Current Topics in Microbiology and Immunology

Volume 325

Series Editors

Richard W. Compans
Emory University School of Medicine, Department of Microbiology and
Immunology, 3001 Rollins Research Center, Atlanta, GA 30322, USA

Max D. Cooper
Department of Pathology and Laboratory Medicine, Georgia Research Alliance,
Emory University, 1462 Clifton Road, Atlanta, GA 30322, USA

Tasuku Honjo
Department of Medical Chemistry, Kyoto University, Faculty of Medicine,
Yoshida, Sakyo-ku, Kyoto 606-8501, Japan

Hilary Koprowski
Thomas Jefferson University, Department of Cancer Biology, Biotechnology
Foundation Laboratories, 1020 Locust Street, Suite M85 JAH, Philadelphia,
PA 19107-6799, USA

Fritz Melchers
Biozentrum, Department of Cell Biology, University of Basel, Klingelbergstr.
50–70, 4056 Basel Switzerland

Michael B.A. Oldstone
Department of Neuropharmacology, Division of Virology, The Scripps Research
Institute, 10550 N. Torrey Pines, La Jolla, CA 92037, USA

Sjur Olsnes
Department of Biochemistry, Institute for Cancer Research, The Norwegian
Radium Hospital, Montebello 0310 Oslo, Norway

Peter K. Vogt
The Scripps Research Institute, Dept. of Molecular & Exp. Medicine, Division of
Oncovirology, 10550 N. Torrey Pines. BCC-239, La Jolla, CA 92037, USA

Thomas E. Shenk • Mark F. Stinski
Editors

Human Cytomegalovirus

Springer

Editors
Thomas E. Shenk
Princeton University
Princeton, NJ 08544-1014
USA
tshenk@princeton.edu

Mark F. Stinski
University of Iowa
Iowa City, IA 52242
USA
mark-stinski@uiowa.edu

Cover Illustration: HCMV infected HFF (3 h p.i.). Green dots outside the blue stained nucleus represent pp71 antibodies bound to viral particles. Larger green dots in the (blue) nucleus represent pp71 antibodies bound to Daxx, which is associated with PML at ND10. The three red domains located beside the pp71 stained ND10 denote the localization of IE2 aggregates. Figure from G. Maul, this volume.

ISBN 978-3-540-77348-1   e-ISBN 978-3-540-77349-8
DOI 10.1007/978-3-540-77349-8

Current Topics in Microbiology and Immunology ISSN 0070-217x

Library of Congress Catalog Number: 2008920670

© 2008 Springer-Verlag Berlin Heidelberg

This work is subject to copyright. All rights reserved, whether the whole or part of the material is concerned, specifically the rights of translation, reprinting, reuse of illustrations, recitation, broadcasting, reproduction on microfilm or in any other way, and storage in data banks. Duplication of this publication or parts thereof is permitted only under the provisions of the German Copyright Law of September, 9, 1965, in its current version, and permission for use must always be obtained from Springer-Verlag. Violations are liable for prosecution under the German Copyright Law.

The use of general descriptive names, registered names, trademarks, etc. in this publication does not imply, even in the absence of a specific statement, that such names are exempt from the relevant protective laws and regulations and therefore free for general use.

Product liability: The publisher cannot guarantee the accuracy of any information about dosage and application contained in this book. In every individual case the user must check such information by consulting the relevant literature.

Cover Design: WMXDesign GmbH, Heidelberg, Germany

Printed on acid-free paper

9 8 7 6 5 4 3 2 1

springer.com

Preface

The earliest observation of cytomegalovirus (CMV) interactions with the host cell was owl eye cytopathology in various tissues. It was recognized in the early 1970s that human CMV caused in utero infections resulting in congenital brain damage and other sensory neurological complications. Events of the 1980s and early 1990s, such as the wide application of solid organ and bone marrow transplantation and the emergence of AIDS, put the spotlight on human CMV. We understood that the virus was an opportunistic agent associated with immunosuppression. The golden age of cytomegalovirus research was ushered in during the late 1970s and early 1980s by a set of powerful new technologies that included restriction enzymes, DNA cloning, DNA sequencing, and open reading frame prediction. The genetic manipulation and propagation of novel CMV strains was accelerated with the application of bacterial artificial chromosome technology.

Today, we still struggle to understand the full spectrum of disease associated with human CMV. To the molecular biologist, CMV is a master of regulation in the eukaryotic cell where it either replicates or remains latent. To the immunologist, CMV is a master of immune evasion with tools to escape both the innate and acquired immune responses. The use of animal models with non-human CMVs has become significantly more sophisticated and tied to a more certain understanding of the interrelationships of non-human and human CMV genes. High-throughput assay technologies are providing exceptionally rich data sets, which provide even greater hope of fully understanding this complex virus. These tools plus others have led to a better understanding of virus replication and the diseases caused by the virus.

This volume has gathered some of the experts in the field to review aspects of our understanding of CMV and to offer perspectives of the current problems associated with CMV. It is our hope that the chapters will lead to a better understanding of the virus that will assist in the development of new and unique antivirals, a protective vaccine, and a full understanding of the involvement of CMV in human disease.

We thank the authors for their terrific contributions! We greatly enjoyed reading their chapters, and we learned a lot in the process. We also are grateful to Anne Clauss at Springer for her support and expertise.

Thomas E. Shenk, Princeton
Mark F. Stinski, Iowa

Contents

Human Cytomegalovirus Genome 1
E. Murphy, T. Shenk

Human Cytomegalovirus microRNAs 21
P. J. Fannin Rider, W. Dunn, E. Yang, F. Liu

Mutagenesis of the Cytomegalovirus Genome 41
Z. Ruzsics, U. H. Koszinowski

Cytomegalovirus Cell Tropism 63
C. Sinzger, M. Digel, G. Jahn

Virus Entry and Innate Immune Activation 85
M. K. Isaacson, L. K. Juckem, T. Compton

**Functions of Human Cytomegalovirus Tegument Proteins
Prior to Immediate Early Gene Expression** 101
R. F. Kalejta

Initiation of Cytomegalovirus Infection at ND10 117
G. G. Maul

**Functional Roles of the Human Cytomegalovirus Essential
IE86 Protein** 133
M. F. Stinski, D. T. Petrik

Nuts and Bolts of Human Cytomegalovirus Lytic DNA Replication 153
G. S. Pari

**Interactions of Human Cytomegalovirus Proteins with
the Nuclear Transport Machinery** 167
T. Stamminger

Structure and Formation of the Cytomegalovirus Virion 187
W. Gibson

Human Cytomegalovirus Modulation of Signal Transduction 205
A. D. Yurochko

**Chemokines and Chemokine Receptors Encoded
by Cytomegaloviruses** 221
P. S. Beisser, H. Lavreysen, C. A. Bruggeman, C. Vink

Subversion of Cell Cycle Regulatory Pathways 243
V. Sanchez, D. H. Spector

**Modulation of Host Cell Stress Responses
by Human Cytomegalovirus** 263
J. C. Alwine

Control of Apoptosis by Human Cytomegalovirus 281
A. L. McCormick

Aspects of Human Cytomegalovirus Latency and Reactivation 297
M. Reeves, J. Sinclair

Murine Model of Cytomegalovirus Latency and Reactivation 315
M. J. Reddehase, C. O. Simon, C. K. Seckert, N. Lemmermann,
N. K. A. Grzimek

Cytomegalovirus Immune Evasion 333
C. Powers, V. DeFilippis, D. Malouli, K. Früh

Cytomegalovirus Vaccine Development 361
M. R. Schleiss

**Cytomegalovirus Infection in the Human Placenta: Maternal
Immunity and Developmentally Regulated Receptors
on Trophoblasts Converge** 383
L. Pereira, E. Maidji

**Mechanisms of Cytomegalovirus-Accelerated Vascular Disease:
Induction of Paracrine Factors That Promote Angiogenesis
and Wound Healing** 397
D. N. Streblow, J. Dumortier, A. V. Moses, S. L. Orloff, J. A. Nelson

**Manifestations of Human Cytomegalovirus Infection:
Proposed Mechanisms of Acute and Chronic Disease** 417
W. Britt

Contributors

J.C. Alwine
Department of Cancer Biology and the Abramson Family Cancer Research Institute, University of Pennsylvania, 314 Biomedical Research Building, 421 Curie Blvd. Philadelphia, PA 19104-6142, USA, alwine@mail.med.upenn.edu

P.S. Beisser
Department of Medical Microbiology, Cardiovascular Research Institute Maastricht (CARIM), University Hospital Maastricht, PO Box 5800, 6202 AZ, Maastricht, The Netherlands

W. Britt
Departments of Pediatrics, Microbiology, and Neurobiology, University of Alabama School of Medicine, Childrens Hospital, Harbor Bldg. 104, 1600 7th Avenue South, Birmingham, AL 35233, USA, wbritt@peds.uab.edu

C.A. Bruggeman
Department of Medical Microbiology, Cardiovascular Research Institute Maastricht (CARIM), University Hospital Maastricht, PO Box 5800, 6202 AZ, Maastricht, The Netherlands

T. Compton
Infectious Diseases, Novartis Institutes for Biomedical Research, Cambridge, MA 02139, USA, teresa.compton@novartis.com

V. DeFilippis
Vaccine and Gene Therapy Institute, Oregon Health and Science University, Portland, OR 97201, USA

M. Digel
Institute of Medical Virology; University of Tübingen; Elfriede-Aulhorn-Str. 6; 72076 Tübingen, Germany

J. Dumortier
Vaccine and Gene Therapy Institute and Department of Molecular Microbiology and Immunology, Oregon Health and Science University, Portland, OR 97201, USA

W. Dunn
Program in Comparative Biochemistry, University of California, Berkeley,
CA 94720, USA

P.J. Fannin Rider
Program in Infectious Diseases and Immunity, School of Public Health, 140
Warren Hall, University of California, Berkeley, CA 94720, USA

K. Früh
Vaccine and Gene Therapy Institute, Oregon Health and Science University,
Portland, OR 97201, USA, fruehk@ohsu.edu

W. Gibson
Department of Pharmacology and Molecular Sciences, Johns Hopkins University
School of Medicine, Baltimore, MD 21205, USA, wgibson@jhmi.edu

N.K.A. Grzimek
Institute for Virology, Johannes Gutenberg-University, Obere Zahlbacher Strasse
67, Hochhaus am Augustusplatz, 55131, Mainz, Germany

M.K. Isaacson
McArdle Laboratory for Cancer Research, University of Wisconsin, Madison,
WI 53706, USA

G. Jahn
Institute of Medical Virology; University of Tübingen; Elfriede-Aulhorn-Str. 6;
72076 Tübingen, Germany

L.K. Juckem
McArdle Laboratory for Cancer Research, University of Wisconsin, Madison,
WI 53706, USA

R.F. Kalejta
Institute for Molecular Virology and McArdle Laboratory for Cancer Research,
University of Wisconsin-Madison, Madison, WI 53706-1596, USA,
rfkalejta@wisc.edu

U.H. Koszinowski
Max von Pettenkofer Institute, Department of Virology, Gene Center,
Ludwig-Maximilians-University, 81377 Munich, Germany,
koszinowski@mvp.uni-muenchen.de

H. Lavreysen
Department of Medical Microbiology, Cardiovascular Research Institute
Maastricht (CARIM), University Hospital Maastricht, PO Box 5800, 6202 AZ,
Maastricht, The Netherlands

N. Lemmermann
Institute for Virology, Johannes Gutenberg-University, Obere Zahlbacher Strasse
67, Hochhaus am Augustusplatz, 55131 Mainz, Germany

List of Contributors

F. Liu
Program in Infectious Diseases and Immunity, School of Public Health, 140 Warren Hall, University of California, Berkeley, CA 94720, USA, liu_fy@uclink4.berkeley.edu

E. Maidji
Department of Cell and Tissue Biology, School of Dentistry, University of California San Francisco, 513 Parnassus, C-734, Box-0640, San Francisco, CA 94143, USA

D. Malouli
Vaccine and Gene Therapy Institute, Oregon Health and Science University, Portland, OR 97201, USA

G.G. Maul
The Wistar Institute, 3601 Spruce Street, Philadelphia, PA 19014, USA, maul@wistar.org

A.L. McCormick
Department of Microbiology and Immunology, Emory Vaccine Center, Emory University, Atlanta, GA 30322, USA, louise.mccormick@emory.edu

A.V. Moses
Vaccine and Gene Therapy Institute and Department of Molecular Microbiology and Immunology, Oregon Health and Science University, Portland, OR 97201, USA

E. Murphy
Department of Molecular Biology, Princeton University, Princeton, NJ 08544-1014, USA

J.A. Nelson
Vaccine and Gene Therapy Institute and Department of Molecular Microbiology and Immunology, Oregon Health and Science University, Portland, OR 97201, USA

S.L. Orloff
Vaccine and Gene Therapy Institute and Department of Molecular Microbiology and Immunology, Oregon Health and Science University, Portland, OR 97201, USA

G.S. Pari
University of Nevada, Reno, School of Medicine, Reno NV 89557, USA, gpari@med.unr.edu

L. Pereira
Department of Cell and Tissue Biology, School of Dentistry, University of California San Francisco, 513 Parnassus, C-734, Box-0640, San Francisco, CA 94143, USA lenore.pereira@ucsf.edu

D.T. Petrik
Interdisciplinary Program in Molecular Biology, University of Iowa, Iowa City, IA 52242, USA

C. Powers
Vaccine and Gene Therapy Institute, Oregon Health and Science University,
Portland, OR 97201, USA

M.J. Reddehase
Institute for Virology, Johannes Gutenberg-University, Obere Zahlbacher Strasse
67, Hochhaus am Augustusplatz, 55131, Mainz, Germany,
Matthias.Reddehase@uni-mainz.de

M.B. Reeves
Infectious Diseases, Novartis Institutes for Biomedical Research, Cambridge,
MA 02139, USA

Z. Ruzsics
Max von Pettenkofer Institute, Department of Virology, Gene Center, Ludwig-
Maximilians-University, 81377 Munich, Germany

V. Sanchez
Department of Microbial and Molecular Pathogenesis, Texas A&M Health
Science Center, College Station, TX 77843-1266, USA

M.R. Schleiss
Division of Pediatric Infectious Diseases, Department of Pediatrics, Center for
Infectious Diseases and Microbiology Translational Research, University of
Minnesota Medical School, 2001 6th Street SE, Minneapolis, MN 55455, USA,
schleiss@umn.edu

C.K. Seckert
Institute for Virology, Johannes Gutenberg-University, Obere Zahlbacher Strasse
67, Hochhaus am Augustusplatz, 55131, Mainz, Germany

T. Shenk
Department of Molecular Biology, Princeton University, Princeton, NJ 08544-1014,
USA, tshenk@princeton.edu

C.O. Simon
Institute for Virology, Johannes Gutenberg-University, Obere Zahlbacher Strasse
67, Hochhaus am Augustusplatz, 55131, Mainz, Germany

J. Sinclair
Department of Medicine, University of Cambridge, Addenbrooke's Hospital,
Cambridge, CB2 2QQ, UK, js@mole.bio.cam.ac.uk

C. Sinzger
Institute of Medical Virology; University of Tübingen; Elfriede-Aulhorn-Str. 6;
72076 Tübingen, Germany christian.sinzger@med.uni-tuebingen.de

T. Stamminger
Institute for Clinical and Molecular Virology, University Hospital, University of
Erlangen-Nürnberg, Schlossgarten 4, 91054 Erlangen, Germany,
thomas.stamminger@viro.med.uni-erlangen.de

List of Contributors

M.F. Stinski
Department of Microbiology and Interdisciplinary Program in Molecular Biology, University of Iowa, Iowa City, IA 52242, USA, mark-stinski@uiowa.edu

D.N. Streblow
Vaccine and Gene Therapy Institute and Department of Molecular Microbiology and Immunology, Oregon Health and Science University, Portland, OR 97201, USA, streblow@ohsu.edu

D.H. Spector
Department of Cellular and Molecular Medicine and The Skaggs School of Pharmacy and Pharmaceutical Sciences, University of California San Diego, La Jolla, CA 92093-0712, USA, dspector@ucsd.edu

C. Vink
Laboratory of Pediatrics, Erasmus Medical Center Rotterdam, PO Box 2040, 3000 CA, Rotterdam, The Netherlands, c.vink@erasmusmc.nl

E. Yang
Program in Comparative Biochemistry, University of California, Berkeley, CA 94720, USA

A.D. Yurochko
Department of Microbiology & Immunology, Center for Molecular and Tumor Virology, Feist-Weiller Cancer Center, Louisiana State University Health Sciences Center, 1501 Kings Highway, Shreveport, LA 71130-3932, USA, ayuroc@lsuhsc.edu

Human Cytomegalovirus Genome

E. Murphy, T. Shenk(✉)

Contents

Introduction 2
Genome Organization and *cis*-Acting Elements 3
Clinical Isolates and Laboratory Strains 4
Protein-Coding ORFs 6
Genomic Organization: Evolution and Function. 16
Perspectives 17
Referencce 17

Abstract Human cytomegalovirus (HCMV) contains a large and complex E-type genome. There are both clinical isolates of the virus that have been passaged minimally in fibroblasts and so-called laboratory strains that have been extensively passaged and adapted to growth in fibroblasts. The genomes of laboratory strains have undergone rearrangements. To date, the genomes of five clinical isolates have been sequenced. We have re-evaluated the coding content of clinical isolates by identifying the set of open reading frames (ORFs) that are conserved in all five sequenced clinical isolates. We have further determined which of these ORFs are present in the chimpanzee cytomegalovirus (CCMV) genome. A total of 173 ORFs are present in all HCMV genomes and the CCMV genome, and we conclude that these ORFs are very likely to be functional. An additional 59 ORFs are present in the genomes of all five HCMV isolates, but not in CCMV. We have discounted 26 of this latter set of ORFs, because they reside in regions of the genome unlikely to encode functional ORFs. The remaining 33 ORFs are potentially functional ORFs that are specific to HCMV.

T. Shenk
Department of Molecular Biology, Princeton University, Princeton, NJ 08544-1014, USA
tshenk@princeton.edu

Introduction

Historically, viruses were assigned to the herpesvirus family based on the architecture of their virions. Herpesvirus virions contain a linear double-stranded DNA genome packaged in an icosahedral (T = 16) capsid. The capsid is surrounded by a structured protein layer known as the tegument, and this, in turn, is enclosed in a glycoprotein-containing lipid bilayer. The capsid of herpesviruses ranges from approximately 115 to 130 nm in diameter, and the complete virion measures about 150-200 nm in diameter.

Biological criteria, such as host range and growth kinetics, were used to assign the herpesviruses to three different subfamilies, the α-, β- and γ-herpesviruses. These groupings have proven to accurately reflect the diversity in organization and gene content of herpesvirus genomes as well. Herpesviruses that infect mammals populate each of the subfamilies. Mammalian herpesviruses have genomes that vary in size by a factor of about two. The α-herpesviruses have the smallest genomes, γ-herpesviruses are intermediate in size and the β-herpesviruses are

Table 1 Sequenced members of the β-herpesvirus subfamily

Virus	Genus	Genome type/size[a]	GenBank accession number
Viruses of humans			
Human herpes virus 5 (human cytomegalovirus)	Cytomegalovirus	E/~235	AD169: X17403
			AC146999
			Towne: AC146851
			Towne: AY315197
			Toledo: AC146905
			FIX: AC146907
			TR: AC146904
			PH: AC146904
			Merlin: AY446894
Human herpes virus 6A	Roseolovirus	A/~165	X83413
Human herpes virus 6B	Roseolovirus	A/~165	AB021506
			AF157706
Human herpes virus 7	Roseolovirus	A/~145	AF037218
			U43400
Viruses of nonhuman primates			
Chimpanzee cytomegalovirus	Cytomegalovirus	E/~241	AF480884
Rhesus cytomegalovirus	Cytomegalovirus	F/~221	AY186194
			DQ120516
Tree shrew herpesvirus	Unassigned	F/~196	AF281817
Viruses of rodents			
Mouse cytomegalovirus	Muromegalovirus	A/~235	U68299
Rat cytomegalovirus	Muromegalovirus	A/~230	AF232689

[a]The type of genome sequence arrangement (Pellet and Roizman 2006) and approximate size (kbp) of the genome are indicated

largest. Packaging a "head full" of viral DNA within a capsid has constrained the maximum and minimum size of herpesvirus genomes.

The β-herpesvirus exhibit relatively lengthy replication cycles, remain substantially cell associated, and generally do not cross host species barriers. This subfamily currently includes three genera (Table 1): cytomegaloviruses, muromegaloviruses and roseoloviruses (Pellett and Roizman 2006). Cytomegaloviruses include human (HCMV), chimpanzee (CCMV) and rhesus cytomegalovirus (RhCMV); muromegaloviruses include mouse (MCMV) and rat cytomegalovirus (RCMV); and the more distantly related roseoloviruses include human herpes viruses 6A, 6B and 7 (HHV-6A, HHV-6B and HHV-7). The sizes of the β-herpesvirus genomes that have been sequenced range from approximately 241 kbp for chimpanzee cytomegalovirus (CCMV) to approximately 145 kbp for the HHV-7 roseolovirus; the human cytomegalovirus (HCMV) genome is approximately 235,000 bp (Table 1).

Genome Organization and *cis*-Acting Elements

As is true for other herpesviruses, β-herpesvirus genomes are linear when isolated from virions, and linear HCMV DNA has been shown to contain a single unpaired base at each end (Tamashiro and Spector 1986). HCMV DNA is replicated by a rolling circle mechanism to generate multiple tandemly linked copies of the viral genome. When it is packaged, the catenated DNA is cleaved to generate unit-length molecules, and the unpaired base at each end facilitates circularization of the genome at the start of the next round of infection.

The genomes of HCMV and CCMV have a so-called class E organization (Pellett and Roizman 2006), with two domains that can be inverted relative to each other, yielding equal amounts of four genomic isomers that can be isolated from virus particles. The two domains are known as the long and short genome segments (L and S), and each domain is comprised of a central unique region (U_L and U_S) flanked by repeated segments that reside at either the ends of the complete genome (TR_L and TR_S) or internally at the intersection of long and short segments (IR_L and IR_S). Consequently, the HCMV and CCMV genomes have the general organization: TR_L-U_L-IR_L-IR_S-U_S-TR_S.

The TR_L region is comprised of a_n and b sequences, the IR_L-IR_S region contains $b'a'_n c'$ sequences and the TR_S regions contains c and a_n sequences, where prime designations signify sequences in reverse orientation relative to sequences without primes. Thus, the repeats are organized as follows: $a_n b$-U_L-$b'a'_n c'$-U_S-ca_n. The terminal a_n sequences can recombine with the internal a'_n sequences, thereby enabling the isomerization of the HCMV and CCMV genomes. In contrast, RhCMV has a class F genome with no known terminal repeated sequences, whereas MCMV and RCMV have class A genomes with terminal but not internal repeats. Consequently, class A and F genomes do not isomerize.

The utility of the isomerization of the HCMV and CCMV genomes is not yet understood. Herpes simplex virus type 1 (HSV-1) mutants that are frozen in individual isomeric orientations replicate perfectly well (Poffenberger et al. 1983; Jenkins and Roizman 1986). Consequently, there is no evidence for a functional consequence of the isomerization, although it remains possible that the multiple isomers are useful in cell types that have not yet been investigated or within infected hosts. It is conceivable that the inverted repeats provide a convenient mechanism for the genome to be segmented by recombination and then maintained in two segments under certain conditions.

Cytomegalovirus genomes contain *cis*-acting elements that direct DNA replication, packaging and transcription. In HCMV the replication origin, ori*Lyt*, has been mapped to an approximately 1500-bp domain (Anders et al. 1992; Masse et al. 1992; Borst and Messerle 2005) located near the middle of the U_L domain. The core of ori*Lyt* includes various repeated elements, transcription factor binding sites and sites at which RNA-DNA structures form (see the chapter by G.S. Pari, this volume). It is possible that a second specific origin functions to initiate replication of the viral genome within latently infected cells, where the HCMV genome is thought to persist as an episome (Bolovan-Fritts et al. 1999). As yet, however, a latency origin has not been identified. It is possible that the cellular DNA replication machinery initiates randomly on latent genomes or that the viral genome is not replicated during latency (see the chapter by M. Reeves and J. Sinclair, this volume). The *a* regions of HCMV are comprised of multiple head-to-tail repeats of *a* sequences (Tamashiro and Spector 1986), which include two *cis*-acting packaging elements, *pac*1 and *pac*2 (Kemble and Mocarski 1989). The number of *a* repeats can vary among strains and among different passages of the same strain due to amplifications and deletions. These relatively short sequences with AT-rich cores flanked by GC-rich sequences are recognized by the viral DNA cleavage and encapsidation machinery (Bogner et al. 1998; W. Gibson, this volume). Cleavage of the sequences adjacent to the *pac*1 and *pac*2 sequences leaves a single 3' base overhang which, as noted above, is utilized for circularization of the linear viral genome upon entry. A variety of transcriptional control regions have been characterized on the HCMV genome, most notably the large and complex major immediate-early promoter/enhancer. It is located within the U_L domain, and it plays a pivotal role in the regulation of viral gene expression at the start of infection (see the chapter by M.F. Stinski and D.T. Petrik, this volume).

Clinical Isolates and Laboratory Strains

Five different HCMVs that are generally referred to as clinical isolates have been sequenced. Four of these viruses were passaged to a limited, but not precisely defined, extent in fibroblasts before they were cloned as bacterial artificial chromosomes (BACs) and then sequenced (FIX, TR, PH and Toledo; Murphy et al. 2003b). Toledo has suffered an inversion event that likely enhanced its growth in fibroblasts, as discussed below. The fifth clinical isolate, Merlin, was subjected to three serial passages in fibroblasts before sequencing of uncloned DNA (Dolan et al. 2004). FIX (Hahn et al. 2002) is a derivative of VR1814 (Grazia Revello et al. 2001), which was isolated from a pregnant woman with a primary HCMV infection; TR is a ganciclovir-resistant ocular isolate from an AIDS patient with retinitis (Smith et al. 1998); PH was isolated from a bone marrow transplant patient (Rice et al. 1984; Fish et al. 1995); Toledo (Quinnan et al. 1984) and Merlin (Tomasec et al. 2000) were isolated from the urine of a congenitally infected child. Thus, the sequenced isolates are derived from a variety of clinical settings. Not surprisingly, these viruses each exhibit the same open reading frame (ORF) organization (Fig. 1, clinical strain). For historical reasons discussed below, the ORFs in the U_L domain at the conventional left end of the map are named RL1-14, and they are followed by UL1-151, IRS1, US1-34 and TRS1.

Two viral genes span from repeated to unique domains. IRS1 is coded across the junction of the repeated IR_S sequence and the unique U_S domain, whereas the TRS1 ORF spans the junction of TR_S and U_S. As a result, the two encoded proteins have nearly identical (differing by one amino acid) amino-terminal halves coded by the repeated sequences and unique carboxy-terminal halves. The advantage to the virus of this curious gene organization is not apparent.

Several strains of HCMV were developed as vaccine candidates. To attenuate these viruses, clinical isolates were serially passaged in cultured fibroblasts. After serial passage, the vaccine candidates proved to be different in both their biological and genomic properties. As they were selected for rapid replication in fibroblasts, they simultaneously lost the ability to efficiently enter and replicate in numerous cell types susceptible to the parental virus, such as epithelial cells, endothelial cells, smooth muscle cells and macrophages. Their inability to enter these cells results from mutations in ORFs that encode constituents of a glycoprotein complex (gH-gL-pUL128-pUL130-pUL131; Wang and Shenk 2005b) present in the virion envelope. These passaged viruses also contain changes in additional genes that constrict their host range (see the chapter by

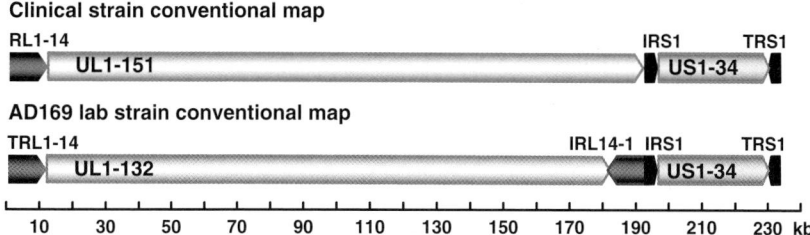

Fig. 1 HCMV ORF organization. *Top*, conventional ORF map of clinical isolates. ORFs are organized using the genome isomer that was originally employed as the conventional map for AD169. *Arrows* portray the relative orientation of the RL1-14, UL1-151, IRS1, US1-34 and TRS1 gene segments. *Bottom*, conventional map of AD169. The RL region is repeated and designated TRL and IRL in clinical isolates

C. Sinzger et al., this volume). Although these passaged viruses have not succeeded as vaccine candidates, they have been widely studied since they grow more rapidly, release more cell-free virus, and generate higher yields in fibroblasts than do clinical strains. Consequently, they are often termed laboratory strains. The improved replication of laboratory strains in fibroblasts results in part from disruption of the glycoprotein complex mentioned above. When the UL131 mutation was repaired in AD169, the repaired virus grew to a reduced yield in fibroblasts (Wang and Shenk 2005a); a similar growth defect was observed when pUL131 was provided to AD169 in *trans* (Adler et al. 2006). It is not clear why an intact gH-gL-pUL128-pUL130-pUL131 complex inhibits the replication of AD169 in fibroblasts.

In addition to numerous more subtle alterations, the genomes of these passaged viruses have undergone major rearrangements, suffering large deletions and concomitant duplications (Cha et al. 1996). Two laboratory strains have been sequenced: AD169 (Chee et al. 1990; Murphy et al. 2003b) and Towne (Dunn et al. 2003; Murphy et al. 2003b). Both viruses have independently lost a multigene segment and acquired a repeated multigene sequence of nearly identical size. The sequence duplication most likely serves to maintain a unit genome length for viral packaging. Even though the two viruses were generated from different clinical isolates in different parts of the world, their substitutions are very similar. AD169 lacks ORFs UL133-UL151 and carries a duplication of ORFs RL1-14 (and termed TRL1-14 and IRL1-14 to discriminate terminal and internal copies of the repeat) (Fig. 1, AD169). The AD169 rearrangement resulted from a recombination event between RL14 and an ORF in a different frame but within UL148 at one end and within the *a* and *b* sequences at the other end. Towne underwent a recombination where sequences within UL1 and the *a* and *b* repeats were duplicated at the expense of ORFs UL144-UL151. Both recombination events disrupted the ORFs that allow for the synthesis of a functioning gH-gL-pUL128-pUL130-pUL131 complex mentioned above, and this appears to be the result of a selective pressure to lose this complex for efficient replication in fibroblasts.

It is noteworthy that Toledo has also, at least partially, adapted to more efficient replication in fibroblasts. In this case, rather than undergoing a deletion/duplication event, an approximately 15-kbp domain has been inverted (Cha et al. 1996). The UL128 ORF was disrupted at one end of the inversion.

Protein-Coding ORFs

How many functional ORFs reside in the HCMV genome? Answering this question is a daunting task. The first estimate came with the original sequence of AD169 (Chee et al. 1990). This annotation predicted that AD169 has the potential to encode 208 ORFs, of which several are repeated (TRL1-14 and IRL1-14). An ORF was considered a coding ORF if it encoded a polypeptide of 100 amino acids or more and did not overlap a larger ORF across more than 60% of its length. A DNA segment containing 19 additional ORFs was discovered in the Toledo clinical strain and shown to be generally present in clinical isolates (Cha et al. 1996). This raised the initial estimate to 227 ORFs, and other studies identified several additional ORFs (Gibson et al. 1996; Mullberg et al. 1999; Kotenko et al. 2000). Subsequent studies utilizing a gene-finding algorithm (Murphy et al. 2003a) and using the sequenced CCMV genome as a comparator (Davison et al. 2003) suggested refinements to the earlier annotations.

More recently, the sequences of clinical HCMV strains were determined. Evaluation of these sequences led to estimates of the number of protein-coding ORFs ranging from a maximum of 252 potentially functional ORFs that are conserved in four different clinical isolates (Murphy et al. 2003b) to a minimum of 165 ORFs that were conserved between one HCMV clinical isolate and CCMV (Davison et al. 2003). These two studies provide a reasonable range for the number of HCMV coding ORFs. The higher estimate was inclusive of all ORFs that might encode a protein. The lower estimate focused on the subset of ORFs for which a strong case can be made for function.

We have revisited these maximal and minimal estimates of potential coding ORFs. We did not try to discover additional ORFs; instead, we reassessed the full set of previously annotated ORFs. Our criteria were simple. If a previously annotated ORF was present in the genomes of five clinical isolates (FIX, TR, PH, Toledo and Merlin), it was considered potentially functional. The subset of annotated ORFs that were also present in the CCMV genome, were considered very likely to be functional, because they have been conserved through 4-4.5 million years of divergent evolution. These criteria closely mimicked those used in the two earlier studies (Davison et al. 2003; Murphy et al. 2003b), and the analysis benefited from combining the two earlier data sets.

To initiate our meta-analysis, MacVector 7.2 (Accelrys, San Diego, CA, USA) was used to identify all start-to-stop ORFs with a coding potential of 80 amino acids or more within each of the genomes. Next, the identified ORFS (>400 per genome) were translated and each polypeptide was used as a query in a BlastP analysis against a database including all previously annotated HCMV ORFs. All ORFs with a local alignment score of 10^{-5} or less were considered matches, and used to generate maps for each of the five genomes with MacVector. Finally, the ORF maps were aligned to determine conservation among the clinical isolates. Due to the substitution of the BAC sequence in four of the five HCMV genome sequences (FIX, TR, PH and Toledo) for viral genes, the IRS1 to US12 region was compiled by conservation between the Merlin sequence (Dolan et al. 2004) and a BAC clone of AD169.

A master map was generated containing all ORFs that met the above criteria and that were conserved in all five clinical isolates (Fig. 2). It contains a total of 232 potentially functional ORFs. Color-coding is used to distinguish ORFs that are known to be essential (red), augmenting, i.e., are required for an optimal yield (yellow), or nonessential for replication in fibroblasts (green) (Dunn et al. 2003; Yu et al. 2003). The 15 gray ORFs have not been tested for a role in replication. The 173 red, yellow, green and gray ORFs are present in all five HCMV genomes and the CCMV genome. The 59 ORFs shown in white are present in the five clinical

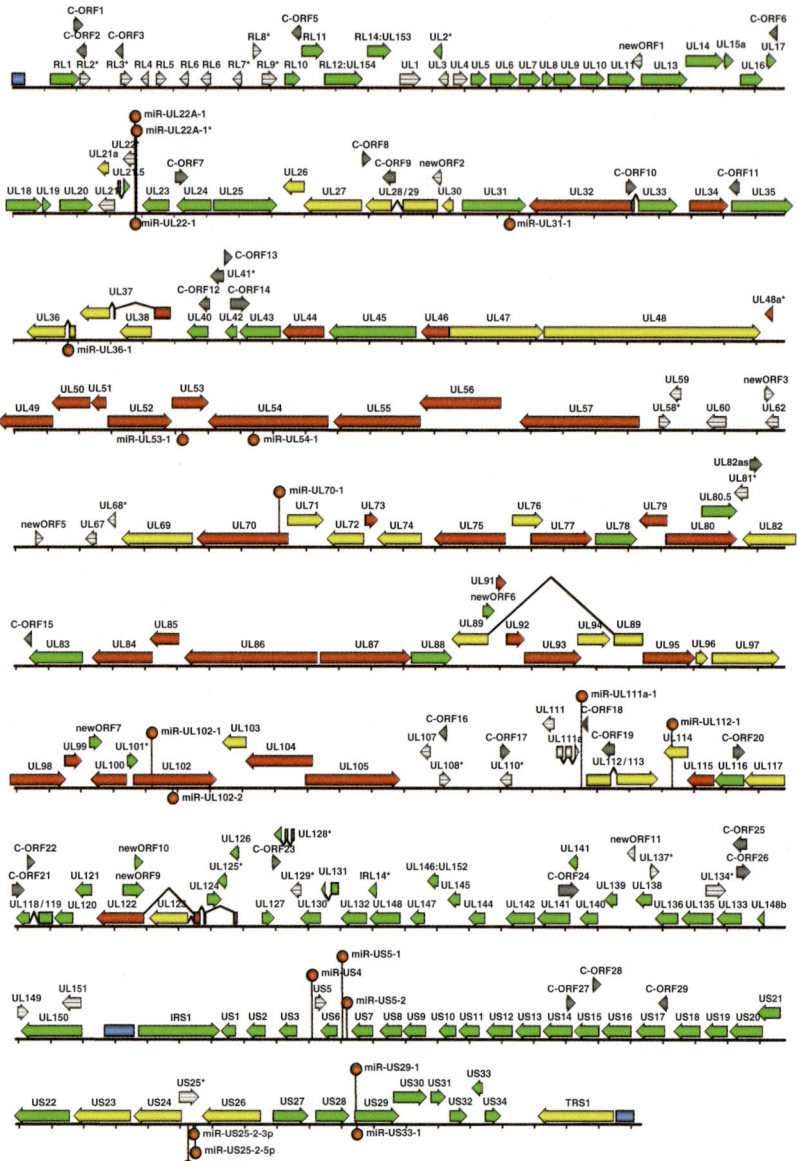

Fig. 2 Genomic arrangement of clinical HCMV strains. HCMV ORFs that are conserved in the five sequenced HCMV strains (FIX, Ph, TR, Toledo and Merlin) are arranged in the conventional HCMV map organization. ORFs are represented as *arrows* demonstrating the relative orientation of ORFs; and, where applicable, *black carat symbols* connect exons. In several cases, e.g., UL37, UL122 and UL123, numerous spliced variants are known, but only one abundant variant is shown on the map. The color codes designate ORFs that are essential (*red arrows*), augmenting (*yellow arrows*) or nonessential (*green arrows*) for replication within cultured fibroblasts. *Gray arrows* represent ORFs that are have not yet been tested for function. *Red, yellow, green* and *gray* ORFs are conserved in CCMV; *white* ORFs are not conserved in CCMV. ORFs with an *asterisk* do not contain an AUG ≥ 80 codons from a stop codon. The three *blue boxes* represent the repeat sequences found at the ends of the unique long and unique short regions. The *orange pins* designate the location of virus-coded miRNAs. Their placement above or below the sequence line designates the strand on which they are encoded. Each *tick mark* on the black sequence line represents 1 kb of DNA

isolates, but are not found in the CCMV genome. Finally, 20 microRNAs (miR-NAs), predicted to be encoded by HCMV (Dunn et al. 2005; Grey et al. 2005; Pfeffer et al. 2005), are identified as orange pins. So far, expression of 14 of the miRNAs has been demonstrated (see the chapter by P.J.F. Rider et al., this volume).

Further details of the full set of 232 ORFs are presented in Table 2, which includes previously annotated ORFs that did not pass our filters for inclusion on the map. As is evident in Table 2, UL147a and UL148a are each missing in only the PH clinical isolate and present in CCMV. Thus, they are likely bona-fide ORFs.

The map in Fig. 2 has numerous uncertainties. Several relate to the filters used previously to qualify an ORF as a potential protein-coding sequence, and, therefore, for inclusion in the database from which we selected ORFs. First, the majority of previously annotated ORFs were required to code polypeptides meeting a minimum size standard, often 80 amino acids or more, as is evident in Table 2. This is an arbitrary cut off, utilized for practical reasons, but, of course, there is no reason to assume that HCMV does not encode smaller polypeptides. As a case in point, analysis of the proteins associated with HCMV virions (Varnum et al. 2004) raises the possibility that the virus encodes some very small polypeptides. In this study, mass spectroscopy was employed to identify proteins in preparations of purified virus particles. The analysis identified 12 tryptic-digestion products corresponding to polypeptides encoded by ORFs that were not previously recognized. Several of the ORFs encode polypeptides of fewer than 80 amino acids, and one has a coding potential of 22 amino acids. This polypeptide might be the result of spurious transcription/translation late after infection, or the polypeptide or a portion of it could be appended to a larger protein as a consequence of splicing. Although it is not possible to conclude that the virus encodes a 22-amino acid polypeptide from this data set, the observation nevertheless serves to reinforce the very likely possibility that the virus encodes small polypeptides that have been overlooked.

A second uncertainty comes from overlap restrictions placed on the pool of previously annotated ORFs. An ORF on one strand can potentially bias the sequence of the opposing strand (Silke 1997; Cebrat et al. 1998), and the high G+C content of HCMV (57%) potentially favors the presence of spurious ORFs since stop codons are A+U-rich. In past annotations, the overlap of the shorter of two overlapping ORFs has been arbitrarily limited to 60% or more or 25% or more, or the overlap has been limited to 396 bp, the longest overlap documented for two HCMV ORFs known to code proteins (UL76 and UL77). It is certainly possible that, in some instances, functional ORFs have evolved with longer overlaps.

Another significant uncertainty to the map in Fig. 2 is our incomplete understanding of HCMV splicing. It is not possible to predict splice donors and acceptors with certainty. A variety of spiced mRNAs have been successfully identified, (e.g., Stenberg et al. 1984; Rawlinson and Barrell 1993; Scott et al. 2002; Adair et al. 2003), but so far, there has been no exhaustive experimental search for spliced HCMV mRNAs. Splicing can, of course, combine ORFs originally assumed to be separate, or utilize small coding regions as a constituent of a larger mRNA.

The majority of the 173 ORFs that are present in all HCMV clinical isolates and in CCMV are extremely likely to encode proteins. Indeed, 130 of these ORFs have

Table 2 Conserved ORFs in clinical HCMV Strains

HCMV ORF	Amino acids[a]	Pheno-type[b]	Expres-sion[c]	Toledo[d]	FIX	PH	TR	Merlin	CCMV	RhCMV
RL1	311	NE	nr	+	+	+	+	+	+	+
C-ORF1	102	nt	nr	+	+	+	+	+	−	−
C-ORF2	103	nt	nr	+	+	+	+	+	−	−
RL2	115	nt	nr	+	+	+	+	+	−	−
C-ORF3	87	nt	nr	+	+	+	+	+	−	−
RL3	114	nt	nr	+	+	+	+	+	−	−
RL4	170	NE	+	+	2	2	2	2	−	−
RL5	114	NE	nr	+	+	+	+	+	−	−
RL6	111	NE	nr	+	2	+	2	+	−	−
RL7	82	NE	nr	+	+	+	+	+	−	−
RL8	129	nt	nr	+	+	+	+	+	−	−
C-ORF4	91	nt	nr	+	+	+	+	+	−	−
RL9	143	NE	nr	+	+	+	+	+	−	−
RL10	171	NE	+	+	+	+	+	+	+	−
C-ORF5	107	nt	nr	+	+	+	+	+	−	−
RL11	234	NE	+	+	+	+	+	+	+	+
RL12	416	NE	nr	+	UL154	+	+	+	+	−
RL13/14	253	NE	+	+	UL153	UL153	+	+	+	−
UL1	224	NE	nr	+	+	+	+	+	−	−
UL2	60	nt	+	+	+	+	+	+	+	−
UL3	105	NE	nr	+	+	+	+	+	−	−
UL4	152	NE	+	+	+	+	+	+	−	+
UL5	166	NE	+	+	+	+	+	+	+	−
UL6	284	NE	nr	+	+	+	+	+	+	+
UL7	222	NE	nr	+	+	+	+	+	+	+
UL8	122	NE	nr	+	+	+	+	+	+	−
UL9	228	NE	+	2	+	+	+	+	+	2
UL10	326	NE	nr	+	+	+	+	+	+	−
UL11	275	NE	+	+	+	+	+	+	+	3
newORF1	84	nt	nr	+	+	+	+	+	·	−
UL12	73	nt	+	+	−	−	−	−	−	−
UL13	473	NE	nr	+	+	+	+	+	+	+
UL14	343	NE	+	+	+	+	+	+	+	+
UL15	322	NE	nr	+	+	+	−	+	−	−
UL15A	101	NE	nr	+	+	+	+	+	+	−
UL16	230	NE	+	+	+	+	+	+	+	−
UL17	104	NE	+	+	+	+	+	+	−	−
C-ORF6	86	nt	nr	+	+	+	+	+	−	−
UL18	368	NE	+	+	+	+	+	+	+	−
UL19	98	NE	nr	+	+	+	+	+	+	+
UL20	340	NE	+	+	+	+	+	+	+	+
UL21	175	nt	+	+	+	+	+	+	−	−
UL21A (UL21)	124	A	+	+	+	+	+	+	+	+
UL21.5 (22A)	104	NE	nr	+	+	+	+	+	−	−
UL22	128	nt	nr	+	+	+	+	+	−	−
UL23	342	NE	+	+	+	+	+	+	+	+
C-ORF7	130	nt	nr	+	+	+	+	+	+	−

(continued)

Table 2 (continued)

HCMV ORF	Amino acids[a]	Pheno-type[b]	Expression[c]	Toledo[d]	FIX	PH	TR	Merlin	CCMV	RhCMV
UL24	358	NE	+	+	+	+	+	+	+	+
UL25	656	NE	+	+	+	+	+	+	+	+
UL26	188	A	+	+	+	+	+	+	+	+
UL27	608	A	+	+	+	+	+	+	+	+
C-ORF8	91	nt	nr	+	+	+	+	+	−	−
C-ORF9	139	nt	nr	+	+	+	+	+	+	+
UL28/29	701	A	+	+	+	+	+	+	+	+
newORF2	105	nt	nr	+	+	+	+	+	−	−
UL30	121	A	+	+	+	+	+	+	+	+
UL31	694	NE	+	+	+	+	+	+	+	+
UL32	1048	E	+	+	+	+	+	+	+	+
C-ORF10	113	nt	nr	+	+	+	+	+	+	+
UL33	390	NE	+	+	+	+	+	+	+	+
UL34	504	E	+	+	+	+	+	+	+	+
C-ORF11	114	nt	nr	+	+	+	+	+	−	−
UL35	640	NE	+	+	+	+	+	+	+	+
UL36	476	NE	+	+	+	+	+	+	+	+
UL37	487	E	+	+	+	+	+	+	+	+
UL38	331	A	+	+	+	+	+	+	+	+
UL39	124	nt	nr	−	+	+	+	−	−	−
UL40	221	NE	+	+	+	+	+	+	+	−
C-ORF12	109	nt	nr	+	+	+	+	+	−	−
UL41	141	nt	nr	+	+	+	+	+	+	−
C-ORF13	86	nt	nr	+	+	+	+	+	−	−
C-ORF14	205	nt	nr	+	+	+	+	+	+	−
UL41a	79	nt	+	−	+	−	+	+	+	−
UL42	157	NE	nr	+	+	+	+	+	+	+
UL43	187	NE	+	+	+	+	+	+	+	+
UL44	433	E	+	+	+	+	+	+	+	+
UL45	906	NE	+	+	+	+	+	+	+	+
UL46	290	E	+	+	+	+	+	+	+	+
UL47	982	A	+	+	+	+	+	+	+	+
UL48	2241	A	+	+	+	+	+	+	+	−
UL48a	76	nt	+	+	+	+	+	+	+	+
UL49	570	E	+	+	+	+	+	+	+	+
UL50	397	E	+	+	+	+	+	+	+	+
UL51	157	E	+	+	+	+	+	+	+	+
UL52	668	E	+	+	+	+	+	+	+	+
UL53	376	E	+	+	+	+	+	+	+	+
UL54	1242	E	+	+	+	+	+	+	+	+
UL55	906	E	+	+	+	+	+	+	+	+
UL56	850	E	+	+	+	+	+	+	+	+
UL57	1235	E	+	+	+	+	+	+	+	+
UL58	124	nt	nr	+	+	+	+	+	−	−
UL59	123	NE	nr	+	+	+	+	+	−	−
UL60	160	NE	+	+	+	+	+	+	−	−
UL61	431	NE	+	+	+	−	+	−	+	−

(continued)

Table 2 (continued)

HCMV ORF	Amino acids[a]	Pheno-type[b]	Expres-sion[c]	Toledo[d]	FIX	PH	TR	Merlin	CCMV	RhCMV
NewORF3	104	nt	nr	+	+	+	+	+	−	−
UL62	217	NE	nr	+	+	+	+	+	−	−
UL63	129	nt	nr	+	+	+	−	−	−	−
NewORF5	87	nt	nr	+	+	+	+	+	−	−
UL64	100	nt	nr	+	+	+	+	−	−	−
UL65	102	nt	+	+	+	−	−	+	−	−
UL66	114	nt	nr	+	+	+	+	−	−	−
UL67	113	NE	nr	+	+	+	+	+	−	−
UL68	110	NE	nr	+	+	+	+	+	−	−
UL69	744	A	+	+	+	+	+	+	+	+
UL70	1062	E	+	+	+	+	+	+	+	+
UL71	411	A	+	+	+	+	+	+	+	+
UL72	388	A	+	+	+	+	+	+	+	+
UL73	138	E	+	+	+	+	+	+	+	+
UL74	466	A	+	+	+	+	+	+	+	+
UL75	743	E	+	+	+	+	+	+	+	+
UL76	325	A	+	+	+	+	+	+	+	+
UL77	642	E	+	+	+	+	+	+	+	+
UL78	431	NE	nr	+	+	+	+	+	+	+
UL79	295	E	+	+	+	+	+	+	+	+
UL80	708	E	+	+	+	+	+	+	+	+
UL80.5	373	NE	+	+	+	+	+	+	+	−
UL81	116	nt	nr	+	+	+	+	+	−	−
UL82	559	A	+	+	+	+	+	+	+	+
UL82as	134	nt	+	+	+	+	+	+	+	−
C-ORF15	86	nt	nr	+	+	+	+	+	+	−
UL83	561	NE	+	+	+	+	+	+	+	2
UL84	586	E	+	+	+	+	+	+	+	+
UL85	306	E	+	+	+	+	+	+	+	+
UL86	1370	E	+	+	+	+	+	+	+	+
UL87	941	E	+	+	+	+	+	+	+	+
UL88	429	NE	+	+	+	+	+	+	+	+
UL89	675	E	+	+	+	+	+	+	+	+
UL90	66	nt	+	+	+	+	−	+	−	−
newORF6	124	nt	nr	+	+	+	+	+	+	+
UL91	111	E	+	+	+	+	+	+	+	+
UL92	201	E	+	+	+	+	+	+	+	+
UL93	594	E	+	+	+	+	+	+	+	+
UL94	345	A	+	+	+	+	+	+	+	+
UL95	531	E	+	+	+	+	+	+	+	+
UL96	115	A	+	+	+	+	+	+	+	+
UL97	707	A	+	+	+	+	+	+	+	+
UL98	584	E	+	+	+	+	+	+	+	+
UL99	190	E	+	+	+	+	+	+	+	+
newORF7	194	nt	nr	+	+	+	+	+	+	−
UL100	372	E	+	+	+	+	+	+	+	+
UL101	115	E	nr	+	+	+	+	+	−	−

(continued)

Table 2 (continued)

HCMV ORF	Amino acids[a]	Pheno-type[b]	Expres-sion[c]	Toledo[d]	FIX	PH	TR	Merlin	CCMV	RhCMV
UL102	798	E	+	+	+	+	+	+	+	+
UL103	249	A	nr	+	+	+	+	+	+	+
UL104	697	E	+	+	+	+	+	+	+	+
UL105	956	E	+	+	+	+	+	+	+	+
UL106	125	NE	nr	−	+	−	−	+	−	−
UL107	150	NE	nr	+	+	+	+	+	−	−
UL108	123	NE	+	+	+	+	+	+	−	−
C-ORF16	89	nt	nr	+	+	+	+	+	−	−
C-ORF17	105	nt	nr	+	+	+	+	+	−	−
UL109	98	NE	nr	+	+	−	+	+	−	−
UL110	127	NE	nr	+	+	+	+	+	−	−
UL111	107	NE	nr	+	+	+	+	+	−	−
UL111a	176	NE	+	+	+	+	+	+	−	−
C-ORF18	83	nt	nr	+	+	+	+	+	+	−
C-ORF19	136	nt	nr	+	+	+	+	+	−	−
UL112/113	268	A	+	+	+	+	+	+	+	+
UL114	250	A	+	+	+	+	+	+	+	+
UL115	278	E	+	+	+	+	+	+	+	+
UL116	344	NE	nr	+	+	+	+	+	+	+
C-ORF20	126	nt	nr	+	+	+	+	+	−	−
UL117	424	A	+	+	+	+	+	+	+	+
C-ORF21	134	nt	nr	+	+	+	+	+	−	−
C-ORF22	88	nt	nr	+	+	+	+	+	+	−
UL118/119	231	NE	+	+	+	+	+	+	+	+
UL120	201	NE	nr	+	+	+	+	+	+	+
UL121	180	NE	nr	+	+	+	+	+	+	+
NewORF9	228	nt	nr	+	+	+	+	+	+	−
NewORF10	93	nt	nr	+	+	+	+	+	+	−
UL122	579	E	+	+	+	+	+	+	+	+
UL123	491	A	+	+	+	+	+	+	+	+
UL124	152	NE	+	+	+	+	+	+	+	−
UL125	102	nt	nr	+	+	+	+	+	−	−
UL126	134	nt	nr	+	+	+	+	+	+	−
UL127	131	nt	nr	+	+	+	+	+	+	−
C-ORF23	92	nt	nr	+	+	+	+	+	+	−
UL128	175	NE	+	+	+	+	+	+	+	−
UL129	116	nt	nr	+	+	+	+	+	−	−
UL130	215	NE	+	+	+	+	+	+	+	+
UL131	129	NE	+	+	+	+	+	+	+	−
UL132	270	NE	+	+	+	+	+	+	+	+
IRL14	183	NE	nr	+	+	+	+	+	+	−
UL148	317	NE	nr	+	+	+	+	+	+	+
UL147a	75	NE	nr	+	+	−	+	+	+	−
UL147	160	NE	+	+	+	+	+	+	+	+
UL146	118	NE	+	+	UL152	UL152	+	+	+	−
UL145	101	NE	nr	+	+	+	+	+	+	+
UL144	177	NE	+	+	+	+	+	+	+	+

(continued)

Table 2 (continued)

HCMV ORF	Amino acids[a]	Pheno-type[b]	Expres-sion[c]	Toledo[d]	FIX	PH	TR	Merlin	CCMV	RhCMV
UL143	93	NE	nr	+	−	+	+	−	−	−
UL142	307	NE	+	+	+	+	+	+	+	−
C-ORF24	214	nt	nr	+	+	+	+	+	+	−
UL141	426	NE	+	+	2	2	2	2	+	+
UL140	115	NE	nr	+	+	+	+	+	+	−
newORF11	89	nt	nr	+	+	+	+	+	−	−
UL139	136	NE	nr	+	+	+	+	+	+	−
UL138	170	NE	+	+	+	+	+	+	+	−
UL137	97	NE	nr	+	+	+	+	+	−	−
UL136	241	NE	nr	+	+	+	+	+	+	−
UL135	329	NE	nr	+	+	+	+	+	+	−
UL134	176	NE	nr	+	+	+	+	+	−	−
UL133	258	NE	nr	+	+	+	+	+	+	−
C-ORF25	149	nt	nr	+	+	+	+	+	−	−
C-ORF26	151	nt	nr	+	+	+	+	+	−	−
UL148a	80	NE	nr	+	+	−	+	+	+	−
UL148b	80	NE	nr	+	+	+	+	+	+	−
UL148c	77	NE	nr	−	+	−	−	+	+	−
UL148d	63	NE	nr	−	+	−	+	+	+	−
UL149	123	NE	nr	+	+	+	+	+	−	−
UL150	643	NE	nr	+	+	+	+	2	+	−
UL151	337	NE	nr	2	2	2	+	+	−	−
IRS1	846	NE	+	+	BAC	BAC	+	+	+	−
US1	212	NE	nr	+	BAC	BAC	+	+	+	+
US2	199	NE	+	BAC	BAC	BAC	BAC	+	+	+
US3	186	NE	+	BAC	BAC	BAC	BAC	+	+	+
US4	119	NE	nr	BAC	BAC	BAC	BAC	+	−	−
US5	126	NE	nr	BAC	BAC	BAC	BAC	+	−	−
US6	183	NE	+	BAC	BAC	BAC	+	+	+	−
US7	225	NE	+	BAC	+	+	+	+	+	−
US8	227	NE	+	BAC	+	+	+	+	+	−
US9	247	NE	+	BAC	+	+	+	+	+	−
US10	185	NE	+	BAC	+	+	+	+	+	−
US11	215	NE	+	BAC	+	+	+	+	+	+
US12	281	NE	nr	+	+	+	+	+	+	−
US13	261	NE	+	+	+	+	+	+	+	−
US14	310	NE	+	+	+	+	+	+	+	3
C-ORF27	97	nt	nr	+	+	+	+	+	−	−
US15	484	NE	nr	+	+	+	+	+	+	−
C-ORF28	93	nt	nr	+	+	+	+	+	+	−
US16	309	NE	nr	+	+	+	+	+	+	−
US17	293	NE	+	+	+	+	+	+	+	+
C-ORF29	97	nt	nr	+	+	+	+	+	+	−
US18	274	NE	+	+	+	+	+	+	+	+
US19	240	NE	+	+	+	+	+	+	+	+
US20	342	NE	+	+	+	+	+	+	+	+
US21	239	NE	nr	+	+	+	+	+	+	+

(continued)

Table 2 (continued)

HCMV ORF	Amino acids[a]	Pheno-type[b]	Expres-sion[c]	Toledo[d]	FIX	PH	TR	Merlin	CCMV	RhCMV
US22	593	NE	+	+	+	+	+	+	+	+
US23	592	A	+	+	+	+	+	+	+	+
US24	500	A	+	+	+	+	+	+	+	+
US25	179	NE	nr	+	+	+	+	+	–	–
US26	603	A	+	+	+	+	+	+	+	+
US27	362	NE	+	+	+	+	+	+	+	–
US28	323	NE	+	+	+	+	+	+	+	5
US29	462	NE	nr	+	+	+	+	+	+	+
US30	349	NE	+	+	+	+	+	+	+	+
US31	197	NE	nr	+	+	+	+	+	+	+
US32	183	NE	nr	+	+	+	+	+	+	+
US33	137	NE	nr	+	+	+	+	+	+	–
US34	163	NE	nr	+	+	+	+	+	+	–
US34a	65	nt	nr	–	+	–	+	+	–	–
US35	109	nt	nr	–	–	+	–	–	–	–
US36	110	nt	nr	–	–	+	–	–	–	–
TRS1	788	A	+	+	+	+	+	+	+	+

[a] Open reading frame length in amino acids (aa) is listed using FIX as the reference strain except where sequence was deleted due to the insertion of the bacterial artificial chromosome. In those regions, the Merlin sequence was used to determine ORF size. Light grey boxes denoting the number of amino acids indicate that more than one HCMV strain lacks a methionine within 80 amino acids from the stop codon

[b] Fibroblast phenotypes of mutant viruses as determined by Yu et al. (2003) are listed. The phenotypes described are nonessential (NE), essential (E), augmenting (A) or not tested (nt)

[c] Reported expression of the ORF was determined if the ORF has been reported to have a growth phenotype when mutated or if confirmed by Western blot or mass spectroscopy. (+) designates that the ORF has been confirmed to be expressed or to have a functional impact on viral replication and (nr) designates no report of expression or functional impact

[d] For each of the seven indicated cytomegaloviruses, BlastP analysis of the translation of all ORFs >80 amino acids against a database of previously identified HCMV ORFS was utilized to determine conservation of the ORF. All translations that had a blast expect value of $<10^{-5}$ was considered a match. Matches are designated by (+) in a white box and nonmatches by (–) in a dark grey box. Light grey boxes with a (+) designate ORFs that lack a methionine within 80 amino acids of the stop codon. A number in the box designates that two neighboring ORFs contain translated peptides that match a single predicted ORF thereby suggesting a frame shift mutation. In strains where a different but highly related ORF is present in place of the listed ORF, the ORF name is listed as is the case with UL154 in place of RL12 in FIX BAC and UL153 in place of RL13/14 and UL152 in place of UL146 in FIX BAC and PH BAC. BAC designates where the bacterial artificial chromosome sequence was inserted in the HCMV genome

either been directly shown to encode proteins or to influence viral replication when they are mutated (Table 2). Perhaps the more intriguing set of ORFs are the 59 ORFs present in all HCMVs, but not in CCMV (Fig. 2, white arrows). Some of these are likely to be spurious. For example, many of the ORFs extending from RL2* through RL9* may not encode proteins. Analysis with a gene-finding algorithm indicated that seven out of nine ORFs tested in this region have low coding potential (Murphy et al. 2003a), and an abundant transcript coded in this region has very recently been shown to function as an RNA (Reeves et al. 2007). Similarly, the ORFs extending from UL107 to UL111 are not likely to encode proteins. Again, the gene-finding algorithm found ORFs in this genome segment to have very low coding potential, and the so-called 5-kb transcript, an intron that functions as an RNA molecule (Kulesza and Shenk 2006), extends through this region. Finally, the ORFs from UL58* through UL68* span a region that includes the HCMV *ori*Lyt. Six of the ORFs in this region were tested by the gene-finding algorithm and classed as unlikely to encode a protein. Collectively, these two RNA-coding regions plus the ori*Lyt* domain are unlikely to contain protein-coding ORFs, removing all or most of 26 ORFs in these regions from further consideration. This leaves 33 ORFs that are present in all five HCMV clinical isolates, but not in CCMV. They are spread across the entire genome, although more are present in terminal domains; they are generally relatively small, but not smaller than known protein-coding ORFs; and some might encode segments that are incorporated into larger proteins by splicing. The entire set are candidates for genes that are unique to the human virus. CCMV likely contains several ORFs that are not present in human viruses (Davison et al. 2003), so it is not surprising that human virus-specific ORFs would also exist. In fact, one member of this set, UL111a, is known to encode a functional IL10 homolog (Kotenko et al. 2000). Consequently, it is likely that additional ORFs present in the human viruses but not CCMV will prove to encode proteins specific to the human virus, but it will be necessary to test each for expression and/or function to be certain.

Genomic Organization: Evolution and Function

As is true for all herpes viruses, the HCMV genome contains a set of evolutionally conserved, herpes virus-common ORFs that encode core functions required for replication of the viral DNA and its assembly into virus particles (Mocarski et al. 2006). The 40 core HCMV genes are primarily located in the central region of the viral genome within the U_L domain. The terminal regions, including the U_S domain, contain genes that are cytomegalovirus-specific and generally nonessential for replication in cultured cells. This organization leads to the view that modern herpes viruses have evolved from a common ancestral virus by the acquisition of specialized, luxury gene functions primarily within their terminal domains. This generalization leads to the suggestion that the apparent noncoding region ranging from RL2* through RL9* (Fig. 2) contains cytomegalovirus-specific functions. Consistent with this prediction, the function of β2.7 RNA, which is coded in this region, is without precedent in other herpes viruses. The RNA binds to mitochondrial enzyme complex I and blocks the induction of apoptosis in response to stress (see the chapter by M. Reeves and J. Sinclair, this volume). Many additional genes within the terminal regions, including much of the U_S domain, have been shown to function in immune evasion (see the chapter by C. Powers et al., this volume).

In HCMV, 12 gene families have been identified (Chee et al. 1990; Rigoutsos et al. 2003; Lesniewski et al. 2006), which have likely arisen by duplication events. One example is the UL12 family, which consists of ten contiguous genes, US12-US21 (Lesniewski et al. 2006). After the primordial duplication event, members of gene families have functionally diverged as the virus has continued to evolve.

In some cases, genes with similar functions reside in close proximity to each other on the viral genome. As noted above, immune evasion genes are near the termini, the key immediate-early transcriptional regulatory proteins are derived from the same transcription unit (UL122, 123) (Stinski and Petrik 2007), genes known to block apoptosis are grouped (UL36-38) (see the chapter by A.L. McCormick, this volume), and three genes important for entry into endothelial and epithelial cells are grouped (UL128-131) (see the chapter by C. Sinzger et al., this volume). Perhaps these genes have evolved to function in an interactive manner, and their groupings allow them to generally travel in functional sets as the viral genome undergoes recombination.

Perspectives

The principal challenge we face in HCMV genomics is to identify the full set of viral gene products. The majority of larger protein-coding ORFs have certainly been identified, but there are very likely additional smaller protein-coding ORFs and functional transcripts that do not serve as mRNAs that remain to be identified.

Finally, it is noteworthy that our knowledge of the HCMV genome relies on the sequences of a relatively small number of clinical isolates. It remains possible that variants will be discovered that contain one or more modified genes or additional genes that will prove to be associated with specific clinical manifestations of viral infection.

References

Adair R, Liebisch GW, Colberg-Poley AM (2003) Complex alternative processing of human cytomegalovirus UL37 pre-mRNA. J Gen Virol 84:3353-3358

Adler B, Scrivano L, Ruzcics Z, Rupp B, Sinzger C, Koszinowski U (2006) Role of human cytomegalovirus UL131A in cell type-specific virus entry and release. J Gen Virol 87:2451-2460

Anders DG, Kacica MA, Pari G, Punturieri SM (1992) Boundaries and structure of human cytomegalovirus oriLyt, a complex origin for lytic-phase DNA replication. J Virol 66:3373-3384

Bogner E, Radsak K, Stinski MF (1998) The gene product of human cytomegalovirus open reading frame UL56 binds the pac motif and has specific nuclease activity. J Virol 72:2259-2264

Bolovan-Fritts CA, Mocarski ES, Wiedeman JA (1999) Peripheral blood CD14(+) cells from healthy subjects carry a circular conformation of latent cytomegalovirus genome. Blood 93:394-398

Borst EM, Messerle M (2005) Analysis of human cytomegalovirus oriLyt sequence requirements in the context of the viral genome. J Virol 79:3615-3626

Cebrat S, Mackiewicz P, Dudek MR (1998) The role of the genetic code in generating new coding sequences inside existing genes. Biosystems 45:165-176

Cha TA, Tom E, Kemble GW, Duke GM, Mocarski ES, Spaete RR (1996) Human cytomegalovirus clinical isolates carry at least 19 genes not found in laboratory strains. J Virol 70:78-83

Chee MS, Bankier AT, Beck S, Bohni R, Brown CM, Cerny R, Horsnell T, Hutchison CA 3rd, Kouzarides T, Martignetti JA et al (1990) Analysis of the protein-coding content of the sequence of human cytomegalovirus strain AD169. Curr Top Microbiol Immunol 154:125-169

Davison AJ, Dolan A, Akter P, Addison C, Dargan DJ, Alcendor DJ, McGeoch DJ, Hayward GS (2003) The human cytomegalovirus genome revisited: comparison with the chimpanzee cytomegalovirus genome. J Gen Virol 84:17-28

Dolan A, Cunningham C, Hector RD, Hassan-Walker AF, Lee L, Addison C, Dargan DJ, McGeoch DJ, Gatherer D, Emery VC, Griffiths PD, Sinzger C, McSharry BP, Wilkinson GW, Davison AJ (2004) Genetic content of wild-type human cytomegalovirus. J Gen Virol 85:1301-1312

Dunn W, Chou C, Li H, Hai R, Patterson D, Stolc V, Zhu H, Liu F (2003) Functional profiling of a human cytomegalovirus genome. Proc Natl Acad Sci USA 100:14223-14228

Dunn W, Trang P, Zhong Q, Yang E, van Belle C, Liu F (2005) Human cytomegalovirus expresses novel microRNAs during productive viral infection. Cell Microbiol 7:1684-1695

Fish KN, Depto AS, Moses AV, Britt W, Nelson JA (1995) Growth kinetics of human cytomegalovirus are altered in monocyte-derived macrophages. J Virol 69:3737-3743

Gibson W, Clopper KS, Britt WJ, Baxter MK (1996) Human cytomegalovirus (HCMV) smallest capsid protein identified as product of short open reading frame located between HCMV UL48 and UL49. J Virol 70:5680-5683

Grazia Revello M, Baldanti F, Percivalle E, Sarasini A, De-Giuli L, Genini E, Lilleri D, Labo N, Gerna G (2001) In vitro selection of human cytomegalovirus variants unable to transfer virus and virus products from infected cells to polymorphonuclear leukocytes and to grow in endothelial cells. J Gen Virol 82:1429-1438

Grey F, Antoniewicz A, Allen E, Saugstad J, McShea A, Carrington JC, Nelson J (2005) Identification and characterization of human cytomegalovirus-encoded microRNAs. J Virol 79:12095-12099

Hahn G, Khan H, Baldanti F, Koszinowski UH, Revello MG, Gerna G (2002) The human cytomegalovirus ribonucleotide reductase homolog UL45 is dispensable for growth in endothelial cells, as determined by a BAC-cloned clinical isolate of human cytomegalovirus with preserved wild-type characteristics. J Virol 76:9551-9555

Jenkins FJ, Roizman B (1986) Herpes simplex virus 1 recombinants with noninverting genomes frozen in different isomeric arrangements are capable of independent replication. J Virol 59:494-499

Kemble GW, Mocarski ES (1989) A host cell protein binds to a highly conserved sequence element (pac-2) within the cytomegalovirus a sequence. J Virol 63:4715-4728

Kotenko SV, Saccani S, Izotova LS, Mirochnitchenko OV, Pestka S (2000) Human cytomegalovirus harbors its own unique IL-10 homolog (cmvIL-10). Proc Natl Acad Sci USA 97:1695-1700

Kulesza CA, Shenk T (2006) Murine cytomegalovirus encodes a stable intron that facilitates persistent replication in the mouse. Proc Natl Acad Sci USA 103:18302-18307

Lesniewski M, Das S, Skomorovska-Prokvolit Y, Wang FZ, Pellett PE (2006) Primate cytomegalovirus US12 gene family: a distinct and diverse clade of seven-transmembrane proteins. Virology 354:286-298

Masse MJ, Karlin S, Schachtel GA, Mocarski ES (1992) Human cytomegalovirus origin of DNA replication (oriLyt) resides within a highly complex repetitive region. Proc Natl Acad Sci USA 89:5246-5250

Mocarski ES, Shenk T, Pass RF (2006) Cytomegaloviruses. Lippincott, Williams and Wilkins, Philadelphia

Mullberg J, Hsu ML, Rauch CT, Gerhart MJ, Kaykas A, Cosman D (1999) The R27080 glycoprotein is abundantly secreted from human cytomegalovirus-infected fibroblasts. J Gen Virol 80:437-440

Murphy E, Rigoutsos I, Shibuya T, Shenk TE (2003a) Reevaluation of human cytomegalovirus coding potential. Proc Natl Acad Sci USA 100:13585-13590

Murphy E, Yu D, Grimwood J, Schmutz J, Dickson M, Jarvis MA, Hahn G, Nelson JA, Myers RM, Shenk TE (2003b) Coding potential of laboratory and clinical strains of human cytomegalovirus. Proc Natl Acad Sci USA 100:14976-14981

Pellett PE, Roizman B (2006) The family Herpesviridae: a brief introduction. Lippincott, Williams and Wilkins, Philadelphia

Pfeffer S, Sewer A, Lagos-Quintana M, Sheridan R, Sander C, Grasser FA, van Dyk LF, Ho CK, Shuman S, Chien M, Russo JJ, Ju J, Randall G, Lindenbach BD, Rice CM, Simon V, Ho DD, Zavolan M, Tuschl T (2005) Identification of microRNAs of the herpesvirus family. Nat Methods 2:269-276

Poffenberger KL, Tabares E, Roizman B (1983) Characterization of a viable, noninverting herpes simplex virus 1 genome derived by insertion and deletion of sequences at the junction of components L and S. Proc Natl Acad Sci USA 80:2690-2694

Quinnan GV Jr, Delery M, Rook AH, Frederick WR, Epstein JS, Manischewitz JF, Jackson L, Ramsey KM, Mittal K, Plotkin SA et al (1984) Comparative virulence and immunogenicity of the Towne strain and a nonattenuated strain of cytomegalovirus. Ann Intern Med 101:478-483

Rawlinson WD, Barrell BG (1993) Spliced transcripts of human cytomegalovirus. J Virol 67:5502-5513

Reeves MB, Davies AA, McSharry BP, Wilkinson GW, Sinclair JH (2007) Complex I binding by a virally encoded RNA regulates mitochondria-induced cell death. Science 316:1345-1348

Rice GP, Schrier RD, Oldstone MB (1984) Cytomegalovirus infects human lymphocytes and monocytes: virus expression is restricted to immediate-early gene products. Proc Natl Acad Sci USA 81:6134-6138

Rigoutsos I, Novotny J, Huynh T, Chin-Bow ST, Parida L, Platt D, Coleman D, Shenk T (2003) In silico pattern-based analysis of the human cytomegalovirus genome. J Virol 77:4326-4344

Scott GM, Barrell BG, Oram J, Rawlinson WD (2002) Characterisation of transcripts from the human cytomegalovirus genes TRL7, UL20a, UL36, UL65, UL94, US3 and US34. Virus Genes 24:39-48

Silke J (1997) The majority of long non-stop reading frames on the antisense strand can be explained by biased codon usage. Gene 194:143-155

Smith IL, Taskintuna I, Rahhal FM, Powell HC, Ai E, Mueller AJ, Spector SA, Freeman WR (1998) Clinical failure of CMV retinitis with intravitreal cidofovir is associated with antiviral resistance. Arch Ophthalmol 116:178-185

Stenberg RM, Thomsen DR, Stinski MF (1984) Structural analysis of the major immediate early gene of human cytomegalovirus. J Virol 49:190-199

Tamashiro JC, Spector DH (1986) Terminal structure and heterogeneity in human cytomegalovirus strain AD169. J Virol 59:591-604

Tomasec P, Braud VM, Rickards C, Powell MB, McSharry BP, Gadola S, Cerundolo V, Borysiewicz LK, McMichael AJ, Wilkinson GW (2000) Surface expression of HLA-E, an inhibitor of natural killer cells, enhanced by human cytomegalovirus gpUL40. Science 287:1031

Varnum SM, Streblow DN, Monroe ME, Smith P, Auberry KJ, Pasa-Tolic L, Wang D, Camp DG 2nd, Rodland K, Wiley S, Britt W, Shenk T, Smith RD, Nelson JA (2004) Identification of proteins in human cytomegalovirus (HCMV) particles: the HCMV proteome. J Virol 78:10960-10966

Wang D, Shenk T (2005a) Human cytomegalovirus UL131 open reading frame is required for epithelial cell tropism. J Virol 79:10330-10338

Wang D, Shenk T (2005b) Human cytomegalovirus virion protein complex required for epithelial and endothelial cell tropism. Proc Natl Acad Sci USA 102:18153-18758

Yu D, Silva MC, Shenk T (2003) Functional map of human cytomegalovirus AD169 defined by global mutational analysis. Proc Natl Acad Sci USA 100:12396-12401

Human Cytomegalovirus microRNAs

P. J. Fannin Rider, W. Dunn, E. Yang, F. Liu(✉)

Contents

Introduction 22
miRNA Biogenesis 23
Location and Conservation of HCMV miRNAs 27
   Mapping of HCMV miRNAs. 27
   HCMV miRNA Conservation 29
   Genomic Arrangement of HCMV miRNAs 30
HCMV miRNA Expression 31
   Kinetic Classes of HCMV miRNAs 31
   Tissue-Specific HCMV miRNA Expression 32
   Latent Versus Lytic Infection. 34
Potential Function of HCMV miRNAs 34
Future Directions 35
References 35

Abstract MicroRNAs (miRNAs) are approximately 22 nucleotide RNAs that mediate the posttranscriptional regulation of gene expression. miRNAs regulate diverse cellular processes such as development, differentiation, cell cycling, apoptosis, and immune responses. More than 400 miRNAs have been identified in humans and it is predicted that over 30% of human gene transcripts are regulated via miRNAs. Since 2004, many viral miRNAs have been described in several families of viruses. More than half of currently known viral miRNAs are encoded by viruses of the human Herepsviridae and 14 miRNAs have been found to be encoded by Human cytomegalovirus (HCMV). Thus far, HCMV is the only betaherpesvirus in which miRNAs have been described and these miRNAs possess many characteristics, including their genomic arrangement and temporal/spatial expression, which distinguish them from the other known herpesvirus miRNAs described. As a herpesvirus,

---

F. Liu
Program in Infectious Diseases and Immunity, School of Public Health, 140 Warren Hall, University of California, Berkeley, CA 94720, USA
liu_fy@uclink4.berkeley.edu

HCMV establishes infection for the life of the host characterized by latent infection with periodic reactivation for production and spread of infectious progeny. This multifaceted life cycle of the herpesvirus requires an abundance of gene products and regulatory elements that makes cytomegalovirus genomes one of the most complex among human viruses. The defining characteristics of the cytomegalovirus and the minimal impact on genome size afforded by miRNAs inform the logic of virus-encoded miRNAs.

Abbreviations miRNA: MicroRNA; HCMV: Human cytomegalovirus; EBV: Epstein Bar Virus; HSV: Herpes simplex virus; miRISC: MicroRNA-induced silencing complex; MHV 68: Murine gammaherpesvirus 68; RRV: Rhesus macaque rhadinovirus; rLCV: Rhesus lymphocryptovirus; MDV: Marek's disease virus

Introduction

In 1993, the discovery of an approximately 22-nucleotide RNA (*lin-4*) responsible for the posttranscriptional regulation of LIN-14 protein levels in *caenorhabditis elegans* represented the beginnings of a paradigm shift in molecular biology (Lee et al. 1993; Wightman et al. 1993). While *lin-4* was discovered in 1993, it was not until 1999 that it was shown to inhibit protein synthesis after the initiation of translation (Olsen and Ambros 1999). *Lin-4* is the founding member of a family of small RNAs, termed microRNAs (miRNA), that now number over 400 in humans (Bentwich et al. 2005). Not 15 years after *lin-4*'s discovery, it has been predicted that over 30% of human gene products are subject to miRNA-mediated regulation (Lewis et al. 2005).

miRNAs are endogenously encoded approximately 22-nt RNAs that are responsible for the temporal and spatial regulation of gene products involved in diverse cellular processes including development, apoptosis, differentiation, cell cycle regulation, and immune response (Ambros 2004; Taganov et al. 2006; O'Connell et al. 2007; Rodriguez et al. 2007; Taganov et al. 2007). Functionally, miRNAs mediate gene silencing by guiding the miRNA-induced gene silencing complex (miRISC) to target mRNAs (Tang 2005). Targeting of mRNA by miRISC leads to translational inhibition or cleavage of the targeted mRNA. The specificity of most animal miRNA–target mRNA complexes is determined by complementarity of a seed region, namely nucleotides 2–7 of the miRNA (Brennecke et al. 2005). When recognition of mRNA by miRNA is mediated primarily by the seed region, translational inhibition is typically the end result, while extensive sequence complementarity between the miRNA and target mRNA results in cleavage of the target mRNA (Ambros 2004). Complementary sequences in the target mRNA usually reside in the 3′ UTR (Lewis et al. 2005). Due to the size of the seed region, miRNAs are predicted to target as many as ten mRNAs. Additionally, multiple miRNAs may target the same mRNA with the 3′-UTRs containing target sites for multiple miRNAs, and in situ adenosine to inosine substitutions in mature miRNAs can also alter targeting (Wightman et al. 1993; Doench and Sharp 2004; Kawahara et al. 2007). Finally, the miRNA sequence outside the seed region has been found to play a role in subcellular localization of miRNAs (Hwang et al. 2007).

Eleven years after the discovery of miRNAs, the first virally encoded miRNAs were reported (Pfeffer et al. 2004). As most algorithms for the identification of miRNAs rely on conservation of sequence, viral miRNA prediction was particularly difficult due to the absence of significant homology to known miRNAs. By sequencing of a small RNA library from a Burkitt's lymphoma cell line latently infected with Epstein Bar Virus (EBV), Tuschl and colleagues identified five miRNAs that originated from the EBV (Pfeffer et al. 2004). Subsequently, many groups, including our own, have contributed to the identification of miRNAs encoded by other human herpesviruses. Additionally, many nonhuman herpesvirus miRNAs have been identified, providing animal models in which to study the function of virally encoded miRNAs. Thus far, 106 of the 108 mature viral miRNAs species in the miRNA registry are encoded by herpesviruses (Griffiths-Jones 2004, 2006; Griffiths-Jones et al. 2006). Currently, it is known that HCMV expresses 14 mature miRNAs from 11 precursor miRNAs (Fig. 1).

A herpesvirus is a large dsDNA virus that replicates in the nucleus of the host cell, and after an initial lytic replication cycle establishes latent infection for the life of the host. Reactivation from latency and initiation of secondary lytic replication occurs periodically. Betaherpesviruses, which include HCMV and herpesvirus-6 and -7, can replicate in a wide variety of cell types, but exhibit strict species specificity. The complex life cyle of herpes viruses illustrates the need for distinctive gene regulation mechanisms that viral miRNAs provide. Utilization of miRNA offers the virus, which possesses limited coding capacity, a means to alter gene expression with relatively minimal impact on genome size; maintenance of latent infection requires limited, nonimmunogenic gene expression; tissue tropism implies an array of gene products suited to the exploitation of specific host-cell types. As such, it is not surprising that the first virally encoded miRNAs were discovered in a herpesvirus, EBV.

miRNA Biogenesis

The progress made in understanding miRNA biogenesis (Fig. 2) stands in stark contrast to the limited understanding of miRNA-mediated regulatory networks. While many fundamental questions about regulation of miRNA biogenesis still need to be addressed, significant progress has been made concerning the process of miRNA maturation.

Many of the subtleties and details concerning miRNA biogenesis are beyond the scope of this review. Rather, it is our aim to provide an overview of the biogenesis process as an aid to understanding HCMV miRNA. There is little evidence to suggest that viral miRNA biogenesis differs from that of cellular miRNAs. Recently, however, aspects of miRNA biogenesis such as substrate

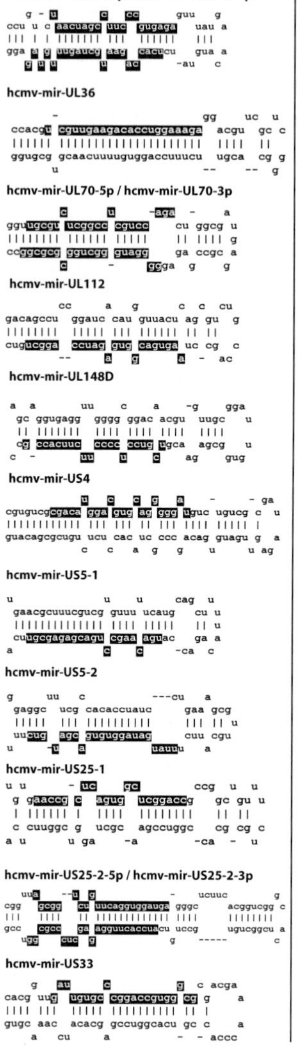

microRNA	Evidence	Expression Kinetics	Position	Reference
hcmv-mir-UL22A-5p / hcmv-mir-UL22A-3p	5p - Cloned 3p - Cloned	IE	16820-16884 (Between UL22-UL23)	Dunn et al. 2005 Pfeffer et al. 2005
hcmv-mir-UL36	Cloned; Northern	IE	38676-38750 (Between UL36-37)	Grey et al. 2005 Pfeffer et al. 2005
hcmv-mir-UL70-5p / hcmv-mir-UL70-3p	5p - Northern 3p - Northern	IE	92890-92955 (Antisense UL70)	Grey et al. 2005
hcmv-mir-UL112	Cloned	unknown	152992-152926 (Antisense UL114)	Pfeffer et al. 2005
hcmv-mir-UL148D	Cloned	unknown		Pfeffer et al. 2005
hcmv-mir-US4	Northern	E	195149 - 195232 (Partial US4)	Grey et al. 2005
hcmv-mir-US5-1	Cloned; Northern	E	196060 - 196125 (Between US6-US7)	Grey et al. 2005 Pfeffer et al. 2005
hcmv-mir-US5-2	Cloned	E	196189 - 196253 (Between US6-US7)	Pfeffer et al. 2005
hcmv-mir-US25-1	Cloned	E	208315-208384 (Between US24-25)	Dunn et al. 2005 Pfeffer et al. 2005
hcmv-mir-US25-2-5p / hcmv-mir-US25-2-3p	5p - Cloned 3p - Cloned	E	208517-208606 (Partial antisense US25)	Pfeffer et al. 2005
hcmv-mir-US33	Cloned	unknown	213547-213616 (Antisense US29)	Pfeffer et al. 2005

Fig. 1 HCMV miRNAs. The eleven HCMV pre-miRNAs found in the miRNA registry and their predicted foldback structures are presented. Mature miRNA sequences are *highlighted*. Information regarding evidence for the indicated miRNA its expression, position and potential targets are given

specificity and Dicer activity have been found to differ during stages of development (Eis et al. 2005; Kim 2005; Lund and Dahlberg 2006). Finally, unique features of the murine gammaherpesvirus-68 (MHV68) miRNAs such as transcription by pol III and short hairpin structure suggest distinct mechanisms of biogenesis

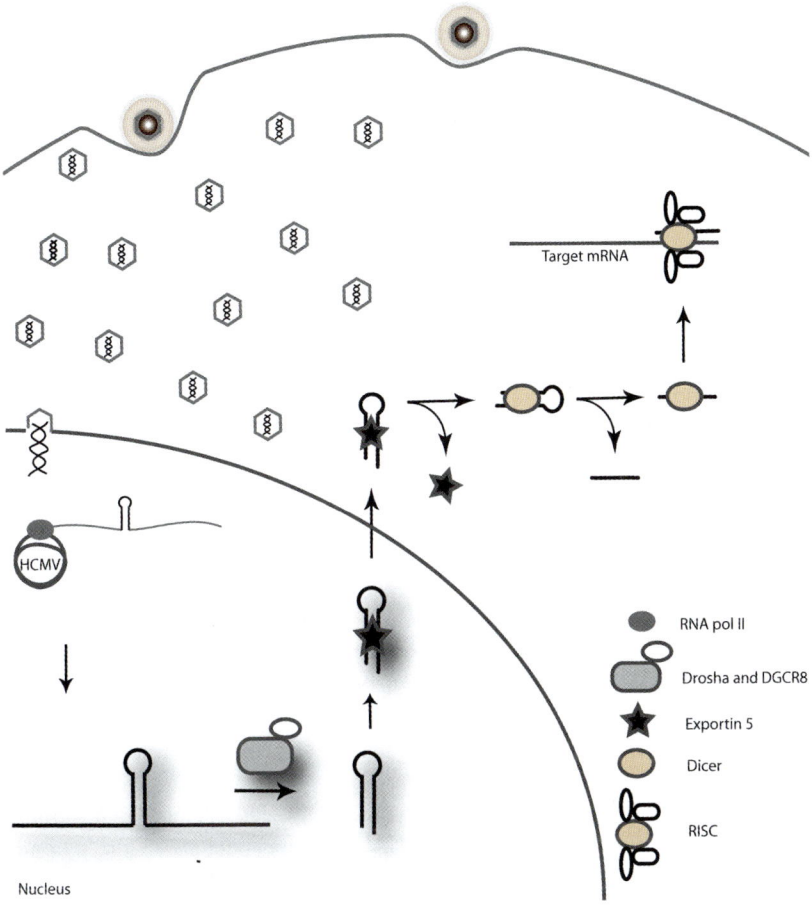

Fig. 2 Biogenesis of miRNAs. Transcription of viral pri-miRNA occurs most often by RNA polymerase II. Pri-miRNAs are then capped and polyadenylated. Nuclear RNase type III endonuclease Drosha then processes the pri-miRNA to pre-miRNA. Pre-miRNAs contain 2-nt 3′ overhangs characteristic of processing by RNase type III nucleases. Pre-miRNA is then exported via exportin-5 from the nucleus. Cytoplasmic type III endonuclease Dicer then cleaves the stem loop of the pre-miRNA to yield a 22-nt dsRNA. The strand with lowest thermodynamic stability at its 5′ end is chosen as the guide strand and incorporated into the RISC complex. The strand not chosen for incorporation is typically degraded. Guide strand miRNA enables recognition of target mRNA by RISC. Target mRNAs are then either cleaved or translation is inhibited

(Pfeffer et al. 2005). As such, the reader is encouraged to pursue the topics included in this section with the host–microbe interaction in mind.

The miRNA transcriptional unit is termed a primary miRNA (pri-miRNA). The pri-miRNA may possess multiple hairpins or single hairpins (Kim 2005). While most miRNAs are located within an intron of a protein-coding transcript, miRNAs can be found in the exons as well as the protein coding potential of the transcriptional unit (Rodriguez et al. 2004). Finally, some pri-miRNAs are found in intergenic regions whereby they contain their own promoters and are independently transcribed (Zeng 2006).

Transcription of most pri-miRNAs is RNA polymerase II-dependent (Lee et al. 2004). Notable exceptions include the MHV68 miRNAs, whose expression is believed to be driven by a pol III promoter (Pfeffer et al. 2005). Pol II-dependent pri-miRNAs possess 5′ cap structures and are polyadenylated (Cullen 2004). Due to the enormity and diversity of pol II-associated transcription factors, pol II-dependent transcription allows for the exquisite temporal and spatial control characteristic of miRNAs.

Following transcription, an approximately 65-nt hairpin with a 2-nt 3′ overhang is excised from the pri-miRNA by the nuclear Microprocessor (Gregory et al. 2004). Microprocessor consists of the RNase III type endonuclease Drosha in complex with the cofactor DGCR8 (Han et al. 2004). These two components of Microprocessor are both necessary and sufficient to affect cleavage in vitro. Using mutagenesis and in vitro processing assays, it was determined that Microprocessor recognition of pri-miRNA is through the ssRNA flanking strands and the structural motif of the hairpin (Han et al. 2006). Recent studies suggest that pri-miRNA processing to mature miRNAs is a regulated step in miRNA biogenesis. Mature miRNAs have been found in fully differentiated cells, while the pri-miRNAs accumulate in undifferentiated cells (Thomson et al. 2006). Additionally, in human tumors pri-miRNAs are expressed at high levels while the formation of mature miRNA is downregulated (Thomson et al. 2006). This suggests that there may be additional cofactors modulating either substrate specificity and/or activity of Microprocessor.

Following conversion of pri-miRNA to pre-miRNA, Exportin-5, in a Ran-GTP-dependant manner, exports the pre-miRNA from the nucleus to the cytoplasm (Lund et al. 2004). A cytoplasmic RNase III type endonuclease, Dicer, then removes the stem loop from the pre-miRNA converting it to a 22-bp dsRNA with 2-nt 3′ overhangs on each strand (Carmell and Hannon 2004). Evidence suggests that pre-miRNA processing to mature miRNAs is also subject to regulation. It is unclear whether this is due to regulation of pre-miRNA export from the nucleus, or due to the presence of cofactors, which may influence substrate specificity, and/or activity of Dicer cleavage (Lund and Dahlberg 2006; Obernosterer et al. 2006).

Typically, one strand of the 22-bp dsRNA, designated the guide strand, is fated to pair with a target mRNA to facilitate translation inhibition or cleavage of the target mRNA. The guide strand is chosen based on lower thermodynamic stability at its 5′ end in the duplex RNA (Khvorova et al. 2003; Schwarz et al. 2003). Incorporation of the guide strand in the miRISC enables identification of target mRNA for either translation inhibition or cleavage.

Mammalian miRISC is at minimum comprised of Dicer, Argonaute proteins, TRBP, and PACT (Rana 2007). In general, translation inhibition occurs if there is imperfect complementarity of the guide strand with the target mRNA. Cleavage occurs if the complementarity between the guide strand and the target mRNA is perfect. Nucleotides 2–7 of the miRNA constitute the seed region of the miRNA and appear to be critical for miRNA-target mRNA recognition (Jackson and Standart 2007).

The major functional consequence of miRNA biogenesis is alteration of protein concentration through either translation inhibition or cleavage of mRNA. It is important to note that operationally miRNA-mediated inhibition of gene expression is not necessarily an all-or-nothing mechanism. Rather, the present consensus is that miRNAs allows a fine tuning of gene expression (Bartel 2004). miRNAs may be seen as a means by which the cell or virus can achieve a spectrum of gene expression levels appropriate to differentiation or developmental states, proliferative signals, and most recently in response to infection. As such, it is not difficult to imagine a role for viral miRNAs, particularly in those viruses that find persistence and/or latency part of their life cycle.

Location and Conservation of HCMV miRNAs

Mapping of HCMV miRNAs

Using a small RNA cloning and sequencing approach, a total number of 14 mature HCMV miRNAs were identified, which arise from 11 pre-miRNAs (Fig. 1) (Dunn et al., 2005: Grey et al., 2005; Pfeffer et al., 2005). In order to correlate our previous work (Dunn et al. 2003), where we functionally profiled the HCMV genome by constructing a deletion mutant library, with the discovery of HCMV miRNAs, we have mapped the HCMV miRNAs to the genome of the HCMV Towne (HCMV Towne-BAC) strain that was cloned as a bacterial artificial chromosome (Fig. 3). Our annotation of open reading frames (ORFs) within the HCMV Towne-BAC reveals that five HCMV pre-miRNAs are intergenic, four are found within ORFs and two partially overlap annotated ORFs. The vector for bacterial propagation of the Towne-BAC replaces ORFs US1–US12. Additionally, a well-characterized spontaneous deletion of UL150 has occurred on passaging of Towne in tissue culture. To map miRNAs in these regions, we have deferred to other publications for their annotation (Grey et al. 2005; Pfeffer et al. 2005). From the map and data generated by the deletion mutant library, we have determined that six of the miRNAs (mir-US4, mir-US5-1, mir-US5-2, mir US25-2, mir-US33 and mir-UL148D) are not essential for growth in tissue culture. These six miRNAs along with their associated ORFs have been deleted from the viral genome with no apparent resulting defect in viral replication in tissue culture. The possibility remains, however, that these miRNA targeted the ORFs in which they are found, and that concomitant deletion of both the miRNA and its target ORF precluded any phenotype we may have observed if only the miRNA had been disrupted. The two deletions that did result in a phenotype (UL70 and UL114) are in ORFs whose gene products are known to be important for viral infection (Pari and Anders 1993; Courcelle et al. 2001). This precludes determining the contribution that the deletion of miRNAs in these ORFs makes to viral growth

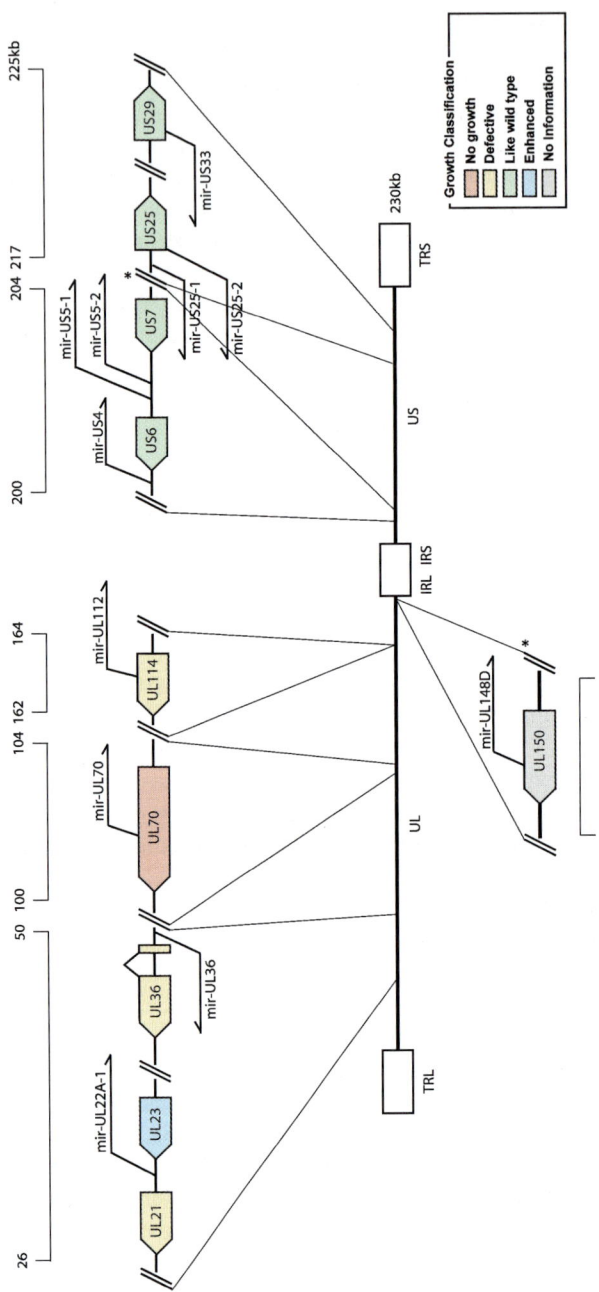

Fig. 3 Map of HCMV miRNAs. Relative positions of HCMV miRNAs were mapped by aligning the sequence of 11 miRNAs found in the miRNA registry with the Towne BAC sequence. The Towne BAC genome (illustrated in the *middle*) consists of two unique regions bracketed by repeat regions. Regions of the genome containing miRNAs have been expanded. The Towne BAC is missing ORFs US1–11 and UL150. To accommodate HCMV miRNAs in these regions, we have used the HCMV Merlin sequence, which possesses a similar arrangement of ORFs as Towne BAC. Growth phenotype of deletion mutants is indicated for ORFs that either contain or immediately surround HCMV miRNAs. Scale in kilobase pairs

in tissue culture. Characterizing the contribution that the remaining HCMV miRNAs (mir-UL22A-1, mir-UL36 and mir-US25-1) make to virus growth in tissue culture awaits the generation of novel recombinant virus.

HCMV is the only betaherpesvirus in which miRNAs have been described, and several aspects of HCMV miRNAs are unique. Firstly, unlike the miRNAs identified in alpha- and gammaherpesviruses, HCMV miRNAs are not found clustered in small regions of the genome, but can be found up to 195 kb apart (Fig. 3). Secondly, due to the diversity of cells permissive to HCMV infection, it is possible to study the expression of viral miRNAs in a number of cell types. Thirdly, because of the lack of a simple in vitro model for HCMV latency, all of the HCMV miRNAs identified in the experiments above have been in cells lytically infected with HCMV. Thus far, all miRNAs identified in prototypic human alpha-(herpes simplex virus 1 [HSV-1] and gamma-Kaposi's sarcoma associated herpesvirus [KSHV] and EBV) herpes viruses have been found in regions of the genome that are transcriptionally active during latent infection. The similarities and differences between HCMV miRNAs and other herpesvirus miRNAs, in terms of their genomic arrangement, expression and conservation will be discussed below.

HCMV miRNA Conservation

Multiple ORFs are conserved throughout the herpesvirus family, which are referred to as core herpesvirus ORFs. As none of the miRNAs identified thus far are conserved among the human herpesviruses, it can be inferred, just as it is for the non-core herpesvirus ORFs, that these miRNAs impart functions unique to the identity and lifestyle of each virus.

Our lab was the first to show the conservation of miRNAs between primate herpesviruses of the same genera, HCMV and chimpanzee cytomegalovirus (CCMV) (Dunn et al. 2005). For purposes of this review, we have extended our analysis to include all HCMV miRNAs identified to date and we report here that at least five of the HCMV miRNAs are 100% conserved in CCMV (Fig. 4). Related herpesviruses are expected to have diverged with their host species. Chimpanzee and human are thought to have diverged approximately 5 million years ago and as such, CCMV is the closest relative of HCMV (Davison et al. 2003).

Similar analysis of the conservation of miRNAs between EBV and the closely related rhesus lymphocryptovirus (rLCV) revealed that at least seven of the EBV miRNAs are conserved in rLCV (Cai et al. 2006). The strength of this report lies in the fact that bioinformatics were not used to identify rLCV miRNAs. Rather, rLCV miRNAs were cloned from infected cells prior to sequence analysis. Interestingly, the rLCV miRNAs that have sequence identity with EBV BART miRNAs possess identical synteny with EBV BART miRNAs (Cai et al. 2006).

Two human herpesviruses, HSV-1 and HSV-2, exhibit close homology. Eight of the miRNAs predicted for HSV-1 were conserved in HSV-2 (Pfeffer et al. 2005; Cui et al. 2006). Of the eight, two of the precursor miRNAs were conserved in HSV-2 (Pfeffer

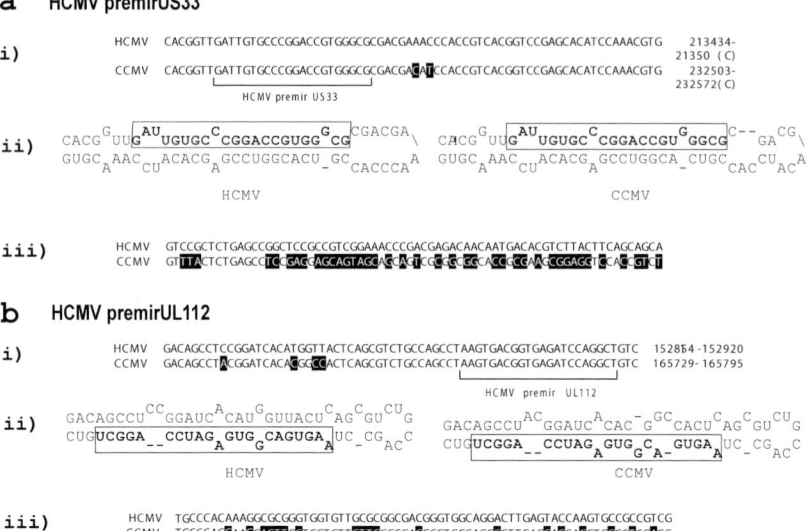

Fig. 4 Conservation of HCMV miRNAs in chimpanzee cytomegalovirus (CCMV). Alignment and secondary structure of two HCMV pre-miRNAs predicted to be conserved in CCMV. HCMV miRNA sequences found in the miRNA registry were aligned with the publicly available CCMV genome sequence (Davison et al. 2003). Sequences for HCMV pre-miRNAs and predicted CCMV pre-miRNAs are shown in (i), secondary structures are compared in (ii), finally alignment of sequences flanking the respective premiRNAs are compared in (iii). Mismatched bases are *highlighted*

et al. 2005; Cui et al. 2006). However, neither of these conserved miRNAs has been demonstrated to be expressed by either virus. Finally, Marek's disease viruses one and two (MDV-1, -2), closely related avian alpha herpesviruses, have been shown to express miRNAs (Burnside et al. 2006; Yao et al. 2007). Interestingly, while these miRNAs are conserved in their genomic location, none of the miRNAs possess sequence homology (Yao et al. 2007). It will be interesting to reevaluate the relatedness of viral miRNAs in terms of their function, as more targets of these miRNAs are identified.

Genomic Arrangement of HCMV miRNAs

As mentioned above, one of the distinguishing features of HCMV miRNAs is that they are widely distributed across the genome (Fig. 3). In contrast, miRNAs identified in all other herpes viruses to date are clustered in short regions of the genome, usually less than 5 kbp. An additional important difference is that all herpesvirus-encoded miRNAs are encoded in regions of the genome that are transcriptionally active during latency. Due to the lack of a simple tissue culture model for HCMV latency, transcriptionally active regions of the viral genome during latency remain controversial.

Genomic arrangement has important implications for miRNA expression. Clustering of miRNAs implies the ability to coordinate miRNA expression. It has recently been shown that the 14 EBV *BART* miRNAs are indeed coordinately expressed, and that expression of the *BART* miRNAs, which lie in the intron(s) of the *BART* mRNA, was correlated with the expression of the *BART* mRNA (Cai et al. 2006). The three *BHRF1* miRNAs are temporally and coordinately expressed (discussed Sect. 4.1 below) predominately by the usage of alternate promoters active in stage III EBV latency (Cai et al. 2006). All twelve KSHV miRNAs identified to date originate from a single cluster approximately 4 kb in length (Gottwein et al. 2006). Of the 12, ten of these miRNAs are intronic, one is in the kaposin ORF, and one is found in the 3′-UTR of the kaposin gene (Cai et al. 2005; Pfeffer et al. 2005). Transcript mapping has identified four different transcripts from which all or only some of the KSHV miRNAs are derived (Cai and Cullen 2006). Interestingly, rLCV and rhesus macaque rhadinovirus (RRV), primate herpesviruses that are distantly related to EBV and KSHV, respectively, possess miRNA clusters that are in similar genomic positions (Schafer et al. 2007).

HCMV pre-miRNAs are mostly intergenic (5) or are coded for on the anti-sense strand of ORFs (4), while the remaining two HCMV pre-miRNAs partially overlap known ORFs (Figs. 1, 3). The distribution of HCMV miRNAs has likely implications for their transcriptional regulation. The clustering of other herpesvirus miRNAs implies their coordinated expression under the control of a master regulatory sequence, while the broad distribution of HCMV miRNAs across the genome implies different mechanisms of transcriptional control. Most likely, each spatially isolated miRNA may have its own regulatory sequence dedicated solely to its transcriptional control and would be operating independently of other miRNAs. Alternatively, expression of HCMV miRNAs may also be coordinated with each other, with an enhancer-like element coordinating expression of multiple miRNAs found far apart on the genome. It will be important to identify the promoter elements and signaling events associated with expression of HCMV miRNAs.

HCMV miRNA Expression

Owing to the complex life cycle exhibited by HCMV, multiple issues regarding the expression of viral miRNAs should be addressed, including expression kinetics, miRNA expression in different cell types and latent vs lytic expression.

Kinetic Classes of HCMV miRNAs

Herpesvirus lytic gene expression is characterized by cascade regulation (Roizman 2001; Roizman and Pellett 2007). As such, there are three kinetic classes of HCMV genes characterized by their dependence on viral protein synthesis and/or viral genome replication. Viral genes that are expressed prior to de novo viral protein synthesis are defined as immediate early genes. Early gene expression requires de novo protein synthesis but not viral genome replication. All other genes are designated as late genes and their expression occurs after viral genome replication.

Work in our lab and the Nelson lab has addressed the kinetics of HCMV miRNA expression. In general, HCMV miRNA levels within the infected cell are observed to increase over the course of the viral replication cycle. Specifically, HCMV miRNA expression is dependent on protein synthesis but independent of viral genome replication and as such, HCMV miRNAs are characterized as early genes (Dunn 2005; Grey 2005). Exceptions include mir-UL22A* (referred to as mir-UL22A-3p in Fig. 1) and mir-UL70. Our experiments show that mir-UL22A* is expressed in cells treated with a protein synthesis inhibitor (Dunn et al. 2005; Grey et al. 2005). This suggests that mir-UL22A* is expressed with immediate early kinetics. Mir-UL70, like mir-UL22A*, is expressed in the absence of protein synthesis (Grey et al. 2005). Curiously however, it was not expressed in the absence of viral genome replication and so does not fall into any of the previously described kinetic classes. miR-UL148D kinetics have not been examined. Mir-US33, reported by Tuschl and colleagues, could not be confirmed by experiments (Grey et al. 2005). A caveat about ascribing kinetics to HCMV miRNAs is that pre-mir-UL36 was shown to accumulate in cells treated with a protein synthesis inhibitor (Grey et al. 2005). Data on other pre-miRNAs in the presence of a protein synthesis inhibitor was not presented. The accumulation of pre-miRNA, and thus the absence of mature miRNA, may be due to the paucity of components in the miRNA biogenesis pathway due to inhibition of protein synthesis. Alternatively, the accumulation of pre-miRNA in cells treated with protein synthesis inhibitors may be due to the absence of host/viral cofactors important for the processing of pre-miRNAs.

Another issue of interest is that most of the HCMV strains used in the studies of HCMV encoded miRNAs were laboratory strains that had been passaged multiple times. Thus, to examine the possibility that miRNA expression from laboratory strains of HCMV may have been altered by their history of extensive passaging, we performed a limited comparison of miRNA expression between the laboratory Towne strain and a HCMV clinical isolate that had undergone very limited passaging in tissue culture (Dunn et al. 2005). Qualitatively speaking, the miRNA expression from the laboratory strain is comparable to that observed in the clinical strain.

Tissue-Specific HCMV miRNA Expression

In animals, miRNA expression is characterized by tissue specific expression patterns. HCMV lytically replicates in a wide variety of cell types both in vivo and in vitro. Viral gene expression has been shown to vary depending on host cell type (Yang et al. 2006; Goodrum et al. 2002, 2007) and work in our lab has identified a number of HCMV gene products that contribute differentially to the replication success of the virus in different cell types (Dunn et al. 2003). Thus, HCMV miRNA expression might also be expected to vary in different cell types.

We have examined HCMV miRNA expression in multiple, clinically relevant cell types (Dunn et al. 2005). These cell types were of epithelial, microglial, and fibroblast origin. We found that at least in these three cell types, all of the HCMV miRNAs tested were expressed (Dunn et al. 2005). Generally, all HCMV miRNA examined were found to be expressed among all the cell types tested. Qualitatively, expression levels of individual HCMV miRNAs did vary between cell types. Specifically, they appeared to be expressed most highly in an astroglial cell line. In infected retinal pigment epithelial cell lines, viral miRNA expression appeared to be the lowest among the cell lines examined. Whether this was due to differential susceptibility of these cell types to viral infection, or in fact a result of cell type-specific differential HCMV miRNA expression is under investigation.

To aid in extending the discussion of tissue-specific expression of herpes virus miRNAs, interesting data have been recently reported concerning the tissue-specific control of viral miRNA expression in EBV. As noted above, EBV has been found to express at least 17 miRNAs. These miRNAs are encoded in two distinct clusters, denoted *BART* and *BHRF1*. While miRNAs derived from the *BART* cluster were detected in all cell lines studied, they were expressed at much higher levels in epithelial cells (Cai et al. 2006). In contrast, the *BHRF1* miRNAs were expressed predominantly in B cells exhibiting stage III EBV latency (Cai et al. 2006). Finally, it was shown that gastric carcinomas harboring EBV expressed the BART miRNA cluster but not the BHRF1 miRNAs (Kim do et al. 2007). Presently, however, it is not clear what role if any the EBV miRNAs play in different cell types and at different stages of viral infection. The impact of specific EBV-encoded miRNAs in infection awaits interference studies, miRNA target identification and analysis of recombinant virus deficient in one or more viral miRNAs.

Prior to the discovery of the MHV68 miRNAs, Stewart and colleagues examined the growth phenotype of a naturally occurring MHV68 variant, MHV-76 (Macrae et al. 2001). MHV76 was isolated from the yellow necked wood mouse (Blaskovic et al. 1980). MHV76 is identical to MHV68 except for a deletion of approximately 10 kbp of genomic DNA from the left end of the unique region (Macrae et al. 2001). This region includes four genes (M1–M4) and all nine MHV68 miRNAs. Identical growth of MHV68 and MHV76 were observed in vitro. However, in vivo growth phenotypes were distinguishable, most notably by a decrease in MHV76 pathogenicity in the spleen and more rapid clearance from the lung (Macrae et al. 2001; Clambey et al. 2002). It is difficult to ascribe the MHV76 in vivo growth to the loss of miRNA and analysis is confounded by the absence of four genes also found in the deleted region. Viruses with deletions in either M2, M3 or M4 were shown to have similar in vivo growth phenotypes (Parry et al. 2000; van Berkel et al. 2000; Jacoby et al. 2002; Geere et al. 2006). Nevertheless, it is interesting that a naturally occurring variant of MHV68, which is lacking the nine MHV68 miRNAs, exists.

Latent Versus Lytic Infection

The finding that, with the exception of HCMV, all human herpesvirus miRNAs originate from regions of the viral genome that are transcriptionally active during latency suggests a role for herpesvirus miRNAs in latency. Indeed, recent data suggest that miRNAs are potentially responsible for maintenance of a latent HSV-1 infection (Gupta et al. 2006). The HSV-1 mir-LAT has been shown to inhibit apoptosis in infected neuronal cells. However, as mir-LAT has not been shown to be expressed during latency, its role in latent infection has yet to be proven.

As mentioned above, there is no simple tissue-culture model in which to study HCMV latency. As such, there are no widely accepted HCMV gene products known to contribute to the establishment and/or maintenance of a latent infection. However, a promising recent cell culture model in which to study HCMV latency has been reported recently(Goodrum et al. 2007). We hope that this will allow future study of a potential role for HCMV miRNAs in latent infection.

Gammaherpesviruses, EBV and KSHV, establish latent infection in B cells. Both can be induced to undergo lytic infection after treatment of infected cells with either TPA or n-butyrate. The expression of EBV and KSHV miRNAs in latent vs lytically infected cells has been studied in such systems.

For KSHV, there was little change in the expression of miRNAs after the switch to lytic infection (Pfeffer et al. 2004, 2005; Cai et al. 2005). The exception from these studies appears to be mir-K12–10, which showed an increase after induction of lytic infection. Mir-K12–10 is located in the 3′-UTR of K12 (Cai et al. 2005; Pfeffer et al. 2005). K12 mRNA expression is increased during lytic infection and thus increased levels of mir-K12–10 are consistent with upregulated expression of K12 mRNA. As mir-K12–10 is the only KSHV miRNA not located in the intron (see Sect. 3.3 above), it is interesting to speculate about whether mir-K12–10's location in the 3′ UTR of K12 and its coincident expression are important to lytic replication.

Potential Function of HCMV miRNAs

Currently, miRNA target prediction lags far behind miRNA identification. As such, just two of the 80 or so viral miRNAs in the miRNA registry have validated targets. Simian Virus 40 (SV40), a polyomavirus, encodes one identified pre-miRNA that yields mature miRNA(s) from each arm of the hairpin (Sullivan et al. 2005). These miRNAs are perfectly complementary to early viral mRNAs and mediate cleavage of transcripts leading to a decrease in viral T antigens (Sullivan et al. 2005). Ganem and co-workers used a mutant virus deficient in the ability to form a pre-miRNA to demonstrate viral miRNA function. Thus, the differential contribution of each arm was not established and we refer to it as a single miRNA with a single function. Interestingly, mutant virus lacking functional miRNAs showed no replicative impairment in vitro and grew to wild type levels (Sullivan et al. 2005). Cells infected with these mutant virus were, however, more susceptible to lysis by cytotoxic T lymphocytes (Sullivan et al. 2005), and thus we would expect that the miRNA mutant virus would be replicatively attenuated in vivo.

Two targets of the HSV-1 mir-LAT were recently identified. Mir-LAT targets the 3'-UTRs of TGF beta 1 and SMAD3 (Gupta et al. 2006), resulting in the disruption of pro-apoptotic pathways (Gupta et al. 2006). Expression of a miRNA from the LAT prompted many to attribute a role for mir-LAT in the maintenance of a latent infection. However, to date mir-LAT has not been studied in the context of a latent infection and its role in the maintenance of HSV-1 latency awaits further verification.

HCMV mir-UL112 has been found to target two viral open reading frames (Grey 2006). Nelson and co-workers reported that mir-UL112 targeted viral immediate early (IE) proteins-1 and -2. IE1 and IE2 are essential for viral infection and possess numerous reported functions, including transactivation and cell cycle inhibition (Mocarski et al. 2007). IE1 is not essential, but a deletion mutant shows a severe growth defect (Mocarski et al. 2007). Mir-UL112 is expressed with immediate early kinetics. The regulation of these two viral proteins is likely to have an important role in viral infection. Further studies are needed to address the functional importance of mir-UL112's negative regulation of IE1 and IE2.

Future Directions

The identification of viral miRNA targets is currently the bottleneck in further understanding the role and function of virally encoded miRNAs. It may be expected that an understanding of HCMV miRNA function will reveal novel mechanisms contributing to the unique replicative success of this herpesvirus. Further, identification of HCMV miRNA targets may well illuminate fundamental and universal components of the host response to microbial infection. Microbes have provided the foundation for many important discoveries in molecular biology. It is our hope that, in addition to expanding therapeutic and prevention strategies, probing the functions of miRNA in the context of viral infection will illuminate fundamental aspects of miRNA-mediated control of gene expression.

Acknowledgements We thank Gerry Abenes, Kihoon Kim, Aaron To, and Eric Sanborn for helpful discussions and comments on the manuscript. Gratitude also goes to Kihoon Kim and Gina Blackledge for contributions to the figures. P.J.R. was supported by a NIH predoctoral training grant (AI007620) and E.Y. acknowledges predoctoral fellowship support from University of California at Berkeley (Graduate Division). The research has been supported by grants from NIH.

References

Ambros V (2004) The functions of animal microRNAs. Nature 431:350–355
Bartel DP (2004) MicroRNAs: genomics, biogenesis, mechanism, and function. Cell 116:281–297

Bentwich I, Avniel A, Karov Y, Aharonov R, Gilad S, Barad O, Barzilai A, Einat P, Einav U, Meiri E, Sharon E, Spector Y, Bentwich Z (2005) Identification of hundreds of conserved and nonconserved human microRNAs. Nat Genet 37:766–770

Blaskovic D, Stancekova M, Svobodova J, Mistrikova J (1980) Isolation of five strains of herpesviruses from two species of free living small rodents. Acta Virol 24:468

Brennecke J, Stark A, Russell RB, Cohen SM (2005) Principles of microRNA-target recognition. PLoS Biol 3:e85

Burnside J, Bernberg E, Anderson A, Lu C, Meyers BC, Green PJ, Jain N, Isaacs G, Morgan RW (2006) Marek's disease virus encodes MicroRNAs that map to meq and the latency-associated transcript. J Virol 80:8778–8786

Cai X, Cullen BR (2006) Transcriptional origin of Kaposi's sarcoma-associated herpesvirus microRNAs. J Virol 80:2234–2242

Cai X, Lu S, Zhang Z, Gonzalez CM, Damania B, Cullen BR (2005) Kaposi's sarcoma-associated herpesvirus expresses an array of viral microRNAs in latently infected cells. Proc Natl Acad Sci USA 102:5570–5575

Cai X, Schafer A, Lu S, Bilello JP, Desrosiers RC, Edwards R, Raab-Traub N, Cullen BR (2006) Epstein-Barr virus microRNAs are evolutionarily conserved and differentially expressed. PLoS Pathog 2:e23

Carmell MA, Hannon GJ (2004) RNase III enzymes and the initiation of gene silencing. Nat Struct Mol Biol 11:214–218

Clambey ET, Virgin HWt, Speck SH (2002) Characterization of a spontaneous 9.5-kilobase-deletion mutant of murine gammaherpesvirus 68 reveals tissue-specific genetic requirements for latency. J Virol 76:6532–6544

Courcelle CT, Courcelle J, Prichard MN, Mocarski ES (2001) Requirement for uracil-DNA glycosylase during the transition to late-phase cytomegalovirus DNA replication. J Virol 75:7592–7601

Cui C, Griffiths A, Li G, Silva LM, Kramer MF, Gaasterland T, Wang XJ, Coen DM (2006) Prediction and identification of herpes simplex virus 1-encoded microRNAs. J Virol 80:5499–5508

Cullen BR (2004) Transcription and processing of human microRNA precursors. Mol Cell 16:861–865

Davison AJ, Dolan A, Akter P, Addison C, Dargan DJ, Alcendor DJ, McGeoch DJ, Hayward GS (2003) The human cytomegalovirus genome revisited: comparison with the chimpanzee cytomegalovirus genome. J Gen Virol 84:17–28

Doench JG, Sharp PA (2004) Specificity of microRNA target selection in translational repression. Genes Dev 18:504–511

Dunn W, Chou C, Li H, Hai R, Patterson D, Stolc V, Zhu H, Liu F (2003) Functional profiling of a human cytomegalovirus genome. Proc Natl Acad Sci USA 100:14223–14228

Dunn W, Trang P, Zhong Q, Yang E, van Belle C, Liu F (2005) Human cytomegalovirus expresses novel microRNAs during productive viral infection. Cell Microbiol 7:1684–1695

Eis PS, Tam W, Sun L, Chadburn A, Li Z, Gomez MF, Lund E, Dahlberg JE (2005) Accumulation of miR-155 and BIC RNA in human B cell lymphomas. Proc Natl Acad Sci USA 102:3627–3632

Geere HM, Ligertwood Y, Templeton KM, Bennet I, Gangadharan B, Rhind SM, Nash AA, Dutia BM (2006) The M4 gene of murine gammaherpesvirus 68 modulates latent infection. J Gen Virol 87:803–807

Goodrum FD, Jordan CT, High K, Shenk T (2002) Human cytomegalovirus gene expression during infection of primary hematopoietic progenitor cells: a model for latency. Proc Natl Acad Sci USA 99:16255–16260

Goodrum F, Reeves M, Sinclair J, High K, Shenk T (2007) Human cytomegalovirus sequences expressed in latently infected individuals promote a latent infection in vitro. Blood 110:937–945

Gottwein E, Cai X, Cullen BR (2006) Expression and function of MicroRNAs encoded by Kaposi's sarcoma-associated herpesvirus. Cold Spring Harb Symp Quant Biol 71:357–364

Gregory RI, Yan KP, Amuthan G, Chendrimada T, Doratotaj B, Cooch N, Shiekhattar R (2004) The Microprocessor complex mediates the genesis of microRNAs. Nature 432:235–240

Grey F, Antoniewicz A, Allen E, Saugstad J, McShea A, Carrington JC, Nelson J (2005) Identification and characterization of human cytomegalovirus-encoded microRNAs. J Virol 79:12095–12099

Grey F, Meyers HL, Nelson JA (2006) HCMV miRNA UL112-1 regulates the expression of two different viral genes (abstract no. 1.03). In: 31st International Herpesvirus Workshop, Seattle, Washington

Griffiths-Jones S (2004) The microRNA Registry. Nucleic Acids Res 32:D109–D111

Griffiths-Jones S (2006) miRBase: the microRNA sequence database. Methods Mol Biol 342:129–138

Griffiths-Jones S, Grocock RJ, van Dongen S, Bateman A, Enright AJ (2006) miRBase: microRNA sequences, targets and gene nomenclature. Nucleic Acids Res 34:D140–D144

Gupta A, Gartner JJ, Sethupathy P, Hatzigeorgiou AG, Fraser NW (2006) Anti-apoptotic function of a microRNA encoded by the HSV-1 latency-associated transcript. Nature 442:82–85

Han J, Lee Y, Yeom KH, Kim YK, Jin H, Kim VN (2004) The Drosha-DGCR8 complex in primary microRNA processing. Genes Dev 18:3016–3027

Han J, Lee Y, Yeom KH, Nam JW, Heo I, Rhee JK, Sohn SY, Cho Y, Zhang BT, Kim VN (2006) Molecular basis for the recognition of primary microRNAs by the Drosha-DGCR8 complex. Cell 125:887–901

Hwang HW, Wentzel EA, Mendell JT (2007) A hexanucleotide element directs microRNA nuclear import. Science 315:97–100

Jackson RJ, Standart N (2007) How do microRNAs regulate gene expression? Sci STKE 2007:re1

Jacoby MA, Virgin HWt, Speck SH (2002) Disruption of the M2 gene of murine gammaherpesvirus 68 alters splenic latency following intranasal, but not intraperitoneal, inoculation. J Virol 76:1790–1801

Kawahara Y, Zinshteyn B, Sethupathy P, Iizasa H, Hatzigeorgiou AG, Nishikura K (2007) Redirection of silencing targets by adenosine-to-inosine editing of miRNAs. Science 315:1137–1140

Khvorova A, Reynolds A, Jayasena SD (2003) Functional siRNAs and miRNAs exhibit strand bias. Cell 115:209–216

Kim do N, Chae HS, Oh ST, Kang JH, Park CH, Park WS, Takada K, Lee JM, Lee WK, Lee SK (2007) Expression of viral microRNAs in Epstein-Barr virus-associated gastric carcinoma. J Virol 81:1033–1036

Kim VN (2005) MicroRNA biogenesis: coordinated cropping and dicing. Nat Rev Mol Cell Biol 6:376–385

Lee RC, Feinbaum RL, Ambros V (1993) The *C. elegans* heterochronic gene lin-4 encodes small RNAs with antisense complementarity to lin-14. Cell 75:843–854

Lee Y, Kim M, Han J, Yeom KH, Lee S, Baek SH, Kim VN (2004) MicroRNA genes are transcribed by RNA polymerase II. EMBO J 23:4051–4060

Lewis BP, Burge CB, Bartel DP (2005) Conserved seed pairing, often flanked by adenosines, indicates that thousands of human genes are microRNA targets. Cell 120:15–20

Lund E, Dahlberg JE (2006) Substrate selectivity of exportin 5 and dicer in the biogenesis of microRNAs. Cold Spring Harb Symp Quant Biol 71:59–66

Lund E, Guttinger S, Calado A, Dahlberg JE, Kutay U (2004) Nuclear export of microRNA precursors. Science 303:95–98

Macrae AI, Dutia BM, Milligan S, Brownstein DG, Allen DJ, Mistrikova J, Davison AJ, Nash AA, Stewart JP (2001) Analysis of a novel strain of murine gammaherpesvirus reveals a genomic locus important for acute pathogenesis. J Virol 75:5315–5327

Mocarski ES, Shenk T, Pass RF (2007) Cytomegaloviruses. In: Fields' virology. Lippincott Williams & Wilkins, Philadelphia, pp 2701–2722

Obernosterer G, Leuschner PJ, Alenius M, Martinez J (2006) Post-transcriptional regulation of microRNA expression. RNA 12:1161–1167

O'Connell RM, Taganov KD, Boldin MP, Cheng G, Baltimore D (2007) MicroRNA-155 is induced during the macrophage inflammatory response. Proc Natl Acad Sci USA 104:1604–1609

Olsen PH, Ambros V (1999) The lin-4 regulatory RNA controls developmental timing in *Caenorhabditis elegans* by blocking LIN-14 protein synthesis after the initiation of translation. Dev Biol 216:671–680

Pari GS, Anders DG (1993) Eleven loci encoding trans-acting factors are required for transient complementation of human cytomegalovirus oriLyt-dependent DNA replication. J Virol 67:6979–6988

Parry CM, Simas JP, Smith VP, Stewart CA, Minson AC, Efstathiou S, Alcami A (2000) A broad spectrum secreted chemokine binding protein encoded by a herpesvirus. J Exp Med 191:573–578

Pfeffer S, Zavolan M, Grasser FA, Chien M, Russo JJ, Ju J, John B, Enright AJ, Marks D, Sander C, Tuschl T (2004) Identification of virus-encoded microRNAs. Science 304:734–736

Pfeffer S, Sewer A, Lagos-Quintana M, Sheridan R, Sander C, Grasser FA, van Dyk LF, Ho CK, Shuman S, Chien M, Russo JJ, Ju J, Randall G, Lindenbach BD, Rice CM, Simon V, Ho DD, Zavolan M, Tuschl T (2005) Identification of microRNAs of the herpesvirus family. Nat Methods 2:269–276

Rana TM (2007) Illuminating the silence: understanding the structure and function of small RNAs. Nat Rev Mol Cell Biol 8:23–36

Rodriguez A, Griffiths-Jones S, Ashurst JL, Bradley A (2004) Identification of mammalian microRNA host genes and transcription units. Genome Res 14:1902–1910

Rodriguez A, Vigorito E, Clare S, Warren MV, Couttet P, Soond DR, van Dongen S, Grocock RJ, Das PP, Miska EA, Vetrie D, Okkenhaug K, Enright AJ, Dougan G, Turner M, Bradley A (2007) Requirement of bic/microRNA-155 for normal immune function. Science 316:608–611

Roizman B, Knipe DM (2001) Herpes simplex viruses and their replication. In: Knipe DM, Howley PM, Griffin DE, Martin MA, Lamb RA, Roizman B, Straus SE (eds) Fields' Virology. Lippincott-Williams &Wilkins, Philadelphia, pp 2399–2460

Roizman B, Pellett PE (2007) The Herpesviridae family: a brief introduction. In: Fields' virology. Lippincott Williams & Wilkins, Philadelphia, pp 2501–2602

Schafer A, Cai X, Bilello JP, Desrosiers RC, Cullen BR (2007) Cloning and analysis of microRNAs encoded by the primate gamma-herpesvirus rhesus monkey rhadinovirus. Virology 364:21–27

Schwarz DS, Hutvagner G, Du T, Xu Z, Aronin N, Zamore PD (2003) Asymmetry in the assembly of the RNAi enzyme complex. Cell 115:199–208

Sullivan CS, Grundhoff AT, Tevethia S, Pipas JM, Ganem D (2005) SV40-encoded microRNAs regulate viral gene expression and reduce susceptibility to cytotoxic T cells. Nature 435:682–686

Taganov KD, Boldin MP, Chang KJ, Baltimore D (2006) NF-kappaB-dependent induction of microRNA miR-146, an inhibitor targeted to signaling proteins of innate immune responses. Proc Natl Acad Sci USA 103:12481–12486

Taganov KD, Boldin MP, Baltimore D (2007) MicroRNAs and immunity: tiny players in a big field. Immunity 26:133–137

Tang G (2005) siRNA and miRNA: an insight into RISCs. Trends Biochem Sci 30:106–114

Thomson JM, Newman M, Parker JS, Morin-Kensicki EM, Wright T, Hammond SM (2006) Extensive post-transcriptional regulation of microRNAs and its implications for cancer. Genes Dev 20:2202–2207

Van Berkel V, Barrett J, Tiffany HL, Fremont DH, Murphy PM, McFadden G, Speck SH, Virgin HI (2000) Identification of a gammaherpesvirus selective chemokine binding protein that inhibits chemokine action. J Virol 74:6741–6747

Wightman B, Ha I, Ruvkun G (1993) Posttranscriptional regulation of the heterochronic gene lin-14 by lin-4 mediates temporal pattern formation in *C. elegans*. Cell 75:855–862

Yang S, Ghanny S, Wang W, Galante A, Dunn W, Liu F, Soteropoulos P, Zhu H (2006) Using DNA microarray to study human cytomegalovirus gene expression. J Virol Methods 131:202–208

Yao Y, Zhao Y, Xu H, Smith LP, Lawrie CH, Sewer A, Zavolan M, Nair V (2007b) Marek's disease virus type 2 (MDV-2)-encoded microRNAs show no sequence conservation with those encoded by MDV-1. J Virol 81:7164–7170

Zeng Y (2006) Principles of micro-RNA production and maturation. Oncogene 25:6156–6162

Mutagenesis of the Cytomegalovirus Genome

Z. Ruzsics, U. H. Koszinowski(✉)

Contents

Introduction 42
CMV Genetics in Cells 42
    Forward (Classical) Genetics. 42
    Reverse Genetics 43
CMV Genetics in Bacteria. 45
    Bacterial Artificial Chromosomes. 45
    Cloning and Maintenance of CMV Genomes as BACs 45
    Allelic Exchange by Shuttle Plasmid Mutagenesis. 46
    Allelic Exchange Using Linear Fragment Mutagenesis 47
    Transposon Mutagenesis for Reverse and Forward Genetics 49
    Mutants, Revertants and the Mutation-Phenotype Connection. 51
Genetic Analysis of Essential Genes. 53
    Comprehensive Mutational Analysis of Essential CMV Genes 53
    Identification and Analysis of Dominant Negative Mutants. 55
Concluding Remarks 57
References 57

Abstract Bacterial artificial chromosomes (BACs) are DNA molecules assembled in vitro from defined constituents and are stably maintained as one large DNA fragment in *Escherichia coli*. Artificial chromosomes are useful for genome sequencing programs, for transduction of DNA segments into eukaryotic cells, and for functional characterization of genomic regions and entire viral genomes such as cytomegalovirus (CMV) genomes. CMV genomes in BACs are ready for the advanced tools of *E. coli* genetics. Homologous and site-specific recombination, or transposon-based approaches allow for the engineering of virtually any kind of genetic change.

U.H. Koszinowski
Max von Pettenkofer Institute, Dept. of Virology, Gene Center,
Ludwig-Maximilians-University, 81377 Munich, Germany
koszinowski@mvp.uni-muenchen.de

Introduction

Animal viruses use a small set of genes to profoundly affect functions of complex hosts. The goal of virus genetics is to understand virus-host interactions at the molecular level. Collection or construction of virus mutants is obligatory for this goal. Until recently, the manipulation of herpesvirus genomes was bound to construction of mutants by homologous recombination in infected cells. Bacterial artificial chromosomes (BACs) are large, circular single-copy episomes of *Escherichia coli* which are suited for maintenance of very large foreign DNA fragments including genomes of large DNA viruses, as first demonstrated by Luckow and colleagues for baculovirus (Luckow et al. 1993). In 1997, we pioneered the BAC-based genetic analysis of herpesviruses with the first example of the cloning and mutagenesis of infectious mouse CMV (MCMV) genome in *E. coli* (Messerle et al. 1997). BAC technology has become a general approach in herpesvirus genetics and beyond for analysis of large and/or unstable viral genomes (Almazan et al. 2000; Ruzsics et al. 2006). This review discusses the different strategies for generation of recombinant herpesviruses and shows the potential of BAC-based herpesvirus genetics. Examples are taken mainly from human CMV (HCMV) and MCMV; reference is also given to work on other herpesviruses.

CMV Genetics in Cells

Forward (Classical) Genetics

The objects of viral genetics are the mutant alleles, where changes in genetic material result in phenotypic alterations that can be analyzed. The frequency of spontaneous mutations in the DNA genomes of herpesviruses ranges between 10^{-8} and 10^{-11} per incorporated nucleotide. Therefore, mutagens have been used to increase the rate of mutations during virus replication by a procedure called in vivo mutagenesis (Schaffer 1975; Schaffer et al. 1984). The characterization of conditional alleles, such as temperature sensitive (*ts*) mutations, has been favored, because both isolation of mutants and their analysis required operational viability. *Ts* mutants are a result of a missense point mutation that alters the primary amino acid sequence of the encoded protein, leading to a loss of function only at a higher (restrictive) temperature. The demanding step is the genetic mapping of the causative mutations. The methods for genetic mapping of *ts* mutations, such as cross-complementation and marker rescue assays, are time-consuming and work-intense. Yet, until recently, *ts* mutants were the only generally applicable tools for studying null phenotypes of essential herpesvirus genes.

Chemical mutagenesis has also been applied to HCMV and MCMV genetics. Early work on HCMV led to the classification of complementation groups

(Yamanishi and Rapp 1977; Ihara et al. 1978). Only few HCMV *ts* mutants could be mapped and associated with open reading frames (ORFs) (Dion et al. 1990; Ihara et al. 1994). The analysis of MCMV *ts* mutants has mainly centered on direct in vivo studies using mutants without mapping (Akel et al. 1993; Akel and Sweet 1993; Bevan et al. 1996; Gill et al. 2000). To our knowledge, none of the *ts* mutations of MCMV has so far been associated with a specific genetic locus. Recently, targeted *ts* mutants for the HCMV IE 2 gene were constructed using BAC technology (Heider et al. 2002).

Reverse Genetics

The ability to clone viral DNA fragments, to propagate and manipulate those recombinant plasmids in *E. coli*, opened several opportunities for the analysis of isolated viral genes. Subcloned viral genes, after mutation and functional analysis in vitro, could be reintroduced into the viral genome to investigate the phenotype in the genomic context (reverse genetics). First, the herpes simplex virus (HSV) genome was used to construct site-directed genome mutants (Mocarski et al. 1980; Smiley 1980; Post and Roizman 1981). After the availability of sequences, site-directed mutagenesis also became an option for CMVs (Spaete and Mocarski 1987; Manning and Mocarski 1988). In principle, a marker gene is introduced into the viral genome by homologous recombination in infected cells thereby disrupting or deleting a viral gene (Fig. 1a). The recombination is under the control of the cellular recombination machinery. It is a rare event and the *wt* virus dominates the resulting progeny pool. Therefore, labeling or even positive selection of the mutants is essential for their isolation by markers such as the β-galactosidase (Spaete and Mocarski 1987; Manning and Mocarski 1988), the neomycin resistance gene (Wolff et al. 1993), and the xanthine guanine phosphoribosyltransferase (gpt) gene (Mulligan and Berg 1981; Greaves et al. 1995).

The advent of cosmid vectors with a capacity to maintain larger DNA fragments (20-50 kbp) in *E. coli* led to cloning of the HSV genome as a set of overlapping cosmid clones (van Zijl et al. 1988). Infectious *wt* virus is reconstituted after co-transfection of the overlapping cosmid set into permissive cells via multiple homologous recombination (Fig. 1b). For mutagenesis, the genetic change is introduced into one of the cosmid fragments. The mutant virus is reconstituted by co-transfection of the mutated cosmid clone with the other cosmids. HSV cloning was followed by cosmid-based reconstruction of HCMV and MCMV (Kemble et al. 1996; Ehsani et al. 2000). The utility of cosmids for generation of mutant CMVs was demonstrated by generation of 17 different UL54 mutants of HCMV (Cihlar et al. 1998). The huge advantage of the cosmid approach over the direct recombination methods is the absence of the *wt* genome. The size of the cloned viral fragment to be mutated allows standard in vitro mutagenesis techniques applicable for plasmids. However, the genetic instability of the system during

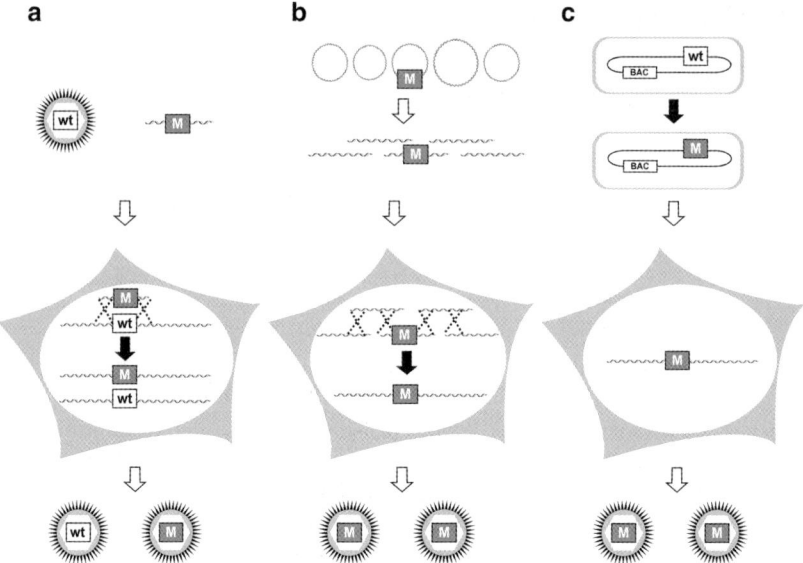

Fig. 1 Different methods of herpesvirus mutagenesis. **a** Site-directed mutagenesis in eukaryotic cells. A linear DNA fragment containing a marker along with the mutation (*M*) flanked by homologies to the viral target sequence is transfected into virus infected cells. By homologous recombination (*dashed lines*) the marker gene and the mutation insert into some of the virus genomes deleting the wild type sequence (wt) Recombinant viruses and wild type viruses need further separation. **b** Cosmid mutagenesis in eukaryotic cells. Overlapping viral fragments spanning the entire genome are cloned as cosmids. A mutation (*M*) is introduced into one fragment. After transfection of the linearized cosmid clones into permissive cells, the virus genome can be reassembled by several homologous recombination steps generating the mutant virus. **c** Principle of the mutagenesis with bacterial artificial chromosomes (BACs) in *E. coli*. Recombinant viral BACs can be generated using various site-directed and random mutagenesis approaches. Recombinant viral BAC DNA with a mutation (*M*) is then transfected into permissive eukaryotic cells and the mutant virus progeny is thereby reconstituted

virus reconstitution became an issue (Horsburgh et al. 1999). Since virus reconstitution relies on several recombination events in eukaryotic cells, changes in the recombinant genomes may occur (Kemble et al. 1996). The cosmid-based mutagenesis is based on recombination in cells and requires the regeneration of a replication-competent virus genome. Therefore, mutant genomes in which an essential gene is affected are difficult, if not impossible, to construct. Unfortunately, revertants cannot be constructed without unreasonable efforts, since the generation of a revertant in its strict sense would require cosmid cloning of the newly generated recombinant genome. Therefore, the usage of this elegant method was restricted to mutagenesis of genes with a known phenotype.

CMV Genetics in Bacteria

Bacterial Artificial Chromosomes

The desire to clone large eukaryotic genomes in order to acquire contiguous physical chromosome maps brought emphasis to cloning vectors of larger insert capacity. Although yeast artificial chromosomes (YACs) can encompass DNA fragments larger than 1000 kbp, YACs are marred by spontaneous rearrangements, insert instability and yeast DNA contamination (Ramsay 1994; Schalkwyk et al. 1995). Stable maintenance of foreign DNA larger than 300 kbp in size was reported using either a fertility factor (F-factor) replicon based bacterial artificial chromosome (BAC) (Shizuya et al. 1992) or a bacteriophage P1 replicon-based cloning system called PAC (Ioannou et al. 1994). In contrast to YACs or cosmids, the BAC clones show surprising sequence stability in appropriate strains of *E. coli*. As a rule, most of the useful BAC hosts are derivatives of DH10B, pointing to the importance of the genetic background of the *E. coli* host strain (Shizuya et al. 1992; Tao and Zhang 1998). Human genome fragments as BACs were maintained over 100 generations in bacteria without detectable changes (Shizuya et al. 1992). The strict control of the F-factor-based replicon keeps one copy of the BAC per cell and reduces recombination events via repetitive DNA elements present in the eukaryotic DNAs. BACs now play a central role in genome research.

Cloning and Maintenance of CMV Genomes as BACs

We pioneered BAC cloning and mutagenesis of MCMV (Messerle et al. 1997). This concept was quickly taken up by several groups. By now, many genomes of many herpesviruses, including different strains of CMVs from various species, have been cloned (for review see Brune et al. 2000). The construction of a herpesvirus BAC starts with conventional mutagenesis procedures. First, the BAC vector sequences flanked with appropriate viral sequences are introduced into the genome by homologous recombination in cells. The linear double-stranded DNA genome of herpesviruses circularizes after infection and these replication intermediates of the BAC-containing herpesvirus genome are transferred by transformation into *E. coli*. This transformation step is needed only once. In *E. coli*, virus functions do not need to be expressed for either genome amplification or mutagenesis procedures. Potential size constraints of the CMV genomes with regard to packaging limits due to the oversize of the inserted BAC cassette can be solved by deletion of nonessential genomic sequences. (Messerle et al. 1997; Borst et al. 1999; Wagner et al. 1999). The deleted sequences can be reinserted after the cloning procedure (Wagner et al. 1999). To regenerate infectious virus, the herpesvirus BACs are transfected into permissive host cells (Fig. 1c). Herpesvirus genomes in BACs are ready for the advanced tools of *E. coli* genetics, which include homologous and site-specific recombination, or transposon-based methods. Here, the techniques have often been pioneered by research labs outside the field of virology and are discussed as they have been adapted to CMV genetics.

Allelic Exchange by Shuttle Plasmid Mutagenesis

Allelic exchange by shuttle plasmids shares aspects of conventional reverse herpesvirus genetics. The desired mutation is cloned into a temperature-sensitive suicide plasmid and flanked by viral sequences homologous to the genomic target site. The shuttle plasmid is then transformed into the BAC carrying *E. coli*, which expresses RecA. The exchange between the mutated and *wt* sequence requires two homologous recombination events (Fig. 2). In the first step, the shuttle plasmid and the BAC recombine via one homology arm resulting in a co-integrate. At the nonpermissive

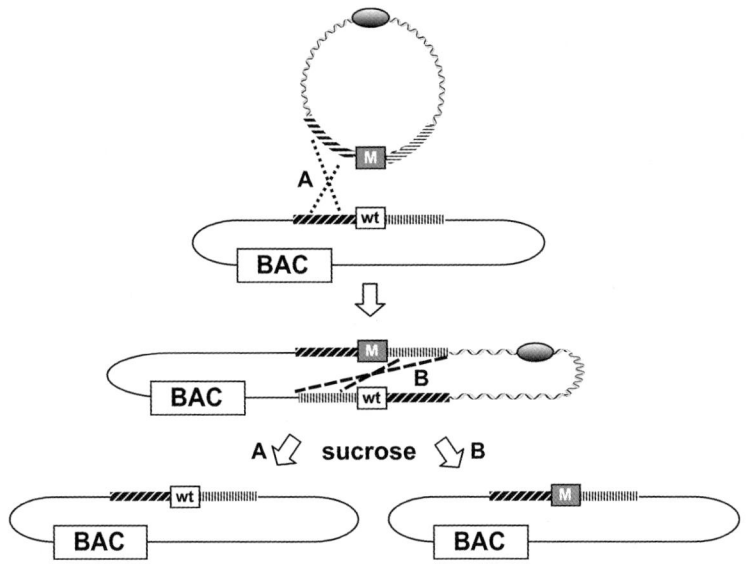

Fig. 2 Allelic exchange using shuttle plasmids. A temperature-sensitive suicide shuttle plasmid that contains the desired mutation (*M*) and viral homologies up- and downstream to the viral target sequence (*dashed lines*) is transformed into *E. coli* carrying the viral BAC. RecA-mediated homologous recombination via one homology arm (*A*) leads to cointegrate formation, which is selected by the antibiotic-resistance genes of the viral BAC and the shuttle plasmid. Free shuttle plasmids are lost at a nonpermissive temperature. In the second recombination step, the cointegrate is resolved. Recombination via the same homology arm (*A*) leads to a wild type (wt) viral BAC (*on the left*), recombination via the other homology arm (*B*) leads to generation of mutant (M) viral BAC (*on the right*) Unresolved cointegrates and shuttle plasmids are eliminated by sucrose counter selection against SacB (*shaded sphere*), which is present in the shuttle plasmid backbone

temperature the nonintegrated shuttle plasmid is lost. Then the selected co-integrate is resolved by a second recombination event (Fig. 2). There are two possibilities: if the second recombination occurs via the homology arm of first step, *wt* BAC is reconstituted. If the second recombination occurs via the other homology arm, a mutant is gained. Only a minority of the co-integrates undergo RecA-mediated resolution (O'Connor et al. 1989; Messerle et al. 1997). The inclusion of sacB (Steinmetz et al. 1983; Blomfield et al. 1991) in the shuttle plasmid allows counter-selection against unresolved co-integrates. Several herpesvirus mutants have been generated using this method (Angulo et al. 1998, 2000a, 2000b; Wagner et al. 1999, 2000; Hobom et al. 2000; Brune et al. 2001a; Sanchez et al. 2002). The homologous recombination mediated by RecA prefers long (1-3 kbp) homologies, which need to be cloned along with the mutation into the shuttle plasmid. Therefore, the construction of the shuttle plasmid may need several cloning steps. Shuttle plasmid-based allelic exchange allows the neat introduction of any kind of mutation (point mutation, deletion, insertion, sequence replacement) into a viral BAC without leaving any operational trace in the genome and represents a method of choice when a complex work is concentrated on one specific genomic region.

Allelic Exchange Using Linear Fragment Mutagenesis

Stewart and colleagues described a one-step mutagenesis method called ET recombination, which uses the recombination functions recET from prophage Rac or the functions redαβ from bacteriophage λ for introduction of mutations into a circular DNA by in vitro-generated linear fragments (Zhang et al. 1998; Muyrers et al. 2000). We and others adapted this method to mutagenesis of viral BACs (Adler et al. 2000; Borst et al. 2001; Kavanagh et al. 2001; Schumacher et al. 2001; Dorange et al. 2002; Rudolph and Osterrieder 2002; Strive et al. 2002; Tischer et al. 2002; Wagner et al. 2002). A linear DNA fragment containing a selectable marker and homologous sequences flanking the target site are transferred into recombination proficient *E. coli* carrying the target BAC. It is important to prevent the degradation of the transformed linear DNA. Therefore, either exonuclease-negative bacteria are used or the exonuclease inhibitor red γ from bacteriophage λ is co-expressed with the recombinases. The selectable marker along with the mutation is introduced into the BAC by a double crossover event (Fig. 3). Compared to the RecA-mediated two-step recombination with shuttle plasmids, ET recombination has advantages. The RecET or redαβ expression allows exact recombination between homologies as short as 25-50 nts. Therefore, the homology arms including the mutated sequence can be provided by synthetic oligonucleotide primers, which are used to amplify the selection cassette. This form of BAC engineering is termed ET cloning, ET recombination, recombinogenic engineering, or recombineering. Many systems have been published that use different or altered recombinases and/or different expression systems controlling their expression. Recombineering facilitates many kinds of genomic experiments that have otherwise been difficult

Fig. 3 Allelic exchange using linear PCR fragments. **a** A selectable marker gene (*open box*) that can be flanked by FRT sites (*black triangle in a square*) is amplified by PCR using a contiguous primer pair (*arrows*) The primers contain homologies of 35-50 nt to the viral target sequence at their 5'-ends (*hatched lines*), a mutation (*M*) and priming regions to the selectable marker gene (*black lines*) **b** The generated linear PCR fragment is transformed into *E. coli* carrying the viral BAC and containing recombinases and an exonuclease inhibitor (red$\alpha\beta\gamma$). The desired mutation along with the selectable marker and the FRT sites is introduced into the viral BAC by double crossover. **c** Additional expression of the site-specific recombinase FLP leads to the excision of the selectable marker reducing the operational sequences to only one FRT site

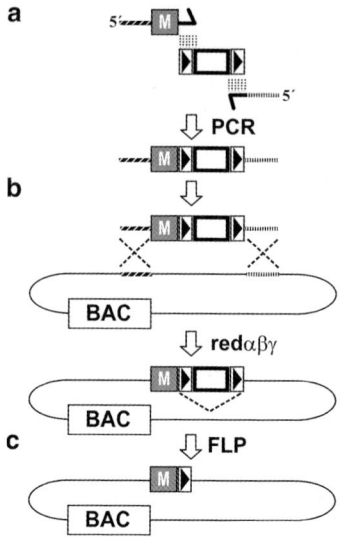

to carry out. The mutagenesis is independent of specific sequence elements; thus the site of mutagenesis can be freely chosen. A risk is the instability of the viral BAC during this mutagenesis procedure since presence of even short repeated sequences in the target genome can lead to unwanted recombination events. Replacing *wt* sequences with a positive selection marker requires only one recombination step and it is easy to create knockout mutants. In addition, it operates practically without background because the positive selection allows the survival of only the desired recombinants and directly sorts out the genome rearrangements induced by the repeat regions. Recently, a comprehensive set of individual deletion mutants of all HCMV genes have been generated by recombineering for functional profiling of the entire genome (Yu et al. 2003). However, this procedure leaves operational trace in the mutated genomes, namely the bacterial selection marker, which is associated with the risk of unpredictable polar effect on usually complex viral transcription units. To lower the size and the risk of the mutational traces, the selectable marker can be flanked with FRT (FLP recognition target) sites (Cherepanov and Wackernagel 1995). These sites allow excision of the marker in a second step by Flp recombinase in *E. coli* (Fig. 3c), leaving only approximately 70-100 nt extra sequence around the introduced mutation (Wagner and Koszinowski 2004).

From the very beginning of BAC recombineering, attempts were made to construct mutants without operational trace. All strategies that are useful for the size of herpesvirus BACs apply two consecutive steps of homologous recombination. First, a combined marker, which allows both positive and counter-selection in *E. coli*, is introduced at the targeted site, resulting in an intermediate that is isolated by positive selection. Next, by a second round of recombination, the markers are replaced with the desired sequence and the right recombinant can be enriched by a pressure against the counter-selection marker. This step, however, is error-prone. First of all, there is no counter-selection system known that works with 100% efficiency. This constitutive leakiness of any counter-selection provides a background that can be critical because the efficiency of the ET/red recombination is low. Second, any counter-selection marker can mutate and the mutated alleles will also appear as a background in large-scale cultures of bacteria proficient for recombination. Third, the selection is applied for the loss of the marker; therefore any unwanted rearrangement induced by repeated sequences in the target genome leading to the loss of the marker will produce selectable recombinants. The first two problems are associated with the counter-selection system used. There are two procedures that seem to be efficient enough for recombineering herpesvirus BACs, namely galK and the I-SceI meganuclease-based counter-selection (Warming et al. 2005; Tischer et al. 2006). The risk of genome rearrangements is controlled at best by the fine-tuning of recombinase expression. However, the instability of the BACs is also influenced by the repeat regions within the specific target genomes (Adler et al. 2000; Warming et al. 2005). Unfortunately, herpesvirus BACs abound in repeats of any kind. Not surprisingly, the approaches of traceless recombineering have been tailored to these genomes only long after the first reports on linear fragment mutagenesis of MCMV BAC (Tischer et al. 2006).

Transposon Mutagenesis for Reverse and Forward Genetics

Transposons (Tns) are mobile genetic elements that insert themselves into a DNA molecule (Craig 1997). After transfer of a Tn-donor plasmid into *E. coli*, the Tn can jump into the viral BAC (Fig. 4). The temperature-dependent suicide Tn donor plasmid is eliminated at the restrictive temperature. Some Tns preferentially insert into the negatively coiled plasmids (e.g., Tn1721) and allow direct isolation of mutated BACs (Brune et al. 1999). Others, like Tn5 or Tn10, are less selective and mutated BACs need to be enriched by a retransformation round (Smith and Enquist 1999). The Tn insertion is determined by sequencing from primer sites within the Tn (Brune et al. 1999) (Fig. 4b). Large libraries of mutant BAC genomes can be established and screened for mutants of specific genes or gene families (Fig. 4c). Even a comprehensive library of transposon mutants classified all known genes of HCMV and guinea pig CMV with regard to their influence on the virus growth in vitro (Yu et al. 2003; McGregor et al. 2004).

A support for genetics applications based on large libraries of randomly generated Tn insertion mutants is the usage of invasive bacteria as vehicles of virus reconstitution. Certain *Salmonella* strains and *E. coli* strains expressing the bacterial gene invasin and listeriolysin can invade mammalian cells and release plasmids. Experimental transfer of a engineered plasmid-encoded transcription units by invasive bacteria to mammalian cells has been shown both in vitro and in vivo (Darji et al. 1997; Grillot-Courvalin et al. 1998). Accordingly, the MCMV-BAC was

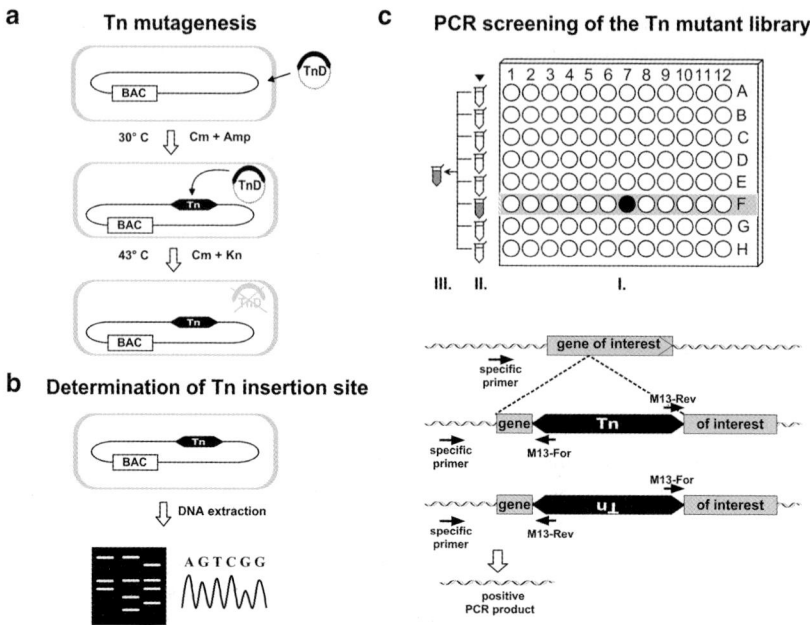

Fig. 4 Random transposon mutagenesis of herpesvirus BACs and screening for the transposon insertion site. **a** Transposon mutagenesis. A temperature-sensitive transposon donor plasmid (TnD) is transformed into *E. coli* carrying the herpesvirus BAC. If a plasmid-selective transposon is used, transposition leads to insertion of the transposon (*Tn*) into the BAC under antibiotic selection for chloramphenicol (*Cm*) and ampicillin (*Amp*) at the permissive temperature (30°C). By following selection with kanamycin (*Kn*) for the transposon and Cm for the BAC at a nonpermissive temperature (43°C), the donor plasmid gets lost and for transposon-inserted BACs can be selected. **b** Determination of the transposon insertion site. From individual *E. coli* clones carrying a viral BAC with Tn insertion, DNA can be extracted and used for sequencing. Forward and reverse M13 primer binding sites at both ends of the transposon allow sequencing from the Tn into the viral sequence, thereby determining the exact nucleotide position of transposon insertion. **c** PCR screening of the transposon mutant library contained in several 96-well plates. Eight rows (A-H) of 12 individual *E. coli* clones (1-12) from each 96-well plate (*I*) are pooled together into eight single vials (*II*). These eight vials are then pooled into one master vial (*III*) containing 96 different *E. coli* clones with mutant BAC DNA. DNA extracted from individual master vials is used for the PCR screening reaction. Here one specific primer binding up- or downstream to the viral gene of interest and the M13 forward (*M13-for*) and reverse (*M13-Rev*) primers binding to the ends of the transposon are used. A clone with a Tn insertion within the gene of interest gives a positive PCR product generated by the specific primer with one of the M13 primers. The master vial with the positive PCR product is selected and the corresponding eight vials are tested using the same primers. After identification of the corresponding positive vial, all 12 individual *E. coli* clones from this vial are tested and the clone with the transposon insertion within the gene of interest is identified (*black circle*)

released under invasive conditions in host cells, leading to virus reconstitution (Brune et al. 2001b). This procedure allows reconstitution of hundreds of random mutant viruses, thereby setting up direct screens for specific phenotypes and selects for nonessential genes because genomes in which Tns disable essential genes do not give rise to progeny (Brune et al. 2001b; Menard et al. 2003; Zimmermann et al. 2005). In its random approach, it is most comparable to the classical herpesvirus genetics by chemical mutagenesis. The advantage over the older method is the immediate access to the mutated position. However, Tn mutants identify nonessential gene functions, whereas *ts* mutants characterize essential genes. Interestingly, *E. coli* carrying the infectious herpesvirus BAC can also reconstitute the infectious virus in the natural host. However, only high bacterial load and the parenteral route gave rise to a barely detectable MCMV progeny in vivo (Cicin-Sain et al. 2003).

Mutants, Revertants and the Mutation-Phenotype Connection

The experimental mutations but also spontaneous mutations elsewhere in the genome - or both - may characterize a phenotype. To date, the size of herpesvirus genomes precludes sequencing as a method to completely exclude the contribution of unwanted mutations. Restriction pattern and Southern blot analysis can exclude only gross genomic alterations. Therefore, different experimental approaches have been developed to confirm the observed mutation/phenotype correlation (Fig. 5).

The local genetic change is considered as the causative principle when at least two independent mutants are evaluated and resulted in the same phenotype. The same conclusion is drawn when a revertant is associated with the *wt* phenotype. Both controls support the direct correlation between the targeted site and the observed phenotype. In the *ts* mutagenesis, independent mutants of the same locus confirm the mutation-phenotype connection. CMV-BAC mutants permit both construction of independent mutants of the same locus and generation of a revertant to the *wt* sequence. Beside the shuttle plasmid method, the recent developments in linear fragment mutagenesis provide convenient access to both kinds of controls.

The genotype-phenotype connection, however, cannot answer the question of whether the phenotype is caused by the mutation of the intended gene of interest, or is a consequence of polar effects on other genetic features. Traditionally, a linkage between a specific mutation and a gene product is proven when the principle of the *wt* phenotype can be restored by providing the wt gene product in trans. *Trans-*complementation of CMV gene products is a plausible approach but is associated with major technical difficulties because of the slow virus growth and the limited access of suitable cell lines for genetic engineering. Toxicity of the viral proteins will also inhibit generation *trans*-complementing cell lines.

Recently, a *trans*-complementation of an HCMV gene product by adenovirus mediated transient transduction has been reported which provides a promising alternative approach to deliver CMV genes to infected cells (Murphy et al. 2000).

Cis-complementation via reinserting the deleted gene product at an ectopic position into the viral genome is another option (Borst et al. 2001). An FRT site is inserted at a position into the genome where it does not affect the *wt* properties. This *wt*-like viral BAC is used for mutagenesis and the induced phenotypes can be analyzed and

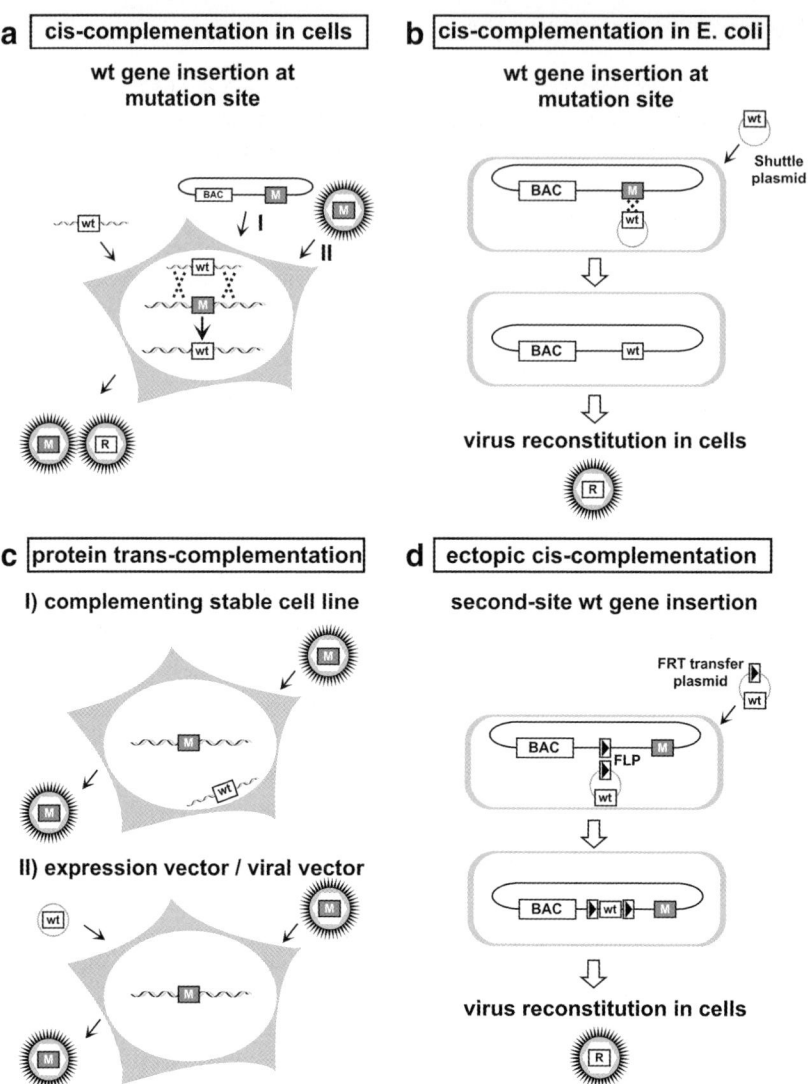

Fig. 5 Different approaches for confirmation of the mutation-phenotype connection. **a** *Cis*-complementation in cells allows reversion of the mutation to the wt sequence. By transfection of cells with the mutant BAC genome (*I*) or infection with the mutant virus (*II*) and co-transfection of a DNA fragment carrying the wild type (wt) sequence and appropriate viral homologies, the mutation (*M*) can be reverted to the wt sequence. Since revertant and mutant viruses need further separation, this approach only works efficiently if one can select for the revertant, e.g., if it has a growth advantage over the mutant virus. **b** *Cis*-complementation of viral BACs in *E. coli* is best performed by shuttle plasmid mutagenesis. The shuttle plasmid carrying the wt sequence and appropriate homologies is introduced into *E. coli* carrying the mutant BAC plasmid. By RecA-mediated homologous recombination, the wt sequence is inserted at the mutation site without leaving any operational sequences. After transfection of the revertant BAC genome into permissive cells, a homogenous revertant population is gained without any further need for selection against mutant viruses. **c** Protein *trans*-complementation in cells. Cells that express the viral wt gene product permanently (*I*) or transiently by an additional expression vector (*II*) are superinfected with the mutant virus. This allows transient complementation of the mutant phenotype if the expression times and levels of the wt gene product are appropriate. **d** Ectopic *cis*-complementation using viral BACs.

compared to *wt*. To generate the phenotypic reversion, first a rescue plasmid is generated by cloning the *wt* gene into an FRT transfer plasmid that carries an FRT site and can only be maintained in special *E. coli* strain by a conditional origin of replication. This rescue plasmid is then reinserted into the mutant genome using the FRT/Flp system. In contrast to the commonly used approaches, in which the FRT sites are located on the same DNA molecule and mediate deletions or inversions, here the FRT sites induce intermolecular recombination, resulting in the unification of the mutant BAC and the rescue plasmid. The recombinants are selected by the antibiotic resistance of the rescue plasmid and the recombinants are reconstituted for analysis of the phenotype. Ectopic reinsertion, like *trans*-complementation, will not restore polar effects at the site of mutation. However, this approach has advantages compared to protein *trans*-complementation by cells: (a) cell toxicity of the viral gene product is not an issue; (b) the gene expression is controlled by the virus life cycle; (c) Complementation works in any cell-type since the complementing gene is expressed from the virus and permits the analysis of in vivo phenotypes; and (d) it does not require cumbersome establishment of *trans*-complementing cell lines. Ectopic *cis*-complementation system has been established in the context of both MCMV and HCMV and was proven for essential and nonessential genes in vitro and in vivo (Borst et al. 2001; Bubeck et al. 2004; Bubic et al. 2004).

Genetic Analysis of Essential Genes

Comprehensive Mutational Analysis of Essential CMV Genes

Detailed genetic analysis of genes involved in DNA replication and packaging, morphogenesis and egress of infectious particles, requires expression of mutant genes in the viral context, where the relevant functions are expressed in operational conditions. High-resolution genetic analysis has been demonstrated by elegant pool screens for genetic foot-printing of viral genes, subgenomic fragments, and even of complete viral genomes up to 10 kbp in size (Laurent et al. 2000; Rothenberg et al. 2001). These screens discriminate between virus mutants that replicate and mutants in which essential functions are affected and therefore cannot be retrieved. The transfection of viral DNA corresponding to the size of a CMV genome is not yet efficient enough for pool reconstitution. Therefore, CMV

◄

Fig. 5 (continued) The wt gene can be introduced into the mutant viral BAC at a neutral second site. By FLP-mediated site-specific recombination of an FRT transfer plasmid carrying the wt gene including regulatory sequences and the mutant BAC genome with an FRT site (*black triangle in a square*), the wt gene product can be expressed from the mutant genome itself. After transfection of permissive cells with this revertant BAC genome, a homogenous population of revertant virus is reconstituted. Only protein *trans*-complementation (**c**) and ectopic *cis*-complementation (**d**) allow the formal confirmation that the mutated gene product (and not possible other *cis*-effects of the mutated sequence) is responsible for the observed phenotype

mutants need to be analyzed individually. The random transposon mutagenesis of the BACs targets the entire genome and null mutants of the respective genes discriminate essential from nonessential genes. Yet, these insertion libraries cannot be used for the functional characterization of a coding sequence. Comprehensive mutant pools of subcloned genes can be obtained through different random mutagenesis procedures. However, a large set of these mutants has to be introduced one by one into the CMV genome lacking the gene of interest to analyze their effect in the context of virus replication. Therefore, we developed a strategy combining a comprehensive Tn7-based linker-scanning mutagenesis of isolated genes (Biery et al. 2000) with fast reinsertion of mutants at an ectopic position into the viral genome by FLP/FRT-mediated site-specific recombination (Fig. 6) as described

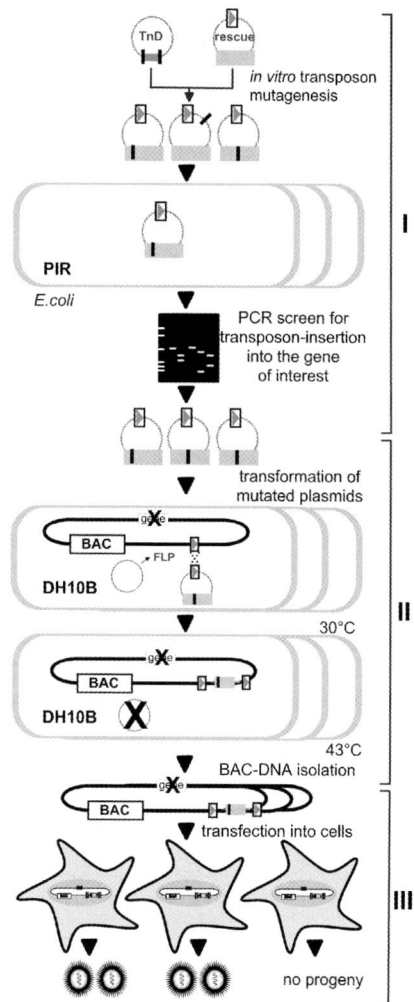

Fig. 6 Scheme of the strategy for random mutagenesis of an essential viral gene in the viral genome context. Part I: In the first step, the viral gene of interest (*gray box*) is subcloned into a rescue plasmid (*rescue*) containing one FRT site (*open box with gray triangle*) This plasmid is subjected to an in vitro Tn7-based random mutagenesis procedure, leading to a mutant library with 15-bp insertions (*black box*) through the target plasmid. This mutant library is transformed into special *E. coli* strain (PIR) that is permissive for the rescue plasmid and single clones are screened by PCR or followed by sequencing to identify insertions within the ORF under study. Part II: To reinsert the gene mutants into the viral genome lacking the gene of interest, the respective deletion mutant-BAC and a FLP recombinase-expressing plasmid (FLP) are maintained in normal *E. coli* strain (DH10B) and transformed with the rescue plasmids. FLP recombinase mediates site-specific recombination between the FRT sites and unifies the BAC and the rescue plasmid. Combined selection identifies the recombinant BACs with the inserted rescue plasmid because the rescue plasmid itself cannot be maintained in normal *E. coli*. The FLP-expressing helper plasmid is removed by elevated temperature. Part III: Subsequently, BAC DNAs are isolated and transfected one by one into eukaryotic cells for virus reconstitution and cells are screened for viral plaques

above for ectopic *cis*-complementation (see Sect. 4.6). The *wt* gene is subcloned in an FRT containing rescue plasmid. This construct is subjected to random Tn mutagenesis, as described for small plasmids. Mutants are sequenced and a comprehensive set of mutants is selected. Recombination between the FRT transfer plasmids carrying the mutants and the BAC lacking the gene of interest, mediated by the FLP recombinase, provides genomes that are *cis*-complemented by the mutant set. The complemented genomes are tested one by one for virus rescue. In combination with standard biochemical or cell biological assays, this procedure allowed genetic analysis of essential gene functions of MCMV at high resolution (Bubeck et al. 2004; Lotzerich et al. 2006). The method easily maps functionally important sites in essential viral proteins.

Identification and Analysis of Dominant Negative Mutants

The function of nonessential genes is studied by gene deletion and by loss-of-function mutants. By *cis*-complementation assays functionally important sites of essential genes can be mapped. Unfortunately, a major target of genetics, the null phenotype of an essential gene is generally hard to come by. This requires the cumbersome establishment and optimization of a *trans*-complementation system for each gene under study. Here, we try to develop a systematic approach. Dominant negative (DN) mutants are special null mutants that induce the null phenotype even in the presence of the *wt* allele. DN mutants of cellular genes have been proven to be a valuable for genetic analysis of complex pathways (Herskowitz 1987). Knowledge of protein structure, protein functions or sequence motifs aid the design of DN mutants (Crowder and Kirkegaard 2005). Unfortunately, the information on the majority of herpesvirus proteins is too limited for knowledge based construction of DN mutants.

Therefore, we set up a random approach to isolate DN mutants of CMV genes. The Tn7-based linker-scanning mutagenesis introduces 5-aa insertions into coding sequences and provides a comprehensive set of subtle insertion mutants of the ORF. Nonfunctional mutants are selected from a library by a *cis*-complementation screen as described above. These mutants can then be introduced into the *wt* MCMV genome. This allows testing their inhibitory potential (Fig. 7a). If the mutant interferes with the function of the essential *wt* allele, virus reconstitution is inhibited. As we showed for both M50 (Rupp et al. 2007) and the M53 (Z. Ruzsics and U.H. Koszinowski, unpublished data), such mutants represent only a small proportion of the null mutants but can be isolated by a standardized procedure.

Transfection of the viral nucleic acid is error-prone and not only a DN function, but also unrelated effects may prevent virus reconstitution. Conditional expression of the inhibitory mutants in the context of the *wt* genome should allow virus reconstitution in the off state and should induce the null phenotype when turned on (Fig. 7b). We constructed a regulated expression system for MCMV (Rupp et al. 2005) in which the constitutive viral expression of the TetR blocks the transcription of the

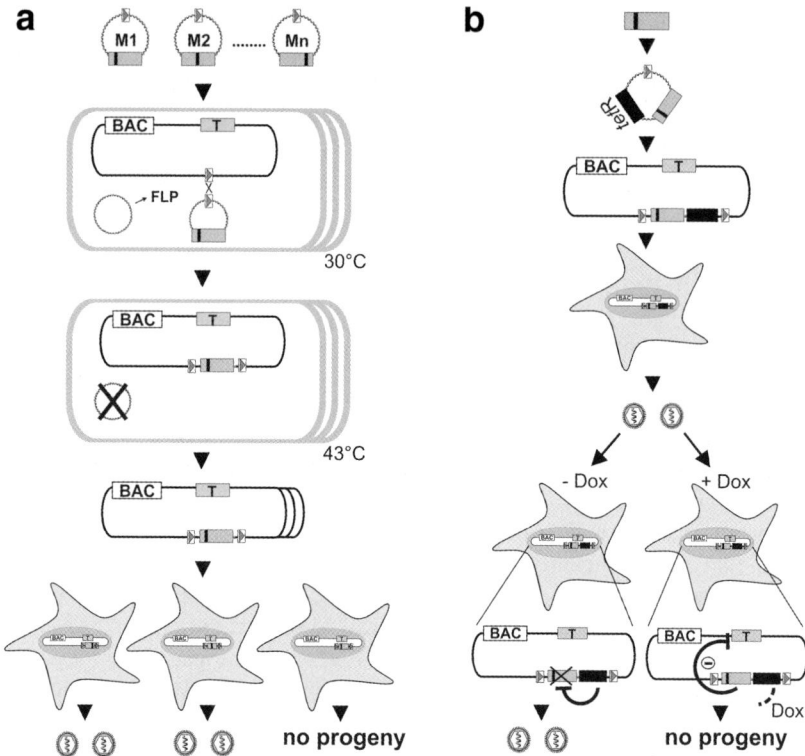

Fig. 7 Screening for and characterization of dominant-negative mutants of essential viral genes. **a** Screening for inhibitory mutants. An essential viral gene, the target gene (*gray box, T*), is subcloned and subjected to a random and comprehensive mutagenesis in vitro leading to a mutant library M1, M2, ... Mn (*small black boxes* indicate mutations). Mutated ORFs are placed under the control of a strong constitutive promoter into an insertion plasmid containing an FRT site (*open box with gray triangle*). The insertion plasmids can only be maintained in a special *E. coli* strain. Normal *E. coli* (*open boxes*) carrying an FRT site-labeled viral bacterial artificial chromosome (*BAC*) and a temperature-sensitive plasmid-expressing FLP recombinase (*FLP*) are transformed with the insertion plasmids carrying different mutants one by one. The FLP recombinase mediates site-specific recombination between the FRT sites in the BAC and the insertion plasmids. This recombinants can then be isolated under combined antibiotic selection for both the BACs and the insertion plasmid. The FLP-expressing helper plasmid is removed by elevated temperature. Then BAC DNA is prepared and permissive cells are transfected with each construct. The mutants that are able to inhibit the virus reconstitution can be selected on the basis of the inability of plaque formation upon transfection. **b** Validation of dominant negative mutants by conditional gene expression. The inhibitory mutants are subcloned under the control of a promoter regulated by the TetR (*black box*) into an insertion plasmid with an FRT site. These constructs are delivered into the viral BAC as described above. Then permissive cells are transfected with the recombinants in order to reconstitute viruses carrying the regulation cassettes for the inhibitory mutants. The inhibitory mutants are not expressed during reconstitution because in the absence of doxycycline (- *Dox*), the constitutively expressed TetR blocks their transcription. The inhibitory function of the mutants can be analyzed upon doxycycline administration (+ *Dox*), which leads to the expression of the inhibitory mutant by releasing the expression cassette from the TetR regulation

regulated gene and induction by doxycycline exposes the viral replication program to the DN mutant. This system allowed us detailed quantitative and qualitative analysis of the effect of DN mutants of both M50 and M53. In addition, the result of the random screen on MCMV M50 aided the construction of a DN mutant of the homolog in HCMV (UL50) (Rupp et al. 2007). We believe that this systematic approach will facilitate the functional analysis of essential CMV genes.

Concluding Remarks

Cloning large DNA sequences as BACs has become the method of choice for mapping, sequencing and manipulation of large eukaryotic genomes. Genetic engineering in BACs is based on homologous recombination and now allows any type of DNA modification. In addition, for herpesviruses, including CMV, these procedures permit the manipulation of the infectious genome as a single plasmid. Mutagenesis is safely carried out in *E. coli* and physical controls can be performed prior to virus reconstitution. These are necessary because CMV genomes contain repetitive sequences that are prone to recombination. In the past, the investigation of CMV gene functions was limited by the laborious and time-consuming generation of virus mutants. Using BAC recombineering, this problem is solved. In the future, more weight has to be placed on careful planning of appropriate controls. The lack of technical limitations allows the production of a plethora of mutants. Traceless mutagenesis permits multiple sequential mutagenesis steps and complex engineering procedures. However, only the local targeted mutations can be thoroughly checked. There is an uncertainty on the numbers of unwanted mutations in other regions that may occur. These mutations do not need to have a growth phenotype in the cell line under study but may mar experiments in other cells or in vivo experiments. According to the law of error propagation, the necessary controls increase with each sequential mutagenesis step. Since at the level of plasmids each step can be controlled by sequencing, as many steps as possible should be done on subcloned target regions. With regard to BAC engineering, it may be advisable to decide on a smaller number of BAC engineering steps rather than on perfect sequence correction.

Acknowledgements The work was supported by the Deutsche Forschungsgemeinschaft through SFB 455, SPP New vaccination strategies, SPP1175 and the VCI.

References

Adler H, Messerle M, Wagner M, Koszinowski UH (2000) Cloning and mutagenesis of the murine gammaherpesvirus 68 genome as an infectious bacterial artificial chromosome. J Virol 74:6964-6974

Akel HM, Sweet C (1993) Isolation and preliminary characterisation of twenty-five temperature-sensitive mutants of mouse cytomegalovirus. FEMS Microbiol Lett 113:253-260

Akel HM, Furarah AM, Sweet C (1993) Further studies of 31 temperature-sensitive mutants of mouse cytomegalovirus: thermal stability, replication and analysis of temperature-sensitive functions by temperature shift. FEMS Microbiol Lett 114:311-316

Almazan F, Gonzalez JM, Penzes Z, Izeta A, Calvo E, Plana-Duran J, Enjuanes L (2000) Engineering the largest RNA virus genome as an infectious bacterial artificial chromosome. Proc Natl Acad Sci USA 97:5516-5521

Angulo A, Messerle M, Koszinowski UH, Ghazal P (1998) Enhancer requirement for murine cytomegalovirus growth and genetic complementation by the human cytomegalovirus enhancer. J Virol 72:8502-8509

Angulo A, Ghazal P, Messerle M (2000a) The major immediate-early gene ie3 of mouse cytomegalovirus is essential for viral growth. J Virol 74:11129-11136

Angulo A, Kerry D, Huang H, Borst EM, Razinsky A, Wu J, Hobom U, Messerle M, Ghazal P (2000b) Identification of a boundary domain adjacent to the potent human cytomegalovirus enhancer that represses transcription of the divergent UL127 promoter. J Virol 74:2826-2839

Bevan IS, Sammons CC, Sweet C (1996) Investigation of murine cytomegalovirus latency and reactivation in mice using viral mutants and the polymerase chain reaction. J Med Virol 48:308-320

Biery MC, Stewart FJ, Stellwagen AE, Raleigh EA, Craig NL (2000) A simple in vitro Tn7-based transposition system with low target site selectivity for genome and gene analysis. Nucleic Acids Res 28:1067-1077

Blomfield IC, Vaughn V, Rest RF, Eisenstein BI (1991) Allelic exchange in *Escherichia coli* using the *Bacillus subtilis* sacB gene and a temperature-sensitive pSC101 replicon. Mol Microbiol 5:1447-1457

Borst EM, Hahn G, Koszinowski UH, Messerle M (1999) Cloning of the human cytomegalovirus (HCMV) genome as an infectious bacterial artificial chromosome in *Escherichia coli*: a new approach for construction of HCMV mutants. J Virol 73:8320-8329

Borst EM, Mathys S, Wagner M, Muranyi W, Messerle M (2001) Genetic evidence of an essential role for cytomegalovirus small capsid protein in viral growth. J Virol 75:1450-1458

Brune W, Menard C, Hobom U, Odenbreit S, Messerle M, Koszinowski UH (1999) Rapid identification of essential and nonessential herpesvirus genes by direct transposon mutagenesis. Nat Biotechnol 17:360-364

Brune W, Messerle M, Koszinowski UH (2000) Forward with BACs: new tools for herpesvirus genomics. Trends Genet 16:254-259

Brune W, Hasan M, Krych M, Bubic I, Jonjic S, Koszinowski UH (2001a) Secreted virus-encoded proteins reflect murine cytomegalovirus productivity in organs. J Infect Dis 184:1320-1324

Brune W, Menard C, Heesemann J, Koszinowski UH (2001b) A ribonucleotide reductase homolog of cytomegalovirus and endothelial cell tropism. Science 291:303-305

Bubeck A, Wagner M, Ruzsics Z, Lötzerich M, Iglesias M, Singh IR, Koszinowski UH (2004) Comprehensive mutational analysis of a herpesvirus gene in the viral genome context reveals a region essential for virus replication. J Virol 78:8026-8035

Bubic I, Wagner M, Krmpotic A, Saulig T, Kim S, Yokoyama WM, Jonjic S, Koszinowski UH (2004) Gain of virulence caused by loss of a gene in murine cytomegalovirus. J Virol 78:7536-7544

Cherepanov PP, Wackernagel W (1995) Gene disruption in *Escherichia coli*: TcR and KmR cassettes with the option of Flp-catalyzed excision of the antibiotic-resistance determinant. Gene 158:9-14

Cicin-Sain L, Brune W, Bubic I, Jonjic S, Koszinowski UH (2003) Vaccination of mice with bacteria carrying a cloned herpesvirus genome reconstituted in vivo. J Virol 77:8249-8255

Cihlar T, Fuller MD, Cherrington JM (1998) Characterization of drug resistance-associated mutations in the human cytomegalovirus DNA polymerase gene by using recombinant mutant viruses generated from overlapping DNA fragments. J Virol 72:5927-5936

Craig NL (1997) Target site selection in transposition. Annu Rev Biochem 66:437-474

Crowder S, Kirkegaard K (2005) Trans-dominant inhibition of RNA viral replication can slow growth of drug-resistant viruses. Nat Genet 37:701-709

Darji A, Guzman CA, Gerstel B, Wachholz P, Timmis KN, Wehland J, Chakraborty T, Weiss S (1997) Oral somatic transgene vaccination using attenuated *S. typhimurium*. Cell 91:765-775

Dion M, Yelle J, Hamelin C (1990) Physical mapping of a temperature-sensitive mutation of human cytomegalovirus by marker rescue. Genet Anal Tech Appl 7:32-34

Dorange F, Tischer BK, Vautherot JF, Osterrieder N (2002) Characterization of Marek's disease virus serotype 1 (MDV-1) deletion mutants that lack UL46 to UL49 genes: MDV-1 UL49, encoding VP22, is indispensable for virus growth. J Virol 76:1959-1970

Ehsani ME, Abraha TW, Netherland-Snell C, Mueller N, Taylor MM, Holwerda B (2000) Generation of mutant murine cytomegalovirus strains from overlapping cosmid and plasmid clones. J Virol 74:8972-8979

Gill TA, Morley PJ, Sweet C (2000) Replication-defective mutants of mouse cytomegalovirus protect against wild-type virus challenge. J Med Virol 62:127-139

Greaves RF, Brown JM, Vieira J, Mocarski ES (1995) Selectable insertion and deletion mutagenesis of the human cytomegalovirus genome using the *Escherichia coli* guanosine phosphoribosyl transferase (gpt) gene. J Gen Virol 76:2151-2160

Grillot-Courvalin C, Goussard S, Huetz F, Ojcius DM, Courvalin P (1998) Functional gene transfer from intracellular bacteria to mammalian cells. Nat Biotechnol 16:862-866

Heider JA, Bresnahan WA, Shenk TE (2002) Construction of a rationally designed human cytomegalovirus variant encoding a temperature-sensitive immediate-early 2 protein. Proc Natl Acad Sci USA 99:3141-3146

Herskowitz I (1987) Functional inactivation of genes by dominant negative mutations. Nature 329:219-222

Hobom U, Brune W, Messerle M, Hahn G, Koszinowski UH (2000) Fast screening procedures for random transposon libraries of cloned herpesvirus genomes: mutational analysis of human cytomegalovirus envelope glycoprotein genes. J Virol 74:7720-7729

Horsburgh BC, Hubinette MM, Qiang D, MacDonald ML, Tufaro F (1999) Allele replacement: an application that permits rapid manipulation of herpes simplex virus type 1 genomes. Gene Ther 6:922-930

Ihara S, Hirai K, Watanabe Y (1978) Temperature-sensitive mutants of human cytomegalovirus: isolation and partial characterization of DNA-minus mutants. Virology 84:218-221

Ihara S, Takekoshi M, Mori N, Sakuma S, Hashimoto J, Watanabe Y (1994) Identification of mutation sites of a temperature-sensitive mutant of HCMV DNA polymerase activity. Arch Virol 137:263-275

Ioannou PA, Amemiya CT, Garnes J, Kroisel PM, Shizuya H, Chen C, Batzer MA, de Jong PJ (1994) A new bacteriophage P1-derived vector for the propagation of large human DNA fragments. Nat Genet 6:84-89

Kavanagh DG, Gold MC, Wagner M, Koszinowski UH, Hill AB (2001) The multiple immune-evasion genes of murine cytomegalovirus are not redundant: m4 and m152 inhibit antigen presentation in a complementary and cooperative fashion. J Exp Med 194:967-978

Kemble G, Duke G, Winter R, Spaete R (1996) Defined large-scale alterations of the human cytomegalovirus genome constructed by cotransfection of overlapping cosmids. J Virol 70:2044-2048

Laurent LC, Olsen MN, Crowley RA, Savilahti H, Brown PO (2000) Functional characterization of the human immunodeficiency virus type 1 genome by genetic footprinting. J Virol 74:2760-2769

Lötzerich M, Ruzsics Z, Koszinowski UH (2006) Functional domains of murine cytomegalovirus nuclear egress protein M53/p38. J Virol 80:73-84

Luckow VA, Lee SC, Barry GF, Olins PO (1993) Efficient generation of infectious recombinant baculoviruses by site-specific transposon-mediated insertion of foreign genes into a baculovirus genome propagated in *Escherichia coli*. J Virol 67:4566-4579

Manning WC, Mocarski ES (1988) Insertional mutagenesis of the murine cytomegalovirus genome: one prominent alpha gene (ie2) is dispensable for growth. Virology 167:477-484

McGregor A, Liu F, Schleiss MR (2004) Identification of essential and non-essential genes of the guinea pig cytomegalovirus (GPCMV) genome via transposome mutagenesis of an infectious BAC clone. Virus Res 101:101-108

Menard C, Wagner M, Ruzsics Z, Holak K, Brune W, Campbell AE, Koszinowski UH (2003) Role of murine cytomegalovirus US22 gene family members in replication in macrophages. J Virol 77:5557-5570

Messerle M, Crnkovic I, Hammerschmidt W, Ziegler H, Koszinowski UH (1997) Cloning and mutagenesis of a herpesvirus genome as an infectious bacterial artificial chromosome. Proc Natl Acad Sci USA 94:14759-14763

Mocarski ES, Post LE, Roizman B (1980) Molecular engineering of the herpes simplex virus genome: insertion of a second L-S junction into the genome causes additional genome inversions. Cell 22:243-255

Mulligan RC, Berg P (1981) Selection for animal cells that express the *Escherichia coli* gene coding for xanthine-guanine phosphoribosyltransferase. Proc Natl Acad Sci USA 78:2072-2076

Murphy EA, Streblow DN, Nelson JA, Stinski MF (2000) The human cytomegalovirus IE86 protein can block cell cycle progression after inducing transition into the S phase of permissive cells. J Virol 74:7108-7118

Muyrers JP, Zhang Y, Stewart AF (2000) ET-cloning: think recombination first. Genet Eng (New York) 22:77-98

O'Connor M, Peifer M, Bender W (1989) Construction of large DNA segments in *Escherichia coli*. Science 244:1307-1312

Post LE, Roizman B (1981) A generalized technique for deletion of specific genes in large genomes: alpha gene 22 of herpes simplex virus 1 is not essential for growth. Cell 25:227-232

Ramsay M (1994) Yeast artificial chromosome cloning. Mol Biotechnol 1:181-201

Rothenberg SM, Olsen MN, Laurent LC, Crowley RA, Brown PO (2001) Comprehensive mutational analysis of the Moloney murine leukemia virus envelope protein. J Virol 75:11851-11862

Rudolph J, Osterrieder N (2002) Equine herpesvirus type 1 devoid of gM and gp2 is severely impaired in virus egress but not direct cell-to-cell spread. Virology 293:356-367

Rupp B, Ruzsics Z, Sacher T, Koszinowski UH (2005) Conditional cytomegalovirus replication in vitro and in vivo. J Virol 79:486-494

Rupp B, Ruzsics Z, Buser C, Adler B, Walther P, Koszinowski UH (2007) Random screening for dominant-negative mutants of the cytomegalovirus nuclear egress protein M50. J Virol 81:5508-5517

Ruzsics Z, Wagner M, Osterlehner A, Cook J, Koszinowski U, Burgert HG (2006) Transposon-assisted cloning and traceless mutagenesis of adenoviruses: Development of a novel vector based on species D. J Virol 80:8100-8113

Sanchez V, Clark CL, Yen JY, Dwarakanath R, Spector DH (2002) Viable human cytomegalovirus recombinant virus with an internal deletion of the IE2 86 gene affects late stages of viral replication. J Virol 76:2973-2989

Schaffer PA (1975) Temperature-sensitive mutants of herpesviruses. Curr Top Microbiol Immunol 70:51-100

Schaffer PA, Weller SK, Pancake BA, Coen DM (1984) Genetics of herpes simplex virus. J Invest Dermatol 83:42s-47s

Schalkwyk LC, Francis F, Lehrach H (1995) Techniques in mammalian genome mapping. Curr Opin Biotechnol 6:37-43

Schumacher D, Tischer BK, Reddy SM, Osterrieder N (2001) Glycoproteins E and I of Marek's disease virus serotype 1 are essential for virus growth in cultured cells. J Virol 75:11307-11318

Shizuya H, Birren B, Kim UJ, Mancino V, Slepak T, Tachiiri Y, Simon M (1992) Cloning and stable maintenance of 300-kilobase-pair fragments of human DNA in *Escherichia coli* using an F-factor-based vector. Proc Natl Acad Sci USA 89:8794-8797

Smiley JR (1980) Construction in vitro and rescue of a thymidine kinase-deficient deletion mutation of herpes simplex virus. Nature 285:333-335

Smith GA, Enquist LW (1999) Construction and transposon mutagenesis in *Escherichia coli* of a full- length infectious clone of pseudorabies virus, an alphaherpesvirus. J Virol 73:6405-6414

Spaete RR, Mocarski ES (1987) Insertion and deletion mutagenesis of the human cytomegalovirus genome. Proc Natl Acad Sci USA 84:7213-7217

Steinmetz M, Le Coq D, Djemia HB, Gay P (1983) Genetic analysis of sacB, the structural gene of a secreted enzyme, levansucrase of Bacillus subtilis Marburg. Mol Gen Genet 191:138-144

Strive T, Borst E, Messerle M, Radsak K (2002) Proteolytic processing of human cytomegalovirus glycoprotein B is dispensable for viral growth in culture. J Virol 76:1252-1264

Tao Q, Zhang HB (1998) Cloning and stable maintenance of DNA fragments over 300 kb in *Escherichia coli* with conventional plasmid-based vectors. Nucleic Acids Res 26:4901-4909

Tischer BK, Schumacher D, Messerle M, Wagner M, Osterrieder N (2002) The products of the UL10 (gM) and the UL49.5 genes of Marek's disease virus serotype 1 are essential for virus growth in cultured cells. J Gen Virol 83:997-1003

Tischer BK, von Einem J, Kaufer B, Osterrieder N (2006) Two-step red-mediated recombination for versatile high-efficiency markerless DNA manipulation in *Escherichia coli*. Biotechniques 40:191-197

van Zijl M, Quint W, Briaire J, de Rover T, Gielkens A, Berns A (1988) Regeneration of herpesviruses from molecularly cloned subgenomic fragments. J Virol 62:2191-2195

Wagner M, Koszinowski UH (2004) Mutagenesis of viral BACs with linear PCR fragments (ET recombination). Methods Mol Biol 256:257-268

Wagner M, Jonjic S, Koszinowski UH, Messerle M (1999) Systematic excision of vector sequences from the BAC-cloned herpesvirus genome during virus reconstitution. J Virol 73:7056-7060

Wagner M, Michel D, Schaarschmidt P, Vaida B, Jonjic S, Messerle M, Mertens T, Koszinowski U (2000) Comparison between human cytomegalovirus pUL97 and murine cytomegalovirus (MCMV) pM97 expressed by MCMV and vaccinia virus: pM97 does not confer ganciclovir sensitivity. J Virol 74:10729-10736

Wagner M, Gutermann A, Podlech J, Reddehase MJ, Koszinowski UH (2002) Major histocompatibility complex class I allele-specific cooperative and competitive interactions between immune evasion proteins of cytomegalovirus. J Exp Med 196:805-816

Warming S, Costantino N, Court DL, Jenkins NA, Copeland NG (2005) Simple and highly efficient BAC recombineering using galK selection. Nucleic Acids Res 33:e36

Wolff D, Jahn G, Plachter B (1993) Generation and effective enrichment of selectable human cytomegalovirus mutants using site-directed insertion of the neo gene. Gene 130:167-173

Yamanishi K, Rapp F (1977) Temperature-sensitive mutants of human cytomegalovirus. J Virol 24:416-418

Yu D, Silva MC, Shenk T (2003) Functional map of human cytomegalovirus AD169 defined by global mutational analysis. Proc Natl Acad Sci USA 100:12396-12401

Zhang Y, Buchholz F, Muyrers JP, Stewart AF (1998) A new logic for DNA engineering using recombination in *Escherichia coli*. Nat Genet 20:123-128

Zimmermann A, Trilling M, Wagner M, Wilborn M, Bubic I, Jonjic S, Koszinowski U, Hengel H (2005) A cytomegaloviral protein reveals a dual role for STAT2 in IFN-{gamma} signaling and antiviral responses. J Exp Med 201:1543-1553

Cytomegalovirus Cell Tropism

C. Sinzger(✉), M. Digel, G. Jahn

Contents

Target Cells of HCMV Infection 64
   Target Cells of HCMV In Vivo 64
   Target Cells of HCMV in Cell Culture 65
   Cell Tropism 67
Pathogenetic Role of Selected Cell Types 68
   Epithelial Cells 70
   Dendritic Cells 70
   Fibroblasts 71
   Smooth Muscle Cells 71
   Endothelial Cells 72
   Leukocytes 72
Cell Biological Basis of HCMV Cell Tropism 73
   Interstrain Differences in Cell Tropism 74
   Viral Genes and Proteins Contributing to Cell Tropism 74
   Critical Events for Replication in Various Cell Types 75
Cell Tropism of Other Cytomegaloviruses 76
Impact of Cell Tropism Analyses 77
References 78

Abstract The human cytomegalovirus (HCMV) can infect a remarkably broad cell range within its host, including parenchymal cells and connective tissue cells of virtually any organ and various hematopoietic cell types. Epithelial cells, endothelial cells, fibroblasts and smooth muscle cells are the predominant targets for virus replication. The pathogenesis of acute HCMV infections is greatly influenced by this broad target cell range. Infection of epithelial cells presumably contributes to interhost transmission. Infection of endothelial cells and hematopoietic cells facilitates systemic spread within the host. Infection of ubiquitous cell types such as fibroblasts and smooth muscle cells provides the platform for efficient proliferation of the virus.

C. Sinzger
Institute of Medical Virology, University of Tübingen, Elfriede-Aulhorn-Str. 6, 72076 Tübingen, Germany
christian.sinzger@med.uni-tuebingen.de

The tropism for endothelial cells, macrophages and dendritic cells varies greatly among different HCMV strains, mostly dependent on alterations within the UL128-131 gene locus. In line with the classification of the respective proteins as structural components of the viral envelope, interstrain differences concerning the infectivity in endothelial cells and macrophages are regulated on the level of viral entry.

Target Cells of HCMV Infection

The question of which cell types in which tissues are targets of HCMV infection derives its relevance from the trivial fact that a virus can only live inside its host cell. The biology of a virus and, even more, its pathogenic effects within the infected organism are therefore inevitably linked to the spectrum of susceptible cell types.

Target Cells of HCMV In Vivo

Generally speaking, HCMV is tightly restricted to humans on the host level, but within the human host it can spread to virtually any tissue due to an exceptionally broad range of target cell types. In fact, it is easier to list the cell types that do not support HCMV replication: despite early reports about some degree of IE gene expression in lymphocytes in cell culture (Rice et al. 1984), we have not found IE antigens in cells of lymphoid origin during extensive immunohistochemical analyses of various organ tissues from acutely infected patients (Sinzger et al. 1995). A similar block of viral replication occurs in polymorphonuclear leukocytes. While these cells can take up virus particles and express IE antigens to some extent, transcripts and proteins of the early and late phase of viral replication are not found (Grefte et al. 1994; Sinzger et al. 1996). These two exceptions are faced by a long list of susceptible cell types, including various cells of ectodermal, mesodermal and endodermal origin. Most prominent examples are epithelial cells of glands and mucosal tissues, connective tissue cells in various organs, smooth muscle cells predominantly in the gastrointestinal tract and vascular endothelial cells (Sinzger et al. 1993; Ng Bautista and Sedmak 1995; Sinzger et al. 1995). Due to the strict host specificity, HCMV infection cannot be studied experimentally in animals, and in vivo data are hence only available from the analysis of diagnostic patient samples or autopsy materials. Dynamic aspects of viral replication and spread have therefore only been addressable within the blood compartment (Emery et al. 1999). Nevertheless, multiple circumstantial evidence strongly suggests successful viral replication in the above-mentioned cell types: Numerous capsids in the nucleus of infected cells as detected by electron microscopy unequivocally represent late-stage infection (Donnellan et al. 1966; Martin and Kurtz 1966; Kasnic et al. 1982; Balazs 1984; Francis et al. 1989; Schwartz et al. 1990; Grefte et al. 1993). Likewise, detection of late structural viral proteins in infected cells argues in the same direction, and the combination of the latter approach with the additional detection of cell marker proteins allowed for a reliable identification of the respective cell types (Sinzger et al. 1993, 1996, 1999a; Digel and Sinzger 2006). Together with the often focal distribution of clusters of infected cells this provided strong evidence that mucosal epithelial cells, connective tissue cells, smooth muscle cells and endothelial cells can produce and transmit viral progeny to their environment (Fig. 1a). HCMV replication can be detected in almost every organ during acute infection under certain conditions, e.g., severe cases of intrauterine infection (Bissinger et al. 2002). Liver, gastrointestinal tract, lung, retina and brain are predominant sites of clinical manifestations of HCMV infections in immunocompromised hosts (Plachter et al. 1996). Within these organs, highly specialized parenchymal cells are frequent targets of HCMV infection, including hepatocytes in the liver (Sinzger et al. 1999a), alveolar epithelial cells in the lung (Ng Bautista and Sedmak 1995; Sinzger et al. 1995), and neuronal cells in retina and brain (Wiley and Nelson 1988; Schmidbauer et al. 1989; Rummelt et al. 1994). In principle, HCMV can thus cause extensive lesions because of its cytolytic nature, which is, however, in most cases limited by a marked cellular immune response (Sinzger and Jahn 1996).

Target Cells of HCMV in Cell Culture

An increasing number of cell culture models almost perfectly reflect the in vivo situation concerning susceptibility of the various cell types. Again, lymphocytes and granulocytes are among the few cell types that were not found to support replication of HCMV in vitro, although they may still act as a passive vehicle for HCMV transmission. On the contrary, the list of susceptible primary cell cultures is long, including skin or lung fibroblasts, vascular smooth muscle cells (Tumilowicz et al. 1985), retina pigment epithelial cells (Tugizov et al. 1996), placental trophoblast cells (Halwachs-Baumann et al. 1998), hepatocytes (Sinzger et al. 1999a), neuronal and glial brain cells (Poland et al. 1990), kidney epithelial cells (Heieren et al. 1988), monocyte-derived macrophages (Ibanez et al. 1991; Lathey and Spector 1991), monocyte-derived dendritic cells (Riegler et al. 2000), and vascular endothelial cells (Ho et al. 1984; Waldman et al. 1989). All of these primary cell types support the complete viral replication cycle, acquire a uniform cytomegalic appearance during the late replication phase and are finally lysed (Fig. 1b). In addition, limited replication can be achieved in a number of immortalized cell lines such as glioblastoma cells, teratocarcinoma cell lines or monocytic cell lines. However, some kind of differentiation is often necessary to render such cell lines supportive of a complete replication cycle (Shelbourn et al. 1989; Ibanez et al. 1991; Lathey and Spector 1991; Spiller et al. 1997; Sinclair and Sissons

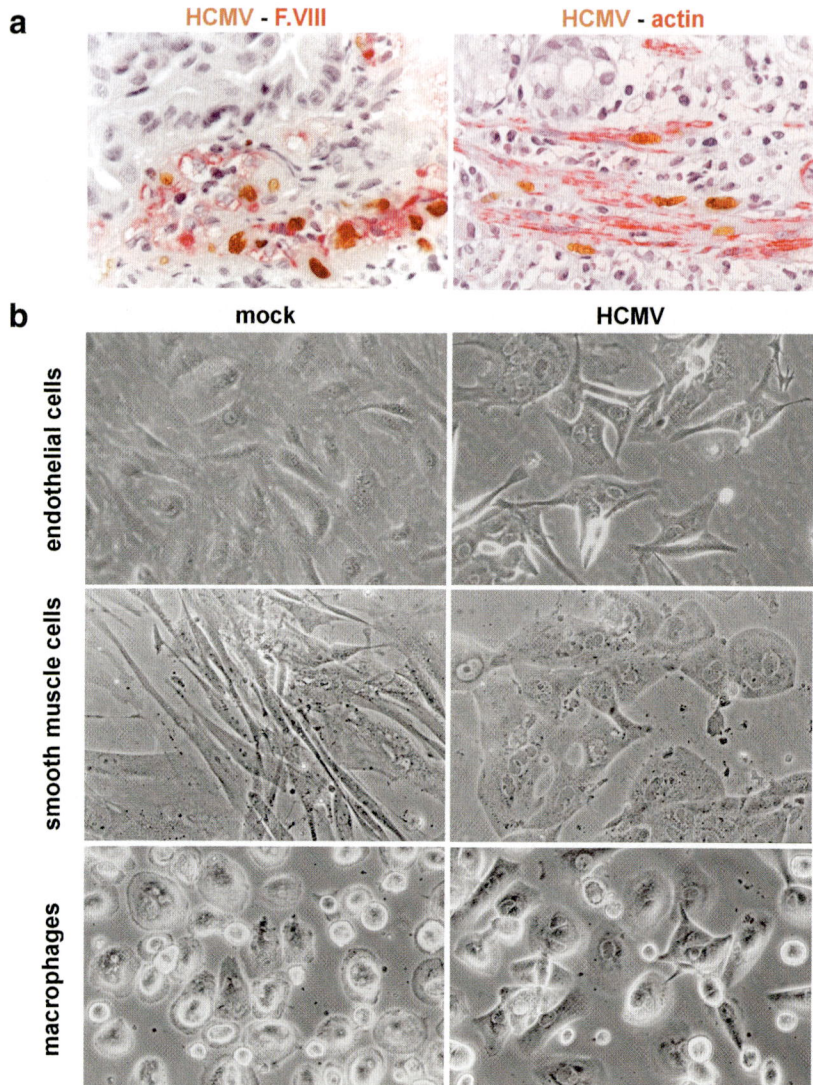

Fig. 1 a Immunohistochemical evidence of productive infection in endothelial cells and smooth muscle cells in vivo, as indicated by focus formation within the respective cell layers. *Brown nuclear signals*, detection of HCMV immediate early antigens by indirect immunoperoxidase labeling; *red cytoplasmic signals*, detection of F.VIII-related antigen (endothelial cells) and actin (smooth muscle cells) by indirect immunoalkaline phosphatase labeling; *blue nuclear signals*, counterstaining with hematoxilin. **b** Phase contrast micrographs of HCMV-infected cell cultures. Irrespective of the great morphological differences between cultured cells prior to infection, HCMV productive replication results in uniform morphological appearance with cytomegaly and nuclear inclusions

2006). While in vivo analyses were apt to descriptively identify the cell types infected by HCMV in its natural host, cell culture models made it possible to address quantitative aspects regarding susceptibility and productivity, thus revealing striking differences between cells of different origin: skin or lung fibroblast have always been the standard cell type for isolation and propagation of HCMV from patient samples and are still the most efficient producer cell line irrespective of the virus strain (Mocarski et al. 2006). For certain HCMV strains, vascular endothelial cells are also sufficiently susceptible and productive to allow long-term propagation of certain virus strains by passaging cell-free supernatants of infected cultures (Digel and Sinzger 2006). Other cell cultures, e.g., monocyte-derived macrophages, are low-level productive (Sinzger et al. 2006) and hardly release sufficient amounts of infectious progeny to maintain the virus during repeated passaging of cell-free supernatant on the respective cell type.

Cell Tropism

Along with the description of quantitative differences regarding susceptibility and productivity, different definitions of cell tropism may be applied to describe different aspects of HCMV-host cell interactions. In a broader sense, cell tropism refers to the simple fact that cells can be entered by the virus and will subsequently express viral genes. More functional definitions of cell tropism may take into account whether the infection is effective. From a biologist's point of view, cell tropism might define those target cells in which the virus can successfully reproduce. From a pathologists view, cell tropism might define those target cells that are damaged by the virus regarding cell survival and/or specific cellular functions.

Thus, even though HCMV infection of polymorphonuclear cells is abortive (Grefte et al. 1994), these cells can obviously contribute to hematogenous spread via the blood stream by carrying internalized virions (Gerna et al. 2000). Similarly, even though generation of viral progeny appears to be rather ineffective in monocytes/macrophages (Sinzger et al. 2006), this might suffice to start infection in an organ after transmigration through the vascular endothelium. Likewise, low-level production in placental trophoblast (Halwachs-Baumann et al. 1998), even though insufficient for long-term propagation of HCMV in trophoblast cultures, can contribute critically to intrauterine infection of the fetus (see also the chapter by L. Pereira and E. Maidji, this volume). Skin fibroblasts, lung fibroblasts and human umbilical vein endothelial cells are suitable for long-term propagation of clinical HCMV isolates, i.e., these cells have a reproductive index greater than 1. During the initial passages when isolates grow strictly cell-associated (Yamane et al. 1983), this would result in an increase of the fraction of infected cells within the culture. After adaptation to a cell-free infection mode, a reproductive index of greater than 1 would mean that progeny production exceeds virus input in the respective cell culture. The cell type used for propagation of an isolate has an effect on the cell tropism of the resulting HCMV strain. Apparently, long-term propagation in fibroblasts selects for HCMV strains with low endothelial cell tropism, whereas long-term propagation in endothelial cells maintains the broader cell tropism characteristic for recent clinical HCMV isolates (Waldman et al. 1991; Sinzger et al. 1999b). Whether propagation in other cell types such as smooth muscle cells or macrophages would also results in a restricted cell tropism has not been tested. Likewise, it is unknown whether an adaptation of HCMV to certain cell types also takes place during natural infection in vivo, i.e., whether tropism variants of HCMV exist within one patient. At present, the assumption of different cell tropism variants in different organ tissues is still speculative, but first hints in that direction come from reports on a strictly localized reactivation of HCMV, e.g., in the lactating breast (Hamprecht et al. 2003). Apart from viral determinants, a tissue-specific immune control might also contribute to differences in the apparent organ tropism of HCMV replication, similar to the apparent tropism of MCMV for the salivary gland (Jonjic et al. 1989). Experimental data from murine cytomegalovirus indicate that under the complex in vivo conditions the apparent cell tropism can be further modified by the microenvironment within a certain tissue. For example, proapoptotic stimuli from surrounding immune cells can limit infection in an otherwise susceptible cell type (Patrone et al. 2003), and this proapoptotic effect might even occur in a cell type-specific manner. For a refined understanding of HCMV's in vivo cell tropism, future work should therefore take into account how the complex organ-typical interactions might influence the susceptibility of target cells for HCMV infection, e.g., by analyzing complex organ tissue cultures (Reinhardt et al. 2003).

In conclusion, the strict host tropism of HCMV is contrasted by a remarkably broad cell tropism within its host, with epithelial cells, endothelial cells, fibroblasts and smooth muscle cells being the predominant targets for virus replication. The discrepancy between in vivo findings and cell culture data has diminished with the introduction of more recent HCMV strains and their application in various primary cell cultures. Since the basic cell culture tools reflecting the in vivo cell tropism of HCMV are now available, the analysis of compound cell culture systems representing the complex composition of organ tissues can be targeted in the future.

Pathogenetic Role of Selected Cell Types

The broad target cell range provides the basis for a highly complex interaction between HCMV and the human host, which can be adapted to many different situations during their lifelong relationship. It should always be kept in mind that HCMV can successfully enter its host, spread within the body, establish latency, reactivate frequently throughout life and be transmitted to other individuals mostly without ever causing clinically apparent disease (for aspects of latency, see the chapters by M. Reeves and J. Sinclair, this volume and M.J. Reddehase et al., this volume). Many of these aspects of HCMV's silent life are still a matter of speculation. More robust information is available on the contribution of certain cell types to viral dissemination and organ infection under conditions when insufficient immune control allows virus replication to exceed the threshold of clinical manifestation (see also the chapter by W. Britt, this volume). Conclusions from analyses of severely ill patients on the behavior of HCMV in the normal host (Fig. 2) are therefore made, with the provision that an intact immune control may modify the apparent cell and organ tropism.

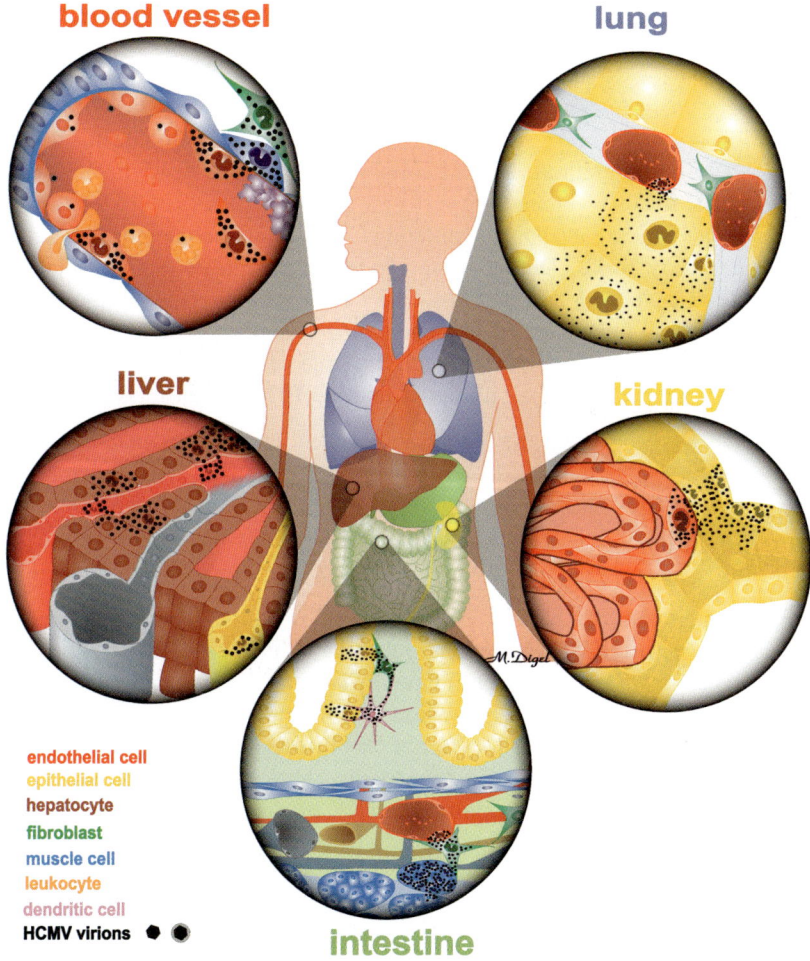

Fig. 2 Hypothetical contribution of various cell types to hematogenous dissemination and organ manifestation as deduced from immunohistochemical findings and cell culture data. *Black dots* represent virus

Epithelial Cells

Epithelial cells are a major target of HCMV infection (Sinzger et al. 1995) and can therefore be assumed to play an important role during host-to-host transmission as they line all external body surfaces. Most likely, HCMV enters a new host by infection of mucosal epithelium. For example, HCMV newborns and infants can be infected by breast milk of a seropositive mother, a highly efficient transmission route which accounts for the majority of HCMV transmissions during early childhood (Stagno and Cloud 1994). More than 95% of seropositive breastfeeding women reactivate HCMV locally, shed cell-free infectivity into the milk, and 30%-40% of them will transmit HCMV to their children (Hamprecht et al. 2001). The infants' mucosal surfaces throughout the gastrointestinal tract are exposed during feeding, and epithelial cells in all parts of the gastrointestinal tract are susceptible and obviously support productive infection. They are the most likely candidates for primary replication of incoming HCMV. However, as these first steps of infection are hardly ever recognized, there are no data available for a direct proof of these considerations. Alternatively, similar to HIV, dendritic cells could also contribute to entry via mucosal surfaces.

More direct data are available supporting a role of epithelial cells in shedding HCMV into body fluids. During acute productive infection, late-stage-infected epithelial cells have been detected in salivary glands, kidney and various parts of the gastrointestinal tract (Variend and Pearse 1986; Sinzger et al. 1995; Bissinger et al. 2002). Undoubtedly, these cells are a source of infectivity detected within saliva, urine and stool and may thus contribute to HCMV transmission via these excretions.

Dendritic Cells

Because of their complex biology, dendritic cells (DCs) may play various roles in the pathogenesis of HCMV infections, resulting in proviral as well as antiviral effects. Immature DCs are resident in virtually all mucosal and epidermal surfaces of the body, controlling for the invasion of foreign organisms. They are well equipped for highly efficient endocytic uptake of material from their environment, e.g., pathogens or remnants from apoptotic cells. Uptake of infectious HCMV into DCs can result in viral replication and release of viral progeny (Riegler et al. 2000). On the other hand, endocytic uptake of HCMV may lead to processing of viral proteins and presentation of viral epitopes by MHC class I and II molecules. This is counteracted to some extent by HCMV-induced downregulation of several immune-stimulatory surface molecules including MHC class I and II (Grigoleit et al. 2002; Moutaftsi et al. 2002; Hertel et al. 2003). As a consequence, HCMV-infection decreases the immune-stimulatory capacity of DCs (Grigoleit et al. 2002; Moutaftsi et al. 2002; Hertel et al. 2003). For immature DCs, antigen uptake is a maturation stimulus, and with maturation DCs downmodulate their endocytic activity and upregulate peptide processing and presentation. Upon maturation, they become mobile and are led by their homing receptors toward lymphatic tissues. Interestingly, immature Langerhans type DCs, which reside in epidermal and mucosal tissues, only become highly susceptible after maturation (Hertel et al. 2003; Reeves et al. 2005), whereas immature interstitial type DCs can be readily infected by HCMV (Riegler et al. 2000; Moutaftsi et al. 2002). It is tempting to speculate that HCMV is endocytosed by immature DCs in mucosal tissues, thus providing a maturation stimulus, which then leads to migration toward the draining lymph node and renders the cells permissive to HCMV replication. In the lymph node, productively infected mature DCs may spread the virus to other cells, whereas their immune-stimulatory capacity may be restricted. To further temper an antiviral immune response, infected mature DCs may directly and indirectly inhibit T cell functions (Raftery et al. 2001). However, despite such immunosuppressive effects, immunocompetent hosts regularly develop a strong T cell response protecting from clinical manifestations of the infection. Cross-presentation of viral antigens by DCs following uptake of apoptotic material from infected cells has been described as an explanation for the well-known robust immune response to HCMV in the normal host (Tabi et al. 2001).

Fibroblasts

Fibroblasts are not only the standard cell culture system for propagation of HCMV to high titers (Mocarski et al. 2006), but they are also among the major targets of HCMV in vivo (Sinzger et al. 1995). Efficient replication in such a ubiquitous cell type opens the possibility for HCMV to replicate in virtually every organ. Consequently, infected connective tissue cells are assumed to contribute to efficient spread of HCMV in organs as different as adrenal glands, bone marrow, heart, kidney, liver, lung, pancreas, placenta, small bowel and spleen (Bissinger et al. 2002). If the particular property of cultured fibroblast to generate and release high titers of viral progeny also applies for infected connective tissue cells in vivo, then they might contribute greatly to the highly dynamic proliferation of HCMV during acute infections (Emery et al. 1999).

Smooth Muscle Cells

Like fibroblasts, smooth muscle cells are also ubiquitously distributed throughout the body. Their basic function is generation of kinetic force by contraction of the actin-myosin skeleton, which may be controlled either by the autonomic nervous system, hormones or stimuli from neighboring cells. Given their spatial organization as multicellular layers in the wall of hollow organs, they regulate the dynamic shape and intraluminal pressure of these organs and help in maintaining organ integrity. Smooth muscle cells are susceptible to productive HCMV infection (Tumilowicz et al. 1985), which may have important pathophysiological consequences. When the host's immune response is severely compromised, focal expansion with subsequent lytic replication in the gastrointestinal tract can result in ulceration (Sinzger et al. 1995) and perforation (Genta et al. 1993), with sometimes fatal outcome. In the immunocompetent host, infection of vascular smooth muscle cells may be pathogenetically important. In these cells, HCMV downregulates extracellular matrix proteins, which may contribute to the development of inflammatory vasculopathies (Reinhardt et al. 2006). In addition, lytic infection of vascular smooth muscle cells might provoke a response to injury reaction, and consequently HCMV is considered a possible pathogenetic (co)factor in the context of atherosclerosis (Stassen et al. 2006).

Endothelial Cells

Speculation on an association of HCMV with vascular damage is additionally supported by the marked endothelial cell tropism of HCMV in vivo. Again, the ubiquitous distribution of small vessels throughout the body is reflected by the detection of HCMV-infected microvascular endothelial cells in various organs, e.g., brain, lung, liver, kidney and the complete gastrointestinal tract (Myerson et al. 1984; Wiley and Nelson 1988; Roberts et al. 1989; Sinzger et al. 1995; Bissinger et al. 2004). They support productive lytic infection and can hence promote hematogenous dissemination HCMV from the circulating blood into organ tissues, often accompanied by a vasculitic response around infected vessel walls (Roberts et al. 1989; Sinzger et al. 1995). Macrovascular endothelial cells are also susceptible to productive lytic infection (Kahl et al. 2000) and combined damage of the endothelial layer and the underlying smooth muscle layer may initiate the cascade of defense reactions finally resulting in vascular lesions. While the contribution of HCMV to atherosclerosis in the general population is still a matter of debate (see also the chapter by D.N. Streblow, this volume), the association is very clear in patients after heart transplantation (Valantine 2004; Potena et al. 2006).

Leukocytes

The disposition of HCMV to systemic dissemination and multiorgan involvement has already been mentioned. Leukocytes are assumed to be a central player with regard to hematogenous spread of the virus, whether by being a target of permissive infection or by passively transporting infectious particles as a vehicle. The latter obviously applies for polymorphonuclear cells, which can take up virus particles and express viral immediate early proteins but do not support the full replicative cycle (Grefte et al. 1994). Even though these cells can not produce viral progeny, they are still capable of transmitting the infection to other cell types, as evidenced by frequent isolation of HCMV from polymorphonuclear cells of immunocompromised patients (Gerna et al. 1992), and this is most likely due to attachment and partial localized fusion of cell membranes with subsequent transfer of engulfed (sub)viral particles, as shown in the opposite direction for the transfer of HCMV from endothelial cells to polymorphonuclear cells (Gerna et al. 2000). In line with these hypotheses, (a) infectivity is predominantly found in the polymorphonuclear fraction of whole blood (Schafer et al. 2000), (b) detection of the viral structural antigen pp65 (pUL83) in polymorphonuclear cells can be clinically used as a marker of acute HCMV infection (The et al. 1990; Gerna et al. 1991) and (c) removal of white blood cells from whole blood prior to transfusion almost completely reduces the risk of HCMV transmission (Gilbert et al. 1989). Monocytes, although a minor target cell with regard to frequency, might also contribute to hematogenous spread of HCMV, particularly as monocyte-derived macrophages support the full replicative cycle (Ibanez et al. 1991; Lathey and Spector 1991). It is tempting to assume a scenario where monocytes rolling along the vascular endothelium take up infectious virus from productively infected endothelial cells at one site of the body, differentiate upon transmigration through an activated endothelial layer at a different site of the body (Waldman et al. 1995), and release virus progeny into the corresponding organ after maturation into tissue macrophages (Sinzger et al. 1996).

Apart from their role in acute HCMV infections, all susceptible cell types may in principle also be sites of viral latency, although experimental data point to a particular role of hematopoietic cells in that context (Sinclair and Sissons 2006 see also the chapters by M. Reeves and J. Sinclair, this volume). Taken together, the pathogenesis of acute HCMV infections is greatly influenced by the broad target cell range of this virus, with hematopoietic cells facilitating systemic spread, ubiquitous cell types like fibroblasts and smooth muscle cells providing the platform for efficient proliferation of the virus and epithelial cells contributing to interhost transmission.

Cell Biological Basis of HCMV Cell Tropism

The longstanding paradox between broad cell tropism of HCMV in vivo and a restricted target cell range of the available virus strains in cell culture has been resolved by the introduction of endothelial-propagated virus strains with a well-preserved natural cell tropism. This enabled recent progress toward the definition of viral genes governing interstrain differences in cell tropism and the first insights into the underlying virus-host interactions.

Interstrain Differences in Cell Tropism

The idea that HCMV strains may differ regarding their reproductive potential in certain cell cultures was already reported in 1980 (Albrecht and Weller 1980). The finding that extended propagation in fibroblasts regularly results in loss of endothelial cell tropism, whereas propagation in endothelial cells maintains a broad cell tropism of the respective strain (Waldman et al. 1991), made the issue of cell tropism accessible to further experimental analyses. Indeed, the phenotypic differences are very pronounced with a 100- to 1,000-fold reduction in endothelial cell tropism of fibroblast-adapted strains irrespective of the origin of endothelial cells (Kahl et al. 2000; Sinzger et al. 2000). Nevertheless, even severely fibroblast-adapted strains, such as AD169 or Towne, can infect endothelial cells to some extent, which may also depend on the origin of the endothelial cell culture. In consideration of this, such HCMV strains are more precisely classified as poorly endotheliotropic rather than nonendotheliotropic.

Interstrain differences in HCMV cell tropism occur as a cell culture artifact, but significant variation has also been described between recent clinical isolates from different patients (Sinzger et al. 1999b). Together with the finding that multiple isolates from the same patient behaved identically with regard to endothelial cell tropism, this suggests a natural interhost variability of HCMV cell tropism. This may contribute to the highly variable clinical course of HCMV infections in various patients (Sinzger et al. 1999b). This notion is further supported by the fact that a high endothelial cell tropism is apparently associated with high infection efficiency also in monocyte-derived macrophages and dendritic cells, cells that are all assumed to mediate hematogenous dissemination of HCMV (Jahn et al. 1999).

Viral Genes and Proteins Contributing to Cell Tropism

Restriction fragment analyses of differentially propagated HCMV strains showed that the restriction of cell tropism during fibroblast adaptation is associated with multiple genetic modifications (Sinzger et al. 1999b). The introduction of BACmid technology and subsequent screening procedures for the effect of genetic deletions led to the identification of several open reading frames involved in endothelial cell and leukocyte tropism. Dunn et al. found that the residual endothelial cell tropism of HCMV strain Towne is further reduced by deletion of the viral tegument UL24-protein, a member of the US22 gene family (Dunn et al. 2003). The highly endotheliotropic phenotype of an endothelial propagated strain was found to depend on the UL128-131 gene region (Hahn et al. 2004). Deletion of either open reading frame within this region strongly reduced endothelial cell tropism, epithelial cell tropism, dendritic cell tropism and the virus transfer rate to granulocytes (Hahn et al. 2004; Wang and Shenk 2005a). UL128, UL130 and also UL131 were found in complex with glycoproteins gH and gL within virion particles (Wang and Shenk 2005b;

Adler et al. 2006). Endothelial cell infection of fibroblast-adapted, poorly endotheliotropic HCMV strains Merlin, Towne and AD169 were rescued by transient expression of intact UL128, UL130 and UL131, respectively (Hahn et al. 2004; Patrone et al. 2005). This suggests that loss of endothelial cell tropism during fibroblast adaptation is frequently caused by alterations in this gene region. In contrast, it is still unclear whether this gene region also contributes to the variability in endothelial cell tropism among naturally occurring HCMV isolates (Sinzger et al. 1999b), as a series of 34 clinical isolates appeared to contain intact copies of these genes (Baldanti et al. 2006).

Critical Events for Replication in Various Cell Types

Regarding the replication steps critical for successful infection of various cell types, interstrain comparisons in endothelial cells showed a particular role of initial postpenetration events. The efficiency of nuclear translocation of incoming virions and subsequent delivery of the viral genome to the nucleus of penetrated endothelial cells is very low with fibroblast-adapted strains (Sinzger et al. 2000), and a similar block occurs in monocyte-derived macrophages (Sinzger et al. 2006). In contrast, these steps are strain-independent in fibroblasts. The demonstration that pUL128-131 are part of glycoprotein complexes with gH and gL in the virion envelope (Wang and Shenk 2005b; Adler et al. 2006) fit well with the finding that endothelial cell tropism of HCMV is determined during entry, as gH is known to be involved in fusion events (see also the chapter by M.K. Isaacson et al., this volume). The particular importance of initial events is further emphasized by recent data suggesting infection of endothelial cells by an endocytic route, in contrast to direct fusion at the plasma membrane of fibroblasts (Sinzger 2008). It appears that, unlike previously assumed (Bodaghi et al. 1999), endocytosis of HCMV in endothelial cells is not necessarily an abortive pathway (Fig. 3). For Epstein Barr virus, different gH/gL complexes are engaged in different cell types, leading either to direct fusion in lymphocytes or endocytosis in epithelial cells. Interestingly, a cell-type-dependent cell-cell fusion activity induced by gH-gL complexes was found in a transient expression system (Kinzler and Compton 2005).

The susceptibility of other cell types may be regulated at later steps of the replication cycle. For example, infection of polymorphonuclear leukocytes is aborted after onset of IE gene expression (Grefte et al. 1994), independent of the virus strain. The exact nature of this block of progression toward the early phase of replication is unknown. In trophoblast cells, hepatocyte or macrophage HCMV can proceed through all phases of the replication cycle, formation and/or release of viral progeny. However, the production of progeny is up to 1,000-fold less efficient than in fibroblasts (Halwachs-Baumann et al. 1998; Sinzger et al. 1999a, 2006) and again the factors contributing to these differences in productivity are not known.

In conclusion, genes UL128-131 classified as nonessential in fibroblast cultures have been shown to contribute to interstrain differences regarding infection of

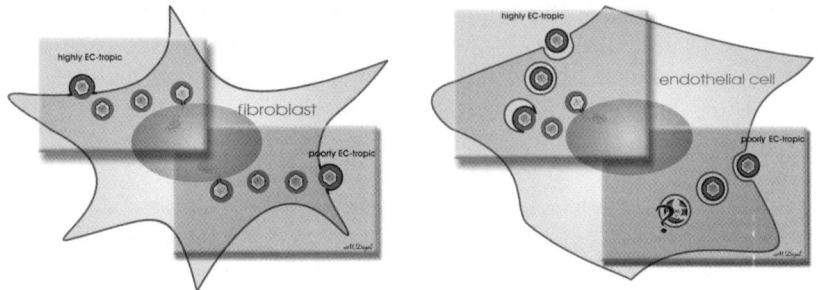

Fig. 3 Hypothetical mechanism mediating interstrain differences in endothelial cell tropism. While all HCMV strains can release their capsids into fibroblasts by direct fusion of their envelope with the plasma membrane, cell type differences are assumed for viral entry into endothelial cells. Both highly endotheliotropic and poorly endotheliotropic strains are internalized by endocytosis, but only highly endotheliotropic can escape from endocytic vesicles and release their capsid into the cytoplasm

endothelial cells, epithelial cells and macrophages. As the respective proteins are structural components of the envelope of virion particles, it is not unexpected that they exert their effects on the level of viral entry. The cellular counterparts mediating the cell-type-specificity of these virion components are to be defined.

Cell Tropism of Other Cytomegaloviruses

The tendency toward systemic dissemination resulting in infection of various organs is not unique to human CMV but has also been reported for animal CMVs. Apparently, a broad organ tropism is a hallmark of cytomegaloviruses, which is based on a similarly broad cell tropism.

Under conditions of severe immunosuppression, murine CMV-infected cells were found in lung, liver, spleen, kidneys, adrenals, gastrointestinal tract, brain, salivary gland, and fibroblasts, epithelial cells, neuronal cells, glial cells, ependymal cells hepatocytes and endothelial cells were identified as predominantly infected cell types within these tissues (Reddehase et al. 1985; Podlech et al. 1998; van Den Pol et al. 1999; Podlech et al. 2000). Likewise, a broad target cell range including fibroblasts, SMC, EC, macrophages was found with rat CMV (Kloover et al. 2000; van der Strate et al. 2003; Streblow et al. 2007).

While the histological distribution of HCMV and MCMV appears almost indistinguishable, the underlying mechanisms regulating cell tropism are apparently not completely conserved between cytomegaloviruses from different species. Deletion of the m45 gene abrogated replication of MCMV in endothelial cell cultures (Brune et al. 2001) by sensitizing infected endothelial cells to apoptosis. In contrast, deletion of UL45 did not influence the endothelial cell tropism of HCMV strain

FIXBAC (Hahn et al. 2002). On the other hand, involvement on US22 family members in cell tropism regulation of both HCMV and MCMV indicated a certain degree of conservation. The contribution of UL24 to endothelial cell tropism of HCMV strain Towne (Dunn et al. 2003) has already been mentioned. US22-family members of MCMV, namely m139, m140 and m141 contribute to macrophage tropism and promote MCMV replication in the spleen of infected mice (Hanson et al. 2001; Menard et al. 2003). M36, another US22 family gene of MCMV, also contributes to efficient replication in macrophages again through its anti-apoptotic function (Menard et al. 2003). Obviously, cell-type-specific inhibition of virus-induced apoptosis is a more general theme with cytomegaloviruses, suggesting that similar tropism-relevant anti-apoptotic genes may also exist in the HCMV genome (see also the chapter by A.L. McCormick, this volume). The HCMV counterpart of M36, UL36, has a known antiapoptotic function (Skaletskaya et al. 2001), which has not yet been tested in the context of cell tropism. UL45, the HCMV homolog of M45, exhibited a weak antiapoptotic activity only upon application of strong proapoptotic stimuli (Patrone et al. 2003) and was dispensable for viral replication in endothelial cells (Hahn et al. 2002). However, in vivo the situation may be different depending on the microenvironment within the infected tissues. In the presence of strong proapoptotic stimuli, UL36 and UL45 may be essential for successful completion of viral replication in a cell-type-dependent fashion, as reported for their murine CMV homologs. Complex cell culture systems reflecting the situation of an inflamed tissue are required to test this hypothesis.

A specific contribution of rat CMV concerns the role of vascular endothelial cells and smooth muscle cells in CMV-associated pathogenesis. Under the well-defined conditions of this animal model, a contribution of vascular CMV infection to the development of atherosclerotic lesions is clearly shown and the molecular mechanisms are partially deciphered, including oxLDL uptake, altering monocyte adhesion or increasing the production of pro-inflammatory cytokines (Stassen et al. 2006). In the human system, it may be impossible to prove the contribution of HCMV to a multifactorial disease such as atherosclerosis under natural clinical conditions. However, similarities between RCMV and HCMV regarding cytopathic effects in vascular cell types nevertheless suggest CMV as a proatherosclerotic agent also in humans.

Impact of Cell Tropism Analyses

The ability of CMV species to infect a variety of different cell types in their respective host appears to be central for successful entry, dissemination, persistence, reactivation and excretion. Analyzing CMV replication in various cell culture systems is therefore an absolute requirement for a comprehensive understanding of their biology and will in itself create additional value. Particularly, many of the genes still classified nonessential with regard to replication in the standard cell culture system will turn out to be essential if tested in other cell types or in complex tissues composed of several interacting cell types.

The practical impact of cell tropism issues may concern the design of cytomegalovirus vaccines. Given the limited success of vaccination with the highly adapted strain Towne, it appears as if a higher level of replication within the vaccinee is necessary for induction of a more robust immune response. Therefore, a more moderate attenuation with partial preservation of endothelial and epithelial cell tropism may be desirable. The introduction of small nonlethal mutations within known tropism genes is one possible approach to achieve such intermediate phenotypes.

Finally, an exact definition of entry pathways in diverse cell types may allow for the development of novel antiviral intervention strategies. At present, all anti-CMV chemotherapies target viral DNA replication. By analogy to HIV, entry inhibitors may complement the available drugs and allow for synergistic effects in combination therapies. Such an approach should certainly consider the possibility of different entry pathways in major target cell types such as fibroblasts and endothelial cells.

Acknowledgements This work was supported by the German research foundation (DFG SI 779/3-2).

References

Adler B, Scrivano L, Ruzcics Z, Rupp B, Sinzger C, Koszinowski U (2006) Role of human cytomegalovirus UL131A in cell type-specific virus entry and release. J Gen Virol 87:2451-2460

Albrecht T, Weller TH (1980) Heterogeneous morphologic features of plaques induced by five strains of human cytomegalovirus. Am J Clin Pathol 73:648-654

Balazs M (1984) Electron microscopic examination of congenital cytomegalovirus hepatitis. Virchows Arch A Pathol Anat Histopathol 405:119-129

Baldanti F, Paolucci S, Campanini G, Sarasini A, Percivalle E, Revello MG, Gerna G (2006) Human cytomegalovirus UL131A, UL130 and UL128 genes are highly conserved among field isolates. Arch Virol 151:1225-1233

Bissinger AL, Sinzger C, Kaiserling E, Jahn G (2002) Human cytomegalovirus as a direct pathogen: correlation of multiorgan involvement and cell distribution with clinical and pathological findings in a case of congenital inclusion disease. J Med Virol 67:200-206

Bissinger AL, Oettle H, Jahn G, Neuhaus P, Sinzger C (2004) Cytomegalovirus infection after orthotopic liver transplantation is restricted by a pre-existing antiviral immune response of the recipient. J Med Virol 73:45-53

Bodaghi B, Slobbe-van Drunen ME, Topilko A, Perret E, Vossen RC, van Dam-Mieras MC, Zipeto D, Virelizier JL, LeHoang P, Bruggeman CA, Michelson S (1999) Entry of human cytomegalovirus into retinal pigment epithelial and endothelial cells by endocytosis. Invest Ophthalmol Vis Sci 40:2598-2607

Brune W, Menard C, Heesemann J, Koszinowski UH (2001) A ribonucleotide reductase homolog of cytomegalovirus and endothelial cell tropism. Science 291:303-305

Digel M, Sinzger C (2006) Determinants of endothelial cell tropism of human cytomegalovirus. In: Reddehase MJ (ed) Cytomegaloviruses: molecular biology and immunology. Caister Academic Press, Norfolk, pp 445-464

Donnellan WL, Chantra-Umporn S, Kidd JM (1966) The cytomegalic inclusion cell. An electron microscopic study. Arch Pathol 82:336-348

Dunn W, Chou C, Li H, Hai R, Patterson D, Stolc V, Zhu H, Liu F (2003) Functional profiling of a human cytomegalovirus genome. Proc Natl Acad Sci USA 100:14223-14228

Emery VC, Cope AV, Bowen EF, Gor D, Griffiths PD (1999) The dynamics of human cytomegalovirus replication in vivo. J Exp Med 190:177-182

Francis ND, Boylston AW, Roberts AH, Parkin JM, Pinching AJ (1989) Cytomegalovirus infection in gastrointestinal tracts of patients infected with HIV-1 or AIDS. J Clin Pathol 42:1055-1064

Gerna G, Zipeto D, Parea M, Revello MG, Silini E, Percivalle E, Zavattoni M, Grossi P, Milanesi G (1991) Monitoring of human cytomegalovirus infections and ganciclovir treatment in heart transplant recipients by determination of viremia, antigenemia, and DNAemia. J Infect Dis 164:488-498

Gerna G, Zipeto D, Percivalle E, Parea M, Revello MG, Maccario R, Peri G, Milanesi G (1992) Human cytomegalovirus infection of the major leukocyte subpopulations and evidence for initial viral replication in polymorphonuclear leukocytes from viremic patients. J Infect Dis 166:1236-1244

Genta RM, Bleyzer I, Cate TR, Tandon AK, Yoffe B (1993) In situ hybridization and immunohistochemical analysis of cytomegalovirus-associated ileal perforation. Gastroenterology 104:1822-1827

Gerna G, Percivalle E, Baldanti F, Sozzani S, Lanzarini P, Genini E, Lilleri D, Revello MG (2000) Human cytomegalovirus replicates abortively in polymorphonuclear leukocytes after transfer from infected endothelial cells via transient microfusion events. J Virol 74:5629-5638

Gilbert GL, Hayes K, Hudson IL, James J (1989) Prevention of transfusion-acquired cytomegalovirus infection in infants by blood filtration to remove leucocytes. Neonatal Cytomegalovirus Infection Study Group. Lancet 1:1228-1231

Grefte A, Blom N, van der Giessen M, van Son W, The TH (1993) Ultrastructural analysis of circulating cytomegalic cells in patients with active cytomegalovirus infection: evidence for virus production and endothelial origin. J Infect Dis 168:1110-1118

Grefte A, Harmsen MC, van der Giessen M, Knollema S, van Son WJ, The TH (1994) Presence of human cytomegalovirus (HCMV) immediate early mRNA but not ppUL83 (lower matrix protein pp65) mRNA in polymorphonuclear and mononuclear leukocytes during active HCMV infection. J Gen Virol 75:1989-1998

Grigoleit U, Riegler S, Einsele H, Laib Sampaio K, Jahn G, Hebart H, Brossart P, Frank F, Sinzger C (2002) Human cytomegalovirus induces a direct inhibitory effect on antigen presentation by monocyte-derived immature dendritic cells. Br J Haematol 119:189-198

Hahn G, Khan H, Baldanti F, Koszinowski UH, Revello MG, Gerna G (2002) The human cytomegalovirus ribonucleotide reductase homolog UL45 is dispensable for growth in endothelial cells, as determined by a BAC-cloned clinical isolate of human cytomegalovirus with preserved wild-type characteristics. J Virol 76:9551-9555

Hahn G, Revello MG, Patrone M, Percivalle E, Campanini G, Sarasini A, Wagner M, Gallina A, Milanesi G, Koszinowski U, Baldanti F, Gerna G (2004) Human cytomegalovirus UL131-128 genes are indispensable for virus growth in endothelial cells and virus transfer to leukocytes. J Virol 78:10023-10033

Halwachs-Baumann G, Wilders-Truschnig M, Desoye G, Hahn T, Kiesel L, Klingel K, Rieger P, Jahn G, Sinzger C (1998) Human trophoblast cells are permissive to the complete replicative cycle of human cytomegalovirus. J Virol 72:7598-7602

Hamprecht K, Maschmann J, Vochem M, Dietz K, Speer CP, Jahn G (2001) Epidemiology of transmission of cytomegalovirus from mother to preterm infant by breastfeeding. Lancet 357:513-518

Hamprecht K, Maschmann J, Vochem M, Speer CP, Jahn G (2003) Transmission of cytomegalovirus to preterm infants by breast-feeding. In: Proesch S, Cinatl J, Scholz M (eds) New aspects of CMV-related immunopathology, vol 24. Karger, Basel, pp 33-42

Hanson LK, Slater JS, Karabekian Z, Ciocco-Schmitt G, Campbell AE (2001) Products of US22 genes M140 and M141 confer efficient replication of murine cytomegalovirus in macrophages and spleen. J Virol 75:6292-6302

Heieren MH, Kim YK, Balfour HH Jr (1988) Human cytomegalovirus infection of kidney glomerular visceral epithelial and tubular epithelial cells in culture. Transplantation 46:426-432

Hertel L, Lacaille VG, Strobl H, Mellins ED, Mocarski ES (2003) Susceptibility of immature and mature Langerhans cell-type dendritic cells to infection and immunomodulation by human cytomegalovirus. J Virol 77:7563-7574

Ho DD, Rota TR, Andrews CA, Hirsch MS (1984) Replication of human cytomegalovirus in endothelial cells. J Infect Dis 150:956-957

Ibanez CE, Schrier R, Ghazal P, Wiley C, Nelson JA (1991) Human cytomegalovirus productively infects primary differentiated macrophages. J Virol 65:6581-6588

Jahn G, Stenglein S, Riegler S, Einsele H, Sinzger C (1999) Human cytomegalovirus infection of immature dendritic cells and macrophages. Intervirology 42:365-372

Jonjic S, Mutter W, Weiland F, Reddehase MJ, Koszinowski UH (1989) Site-restricted persistent cytomegalovirus infection after selective long-term depletion of CD4+ T lymphocytes. J Exp Med 169:1199-1212

Kahl M, Siegel-Axel D, Stenglein S, Jahn G, Sinzger C (2000) Efficient lytic infection of human arterial endothelial cells by human cytomegalovirus strains. J Virol 74:7628-7635

Kasnic G Jr, Sayeed A, Azar HA (1982) Nuclear and cytoplasmic inclusions in disseminated human cytomegalovirus infection. Ultrastruct Pathol 3:229-235

Kinzler ER, Compton T (2005) Characterization of human cytomegalovirus glycoprotein-induced cell-cell fusion. J Virol 79:7827-7837

Kloover JS, Hillebrands JL, de Wit G, Grauls G, Rozing J, Bruggeman CA, Nieuwenhuis P (2000) Rat cytomegalovirus replication in the salivary glands is exclusively confined to striated duct cells. Virchows Arch 437:413-421

Lathey JL, Spector SA (1991) Unrestricted replication of human cytomegalovirus in hydrocortisone-treated macrophages. J Virol 65:6371-6375

Martin AM Jr, Kurtz SM (1966) Cytomegalic inclusion disease. An electron microscopic histochemical study of the virus at necropsy. Arch Pathol 82:27-34

Menard C, Wagner M, Ruzsics Z, Holak K, Brune W, Campbell AE, Koszinowski UH (2003) Role of murine cytomegalovirus US22 gene family members in replication in macrophages. J Virol 77:5557-5570

Mocarski ES, Shenk T, Pass RF (2006) Cytomegaloviruses. In: Knipe DM (ed) Fields virology. Lippincott Williams and Wilkins, Philadelphia, pp 2701-2772

Moutaftsi M, Mehl AM, Borysiewicz LK, Tabi Z (2002) Human cytomegalovirus inhibits maturation and impairs function of monocyte-derived dendritic cells. Blood 99:2913-2921

Myerson D, Hackman RC, Nelson JA, Ward DC, McDougall JK (1984) Widespread presence of histologically occult cytomegalovirus. Hum Pathol 15:430-439

Ng Bautista CL, Sedmak DD (1995) Cytomegalovirus infection is associated with absence of alveolar epithelial cell HLA class II antigen expression. J Infect Dis 171:39-44

Patrone M, Percivalle E, Secchi M, Fiorina L, Pedrali-Noy G, Zoppe M, Baldanti F, Hahn G, Koszinowski UH, Milanesi G, Gallina A (2003) The human cytomegalovirus UL45 gene product is a late, virion-associated protein and influences virus growth at low multiplicities of infection. J Gen Virol 84:3359-3370

Patrone M, Secchi M, Fiorina L, Ierardi M, Milanesi G, Gallina A (2005) Human cytomegalovirus UL130 protein promotes endothelial cell infection through a producer cell modification of the virion. J Virol 79:8361-8373

Plachter B, Sinzger C, Jahn G (1996) Cell types involved in replication and distribution of human cytomegalovirus. Adv Virus Res 46:195-261

Podlech J, Holtappels R, Wirtz N, Steffens HP, Reddehase MJ (1998) Reconstitution of CD8 T cells is essential for the prevention of multiple-organ cytomegalovirus histopathology after bone marrow transplantation. J Gen Virol 79:2099-2104

Podlech J, Holtappels R, Pahl-Seibert MF, Steffens HP, Reddehase MJ (2000) Murine model of interstitial cytomegalovirus pneumonia in syngeneic bone marrow transplantation: persistence of protective pulmonary CD8-T-cell infiltrates after clearance of acute infection. J Virol 74:7496-7507

Poland SD, Costello P, Dekaban GA, Rice GP (1990) Cytomegalovirus in the brain: in vitro infection of human brain-derived cells. J Infect Dis 162:1252-1262

Potena L, Holweg CT, Chin C, Luikart H, Weisshaar D, Narasimhan B, Fearon WF, Lewis DB, Cooke JP, Mocarski ES, Valantine HA (2006) Acute rejection and cardiac allograft vascular disease is reduced by suppression of subclinical cytomegalovirus infection. Transplantation 82:398-405

Raftery MJ, Schwab M, Eibert SM, Samstag Y, Walczak H, Schonrich G (2001) Targeting the function of mature dendritic cells by human cytomegalovirus: a multilayered viral defense strategy. Immunity 15:997-1009

Reddehase MJ, Weiland F, Munch K, Jonjic S, Luske A, Koszinowski UH (1985) Interstitial murine cytomegalovirus pneumonia after irradiation: characterization of cells that limit viral replication during established infection of the lungs. J Virol 55:264-273

Reeves MB, Lehner PJ, Sissons JG, Sinclair JH (2005) An in vitro model for the regulation of human cytomegalovirus latency and reactivation in dendritic cells by chromatin remodelling. J Gen Virol 86:2949-2954

Reinhardt B, Vaida B, Voisard R, Keller L, Breul J, Metzger H, Herter T, Baur R, Luske A, Mertens T (2003) Human cytomegalovirus infection in human renal arteries in vitro. J Virol Methods 109:1-9

Reinhardt B, Winkler M, Schaarschmidt P, Pretsch R, Zhou S, Vaida B, Schmid-Kotsas A, Michel D, Walther P, Bachem M, Mertens T (2006) Human cytomegalovirus-induced reduction of extracellular matrix proteins in vascular smooth muscle cell cultures: a pathomechanism in vasculopathies? J Gen Virol 87:2849-2858

Rice GP, Schrier RD, Oldstone MB (1984) Cytomegalovirus infects human lymphocytes and monocytes: virus expression is restricted to immediate-early gene products. Proc Natl Acad Sci USA 81:6134-6138

Riegler S, Hebart H, Einsele H, Brossart P, Jahn G, Sinzger C (2000) Monocyte-derived dendritic cells are permissive to the complete replicative cycle of human cytomegalovirus. J Gen Virol 81:393-399

Roberts WH, Sneddon JM, Waldman J, Stephens RE (1989) Cytomegalovirus infection of gastrointestinal endothelium demonstrated by simultaneous nucleic acid hybridization and immunohistochemistry. Arch Pathol Lab Med 113:461-464

Rummelt V, Rummelt C, Jahn G, Wenkel H, Sinzger C, Mayer UM, Naumann GO (1994) Triple retinal infection with human immunodeficiency virus type 1, cytomegalovirus, and herpes simplex virus type 1. Light and electron microscopy, immunohistochemistry, and in situ hybridization. Ophthalmology 101:270-279

Schafer P, Tenschert W, Cremaschi L, Schroter M, Gutensohn K, Laufs R (2000) Cytomegalovirus cultured from different major leukocyte subpopulations: association with clinical features in CMV immunoglobulin G-positive renal allograft recipients. J Med Virol 61:488-496

Schmidbauer M, Budka H, Ulrich W, Ambros P (1989) Cytomegalovirus (CMV) disease of the brain in AIDS and connatal infection: a comparative study by histology, immunocytochemistry and in situ DNA hybridization. Acta Neuropathol (Berl) 79:286-293

Schwartz DA, Walker B, Furlong B, Breding E, Someren A (1990) Cytomegalovirus in a macerated second trimester fetus: persistent viral inclusions on light and electron microscopy. South Med J 83:1357-1358

Shelbourn SL, Sissons JG, Sinclair JH (1989) Expression of oncogenic ras in human teratocarcinoma cells induces partial differentiation and permissiveness for human cytomegalovirus infection. J Gen Virol 70:367-374

Sinclair J, Sissons P (2006) Latency and reactivation of human cytomegalovirus. J Gen Virol 87:1763-1779

Sinzger C (2008) Entry route of HCMV into endothelial cells. J Clin Virol 41:174–179

Sinzger C, Muntefering H, Loning T, Stoss H, Plachter B, Jahn G (1993) Cell types infected in human cytomegalovirus placentitis identified by immunohistochemical double staining. Virchows Arch A Pathol Anat Histopathol 423:249-256

Sinzger C, Plachter B, Grefte A, The TH, Jahn G (1996) Tissue macrophages are infected by human cytomegalovirus in vivo. J Infect Dis 173:240-245

Sinzger C, Grefte A, Plachter B, Gouw AS, The TH, Jahn G (1995) Fibroblasts, epithelial cells, endothelial cells and smooth muscle cells are major targets of human cytomegalovirus infection in lung and gastrointestinal tissues. J Gen Virol 76:741-750

Sinzger C, Jahn G (1996) Human cytomegalovirus cell tropism and pathogenesis. Intervirology 39:302-319

Sinzger C, Bissinger AL, Viebahn R, Oettle H, Radke C, Schmidt CA, Jahn G (1999a) Hepatocytes are permissive for human cytomegalovirus infection in human liver cell culture and In vivo. J Infect Dis 180:976-986

Sinzger C, Schmidt K, Knapp J, Kahl M, Beck R, Waldman J, Hebart H, Einsele H, Jahn G (1999b) Modification of human cytomegalovirus tropism through propagation in vitro is associated with changes in the viral genome. J Gen Virol 80:2867-2877

Sinzger C, Kahl M, Laib K, Klingel K, Rieger P, Plachter B, Jahn G (2000) Tropism of human cytomegalovirus for endothelial cells is determined by a post-entry step dependent on efficient translocation to the nucleus. J Gen Virol 81:3021-3035

Sinzger C, Eberhardt K, Cavignac Y, Weinstock C, Kessler T, Jahn G, Davignon JL (2006) Macrophage cultures are susceptible to lytic productive infection by endothelial-cell-propagated human cytomegalovirus strains and present viral IE1 protein to CD4+ T cells despite late downregulation of MHC class II molecules. J Gen Virol 87:1853-1862

Skaletskaya A, Bartle LM, Chittenden T, McCormick AL, Mocarski ES, Goldmacher VS (2001) A cytomegalovirus-encoded inhibitor of apoptosis that suppresses caspase-8 activation. Proc Natl Acad Sci USA 98:7829-7834

Spiller OB, Borysiewicz LK, Morgan BP (1997) Development of a model for cytomegalovirus infection of oligodendrocytes. J Gen Virol 78:3349-3356

Stagno S, Cloud GA (1994) Working parents: the impact of day care and breast-feeding on cytomegalovirus infections in offspring. Proc Natl Acad Sci USA 91:2384-2389

Stassen FR, Vega-Cordova X, Vliegen I, Bruggeman CA (2006) Immune activation following cytomegalovirus infection: more important than direct viral effects in cardiovascular disease? J Clin Virol 35:349-353

Streblow DN, van Cleef KW, Kreklywich CN, Meyer C, Smith P, Defilippis V, Grey F, Fruh K, Searles R, Bruggeman C, Vink C, Nelson JA, Orloff SL (2007) Rat cytomegalovirus gene expression in cardiac allograft recipients is tissue specific and does not parallel the profiles detected in vitro. J Virol 81:3816-3826

Tabi Z, Moutaftsi M, Borysiewicz LK (2001) Human cytomegalovirus pp65- and immediate early 1 antigen-specific HLA class I-restricted cytotoxic T cell responses induced by cross-presentation of viral antigens. J Immunol 166:5695-5703

The TH, van der Bij W, van den Berg AP, van der Giessen M, Weits J, Sprenger HG, van Son WJ (1990) Cytomegalovirus antigenemia. Rev Infect Dis 12 Suppl 7: S734-S744

Tugizov S, Maidji E, Pereira L (1996) Role of apical and basolateral membranes in replication of human cytomegalovirus in polarized retinal pigment epithelial cells. J Gen Virol 77:61-74

Tumilowicz JJ, Gawlik ME, Powell BB, Trentin JJ (1985) Replication of cytomegalovirus in human arterial smooth muscle cells. J Virol 56:839-845

Valantine HA (2004) The role of viruses in cardiac allograft vasculopathy. Am J Transplant 4:169-177

van Den Pol AN, Mocarski E, Saederup N, Vieira J, Meier TJ (1999) Cytomegalovirus cell tropism, replication, and gene transfer in brain. J Neurosci 19:10948-10965

van der Strate BW, Hillebrands JL, Lycklama a Nijeholt SS, Beljaars L, Bruggeman CA, Van Luyn MJ, Rozing J, The TH, Meijer DK, Molema G, Harmsen MC (2003) Dissemination of rat cytomegalovirus through infected granulocytes and monocytes in vitro and in vivo. J Virol 77:11274-11278

Variend S, Pearse RG (1986) Sudden infant death and cytomegalovirus inclusion disease. J Clin Pathol 39:383-386

Waldman WJ, Sneddon JM, Stephens RE, Roberts WH (1989) Enhanced endothelial cytopathogenicity induced by a cytomegalovirus strain propagated in endothelial cells. J Med Virol 28:223-230

Waldman WJ, Roberts WH, Davis DH, Williams MV, Sedmak DD, Stephens RE (1991) Preservation of natural endothelial cytopathogenicity of cytomegalovirus by propagation in endothelial cells. Arch Virol 117:143-164

Waldman WJ, Knight DA, Huang EH, Sedmak DD (1995) Bidirectional transmission of infectious cytomegalovirus between monocytes and vascular endothelial cells: an in vitro model. J Infect Dis 171:263-272

Wang D, Shenk T (2005a) Human cytomegalovirus UL131 open reading frame is required for epithelial cell tropism. J Virol 79:10330-10338

Wang D, Shenk T (2005b) Human cytomegalovirus virion protein complex required for epithelial and endothelial cell tropism. Proc Natl Acad Sci USA 102:18153-18158

Wiley CA, Nelson JA (1988) Role of human immunodeficiency virus and cytomegalovirus in AIDS encephalitis. Am J Pathol 133:73-81

Yamane Y, Furukawa T, Plotkin SA (1983) Supernatant virus release as a differentiating marker between low passage and vaccine strains of human cytomegalovirus. Vaccine 1:23-25

Virus Entry and Innate Immune Activation

M. K. Isaacson, L. K. Juckem, T. Compton(✉)

Contents

Introduction to Virus Entry	86
Cellular Receptors Proposed for HCMV Entry	86
Envelope and Membrane Fusion	90
Introduction to Activation of Innate Immunity	91
Activation of Inflammatory Cytokines	91
Activation of Interferon Responses	93
Coordination of Entry Events and Innate Immune Activation Steps	94
Perspectives	96
References	96

Abstract Human cytomegalovirus (HCMV) exhibits an exceptionally broad cellular tropism as it is capable of infecting most major organ systems and cell types. Definitive proof of an essential role for a cellular molecule that serves as an entry receptor has proven very challenging. It is widely hypothesized that receptor utilization, envelope glycoprotein requirements and entry pathways may all vary according to cell type, which is partially supported by the data. What has clearly emerged in recent years is that virus entry is not going undetected by the host. Robust and rapid induction of innate immune response is intimately associated with entry-related events. Here we review the state of knowledge on HCMV cellular entry mediators confronting the scientific challenges by accruing a definitive data set. We also review the roles of pattern recognition receptors such as Toll-like receptors in activation of specific innate immune response and discuss how entry events are tightly coordinated with innate immune initiation steps.

T. Compton
Infectious Diseases, Novartis Institutes for Biomedical Research, Cambridge, MA 02139, USA
teresa.compton@novartis.com

Introduction to Virus Entry

The ability of a virus to enter a host cell and deliver its genome for replication represents the essential, first step in the replication cycle of the virus. Human cytomegalovirus (HCMV), like most herpesviruses, enters cells via direct fusion of the viral envelope with the plasma membrane at neutral pH (Compton et al. 1992). However, in specific cell lines such as retinal pigment epithelial and endothelial cells, HCMV enters by receptor-mediated endocytosis, requiring low-pH (Bodaghi et al. 1999; Ryckman et al. 2006). Regardless, the HCMV entry process is highly complex, requiring multiple envelope glycoproteins, which interact with a series of cellular receptors.

The HCMV genome is predicted to encode over 50 glycoproteins and the virion itself has been determined by mass spectrometry to consist of at least 19 different envelope proteins (Varnum et al. 2004; reviewed in Mocarski et al. 2007). However, of these, only five glycoproteins are essential for virus replication in vitro (reviewed in Mocarski et al. 2007). These include glycoprotein B (gB; UL55), gM/gN (UL100/UL73), and gH/gL (UL75/UL115) (reviewed in Mocarski et al. 2007). Glycoprotein M, the most abundant glycoprotein, accounting for 10% of the virion mass, is complex with gN and acts as an attachment receptor (Kari and Gehrz 1992; Compton et al. 1993; Mach et al. 2000; Varnum et al. 2004). Glycoprotein B, the second most abundant glycoprotein, is both an attachment and fusion receptor (Kari and Gehrz 1992; Compton et al. 1993; Navarro et al. 1993; Bold et al. 1996; Varnum et al. 2004). Glycoprotein H is a fusion receptor, while gL is a chaperone required for gH localization; both are complexed with gO (UL74), a glycoprotein unique to HCMV, which enhances virus entry (Keay and Baldwin 1991; Huber and Compton 1997, 1998).

Alternatively, the gH/gL complex associates with pUL128 and pUL130, conferring epithelial and endothelial cell tropism to clinical isolates of HCMV (Wang and Shenk 2005). The UL131–128 locus is consistently mutated in laboratory-adapted strains such as AD169 and Towne, which do not replicate well in endothelial or epithelial cells (reviewed in Mocarski et al. 2007). An AD169 strain in which the UL131 locus has been repaired (BAD*r*UL131) forms both gH/gL/gO and gH/gL/pUL128/pUL130 complexes and is also restored for its ability to infect epithelial and endothelial cells (Wang and Shenk 2005).

Cellular Receptors Proposed for HCMV Entry

One characteristic of HCMV entry is its broad cellular tropism, leading to the manifestation of disease in almost every tissue type and organ system in the human host. In vitro, HCMV is able to enter a wide variety of cell types, including dendritic, endothelial, epithelial, fibroblast, and monocyte/macrophage cells (Mocarski et al. 2007). The ability of HCMV to enter such a wide variety of cell types indicates the presence of a large number of cellular receptors and/or ubiquitously expressed receptors, a fact that has challenged researchers in the search for cellular receptors.

It is known that HCMV entry begins with an initial tethering step to heparan sulfate proteoglycans (HSPGs) on the cell surface, mediated by gB and gM (Kari and Gehrz 1992, 1993; Compton et al. 1993). At least in vitro, binding to HSPGs is an essential step in the HCMV entry process and is thought to help stabilize the virion at the cell surface until other downstream receptors are engaged (Compton et al. 1993). This is supported by the biphasic binding properties of a soluble form of gB, which in HSPG-null cells reverts to single-kinetic binding (Boyle and Compton 1998). Additionally, HCMV virions are initially dissociable from the cell surface by soluble heparin; however, the virus quickly moves to a heparin-resistant binding state (Compton et al. 1993). These data suggest that an additional receptor(s) is being engaged by HCMV, most likely via gB, following attachment to HSPGs.

Over the past 25 years, several HCMV receptors have been initially identified, only upon more stringent testing, to discover that they did not fulfill the requirements of an entry receptor (Table 1). It was discovered that HCMV associated with β_2-microglobulin (β_2m), which led to the hypothesis that β_2m-coated HCMV particles bound the α chain of the HLA class I antigens, displacing endogenous β_2m (McKeating et al. 1986; Grundy et al. 1987a, 1987b; McKeating et al. 1987). However, β_2m did not bind envelope proteins as expected, but instead associated with the tegument (Stannard 1989). Additionally, preincubating cells with antibodies to MHC class I molecules did not inhibit HCMV infectivity and cell lines differentially expressing MHC class I molecules or β_2m showed no correlation with HCMV infection (Beersma et al. 1990, 1991, 1992; Wu et al. 1994).

A 30-kDa HCMV receptor was identified by binding radiolabeled HCMV particles to cell lysate blots (Adlish et al. 1990; Taylor and Cooper 1990). Virus attachment correlated with the abundance of the receptor but penetration did not, suggesting the receptor acted at the attachment stage only (Nowlin et al. 1991). The receptor was eventually identified as annexin II, which is known to interact with phospholipid membranes and has been implicated in the bridging and fusion of membranes (Wright et al. 1994). HCMV virions were found to bind annexin II, via gB, an event which increased virus binding and fusion (Pietropaolo and

Table 1 Proposed HCMV cellular receptors

Cellular receptor	Interacting HCMV glycoprotein	Currently considered HCMV entry receptor?
HSPG	gB and gM	Yes
MHC class I molecules	None identified	Unlikely
Annexin II	gB	Accessory role?
CD13	None identified	Unlikely
92.5 KDas receptor	gH	More data needed
EGFR	gB	More data needed
α2β1, α6β1, αVβ3 integrins	gB and gH	Yes

Compton 1997; Raynor et al. 1999). Antiserum against annexin II was also able to inhibit plaque formation in fibroblasts; however, cells lacking annexin II are completely permissive for HCMV infection (Wright et al. 1995; Pietropaolo and Compton 1999). Therefore, it is unlikely that annexin II is an HCMV entry receptor, but may enhance entry, cell-to-cell spread, and/or viral egress.

It was discovered that peripheral blood mononuclear cells that expressed CD13 (human aminopeptidase N), were permissive for HCMV infection (Soderberg et al. 1993b; Larsson et al. 1998). Antibodies to CD13 inhibited infection as well as virus attachment to cells but it was later determined that these antibodies also prevented entry into CD13-null cells when the antibody was incubated with the virus, suggesting that the antibodies are binding directly to HCMV virions. (Soderberg et al. 1993a; Larsson et al. 1998). Although CD13 may not be involved in virus entry, it was recently discovered that an HCMV-CD13 interaction inhibits macrophage differentiation from monocytes (Gredmark et al. 2004).

A 92.5-kDa cellular receptor for HCMV was identified for gH through the use of anti-idiotype antibodies that bear the image of gH and are able to specifically bind the 92.5-kDa protein (reviewed in Keay and Baldwin 1995). These antibodies inhibit plaque formation but not virus attachment to cells (reviewed in Keay and Baldwin 1995). The 92.5-kDa receptor has been identified as a phosphorylated glycoprotein that mediates a release of intracellular calcium upon virus binding (reviewed in Keay and Baldwin 1995; reviewed in Keay et al. 1995). Partial cloning and sequencing of the 92.5-kDa receptor have not revealed homology to any known protein; therefore, its identity remains unknown (Baldwin et al. 2000).

More recently, both the epidermal growth factor receptor (EGFR) and a specific subset of cellular integrins have been identified as HCMV entry and signaling receptors. It was noticed that many downstream signaling events initiated by HCMV, including Akt, phosphatidylinositol-3-OH kinase, and phospholipase C-γ activation, as well as the mobilization of intracellular Ca^{2+}, are indicative of EGFR activation (Wang et al. 2003). It was demonstrated that HCMV robustly activates the EGFR kinase, suggesting that EGFR acts at least as a signaling receptor for HCMV (Wang et al. 2003). It was also noted that HCMV initiates gene expression in a breast cancer cell line overexpressing EGFR, but gene expression was not detectable in another EGFR-null cell line (Wang et al. 2003). Additionally, a specific EGFR kinase inhibitor blocked HCMV gene expression in the EGFR-expressing cell line (Wang et al. 2003). The first concrete evidence that EGFR may act as an entry receptor was that in fibroblasts an EGFR neutralizing antibody inhibited HCMV attachment and entry (Wang et al. 2005). Last, the authors demonstrated through cross-linking experiments that HCMV gB can interact with EGFR (Wang et al. 2003). These results were exciting, as the major receptor for HCMV entry had yet to be discovered, with the exception of cellular integrins, whose contribution to virus entry was being examined at the same time as EGFR was proposed to be an entry receptor.

However, it was also known that EGFR, although widely expressed, is not found on all cell types permissive for HCMV infection, including hematopoietic cells, suggesting that EGFR may not be a universal receptor for HCMV (Real et al.

1986). Previous work also demonstrated that HCMV infection leads to EGFR downregulation from the cell surface but not through binding and activation of the receptor early in infection, as UV-inactivated virus does not induce downregulation (Fairley et al. 2002; Beutler et al. 2003). Last, when the contribution of CD13 to HCMV entry was being investigated, an EGFR polyclonal antibody was used as a control to demonstrate that blocking the receptor had no effect on HCMV entry (Soderberg et al. 1993a).

More recently, two other studies have been conducted, testing the role of EGFR in HCMV entry and signaling. It was demonstrated that HCMV did not induce activation of EGFR as measured by receptor autophosphorylation and an EGFR kinase inhibitor was not able to inhibit HCMV gene expression in fibroblast cells (Cobbs et al. 2007; Isaacson et al. 2007). Additionally, the same EGFR neutralizing monoclonal antibody that was previously found by Wang et al. to inhibit HCMV attachment and entry was used to pretreat fibroblasts cells, but led to no reduction in virus entry (Isaacson et al. 2007). EGFR-null fibroblasts were also found to be productive for IE gene expression and breast cancer cell lines differentially expressing EGFR were infected with HCMV but surprisingly, IE gene expression could not be detected in either cell line, regardless of EGFR expression (Cobbs et al. 2007; Isaacson et al. 2007). Additionally, epithelial and endothelial cells were pretreated with EGFR-neutralizing antibodies and then challenged with a clinical isolates of HCMV but again, no decrease in virus gene expression was detected (Isaacson et al. 2007). Given the disparity between these EGFR studies, it is clear that more work needs to be done to discover the contribution of EGFR to HCMV entry.

Several integrin heterodimers have proven to be HCMV entry receptors (Feire et al. 2004; Wang et al. 2005). It has often been observed that following HCMV infection, cells round up, suggesting cytoskeleton rearrangements are occurring upon virus penetration. In fact, the actin cytoskeleton is dramatically altered following HCMV entry and both integrins and focal adhesion kinase become activated (Feire et al. 2004). Through the use of neutralizing integrin antibodies, it was determined that the integrin heterodimers $\alpha2\beta1$, $\alpha6\beta1$, $\alpha V\beta3$ acted as cellular receptors for HCMV entry (Feire et al. 2004). Beta 1 integrin null fibroblast cells demonstrated diminished HCMV entry but restoration of $\beta1$ integrin expression also restored HCMV entry (Feire et al. 2004). A highly conserved integrin-binding domain, the disintegrin-like domain (DLD), was also discovered on gB, which was found to mediate binding to $\beta1$ integrins (Feire et al. 2004, Feire et al., unpublished results). Additionally, a protein fragment encompassing the DLD region and antibodies raised against this protein were able to specifically inhibit HCMV entry (Feire et al., unpublished results). The role of the $\alpha V\beta3$ integrin heterodimer as an HCMV entry receptor was further supported by its interaction with gH; however, an integrin-binding domain has yet to be discovered in this glycoprotein (Wang et al. 2005). Additionally, coordination between $\alpha V\beta3$ and EGFR has been proposed, suggesting that gB and gH independently engage EGFR and $\alpha v\beta3$, respectively, followed by $\alpha V\beta3$ movement into lipid rafts, where it interacts with EGFR to coordinate virus entry and signaling (Wang et al. 2005). This is an interesting model that is supported by HCMV entry at cholesterol-rich microdomains; however, it remains to be verified given the uncertain status of EGFR as an HCMV entry receptor (Wang et al. 2005).

Envelope and Membrane Fusion

The exact mechanism of the fusion step is not clear. Over the past several years, it has become increasingly apparent that amino acid heptad repeat (HR) motifs, which encode alpha-helical coiled-coils, are playing a role in the fusion process of HCMV. These coils are very well characterized in more simple viral fusogenic systems, such as with influenza, HIV, and Ebola viruses, but they are only beginning to be understood in the more complex herpesviruses (Weissenhorn et al. 1999). It is known that synthetic peptides from the HR region of HCMV gB and gH and β amino acid oligomers derived from the HR region of gB specifically inhibit HCMV entry (Lopper and Compton 2004; English et al. 2006). Additionally, the HR region of gB appears to be important because mutation of hydrophobic amino acid residues in this region results in a replication-deficient virus (M.K. Isaacson and T. Compton, unpublished results).

Our current model for HCMV entry (Fig. 1) consists of an initial tethering step to HSPGs on the cell surface mediated by gB and the gM/gN complex (Kari and Gehrz 1992, 1993; Compton et al. 1993). The virus then quickly moves to a more stable binding step, most likely mediated by gB given its biphasic binding properties

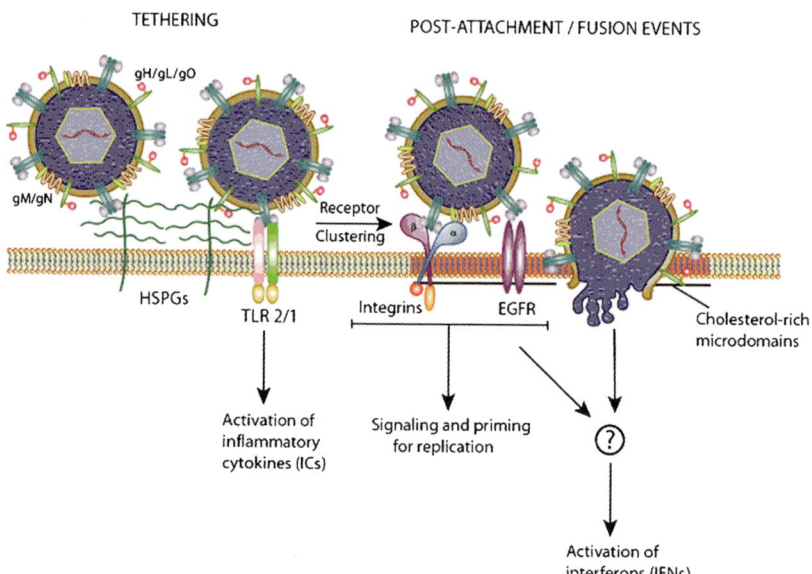

Fig. 1 HCMV entry model

(Boyle and Compton 1998). This secondary binding interaction could be through a gB-EGFR, gB-β1 integrin interaction, or both. However, the role of EGFR in HCMV entry remains uncertain and integrins may act at a post-binding/fusion step; therefore, there is most likely an additional cellular receptor(s) yet to be discovered. It is believed that receptor clustering and signaling may occur following the engagement of cellular integrins by gB and gH, leading to many downstream signaling events necessary for virus entry and/or gene expression (Feire et al. 2004; Wang et al. 2005). Last, gH/gL/gO along with gB mediate the fusion of the viral and cellular membranes, most likely through gB and gH interactions with cellular integrins (Keay and Baldwin 1991; Navarro et al. 1993; Bold et al. 1996; Feire et al. 2004; Wang et al. 2005).

Introduction to Activation of Innate Immunity

Early events in HCMV infection cause a global reprogramming of cellular transcription (Zhu et al. 1997, 1998; Browne et al. 2001; Simmen et al. 2001). Binding and entry of HCMV are known to cause physiological changes such as transient influx of Ca^{2+}, activation of phospholipases C and A2, and the stimulation of arachidonic A metabolism (reviewed in Fortunato et al. 2000). Binding of HCMV envelope glycoproteins to cellular receptors can initiate signal transduction pathways leading to the activation of cellular transcription factors such as NFκ-B, Sp1, and interferon regulatory factor 3 (IRF3) (Yurochko et al. 1997; Navarro et al. 1998; Boehme et al. 2004, 2006). Large-scale studies have revealed that hundreds of cellular genes are affected by early events in HCMV entry and the most strongly induced are antiviral genes belonging to the inflammatory cytokine (IC) family and the interferon-stimulated gene (ISG) family, such as those for RANTES, interleukin-6 (IL-6), IL-8, ISG-54-kDa protein, and IRF-7 (Zhu et al. 1997, 1998; Browne et al. 2001; Simmen et al. 2001). Both of these classes of molecules are hallmarks of innate immunity and contribute significantly to control infection (Stark et al. 1998; Sen 2001). The robust induction of innate immune responses by HCMV does not require virus replication or cellular protein synthesis, suggesting that structural components of the virus are responsible for the changes in gene expression during virus–cell contact and/or entry (Zhu et al. 1997; Browne et al. 2001). Treatment of fibroblasts with glycoprotein B results in a strikingly similar gene expression profile as cells treated with recombinant interferons (Simmen et al. 2001). Together these innate immune responses serve to limit viral replication early during infection as well as activate and promote adaptive immune responses that will ultimately contain or clear the infection.

Activation of Inflammatory Cytokines

The IC branch of the innate immune response is defined by the activation of NFκ-B, which is responsible for the transcription of genes encoding many pro-inflammatory cytokines and chemokines (reviewed in Hayden et al. 2006). In response to stimuli such as cytokines or viruses, activation of the canonical NFκ-B pathway occurs via signal transduction cascades that promote the phosphorylation and degradation of inhibitor of NFκ-B proteins (IκBs), thereby releasing the NFκ-B heterodimer. The activated heterodimer of p50 and p65/Rel A is able to translocate to the nucleus and drive the expression of target genes (reviewed in Baldwin 1996). Gene inactivation experiments of individual NFκ-B family members have revealed that Rel A proteins are required for lymphocyte activation by controlling proliferation, immunoglobulin isotype switching, and the expression of cytokines and their receptors (Attar et al. 1997; Gerondakis et al. 1998). Inflammatory cytokines, including TNF-α, IL-1, IL-6, IL-8, IL-12, and IL-18 have a wide range of biological effects on tissues and cells and are believed to be critical for the recruitment and activation of phagocytic leukocytes to the sites of infection (Laroux 2004). Activation of the IC branch of innate immune signaling is critical for the propagation and elaboration of cytokine responses.

Fibroblasts and monocytes infected with HCMV exhibit activated NFκ-B, as evidenced by its nuclear translocation and increased DNA-binding activity (Yurochko et al. 1995; Yurochko and Huang 1999). HCMV-induced activation of NFκ-B occurs with rapid kinetics that are suggestive of a receptor–ligand interaction (Yurochko et al. 1997). Pretreatment of HCMV with neutralizing antibodies to gB and gH inhibit the induction of the transcription factors NFκ-B and Sp1. Moreover, the use of purified gB and an anti-idiotypic antibody to mimic gH caused the activation of NFκ-B and Sp1 (Yurochko et al. 1997). These data coupled with the ability of HCMV to activate innate immune signaling in the absence of virus replication and cellular protein synthesis led to the hypothesis that viral glycoproteins are initiating the IC response through receptor-binding interactions during the entry process.

The activation of NFκ-B is an immediate host defense mechanism, activated in response to a myriad of stimuli, including ligand stimulation of the Toll-like receptors (TLRs), a class of key pattern recognition receptors. The primary consequences of TLR activation are IC secretion, expression of immune co-stimulatory molecules, dendritic cell maturation, and for a subset of TLRs, interferon α/β activation (reviewed in Takeda and Akira 2003). Together these factors limit viral replication to the site of infection, elicit the infiltration of immune cells to the site of infection, and initiate and modulate adaptive immune responses by T and B cells. To date 12 members of the TLR family have been identified in humans (reviewed in Akira et al. 2006). TLRs are expressed at high levels on phagocytic cells such as dendritic cells or macrophages; however, all cells express at least a subset of these receptors (Hornung et al. 2002; Zarember and Godowski 2002). TLRs recognize microbial pathogens on the basis of structural motifs, termed pathogen-associated molecular patterns (PAMPs), that differ from those found in the host cell (reviewed in Janeway and Medzhitov 2002). Examples of PAMPs include lipopolysaccharide (TLR4), unmethylated CpG DNA (TLR9), and dsRNA (TLR3). Components of a viral envelope can also act as PAMPs to trigger TLR-mediated innate immune signaling. For example, the fusion protein from respiratory syncytial virus and the envelope protein of mouse mammary tumor virus are sensed by TLR4 (Kurt-Jones et al. 2000; Burzyn et al. 2004). Similarly, HCMV activates TLR2 through a physical interaction with gB and gH, to mediate the cellular IC response (Compton et al. 2003; Boehme et al. 2006). TLR2 functional blocking antibodies diminish HCMV-induced IC signaling and a soluble version of gB (gBs) has an inherent ability to induce IL-6 and IL-8 secretion in a TLR2-specific manner (Boehme et al. 2006). Recently we found that the HCMV-mediated IC response is independent of the organization of cholesterol-rich microdomains within the plasma membrane and is not blocked by postbinding HCMV entry inhibitors (L. Juckem and T. Compton, unpublished results). This suggests that the IC response is initiated through TLR2 by outright sensing of the virus during the earliest events in the entry process. The ability of viral glycoproteins to serve as PAMPs is somewhat surprising since the virus-encoded proteins are synthesized by host machinery and ultimately bear protein modifications that are reflective of the host. However, all known viral envelope proteins that are recognized by TLRs share a mutual role in virus entry, suggesting that they may possess unique structural conformations that are recognized by the TLRs.

Recently a parallel mechanism of HCMV-induced NFκ-B activation in response to HCMV has been proposed. It was found that HCMV packages the cellular kinase, casein kinase 2 (CK2), within its tegument during assembly and egress (Varnum et al. 2004). It has been hypothesized that virus–cell fusion deposits the additional constitutively active CK2 molecules into the cytoplasm where they can directly phosphorylate inhibitory IκBα proteins, thereby activating NFκ-B (Nogalski et al. 2007). The contribution of CK2 to the HCMV-induced IC response remains to be fully elucidated.

Activation of Interferon Responses

The type I interferon response, consisting of interferon β and multiple forms of interferon α, is produced in response to viral infection and restricts replication at the earliest stages (reviewed in Stark et al. 1998). Interferon activation is accompanied by the induction of interferon-stimulated genes (ISGs), a subset of cellular genes that carry out many of the antiviral functions of interferon (reviewed in Theofilopoulos et al. 2005). The type I interferon response to virus infection can be divided into two phases: the activation phase and the amplification phase. Virus infection, but not interferon treatment, activates the initial activation phase through the key regulatory transcription factor, IRF3. Signal transduction pathways, which remain incompletely characterized, lead to virus-induced phosphorylation of IRF3 on the carboxyl terminal serine residues by the related kinases TBK1 and IKKε (Servant et al. 2001; Fitzgerald et al. 2003; Sharma et al. 2003). This results in the homodimerization and translocation of IRF3 to the nucleus where it can interact with the co-activators, CREB binding protein (CBP) and p300, and form a complex that drives the transcription of interferon β and a subset of ISGs (reviewed in Taniguchi and Takaoka 2002). The nascent interferon is then able to act in an autocrine and paracrine manner to initiate signaling through the cellular α/β interferon receptor leading to the activation of the janus kinases and signal activators of transcription (JAK/STAT) signal transduction cascade. This amplification phase induces the robust expression of a broad panel of ISGs, which further assists the cell in establishing an antiviral state (reviewed in Stark et al. 1998).

The mechanisms by which cells detect viruses and activate the interferon response have not been completely defined. One of the known mechanisms of IFN activation is through a subset of TLRs (3, 4, 7, 8, 9). The majority of these reside in intracellular compartments, with the exception of TLR4, which resides on the cell surface (reviewed in Akira et al. 2006). The intracellular TLRs allow for rapid recognition of viral ligands during uncoating or degradation processes and require low pH for their activation (Ahmad-Nejad et al. 2002). Herpes simplex virus-1 and -2 and mouse cytomegalovirus are recognized by TLR9 in plasmacytoid dendritic cells (pDCs) and/or dendritic cells (DCs) via their CpG-rich genomes (Lund et al. 2003; Krug et al. 2004a, 2004b). As a global approach to test for the role of intracellular localized TLRs in permissive fibroblasts, we looked at HCMV-induced interferon signaling in the presence of an endosomal acidification inhibitor, bafilomycin, and observed no difference. We also specifically tested TLR3 and TLR4 dominant negative constructs and found no role in HCMV-induced interferon signaling (L.K. Juckem and T. Compton, unpublished results). Since TLR2 is a critical mediator of IC signaling, we tested the ability of HCMV to initiate IFN signaling in cells expressing a dominant negative form of TLR2 (DN-TLR2). We found that the antiviral state was intact in the DN-TLR2 cells, suggesting that it is not involved in HCMV-induced IFN signaling and that the host cell has developed at least two mechanisms to initiate innate immune responses to HCMV infection (K.W. Boehme and T. Compton, unpublished results).

Coordination of Entry Events and Innate Immune Activation Steps

The mechanism of HCMV-induced IFN signaling has only recently begun to be defined. HCMV and gBs are able to activate IRF3 (Navarro et al. 1998; Preston et al. 2001; Boehme et al. 2004). Recent evidence using several strategies to deplete cellular IRF3 confirm its requirement and propose that it is the primary transcription factor mediating HCMV-induced interferon signaling (DeFilippis et al. 2006). The ability of gBs to activate IFN signaling suggests that gB binding to a cell-surface receptor during virus entry is sufficient for antiviral responses. Interestingly, a small molecule HCMV entry inhibitor that targets gB, as well as neutralizing antibodies to both gB and gH, inhibit ISG accumulation (Netterwald et al. 2004). These results provide a link between HCMV-induced interferon activation and early steps in the entry process. In an attempt to dissect the early events in entry, we investigated the function of cholesterol-rich microdomains or lipid rafts. We found that both IE gene expression and pp65 tegument delivery was decreased by pretreatment of cells with a cholesterol-depleting reagent. Similarly, HCMV-induced IFN signaling was also reduced following cholesterol depletion. To confirm a link between entry and IFN activation, we used several novel inhibitors of HCMV entry including a small protein fragment encompassing the disintegrin-like domain of gB and a beta amino acid oligomer that mimics the heptad repeat region of gB (A.L. Feire and T. Compton, unpublished results; English et al. 2006). HCMV-induced interferon signaling was diminished in the presence of these inhibitors, providing further evidence that the IFN activation pathway is connected to downstream events in the entry pathway (L. K. Juckem and T. Compton, unpublished results).

The rapid and direct induction of innate immune responses by HCMV suggests that they are triggered during virus binding and entry into cells. The ability of the virus to activate innate immune signaling in the absence of virus replication and de novo protein synthesis further emphasizes the importance of structural components of the virus. Understanding the coordination of innate immune activation with virus entry is a daunting task due in part to the high level of complexity associated with the molecular events of HCMV entry. Multiple copies of envelope glycoproteins decorate HCMV virions and interact with several cellular receptors to mediate entry into cells (Wang et al. 2003, 2005; Feire et al. 2004). The known interaction between receptors and viral glycoproteins is summarized in Table 2. The importance of gB and gH is exemplified by their roles in entry and both branches of innate immune activation. To date we have found no role for TLR2/1 heterodimer as an entry receptor (K.W. Boehme and T. Compton, unpublished results). Our recent work comparing the requirements for HCMV-induced IC and IFN activation revealed differential regulation. We were able to explore the early events in the HCMV entry process and determine an order to the innate immune activation. The IC response is initiated first by outright sensing of the virus through TLR2 and occurs even if tegument delivery of pp65 is blocked. This activation may be coincident with transfer of HCMV from an initial tethering to a more stable docking step with other cellular receptors. The HCMV-induced IFN response occurs by a postbinding or fusion-dependent mechanism that is dependent on the organization of cholesterol-rich microdomains and is diminished in the presence of HCMV entry inhibitors (L.K. Juckem and T. Compton, unpublished results). Much more work remains to be done to elucidate the underlying mechanism of HCMV-mediated IFN signaling, which may be triggered through: (a) stable binding to cellular receptors, (b) the physical fusion event, and (c) delivery of virion contents into the cytoplasm.

Table 2 Interaction between receptors and viral glycoproteins

Pathway	Cellular component	Virus component	Coordination
IC activation	TLR2/1	gB gH CK2 (packaged)	Outright sensing prior to entry
IFN activation	Unknown	gB gH Fusion event?	Linked to entry pathway

The order of innate immune activation is likely important for the host response to HCMV infection. HCMV has developed an intimate relationship with the host immune response and often the outcome of HCMV infection correlates with the immune status of the host. Rapid recognition of incoming virus could provide a temporal advantage to the cell and prove to be extremely beneficial for combating viral infection. The immediate activation of IC signaling would allow the host to initiate the positive effects of NFκ-B activation such as the infiltration of professional immune cells. However, HCMV may also benefit from innate immune activation. HCMV contains NFκ-B elements within its major immediate early promoter and IC activation even prior to virus entry would prime the cell for viral replication (DeMeritt et al. 2004). Once HCMV has committed to entering the cell, the interferon response is induced, which promotes an antiviral state to protect neighboring cells. The ability of the host to detect HCMV in concert with virus–cell contact and/or entry further highlights the complexity and sophistication of the host innate immune response at the earliest points in HCMV infection.

Perspectives

The exceptionally broad cellular and tissue tropism is a hallmark of CMV pathogenesis. Clearly, CMV exploits is generous glycoprotein coding capacity to form numerous and, in some cases, modular envelope complexes that facilitate interactions with multiple cellular receptors. Mapping the molecular determinants of these interactions will lay the foundation for better insights into the basis of disease and provide targets for therapeutic intervention. Yet, despite the apparent cleverness of CMV to utilize a range of receptors for entry and spread, these events are not silent and do not go undetected by the host. The primal alarm system sounds from the earliest point of virus–cell contact representing yet another avenue for therapeutic manipulation.

References

Adlish JD, Lahijani RS, St Jeor SC (1990) Identification of a putative cell receptor for human cytomegalovirus. Virology 176:337–345

Ahmad-Nejad P, Hacker H, Rutz M, Bauer S, Vabulas RM, Wagner H (2002) Bacterial CpG-DNA and lipopolysaccharides activate Toll-like receptors at distinct cellular compartments. Eur J Immunol 32:1958–1968

Akira S, Uematsu S, Takeuchi O (2006) Pathogen recognition and innate immunity. Cell 124:783–801

Attar RM, Caamano J, Carrasco D, Iotsova V, Ishikawa H, Ryseck RP, Weih F, Bravo R (1997) Genetic approaches to study Rel/NF-kappa B/I kappa B function in mice. Semin Cancer Biol 8:93–101

Baldwin AS Jr (1996) The NF-kappa B and I kappa B proteins: new discoveries and insights. Annu Rev Immunol 14:649–683

Baldwin BR, Zhang CO, Keay S (2000) Cloning and epitope mapping of a functional partial fusion receptor for human cytomegalovirus gH. J Gen Virol 81:27–35

Beersma MF, Wertheim-van Dillen PM, Feltkamp TE (1990) The influence of HLA-B27 on the infectivity of cytomegalovirus for mouse fibroblasts. Scand J Rheumatol Suppl 87:102–103

Beersma MF, Wertheim-van Dillen PM, Geelen JL, Feltkamp TE (1991) Expression of HLA class I heavy chains and beta 2-microglobulin does not affect human cytomegalovirus infectivity. J Gen Virol 72:2757–2764

Beersma MF, Bijlmakers MJ, Geelen JL, Feltkamp TE (1992) HLA-B27 as a receptor for cytomegalovirus. Curr Eye Res 11 Suppl:141–146

Beutler T, Hoflich C, Stevens PA, Kruger DH, Prosch S (2003) Downregulation of the epidermal growth factor receptor by human cytomegalovirus infection in human fetal lung fibroblasts. Am J Respir Cell Mol Biol 28:86–94

Bodaghi B, Slobbe-van Drunen ME, Topilko A, Perret E, Vossen RC, van Dam-Mieras MC, Zipeto D, Virelizier JL, LeHoang P, Bruggeman CA, Michelson S (1999) Entry of human cytomegalovirus into retinal pigment epithelial and endothelial cells by endocytosis. Invest Ophthalmol Vis Sci 40:2598–2607

Boehme KW, Singh J, Perry ST, Compton T (2004) Human cytomegalovirus elicits a coordinated cellular antiviral response via envelope glycoprotein B. J Virol 78:1202–1211

Boehme KW, Guerrero M, Compton T (2006) Human cytomegalovirus envelope glycoproteins B and H are necessary for TLR2 activation in permissive cells. J Immunol 177:7094–7102

Bold S, Ohlin M, Garten W, Radsak K (1996) Structural domains involved in human cytomegalovirus glycoprotein B-mediated cell-cell fusion. J Gen Virol 77:2297–2302

Boyle KA, Compton T (1998) Receptor-binding properties of a soluble form of human cytomegalovirus glycoprotein B. J Virol 72:1826–1833

Browne EP, Wing B, Coleman D, Shenk T (2001) Altered cellular mRNA levels in human cytomegalovirus-infected fibroblasts: viral block to the accumulation of antiviral mRNAs. J Virol 75:12319–12330

Burzyn D, Rassa JC, Kim D, Nepomnaschy I, Ross SR, Piazzon I (2004) Toll-like receptor 4-dependent activation of dendritic cells by a retrovirus. J Virol 78:576–584

Cobbs CS, Soroceanu L, Denham S, Zhang W, Britt WJ, Pieper R, Kraus MH (2007) Human cytomegalovirus induces cellular tyrosine kinase signaling and promotes glioma cell invasiveness. J Neurooncol Jun 23; [Epub ahead of print]

Compton T, Nepomuceno RR, Nowlin DM (1992) Human cytomegalovirus penetrates host cells by pH-independent fusion at the cell surface. Virology 191:387–395

Compton T, Nowlin DM, Cooper NR (1993) Initiation of human cytomegalovirus infection requires initial interaction with cell surface heparan sulfate. Virology 193:834–841

Compton T, Kurt-Jones EA, Boehme KW, Belko J, Latz E, Golenbock DT, Finberg RW (2003) Human cytomegalovirus activates inflammatory cytokine responses via CD14 and Toll-like receptor 2. J Virol 77:4588–4596

DeFilippis VR, Robinson B, Keck TM, Hansen SG, Nelson JA, Fruh KJ (2006) Interferon regulatory factor 3 is necessary for induction of antiviral genes during human cytomegalovirus infection. J Virol 80:1032–1037

DeMeritt IB, Milford LE, Yurochko AD (2004) Activation of the NF-kappaB pathway in human cytomegalovirus-infected cells is necessary for efficient transactivation of the major immediate-early promoter. J Virol 78:4498–4507

English EP, Chumanov RS, Gellman SH, Compton T (2006) Rational development of beta-peptide inhibitors of human cytomegalovirus entry. J Biol Chem 281:2661–2667

Fairley JA, Baillie J, Bain M, Sinclair JH (2002) Human cytomegalovirus infection inhibits epidermal growth factor (EGF) signalling by targeting EGF receptors. J Gen Virol 83:2803–2810

Feire AL, Koss H, Compton T (2004) Cellular integrins function as entry receptors for human cytomegalovirus via a highly conserved disintegrin-like domain. Proc Natl Acad Sci USA 101:15470–15475

Fitzgerald KA, McWhirter SM, Faia KL, Rowe DC, Latz E, Golenbock DT, Coyle AJ, Liao SM, Maniatis T (2003) IKKepsilon and TBK1 are essential components of the IRF3 signaling pathway. Nat Immunol 4:491–496

Fortunato EA, McElroy AK, Sanchez I, Spector DH (2000) Exploitation of cellular signaling and regulatory pathways by human cytomegalovirus. Trends Microbiol 8:111–119

Gerondakis S, Grumont R, Rourke I, Grossmann M (1998) The regulation and roles of Rel/NF-kappa B transcription factors during lymphocyte activation. Curr Opin Immunol 10:353–359

Gredmark S, Britt WB, Xie X, Lindbom L, Soderberg-Naucler C (2004) Human cytomegalovirus induces inhibition of macrophage differentiation by binding to human aminopeptidase N/CD13. J Immunol 173:4897–4907

Grundy JE, McKeating JA, Griffiths PD (1987a) Cytomegalovirus strain AD169 binds beta 2 microglobulin in vitro after release from cells. J Gen Virol 68:777–784

Grundy JE, McKeating JA, Ward PJ, Sanderson AR, Griffiths PD (1987b) Beta 2 microglobulin enhances the infectivity of cytomegalovirus and when bound to the virus enables class I HLA molecules to be used as a virus receptor. J Gen Virol 68):793–803

Hayden MS, West AP, Ghosh S (2006) NF-kappaB and the immune response. Oncogene 25:6758–6780

Hornung V, Rothenfusser S, Britsch S, Krug A, Jahrsdorfer B, Giese T, Endres S, Hartmann G (2002) Quantitative expression of toll-like receptor 1–10 mRNA in cellular subsets of human peripheral blood mononuclear cells and sensitivity to CpG oligodeoxynucleotides. J Immunol 168:4531–4537

Huber MT, Compton T (1997) Characterization of a novel third member of the human cytomegalovirus glycoprotein H-glycoprotein L complex. J Virol 71:5391–5398

Huber MT, Compton T (1998) The human cytomegalovirus UL74 gene encodes the third component of the glycoprotein H-glycoprotein L-containing envelope complex. J Virol 72:8191–8197

Isaacson MK, Feire AL, Compton T (2007) The epidermal growth factor receptor is not required for human cytomegalovirus entry or signaling. J Virol 81:6241–6247

Janeway CA Jr, Medzhitov R (2002) Innate immune recognition. Annu Rev Immunol 20:197–216

Kari B, Gehrz R (1992) A human cytomegalovirus glycoprotein complex designated gC-II is a major heparin-binding component of the envelope. J Virol 66:1761–1764

Kari B, Gehrz R (1993) Structure, composition and heparin binding properties of a human cytomegalovirus glycoprotein complex designated gC-II. J Gen Virol 74:255–264

Keay S, Baldwin B (1991) Anti-idiotype antibodies that mimic gp86 of human cytomegalovirus inhibit viral fusion but not attachment. J Virol 65:5124–5128

Keay S, Baldwin B (1995) Update on the 92.5 kDa putative HCMV fusion receptor. Scand J Infect Dis Suppl 99:32–33

Keay S, Baldwin BR, Smith MW, Wasserman SS, Goldman WF (1995) Increases in [Ca2+]i mediated by the 92.5-kDa putative cell membrane receptor for HCMV gp86. Am J Physiol 269: C11–C21

Krug A, French AR, Barchet W, Fischer JA, Dzionek A, Pingel JT, Orihuela MM, Akira S, Yokoyama WM, Colonna M (2004a) TLR9-dependent recognition of MCMV by IPC and DC generates coordinated cytokine responses that activate antiviral NK cell function. Immunity 21:107–119

Krug A, Luker GD, Barchet W, Leib DA, Akira S, Colonna M (2004b) Herpes simplex virus type 1 activates murine natural interferon-producing cells through toll-like receptor 9. Blood 103:1433–1437

Kurt-Jones EA, Popova L, Kwinn L, Haynes LM, Jones LP, Tripp RA, Walsh EE, Freeman MW, Golenbock DT, Anderson LJ, Finberg RW (2000) Pattern recognition receptors TLR4 and CD14 mediate response to respiratory syncytial virus. Nat Immunol 1:398–401

Laroux FS (2004) Mechanisms of inflammation: the good, the bad and the ugly. Front Biosci 9:3156–3162

Larsson S, Soderberg-Naucler C, Moller E (1998) Productive cytomegalovirus (CMV) infection exclusively in CD13-positive peripheral blood mononuclear cells from CMV-infected individuals: implications for prevention of CMV transmission. Transplantation 65:411–415

Lopper M, Compton T (2004) Coiled-coil domains in glycoproteins B and H are involved in human cytomegalovirus membrane fusion. J Virol 78:8333–8341

Lund J, Sato A, Akira S, Medzhitov R, Iwasaki A (2003) Toll-like receptor 9-mediated recognition of Herpes simplex virus-2 by plasmacytoid dendritic cells. J Exp Med 198:513–520

Mach M, Kropff B, Dal Monte P, Britt W (2000) Complex formation by human cytomegalovirus glycoproteins M (gpUL100) and N (gpUL73). J Virol 74:11881–11892

McKeating JA, Grundy JE, Varghese Z, Griffiths PD (1986) Detection of cytomegalovirus by ELISA in urine samples is inhibited by beta 2 microglobulin. J Med Virol 18:341–348

McKeating JA, Griffiths PD, Grundy JE (1987) Cytomegalovirus in urine specimens has host beta 2 microglobulin bound to the viral envelope: a mechanism of evading the host immune response? J Gen Virol 68:785–792

Mocarski ES, Shenk T, Pass RF (2007) Cytomegaloviruses. In: Knipe DM, Howley PM (eds) Fields virology, vol 2, Lippincott Williams & Wilkins, Philadelphia, pp 2701–2772

Navarro D, Paz P, Tugizov S, Topp K, La Vail J, Pereira L (1993) Glycoprotein B of human cytomegalovirus promotes virion penetration into cells, transmission of infection from cell to cell, and fusion of infected cells. Virology 197:143–158

Navarro L, Mowen K, Rodems S, Weaver B, Reich N, Spector D, David M (1998) Cytomegalovirus activates interferon immediate-early response gene expression and an interferon regulatory factor 3-containing interferon-stimulated response element-binding complex. Mol Cell Biol 18:3796–3802

Netterwald JR, Jones TR, Britt WJ, Yang SJ, McCrone IP, Zhu H (2004) Postattachment events associated with viral entry are necessary for induction of interferon-stimulated genes by human cytomegalovirus. J Virol 78:6688–6691

Nogalski MT, Podduturi JP, Demeritt IB, Milford LE, Yurochko AD (2007) The HCMV virion possesses an activated CKII that allows for the rapid phosphorylation of the inhibitor of NF-{kappa}B, I{kappa}B{alpha}. J Virol 81:5305–5314

Nowlin DM, Cooper NR, Compton T (1991) Expression of a human cytomegalovirus receptor correlates with infectibility of cells. J Virol 65:3114–3121

Pietropaolo RL, Compton T (1997) Direct interaction between human cytomegalovirus glycoprotein B and cellular annexin II. J Virol 71:9803–9807

Pietropaolo R, Compton T (1999) Interference with annexin II has no effect on entry of human cytomegalovirus into fibroblast cells. J Gen Virol 80:1807–1816

Preston CM, Harman AN, Nicholl MJ (2001) Activation of interferon response factor-3 in human cells infected with herpes simplex virus type 1 or human cytomegalovirus. J Virol 75:8909–8916

Raynor CM, Wright JF, Waisman DM, Pryzdial EL (1999) Annexin II enhances cytomegalovirus binding and fusion to phospholipid membranes. Biochemistry 38:5089–5095

Real FX, Rettig WJ, Chesa PG, Melamed MR, Old LJ, Mendelsohn J (1986) Expression of epidermal growth factor receptor in human cultured cells and tissues: relationship to cell lineage and stage of differentiation. Cancer Res 46:4726–4731

Ryckman BJ, Jarvis MA, Drummond DD, Nelson JA, Johnson DC (2006) Human cytomegalovirus entry into epithelial and endothelial cells depends on genes UL128 to UL150 and occurs by endocytosis and low-pH fusion. J Virol 80:710–722

Sen GC (2001) Viruses and interferons. Annu Rev Microbiol 55:255–281

Servant MJ, ten Oever B, LePage C, Conti L, Gessani S, Julkunen I, Lin R, Hiscott J (2001) Identification of distinct signaling pathways leading to the phosphorylation of interferon regulatory factor 3. J Biol Chem 276:355–363

Sharma S, TenOever BR, Grandvaux N, Zhou GP, Lin R, Hiscott J (2003) Triggering the interferon antiviral response through an IKK-related pathway. Science 300:1148–1151

Simmen KA, Singh J, Luukkonen BG, Lopper M, Bittner A, Miller NE, Jackson MR, Compton T, Fruh K (2001) Global modulation of cellular transcription by human cytomegalovirus is initiated by viral glycoprotein B. Proc Natl Acad Sci USA 98:7140–7145

Soderberg C, Giugni TD, Zaia JA, Larsson S, Wahlberg JM, Moller E (1993a) CD13 (human aminopeptidase N) mediates human cytomegalovirus infection. J Virol 67:6576–6585

Soderberg C, Larsson S, Bergstedt-Lindqvist S, Moller E (1993b) Definition of a subset of human peripheral blood mononuclear cells that are permissive to human cytomegalovirus infection. J Virol 67:3166–3175

Stannard LM (1989) Beta 2 microglobulin binds to the tegument of cytomegalovirus: an immunogold study. J Gen Virol 70:2179–2184

Stark GR, Kerr IM, Williams BR, Silverman RH, Schreiber RD (1998) How cells respond to interferons. Annu Rev Biochem 67:227–264

Takeda K, Akira S (2003) Toll receptors and pathogen resistance. Cell Microbiol 5:143–153

Taniguchi T, Takaoka A (2002) The interferon-alpha/beta system in antiviral responses: a multimodal machinery of gene regulation by the IRF family of transcription factors. Curr Opin Immunol 14:111–116

Taylor HP, Cooper NR (1990) The human cytomegalovirus receptor on fibroblasts is a 30-kilodalton membrane protein. J Virol 64:2484–2490

Theofilopoulos AN, Baccala R, Beutler B, Kono DH (2005) Type I interferons (alpha/beta) in immunity and autoimmunity. Annu Rev Immunol 23:307–336

Varnum SM, Streblow DN, Monroe ME, Smith P, Auberry KJ, Pasa-Tolic L, Wang D, Camp DG 2nd, Rodland K, Wiley S, Britt W, Shenk T, Smith RD, Nelson JA (2004) Identification of proteins in human cytomegalovirus (HCMV) particles: the HCMV proteome. J Virol 78:10960–10966

Wang D, Shenk T (2005) Human cytomegalovirus UL131 open reading frame is required for epithelial cell tropism. J Virol 79:10330–10338

Wang X, Huong SM, Chiu ML, Raab-Traub N, Huang ES (2003) Epidermal growth factor receptor is a cellular receptor for human cytomegalovirus. Nature 424:456–461

Wang X, Huang DY, Huong SM, Huang ES (2005) Integrin alphavbeta3 is a coreceptor for human cytomegalovirus. Nat Med 11:515–521

Weissenhorn W, Dessen A, Calder LJ, Harrison SC, Skehel JJ, Wiley DC (1999) Structural basis for membrane fusion by enveloped viruses. Mol Membr Biol 16:3–9

Wright JF, Kurosky A, Wasi S (1994) An endothelial cell-surface form of annexin II binds human cytomegalovirus. Biochem Biophys Res Commun 198:983–989

Wright JF, Kurosky A, Pryzdial EL, Wasi S (1995) Host cellular annexin II is associated with cytomegalovirus particles isolated from cultured human fibroblasts. J Virol 69:4784–4791

Wu QH, Trymbulak W, Tatake RJ, Forman SJ, Zeff RA, Shanley JD (1994) Replication of human cytomegalovirus in cells deficient in beta 2-microglobulin gene expression. J Gen Virol 75:2755–2759

Yurochko AD, Huang ES (1999) Human cytomegalovirus binding to human monocytes induces immunoregulatory gene expression. J Immunol 162:4806–4816

Yurochko AD, Kowalik TF, Huong SM, Huang ES (1995) Human cytomegalovirus upregulates NF-kappa B activity by transactivating the NF-kappa B p105/p50 and p65 promoters. J Virol 69:5391–5400

Yurochko AD, Hwang ES, Rasmussen L, Keay S, Pereira L, Huang ES (1997) The human cytomegalovirus UL55 (gB) and UL75 (gH) glycoprotein ligands initiate the rapid activation of Sp1 and NF-kappaB during infection. J Virol 71:5051–5059

Zarember KA, Godowski PJ (2002) Tissue expression of human Toll-like receptors and differential regulation of Toll-like receptor mRNAs in leukocytes in response to microbes, their products, and cytokines. J Immunol 168:554–561

Zhu H, Cong JP, Shenk T (1997) Use of differential display analysis to assess the effect of human cytomegalovirus infection on the accumulation of cellular RNAs: induction of interferon-responsive RNAs. Proc Natl Acad Sci USA 94:13985–13990

Zhu H, Cong JP, Mamtora G, Gingeras T, Shenk T (1998) Cellular gene expression altered by human cytomegalovirus: global monitoring with oligonucleotide arrays. Proc Natl Acad Sci USA 95:14470–14475

Functions of Human Cytomegalovirus Tegument Proteins Prior to Immediate Early Gene Expression

R. F. Kalejta

Contents

Introduction	101
Tegument Proteins Known to Act at the Very Start of HCMV Infection	104
Delivery of the Genome to the Nucleus	105
Initiating Viral IE Gene Expression	107
Model for Postfusion, Preimmediate Early Events	110
Perspectives	111
References	112

Abstract Proteins within the tegument layer of herpesviruses such as human cytomegalovirus (HCMV) are released into the cell upon entry when the viral envelope fuses with the cell membrane. These proteins are fully formed and active, and they mediate key events at the very start of the lytic infectious cycle, including the delivery of the viral genome to the nucleus and the initiation of viral gene expression. This review examines what is known about tegument protein function prior to the immediate early (IE) phase of the viral lytic replication cycle and identifies key questions that need to be answered to better understand how these proteins promote HCMV infection so that antiviral treatments that target these important viral regulators can be developed.

Introduction

Human cytomegalovirus (HCMV) is a significant human pathogen that infects the majority of the world's population. Viral infection causes birth defects and severe disease in patients with suppressed immune function and is associated with age-related immunosenescence, cancer, and cardiovascular disease (Mocarski

R.F. Kalejta
Institute for Molecular Virology and McArdle Laboratory for Cancer Research, University of Wisconsin-Madison, Madison, WI 53706-1596, USA
rfkalejta@wisc.edu

et al. 2007). Within mature virions, the HCMV genome is housed in an icosahedral protein capsid that is surrounded by a layer of proteins called the tegument, which in turn is enclosed within a lipid membrane termed the envelope. Virally encoded glycoproteins in the envelope function as mediators of viral entry through a membrane fusion event (see the chapter by M.K. Isaacson et al., this volume) that releases both the DNA-containing capsids and the tegument proteins into the cell (Fig. 1).

As many as 59 viral proteins have been found in the viral tegument, although only about 35 are incorporated at significant levels (Baldick et al. 1996; Varnum et al. 2004). Virions also contain a sampling of cellular proteins (Varnum et al. 2004), as well as viral and cellular RNA molecules (Terhune et al. 2004). Bioinformatic and experimental approaches have failed to detect a tegument localization signal (i.e., a sequence necessary and sufficient to direct macromolecules into the tegument) on either proteins or RNAs. The process of assembling the tegument upon viral egress, as well as the disassembly of the tegument upon viral entry into cells, is poorly understood. Likewise, the structure of the tegument within the virion is not known. Although mostly amorphous, there appears to be some structuring of tegument proteins that are closely associated with the capsid (Chen et al. 1999; Trus et al. 1999). Many tegument proteins are phosphorylated (Irmiere and

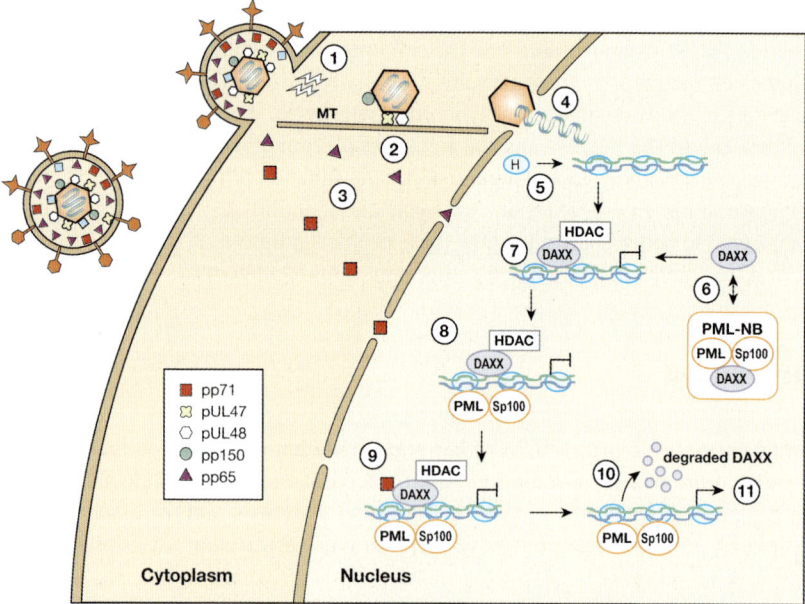

Fig. 1 Postfusion, preimmediate early events during lytic replication of human cytomegalovirus. Schematic representation of delivery of viral genomes and tegument proteins to the nucleus (1.1–1.4), the generation of a silencing complex (PML-NB) on infecting viral genomes (1.5–1.8), and the initial step in the destruction of that complex by tegument-delivered pp71 (1.9–1.11). See the text for further details

Gibson 1983), but the significance of this or other posttranslational modifications to these proteins remains largely unexplored.

Table 1 Human cytomegalovirus tegument proteins with known or predicted functions

Gene, protein	Phenotype	Function(s)	References
UL26	A	Increases stability of virion proteins	Munger et al. 2006 Lorz et al. 2006
UL32 - pp150	E	Directs capsid to site of final envelopment	AuCoin et al. 2006
UL35	A	Activates viral gene expression	Schering et al. 2005
UL36	D	Inhibits apoptosis	Skaletskaya et al. 2001
UL38	A	Inhibits apoptosis	Terhune et al. 2007
UL45	A	Inactive (?) ribonucleotide reductase subunit	Patrone et al. 2003
UL47	A	Release of viral DNA from capsid	Bechtel and Shenk 2002
UL48	E	Deubiquitinating protease	Wang et al. 2006
		Release of viral DNA from capsid	Bechtel and Shenk 2002
UL69	A	Nuclear export of unspliced mRNAs	Lischka et al. 2006
		Arrests cell cycle in G1	Lu and Shenk 1999
UL77	E	Putative pyruvoyl decarboxylase	Yoakum 1993
UL82 - pp71	A	Degrades Daxx, facilitates IE gene expression	Saffert and Kalejta 2006
		Degrades Rb, stimulates cell cycle progression	Kalejta et al. 2003
		Prevents cell surface expression of MHC	Trgovcich et al. 2006
UL83 - pp65	D	Endogenous kinase activity	Yao et al. 2001
		Associated kinase activity	Gallina et al. 1999
		Evasion of adaptive immunity	Gilbert et al. 1996
		Evasion of innate immunity	Arnon et al. 2005
UL94	A/E	Putative DNA-binding protein	Wing et al. 1996
		Similar to autoantigen in systemic sclerosis	Lunardi et al. 2000
UL97	A	Kinase that phosphorylates ganciclovir	Littler et al. 1992 Sullivan et al. 1992
		Stimulates DNA replication, assembly/egress	M. Prichard et al. 1999
UL99 - pp28	E	Directs enclosure of enveloped particles	Silva et al. 2003
IRS1/TRS1	A/E	Inhibits PKR antiviral response	Hakki et al. 2006
		Virion assembly	Adamo et al. 2003
US24	A	Activates viral gene expression	Feng et al. 2006

Genes that encode tegument proteins along with commonly accepted protein names (if applicable) are shown in column 1. Phenotype (column 2), listed as augmenting (A), dispensable (D) or essential (E), refers to the requirement of the gene for lytic replication in human fibroblast cells in vitro as determined in either the provided reference (column 4), the two global mutational analyses of human cytomegalovirus (Dunn et al. 2003; Yu et al. 2003) or as described in a recent review (Mocarski et al. 2007). Column 3 displays either demonstrated or inferred functions for these proteins

Activities for less than half of the tegument proteins have been determined or suggested (Table 1). The phenotypes of recombinant viruses with null mutations in genes encoding tegument proteins demonstrate that some are absolutely essential for viral replication, others are required for efficient replication (often termed augmenting genes), while still others are dispensable for lytic replication in vitro (Mocarski et al. 2007; Dunn et al. 2003; Yu et al. 2003). Many tegument proteins play important roles during the later stages of viral replication, and are required for proper viral assembly and egress (see the chapter by W. Gibson, this volume). This review focuses on the functions of tegument proteins during the initial intracellular events of a lytic infection, after fusion of the viral and cellular envelopes, but prior to the transcription of the first viral genes to be expressed from the infecting viral genome, the immediate early (IE) genes. Thus, what are described here are called postfusion, preimmediate early events.

Tegument Proteins Known to Act at the Very Start of HCMV Infection

The pp65 phospho-protein is the major constituent of HCMV particles (Irmiere and Gibson 1983) and is delivered to the nucleus of permissive cells after fusion of the viral and cellular membranes (Revello et al. 1992). pp65 is either itself a protein kinase (Yao et al. 2001) or associates with a cellular kinase (Gallina et al. 1999) or perhaps both. If pp65 does have intrinsic kinase activity, it is an unusual example of a kinase because it shows poor homology to the catalytic domain sequences of other kinases (Yao et al. 2001). The UL83 gene that encodes pp65 is completely dispensable for replication in cultured fibroblasts (Schmolke et al. 1995), but is likely maintained despite the presence of strong, targeted immune response (Grefte et al. 1992; Wills et al. 1996) because of the ability of pp65 to modulate multiple levels of immune surveillance (Mocarski et al. 2007). Monitoring the delivery of tegument-incorporated pp65 to the nucleus is a common method used to assay for viral entry. Recently, pp150, the second most abundant tegument protein, has also been used to track viral entry (see Sect. 3 below). Encoded by UL32, a gene absolutely essential for lytic replication in vitro (Dunn et al. 2003; Yu et al 2003), pp150 interacts with preformed capsids (Baxter and Gibson 2001), and appears to be required for the incorporation of capsids into forming virions, perhaps because of its ability to stabilize capsids and/or direct their movement within the cytoplasm (Aucoin et al. 2006).

The UL47 and UL48 tegument proteins form a complex with each other and the major capsid protein (Bechtel and Shenk 2002) that appears to play a prominent role during viral entry (see Sect. 3 below). pUL47 has no known enzymatic activity, but pUL48 is a deubiquitinating protease (Wang et al. 2006). While the deubiquitinating activity is not absolutely essential for viral replication, clones with active site mutations in pUL48 show temporal delays in virion release (Wang et al. 2006). Thus, along with their roles in viral entry, pUL47 and pUL48 also likely function during viral maturation and/or egress. UL47 is an augmenting gene, disruption of which results in a 100-fold reduction in viral titers after infections at either high or low multiplicities (Bechtel and Shenk 2002). The UL48 gene has been classified as either augmenting (Yu et al. 2003) or essential (Dunn et al. 2003).

The UL82 gene encodes the pp71 tegument protein that localizes to the nucleus in both HCMV-infected and UL82-transfected cells (Hensel et al. 1996). Although pp71 is not absolutely essential, it is required for efficient viral replication (Bresnahan and Shenk 2000) because of its ability to facilitate viral IE gene expression (see Sect. 4 below). Other tegument proteins such as pUL26 and pUL35 may assist in the pp71-mediated activation of IE gene expression (see Sect. 4 below). pp71 also targets the hypophosphorylated forms of the Rb family of tumor suppressors for proteasome-dependent, ubiquitin-independent degradation, leading to cell cycle stimulation (Kalejta et al. 2003; Kalejta and Shenk 2003), and decreases the cell surface expression of MHC class I proteins by slowing their intracellular transport (Trgovcich et al. 2006).

Delivery of the Genome to the Nucleus

Once in the cytoplasm, HCMV genome-containing capsids and some tegument proteins must make their way to the nucleus. Although a seemingly simple task, this journey is a difficult one due to the size of the viral particle and the density of the cytoplasm. HCMV overcomes these obstacles using strategies that are also employed by other viruses (Dohner et al. 2005; Greber and Way 2006), namely hijacking the intracellular transport machinery. Cells contain an organized network of microtubules (MTs) that extend from the microtubule-organizing center (MTOC) near the nucleus all the way to the periphery, ending near the cell membrane. This network, along with other mechanisms, allows for the temporal and spatial control of the transport of large cargoes to help establish and maintain cell polarity as well as the uneven distributions of proteins, RNAs, and organelles (Welte 2004). MTs are composed of ordered, head-to-tail associations of tubulin monomers, and thus have a distinct polarity, with their negative ends near the MTOC and their positive ends near the cell surface. Cytoplasmic dynein is a minus-end-directed motor protein, which, along with dynactin, uses power generated from ATP hydrolysis to transport cargo along microtubules toward the MTOC (Malik and Gross 2004).

An intact microtubule network is required for the transport to the nucleus of capsids deposited in cells upon HCMV infection (Ogawa-Goto et al. 2003). Nocodazole, a drug that de-polymerizes microtubules, inhibits IE gene expression, likely by preventing infecting HCMV from depositing its DNA in the nucleus. Entering capsids in transit to the nucleus can be localized by detecting the tightly associated tegument protein pp150 through indirect immunofluorescence (Sinzger et al. 2000). In the absence of nocodazole, incoming pp150 is found associated with MTs and concentrated near the nucleus, but is diffusely distributed in the cytoplasm in the presence of the drug (Ogawa-Goto et al. 2003). Transmission electron microscopy (TEM) also showed entering DNA-containing capsids in the cytoplasm that co-localized with MTs (Ogawa-Goto et al. 2003). These capsids displayed a dense outer layer that likely represents tightly associated tegument proteins. This compelling study strongly suggests that entering HCMV capsids and tightly associated tegument proteins travel through the cytoplasm on MTs toward the nucleus, and that this process is important for viral gene expression and replication. One of these tightly associated tegument proteins appears to be pp150, and the recent generation of infectious virus containing a pp150-GFP fusion protein (Sampaio et al. 2005) should allow for the visualization and quantitation of capsid transport to the nucleus in live cells.

Additional candidates for HCMV tegument proteins tightly associated with entering capsids are pUL47 and pUL48. Circumstantial evidence suggests that the UL47/UL48 protein complex (Bechtel and Shenk 2002) may play significant roles in the transport of infecting HCMV capsids to the nucleus and/or in the injection of viral DNA into the nucleus through the nuclear pore complex. The pseudorabiesvirus (PRV) orthologs of pUL47 and pUL48 (pUL37 and pUL36, also called VP1/2, respectively) also form a complex (Klupp et al. 2002) and, as visualized in live cells, travel with entering capsids toward the nucleus and accumulate with them at the nuclear rim (Luxton et al. 2005). The herpes simplex virus type 1 (HSV-1) orthologs (pUL37 and pUL36, also called VP1/2, respectively) also interact with each other (Vittone et al. 2005) and may be the tegument proteins required for the in vitro (and presumably in vivo) transport of viral capsids along MTs (Wolfstein et al. 2006). Interestingly, a temperature-sensitive mutant in HSV-1 UL36 docks at the nuclear pore complex during entry but fails to release viral DNA into the nucleus (Batterson et al. 1983).

Direct evidence of a role for pUL47 and pUL48 in the delivery of viral genomes to the nucleus also exists. Experiments with an HCMV UL47-null mutant revealed a decrease in the overall accumulation of the UL48 protein and in its incorporation into virions (Bechtel and Shenk 2002). Therefore, the UL47-null virus is also hypomorphic for pUL48. Upon infection of permissive fibroblasts with a UL47-null virus, viral immediate early gene expression is delayed, but entry, as assayed by the delivery of the tegument proteins pp65 and pp71 to the nucleus, appears to be normal (Bechtel and Shenk 2002). Thus, viruses lacking UL47 have a defect that is postfusion, but prior to immediate early gene expression. A model in which the HCMV pUL47/pUL48 complex binds to viral capsids and perhaps microtubule motors to mediate the delivery of the capsid to the nuclear pore with the subsequent release of the viral DNA into the nucleus appears to be consistent with the current data. The pp150 protein may also participate in this process. A similar scenario likely occurs for PRV and HSV-1.

Many intriguing questions remain about how tegument proteins and DNA-containing capsids are delivered to nuclear pores, and how the viral genome enters the nucleus. For example, how does the tegument disassemble before, during or after entry? There is a clear example of one tegument protein (pp65) that transits into the nucleus without an intact MT network (Ogawa-Goto et al. 2003), and one, pp150, that remains tightly associated with the capsid (Sinzger et al. 2000) and fails to migrate toward the nucleus in the absence of MTs (Ogawa-Goto et al. 2003). The subcellular localization of other tegument proteins including pp71, pUL47, pUL48 and pUL69 should also be examined during viral entry in both the presence and absence of MT inhibitors to determine which tegument-delivered proteins remain associated with the capsid and thus travel along MTs during viral entry. In addition, a closer examination of the UL26-null virus that appears to have a general defect in tegument formation or stability (Lorz et al. 2006; Munger et al. 2006) could be informative. Is dynein required for the cytoplasmic transport of HCMV capsids toward the nucleus as it is for HSV-1 (Dohner et al. 2002)? If so, to which tegument protein does it bind? Also, how is the viral DNA released from the capsid, through the nuclear pore and into the nucleus? Empty HCMV capsids docked at the nuclear pore complex that presumably had already released their DNA into the nucleus have been observed by TEM (Ogawa-Goto et al. 2003). For HSV-1, the pUL36 protein that is required for DNA release also binds to viral DNA (Chou and Roizman 1989), but it is not clear if this binding plays a role during the release of the genome into the nucleus. Cellular proteins may also play a role in this process, as they do for other viruses (Greber and Fassati 2003). Finally, does one or more of the many cellular signaling pathways induced upon HCMV infection (see the chapter by A. Yurochko, this volume) help during the transport of capsids to the nucleus? Kaposi's sarcoma associated herpesvirus (KSHV) and adenovirus also induce host cell-signaling pathways upon infection, and the activation of these pathways facilitates the MT-directed transport of viral capsids to the nucleus (Naranatt et al. 2005; Suomalainen et al. 2001). Most of the work on the very early stages of HCMV infection has focused on prefusion membrane events or mechanisms of IE gene expression. Although those are defining events in the viral life cycle and certainly merit intense investigation, the time period in between them represents an important, understudied stage of HCMV infection.

Initiating Viral IE Gene Expression

Once viral genomes enter the nucleus, a subset of them associate with subnuclear structures (Ishov et al. 1997) called PML nuclear bodies (PML-NBs), which are sometimes called PODs for PML oncogenic domains or ND10 for nuclear domain 10 (see the chapter by G. Maul, this volume). PML-NBs are visualized as numerous dot-like structures in nuclei, and are built around the PML (promyelocytic leukemia) protein. Other prominent PML-NB proteins include Sp100 and Daxx (Everett and Chelbi-Alix 2007). In the absence of PML, other PML-NB proteins do not co-localize with each other, but are dispersed throughout the nucleus, indicating that the PML protein is required for the integrity of PML-NBs (Ishov et al. 1999). The role of these structures in HCMV-infected cells is beginning to emerge.

Only the HCMV genomes located next to PML-NBs appear to be transcribed, leading to the hypothesis that PML-NBs represent a preferred site for viral gene expression (Ishov et al 1997). However, the proteins that localize to PML-NBs act as transcriptional repressors (Everett and Chelbi-Alix 2007), and many viruses, including HCMV, disrupt PML-NB structures at very early times after infection

(Maul et al. 1993; Korioth et al. 1996). These findings indicate that PML-NBs may actually be detrimental, not helpful, to viral infections. Thus a controversy exists as to whether PML-NBs are pro-virus or anti-virus. Several recent studies from multiple laboratories summarized below paint an incomplete yet quickly resolving portrait of the nuclear events that precede HCMV IE gene expression and argue that PML-NBs are not preferred sites for viral transcription but represent repressive subnuclear domains that are sequentially dismantled during HCMV infection.

Most fluorescent images of PML-NBs give the impression that they are static structures. However, the major constituent proteins of PML-NBs, PML, Sp100 and especially Daxx, actually have a dynamic association with these structures, with high rates of association and disassociation (Wiesmeijer et al. 2002; Everett and Murray 2005). Using synchronized and polarized infections with HSV-1, an elegant series of experiments (Everett and Murray 2005) showed that new PML-NBs are formed de novo around infecting viral genomes (the expected result if PML-NB proteins represent a cellular antiviral defense), and argue against viral genomes migrating through the nucleus to sites of preformed PML-NBs (the expected result if localization to these sites provided an advantage to the virus). The rapidity with which Daxx enters and leaves PML-NBs makes it an obvious candidate for the initial cellular sensor of infecting viral genomes, and this appears to be the case for HCMV. In cells in which the level of PML has been reduced by RNA interference, there are no PML-NBs, and the Daxx and Sp100 proteins are diffusely distributed throughout the nucleus. However, Daxx and Sp100 co-localize to form punctate spots reminiscent of PML-NBs upon HCMV infection (Tavalai et al. 2006). The newly synthesized viral IE2 protein is also found in these aggregates, implying that transcriptionally active viral DNA is located there as well. This significant study showed that in the absence of PML-NBs, Daxx and Sp100 can sense and apparently migrate to infecting HCMV genomes (Tavalai et al. 2006). Any effects of Sp100 on HCMV infection have yet to be established, but it is becoming increasingly clear that Daxx inhibits HCMV infection, and is the very first PML-NB component whose antiviral activities must be neutralized in order for HCMV to express its immediate early genes.

At the low multiplicities presumed to mimic an in vivo infection, the viral pp71 protein, which is delivered from the tegument to the nucleus of infected cells (Hensel et al. 1996), is required for immediate early gene expression and subsequent viral replication (Bresnahan and Shenk 2000). pp71 binds to Daxx through two Daxx-interaction-domains, termed DIDs (Hoffman et al. 2002), and through this interaction partially and transiently localizes to PML-NBs (Hoffman et al. 2002; Marshall et al. 2002; Ishov et al. 2002). Recombinant HCMVs expressing DID-mutant pp71 proteins (and not wild type) have the same phenotype as the pp71-null mutant, indicating that pp71 binding to Daxx is required for efficient viral IE gene expression (Cantrell and Bresnahan 2005). This series of experiments was important because it defined the role of a single function of the multifunctional pp71 protein during viral infection by examining the phenotype of a recombinant virus expressing a mutant pp71 protein, and because it identified Daxx as a critical determinant of HCMV IE gene expression.

The finding that the ability of pp71 to bind Daxx was required for viral IE gene expression at the start of an HCMV infection did not distinguish between two disparate models for pp71 action: the cooperation of pp71 and Daxx to activate IE gene expression (expected if PML-NBs were pro-virus) or the relief of Daxx-mediated repression by pp71 (expected if PML-NBs were anti-virus). The subsequent observation that Daxx levels are dramatically reduced after HCMV infection due to pp71-mediated proteasomal degradation (Saffert and Kalejta 2006) strongly implicated Daxx as a repressor of HCMV IE gene expression and PML-NBs as anti-virus. Multiple approaches by several laboratories have confirmed this. Inhibition of Daxx degradation by a proteasome inhibitor (Saffert and Kalejta 2006) or overexpression of Daxx (Cantrell and Bresnahan 2006; Woodhall et al. 2006) inhibit IE gene expression in HCMV-infected cells, and knockdown of Daxx by RNA interference (Cantrell and Bresnahan 2006; Preston and Nicholl 2006; Saffert and Kalejta 2006; Woodhall et al. 2006) enhances IE gene expression in HCMV-infected cells, especially when pp71 activity is absent or inhibited.

Exactly how Daxx inhibits IE gene expression is still unclear. Daxx is not known to bind directly to DNA, but is recruited to promoters by DNA-binding transcription factors (Salomoni and Khelifi 2006). The cellular factors that mediate the association of Daxx with infecting HCMV genomes and the viral sequences required to recruit Daxx need to be identified. Also, while it is clear that Daxx binds to histone deacetylases (HDACs) and that Daxx-mediated antiviral effects against HCMV requires HDAC activity (Saffert and Kalejta 2006; Woodhall et al. 2006), the specific HDAC or HDACs utilized by Daxx to silence HCMV IE gene expression have not been identified. Also, as chromatin structure plays an important role in HCMV gene expression, the process of chromatin assembly on infecting viral genomes should be explored. In the virion, viral DNA is not associated with histones, but becomes rapidly bound by histones after entering the nucleus (Woodhall et al. 2006). Determining when and how viral genomes acquire a chromatin structure could reveal further insights into the regulation of IE gene expression. For example, chromatin structure may explain why only the viral genomes located at PML-NBs appear to be transcribed (Ishov et al. 1997). Perhaps only a subset of infecting viral genomes becomes properly chromatinized, and only those can be transcribed. By sending PML-NB proteins only to infecting viral genomes competent for transcription, the cell may focus these valuable resources where they are most needed.

Additional tegument proteins may cooperate directly or indirectly with pp71 to stimulate IE gene expression. The pUL35 protein interacts with pp71 (Schierling et al. 2004), and UL35-null viruses have delayed IE gene expression and dramatically reduced production of the early UL44 protein (Schierling et al. 2005), perhaps indicating that while pUL35 has a modest effect on IE gene expression, it may have a much more significant effect on early gene expression. Interestingly, an interaction between pUL35 and pp71 may play a prominent role during viral egress. In cells infected with the UL35-null virus, pp71 (and pp65) remain in the nucleus at late times during infection and do not enter the cytoplasm with egressing capsids

(Schierling et al. 2005). Other assembly/egress defects were also noted. Thus, pUL35 may control the incorporation of other viral proteins (such as pp71) into the tegument. Lower levels of pp71 in UL35-null virions could explain the delay in IE gene expression observed after infection with the mutant virus. A quantitative comparison of tegument proteins incorporated into wild type and UL35-null virions is thus an essential experiment. As pUL35 is only a minor component of the tegument (Varnum et al. 2004), it may have a catalytic (as opposed to stoichiometric) role in tegument assembly, perhaps by influencing nucleocytoplasmic transport pathways, as has been hypothesized (Schierling et al. 2005).

pUL26 is also a minor component of virions (Varnum et al. 2004) that is required for efficient viral replication (Dunn et al. 2003; Yu et al. 2003; Lorz et al. 2006; Munger et al. 2006) and may have an indirect role in the activation of IE gene expression by pp71. In the absence of pUL26, the phosphorylation and stability of at least one tegument protein (Munger et al. 2006) and the stability of virions themselves (Lorz et al. 2006) is reduced. Thus pUL26 appears to play a role in tegument assembly/disassembly, and/or tegument protein/virion stability. How pUL26 may modulate the functions of tegument proteins that act at postfusion, preimmediate early times such as pUL47, pUL48, and pp71 remains to be determined.

In addition to Daxx, the PML protein itself also inhibits HCMV IE gene expression (Tavalai et al. 2006). However, the newly synthesized IE1 protein disrupts PML-NBs (Korioth et al. 1996) and neutralizes the repressive effects of PML and perhaps other PML-NB proteins (Tavalai et al. 2006). Interesting points for further study include determining if PML can be recruited to infecting HCMV genomes in the absence of Daxx, and if Sp100 or any other PML-NB protein also represses HCMV gene expression. We know that at least two PML-NB proteins (Daxx and PML) can repress HCMV IE gene expression, that these structures are sequentially dismantled during HCMV infection, and that HCMV replicates to higher titers in the absence of at least two PML-NB proteins (Daxx and PML), which strongly argues that PML-NBs are not preferred sites of viral transcription and replication, but that the proteins that localize to these structures have antiviral functions. Because these proteins are constitutively expressed, they have been characterized as mediators of intrinsic immunity against HCMV (Saffert and Kalejta 2006; Tavalai et al. 2006), analogous in their effects to retroviral restriction factors (Bieniasz 2004).

Model for Postfusion, Preimmediate Early Events

This section describes an overly simplified model for postfusion, preimmediate early events during HCMV lytic infection of fully permissive fibroblasts (Fig. 1). The model is based on experiments with HCMV, data from other herpesviruses, and a certain amount of speculation on the part of the author. It is not meant as a comprehensive, definitive picture, but as a working model that needs to be refined and built upon.

Figure 1 illustrates that signals initiated on receptor binding may prime the cell for subsequent events during viral entry (1). Mediated by tightly associated tegument proteins such as pUL47, pUL48 and perhaps pp150, capsids are transported along microtubules (MTs) toward the nucleus (2). Cellular motor proteins such as dynein likely assist this transport. Other tegument proteins not directly associated with capsids (such as pp65 and pp71) are transported independently to the nucleus (3). Through unknown mechanisms, capsids dissociate from MTs, dock at nuclear pores, and release their DNA into the nucleus (4). Viral genomes associate with cellular histones (H; 5) and are packaged into chromatin. The Daxx protein, which rapidly dissociates from and reassociates with PML-NBs (6) interacts, in an uncharacterized way, with the viral genome (7), presumably at the major immediate early promoter (MIEP) and perhaps other IE promoters as well. Daxx recruits an HDAC and silences viral gene expression (8) by establishing a repressive chromatin structure. Other PML-NB components are also recruited and participate in the silencing of viral gene expression (8). pp71 binds to Daxx in the newly formed PML-NBs that are silencing infecting viral genomes (9) and induces Daxx degradation (10), thus de-repressing viral IE gene expression (11). The viral IE1 gene product subsequently dismantles PML-NBs and neutralizes the repressive effects of one or more of the proteins that localize to these structures (not shown).

Perspectives

Future work should focus on identifying how, where and when during entry the tegument disassembles, the process of histone association with viral genomes, how Daxx and other PML-NB proteins are recruited to viral genomes, how other tegument proteins cooperate with pp71 (and then IE1) to inactivate the cellular defenses mediated by the PML-NB proteins and how signal transduction cascades induced upon viral entry impact on each of these processes.

Additionally, the question as to whether PML-NBs are either pro-virus or anti-virus needs to be answered. Interestingly, recent evidence suggests that the real answer may be that PML-NBs have both negative and positive effects on the HCMV life cycle. While these proteins clearly inhibit lytic replication (and thus are anti-virus), recent evidence suggests that Daxx may be absolutely necessary to silence expression from the viral genome when latency is established and thus avoid an abortive infection in undifferentiated cells where productive lytic replication cannot be completed (Saffert and Kalejta 2007). Thus, for latent HCMV infections, Daxx could be considered to be pro-virus as well. More work is needed to explore the possibility that HCMV uses the same cellular defense to establish latency that it easily and systematically inactivates at the start of lytic infections.

Acknowledgements My thanks are extended to Leanne Olds for the illustration in Fig. 1, and to my students Ryan Saffert, Adam Hume, and Jiwon Hwang for stimulating discussions and comments on the manuscript. Work in my lab is supported by the American Heart Association

(Scientist Development Grant), the Wisconsin Partnership Fund for a Healthy Future and the Burroughs Wellcome Fund (Investigator in Pathogenesis Award).

References

Adamo JE, Schroer J, Shenk T (2004) Human cytomegalovirus TRS1 protein is required for efficient assembly of DNA-containing capsids. J Virol 78:10221–10229
Arnon TI, Achdout H, Levi O, Markel G, Saleh N, Katz G, Gazit R, Gonen-Gross T, Hanna J, Nahari E, Porgador A, Honigman A, Plachter B, Mevorach D, Wolf DG, Mandelboim O (2005) Inhibition of the NKp30 activating receptor by pp65 of human cytomegalovirus. Nat Immunol 6:515–523
AuCoin DP, Smith GB, Meiering CD, Mocarski ES (2006) Betaherpesvirus-conserved cytomegalovirus tegument protein ppUL32 (pp150) controls cytoplasmic events during virion maturation. J Virol 80:8199–8210
Baldick CJ, Shenk T (1996) Proteins associated with purified human cytomegalovirus particles. J Virol 70:6097–6105
Batterson W, Furlong D, Roizman B (1983) Molecular genetics of herpes simplex virus. VIII. Further characterization of a temperature-sensitive mutant defective in the release of viral DNA and in other stages of the viral reproductive cycle. J Virol 45:397–407
Baxter MK, Gibson W (2001) Cytomegalovirus basic phosphoprotein (pUL32) binds to capsids in vitro through its amino one-third. J Virol 75:6865–6873
Bechtel JT, Shenk T (2002) Human cytomegalovirus UL47 tegument protein functions after entry and before immediate-early gene expression. J Virol 76:1043–1050
Bieniasz PD (2004) Intrinsic immunity: a front-line defense against viral attack. Nat Immunol 5:1109–1115
Bresnahan WA, Shenk T (2000) UL82 virion protein activates expression of immediate early viral genes in human cytomegalovirus-infected cells. Proc Natl Acad Sci USA 97:14506–14511
Cantrell SR, Bresnahan WA (2005) Interaction between the human cytomegalovirus UL82 gene product (pp71) and hDaxx regulates immediate-early gene expression and viral replication. J Virol 79:7792–7802
Cantrell SR, Bresnahan WA (2006) Human cytomegalovirus (HCMV) UL82 gene product (pp71) relieves hDaxx-mediated repression of HCMV replication. J Virol 80:6188–6191
Chen DH, Jiang H, Lee M, Liu F, Zhou ZH (1999) Three-dimensional visualization of tegument/capsid interactions in the intact human cytomegalovirus. Virology 260:10–16
Chou J, Roizman B (1989) Characterization of DNA sequence-common and sequence-specific proteins binding to cis-acting sites for cleavage of the terminal a sequence of the herpes simplex virus 1 genome. J Virol 63:1059–1068
Dohner K, Wolfstein A, Prank U, Echeverri C, Dujardin D, Vallee R, Sodeik B (2002) Function of dynein and dynactin in herpes simplex virus capsid transport. Mol Biol Cell 13:2795–2809
Dohner K, Nagel CH, Sodeik B (2005) Viral stop-and-go along microtubules: taking a ride with dynein and kinesins. Trends Microbiol 13:320–327
Dunn W, Chou C, Li H, Hai R, Patterson D, Stolc V, Zhu H, Liu F (2003) Functional profiling of a human cytomegalovirus genome. Proc Natl Acad Sci USA 11:14223–14228
Everett RD, Chelbi-Alix MK (2007) PML and PML nuclear bodies: implications in antiviral defense. Biochimie 89:819–830
Everett RD, Murray J (2005) ND10 components relocate to sites associated with herpes simplex virus type 1 nucleoprotein complexes during virus infection. J Virol 79:5078–5089
Feng X, Schroer J, Yu D, Shenk T (2006) Human cytomegalovirus pUS24 is a virion protein that functions very early in the replication cycle. J Virol 80:8371–8378
Gallina A, Simoncini L, Garbelli S, Percivalle E, Pedrali-Noy G, Lee KS, Erikson RL, Plachter B, Gerna G, Milanesi G (1999) Polo-like kinase 1 as a target for human cytomegalovirus pp65 lower matrix protein. J Virol 73:1468–1478

Gilbert MJ, Riddell SR, Plachter B, Greenberg PD (1996) Cytomegalovirus selectively blocks antigen processing and presentation of its immediate-early gene product. Nature 383:720–722

Greber UF, Fassati A (2003) Nuclear import of viral DNA genomes. Traffic 4:136–143

Greber UF, Way M (2006) A superhighway to virus infection. Cell 124:741–754

Grefte JM, van der Gun BT, Schmolke S, van der Giessen M, van Son WJ, Plachter B, Jahn G, The TH (1992) The lower matrix protein pp65 is the principal viral antigen present in peripheral blood leukocytes during an active cytomegalovirus infection. J Gen Virol 73:2923–2932

Hakki M, Marshall EE, De Niro KL, Geballe AP (2006) Binding and nuclear relocalization of protein kinase R by human cytomegalovirus TRS1. J Virol 80:11817–11826

Hensel GM, Meyer HH, Buchmann I, Pommerehne D, Schmolke S, Plachter B, Radsak K, Kern HF (1996) Intracellular localization and expression of the human cytomegalovirus matrix phosphoprotein pp71 (ppUL82): evidence for its translocation into the nucleus. J Gen Virol 77:3087–3097

Hoffman H, Sindre H, Stamminger T (2002) Functional interaction between the pp71 protein of human cytomegalovirus and the PML-interacting protein human Daxx. J Virol 76:5769–5783

Irmiere A, Gibson W (1983) Isolation and characterization of a noninfectious virion-like particle released from cells infected with human strains of cytomegalovirus. Virology 130:118–133

Ishov AM, Stenberg RM, Maul GG (1997) Human cytomegalovirus immediate early interaction with host nuclear structures: definition of an immediate transcript environment. J Cell Biol 138:5–16

Ishov AM, Sotnikov AG, Negorev D, Vladimirova OV, Neff N, Kamitani T, Yeh ET, Strauss JF, Maul GG (1999) PML is critical for ND10 formation and recruits the PML-interacting protein daxx to this nuclear structure when modified by SUMO-1. J Cell Biol 147:221–234

Ishov AM, Vladimirova OV, and Maul GG (2002) Daxx-mediated accumulation of human cytomegalovirus tegument protein pp71 at ND10 facilitates initiation of viral infection at these nuclear domains. J Virol 76:7705–7712

Kalejta RF, Shenk T (2003) Proteasome-dependent, ubiquitin-independent degradation of the Rb family of tumor suppressors by the human cytomegalovirus pp71 protein. Proc Natl Acad Sci USA 100:3263–3268

Kalejta RF, Bechtel JT, Shenk T (2003) Human cytomegalovirus pp71 stimulates cell cycle progression by inducing the proteasome-dependent degradation of the retinoblastoma family of tumor suppressors. Mol Cell Biol 23:1885–1895

Klupp BG, Fuchs W, Granzow H, Nixdorf R, Mettenleiter TC (2002) Pseudorabies virus UL36 tegument protein physically interacts with the UL37 protein. J Virol 76:3065–3071

Korioth F, Maul GG, Plachter B, Stamminger T, Frey J (1996) The nuclear domain 10 (ND10) is disrupted by the human cytomegalovirus gene product IE1. Exp Cell Res 229:155–158

Lischka P, Toth Z, Thomas M, Mueller R, Stamminger T (2006) The UL69 transactivator protein of human cytomegalovirus interacts with DEXD/H-box RNA helicase UAP56 to promote cytoplasmic accumulation of unspliced RNA. Mol Cell Biol 26:1631–1643

Littler E, Stuart AD, Chee MS (1992) Human cytomegalovirus UL97 open reading frame encodes a protein that phosphorylates the antiviral nucleoside analog ganciclovir. Nature 358:160–162

Lorz K, Hofmann H, Berndt A, Tavalai N, Mueller R, Schlotzer-Schrehardt U, Stamminger T (2006) Deletion of open reading frame UL26 from the human cytomegalovirus genome results in reduced viral growth, which involves impaired stability of viral particles. J Virol 80:5423–5434

Lu M, Shenk T (1999) Human cytomegalovirus UL69 protein induces cells to accumulate in G1 phase of the cell cycle. J Virol 73:676–683

Lunardi C, Bason C, Navone R, Millo E, Damonte G, Corrocher R, Puccetti A (2000) Systemic sclerosis immunoglobulin G autoantibodies bind the human cytomegalovirus late protein UL94 and induce apoptosis in human endothelial cells. Nat Med 6:1183–1186

Luxton GWG, Haverlock S, Coller, KE, Antinone SE, Pincetic A, Smith GA (2005) Targeting of herpesvirus capsid transport in axons is coupled to association with specific sets of tegument proteins. Proc Natl Acad Sci USA 102:5832–5837

Malik R, Gross SP (2004) Molecular motors: strategies to get along. Curr Biol 14:971–982

Marshall KR, Rowley KV, Rinaldi A, Nicholson IP, Ishov AM, Maul GG, Preston CM (2002) Activity and intracellular localization of the human cytomegalovirus protein pp71. J Gen Virol 83:1601–1612

Maul GG, Guldner HH, Spivack JG (1993) Modification of discrete nuclear domains induced by herpes simplex virus type 1 immediate early gene 1 product (ICP0). J Gen Virol 74:2679–2690

Mocarski ES, Shenk T, Pass RF (2007) Cytomegaloviruses. In: DM Knipe and PM Howley (eds) Fields virology. Lippincott Williams and Wilkins, Philadelphia, pp 2701–2772

Munger J, Yu D, Shenk T (2006) UL26-deficient human cytomegalovirus produces virions with hypophosphorylated pp28 tegument protein that is unstable within newly infected cells. J Virol 80:3541–3548

Naranatt PP, Krishnan HH, Smith MS, Chandran B (2005) Kaposi's sarcoma-associated herpesvirus modulates microtubule dynamics via RhoA-GTP-Diaphanous 2 signaling and utilizes the dynein motors to deliver its DNA to the nucleus. J Virol 79:1191–1206

Ogawa-Goto K, Tanaka K, Gibson W, Moriishi E, Miura Y, Kurata T, Irie S, Sata T (2003) Microtubule network facilitates nuclear targeting of human cytomegalovirus capsid. J Virol 77:8541–8547

Patrone M, Percivalle E, Secchi M, Fiorina L, Pedrali-Noy G, Zoppe M, Baldanti F, Hahn G, Koszinowski UH, Milanesi G, Galina A (2003) The human cytomegalovirus UL45 gene product is a late, virion-associated protein and influences virus growth at low multiplicities of infection. J Gen Virol 84:3359–3370

Preston CM, Nicholl MJ (2006) Role of the cellular protein hDaxx in human cytomegalovirus immediate-early gene expression. J Gen Virol 87:1113–1121

Prichard MN, Gao N, Jairath S, Mulamba G, Krosky P, Coen DM, Parker BO, Pari GS (1999) A recombinant human cytomegalovirus with a large deletion in UL97 has a severe replication deficiency. J Virol 73:5663–5670

Revello MG, Percivalle E, Di Matteo A, Morini F, Gerna G (1992) Nuclear expression of the lower matrix protein of human cytomegalovirus in peripheral blood leukocytes of immunocompromised viraemic patients. J Gen Virol 73:437–442

Saffert RT, Kalejta RF (2006) Inactivating a cellular intrinsic immune defense mediated by Daxx is the mechanism through which the human cytomegalovirus pp71 protein stimulates viral immediate early gene expression. J Virol 80:3863–3871

Saffert RT, Kalejta RF (2007) Human cytomegalovirus gene expression is silenced by Daxx-mediated intrinsic immune defense in model latent infections established in vitro. J Virol 81:9109–9120

Salomoni P, Khelifi, AF (2006) Daxx: death or survival protein? Trends Cell Biol 16:97–104

Sampaio KL, Cavignac Y, Stierhof YD, Sinzger C (2005) Human cytomegalovirus labeled with green fluorescent protein for live analysis of intracellular particle movements. J Virol 79:2754–2767

Schierling K, Stamminger T, Mertens T, Winkler M (2004) Human cytomegalovirus tegument proteins ppUL82 (pp71) and ppUL35 interact and cooperatively activate the major immediate-early enhancer. J Virol 78:9512–9523

Schierling K, Buser C, Mertens T, Winkler M (2005) Human cytomegalovirus tegument protein ppUL35 is important for viral replication and particle formation. J Virol 79:3084–3096

Schmolke S, Kern HF, Drescher P, Jahn G, Plachter B (1995) The dominant phosphoprotein pp65 (UL83) of human cytomegalovirus is dispensable for growth in cell culture. J Virol 69:5959–5968

Silva MC, Yu QC, Enquist L, Shenk T (2003) Human cytomegalovirus UL99-encoded pp28 is required for the cytoplasmic envelopment of tegument-associated capsids. J Virol 77:10594–10605

Sinzger C, Kahl M, Laib K, Klingel K, Rieger P, Plachter B, Jahn G (2000) Tropism of human cytomegalovirus for endothelial cells is determined by a post-entry step dependent on efficient translocation to the nucleus. J Gen Virol 81:3021–3035

Skaletskaya A, Bartle LM, Chittenden T, McCormick AL, Mocarski ES, Goldmacher VS (2001) A cytomegalovirus-encoded inhibitor of apoptosis that suppresses caspace-8 activation. Proc Natl Acad Sci USA 98:7829–7834

Suomalainen M, Nakano MY, Boucke K, Keller S, Greber UF (2001) Adenovirus-activated PKA and p38/MAPK pathways boos microtubule-mediated nuclear targeting of virus. EMBO J 20:1310–1319

Sullivan V, Talarico CL, Stanat SC, Davis M, Coen DM, Biron KK (1992) A protein kinase homologue controls phosphorylation of ganciclovir in human cytomegalovirus-infected cells. Nature 358:162–164

Tavalai N, Papior P, Rechter S, Leis M, Stamminger T (2006) Evidence for a role of the cellular ND10 protein PML in mediating intrinsic immunity against human cytomegalovirus infections. J Virol 80:8006–8018

Terhune SS, Schroer J, Shenk T (2004) RNAs are packaged into human cytomegalovirus virions in proportion to their intracellular concentration. J Virol 78:10390–10398

Terhune S, Torigoi E, Moorman N, Silva M, Qian Z, Shenk T, Yu D (2007) Human cytomegalovirus UL38 protein blocks apoptosis. J Virol 81:3109–3123

Trgovcich J, Cebulla C, Zimmerman P, Sedmak DD (2006) Human cytomegalovirus protein pp71 disrupts major histocompatibility complex class I cell surface expression. J Virol 80:951–963

Trus BL, Gibson W, Cheng N, Steven AC (1999) Capsid structure of simian cytomegalovirus from cryoelectron microscopy: evidence for tegument attachment sites. J Virol 73:2181–2192

Varnum SM, Streblow DN, Monroe ME, Smith P, Auberry KJ, Pasa-Tolic L, Wang D, Camp DG, Rodland K, Wiley S, Britt W, Shenk T, Smith RD, Nelson J (2004) Identification of proteins in human cytomegalovirus (HCMV) particles: the HCMV proteome. J Virol 78:10960–10966

Vittone V, Diefenbach E, Triffett D, Douglas MW, Cunningham AL, Diefenbach RJ (2005) Determination of interactions between tegument proteins of herpes simplex virus type 1. J Virol 79:9566–9571

Wang J, Loveland AN, Kattenhorn LM, Ploegh HL, Gibson W (2006) High-molecular-weight protein (pUL48) of human cytomegalovirus is a competent deubiquitinating protease: mutant viruses altered in its active-site cysteine or histidine are viable. J Virol 80:6003–6012

Welte MA (2004) Bidirectional transport along microtubules. Curr Biol 14:525–537

Wiesmeijer K, Molenaar C, Bekeer IM, Tanke HJ, Dirks RW (2002) Mobile foci of SP100 do not contain PML: PML bodies are immobile but PML and SP100 proteins are not. J Struct Biol 140:180–188

Wills MR, Carmichael AJ, Mynard K, Jin X, Weeks MP, Plachter B, Sissons JGP (1996) The human cytotoxic T-lymphocyte (CTL) response to cytomegalovirus is dominated by structural protein pp65: frequency, specificity, and T-cell receptor usage of pp65-specific CTL. J Virol 70:7569–7579

Wolfstein A, Nagel CH, Radtke K, Dohner K, Allan VJ, Sodeik B (2006) The inner tegument promotes herpes simplex virus capsid motility along microtubules in vitro. Traffic 7:227–237

Woodhall DL, Groves IJ, Reeves MB, Wilkinson G, Sinclair JH (2006) Human Daxx-mediated repression of human cytomegalovirus gene expression correlates with a repressive chromatin structure around the major immediate early promoter. J Biol Chem 49:37652–37660

Wing BA, Lee GCY, Huang E-S (1996) The human cytomegalovirus UL94 open reading frame encodes a conserved herpesvirus capsid/tegument-associated virion protein that is expressed with true late kinetics. J Virol 70:3339–3345

Yao ZQ, Gallez-Hawkins G, Lomeli NA, Li X, Molinder KM, Diamond DJ, Zaia JA (2001) Site-directed mutation in a conserved kinase domain of human cytomegalovirus-pp65 with preservation of cytotoxic T lymphocyte targeting. Vaccine 19:1628–1635

Yoakum GH (1993) Mapping a putative pyruvoyl decarboxylase active site to human cytomegalovirus open reading frame UL77. Biochem Biophys Res Comm 194:1207–1215

Yu D, Silva MC, Shenk T (2003) Functional map of human cytomegalovirus AD169 defined by global mutagenesis analysis. Proc Natl Acad Sci USA 100:12396–12401

Initiation of Cytomegalovirus Infection at ND10

G. G. Maul

Contents

Structural Observations in the First Hours After CMV Infection and Their Limits
   in Interpretative Value 118
Are ND10 Really the Start Sites of CMV Transcription?........................... 121
Structural and Functional Aspects of IE1 122
IE1 Counteracts the Host Cell's Silencing Mechanisms............................ 123
Effect of ND10-Associated Proteins on CMV 125
   PML... 125
   Daxx... 126
   ATRX.. 127
   Sp100.. 127
   HDAC.. 128
Perspectives 128
References 129

Abstract As a large double-stranded DNA virus, CMV replicates in the nucleus, a highly structured environment. Diffusional and solid phases exist as interdependent sets of interactions between many components that determine either replicative success of an infecting virus or the defensive success of the host cell. In their extremes, cell death may be part of the lytic release of viral particles, or, in defense terms, the ultimate sacrifice preventing virus release. Between these extremes exists an evolutionarily derived standoff between virus and cell. Exogenous shifts in homeostasis can disturb this balance, diminishing the cell's defensive powers and reactivating the silenced viral genome. Many of the solid-phase aspects of this process can be seen in situ and analyzed. This review evaluates structural information derived from CMV-infected cells in situ at very early times of infection and the conceptional advances derived from them, mostly centering on the major immediate early gene products, specifically IE1. A scientific basis for considering the major immediate early proteins as potential targets in suppressing CMV disease is discussed.

G.G. Maul
The Wistar Institute, 3601 Spruce Street, Philadelphia, PA 19014, USA
maul@wistar.org

Structural Observations in the First Hours After CMV Infection and Their Limits in Interpretative Value

The last physical hurdle for the infecting viral genome appears to be the nuclear pore complex. How viral DNA enters the host nucleus is not entirely clear, nor is it clear how it is protected from degradation by endonucleases once released from the capsid. One must assume that the viral DNA is first neutralized before or as it enters the host nucleus. This could be mediated either by the host's histones – specifically during S-phase when free histones are abundant – forming an easily silenced chromatin package or by positively charged polyamines neutralizing the virus DNA's negative charge, as has been reported for HSV1 (Gibson and Roizman 1971). Spermidine may still be associated with the viral genome when it leaves the capsid and enters the host nucleus. Because the nucleus is a highly structured environment, viral genomes are excluded from certain domains, such as the tightly packed nucleolus. Any observed nonrandom distribution of viral genomes in the nucleus may therefore be due to exclusion.

We do not know the physical dimensions of the large viral genome of CMV in the nucleus. A tightly coiled, chromatinized genome may be slightly larger than an encapsidated genome. Nor do we know whether the viral genome can move through the nucleus, either by passive diffusion or by active transport. This question has been difficult to address, because we cannot see the virus in real time and must therefore construct a likely sequence of events from observations of fixed material. The large size of CMV genomes aids microscopic identification and localization by in situ hybridization, which allows individual viral genomes to be visualized (Ishov et al. 1997). The size of these signals is close to that obtained by imaging tegument proteins like pp71 (Fig. 1a and b), i.e., the diffraction point size, which is limited by the wavelength of light. Signals can only become weaker when originating from smaller sources. Comparison of encapsidated viruses and in situ hybridization signals representing CMV genomes suggests that the infected host nucleus contains genomes that are highly condensed (Ishov et al. 1997). One caveat is that the completely extended genome (78 µm) would not be visible using this technique, because signals from any point along the approximately 260-Mb genomes would be too weak to register with our current techniques. Therefore, the detectable size, as estimated from what is visible, is equal to or smaller than the wavelength of light (~300 nm).

Fig. 1 (continued) early transcripts (*green*), for IE2 (*blue*) and PML (*red*). IE2 is located like a collar around the emerging transcripts. **d** 3T3 cell infected with MCMV (24 h p.i.). The cell is labeled with antibodies to the 112/113 gene product (*red*) and for viral genomes by in situ hybridization (*green*). Viral genomes are seen on the outside of the 112/113 labeled prereplication domains. **e** HFF 3 h p.i. by HCMV. Cell is triple-labeled for viral immediate early transcripts (*green*), for the splicing compartment delineating SC35 (*blue*) and the PML defining ND10 (*red*). The nucleus is outlined in *blue*. **f** MCMV infected 3T3 cells (24 h p.i.). Cells were probed for viral DNA by in situ hybridization (*green*) and for the 112/113 gene product by antibodies (*red*). The *hollow red spheres* inside the nucleus represent the replication compartments. The nucleus is outlined in *blue*

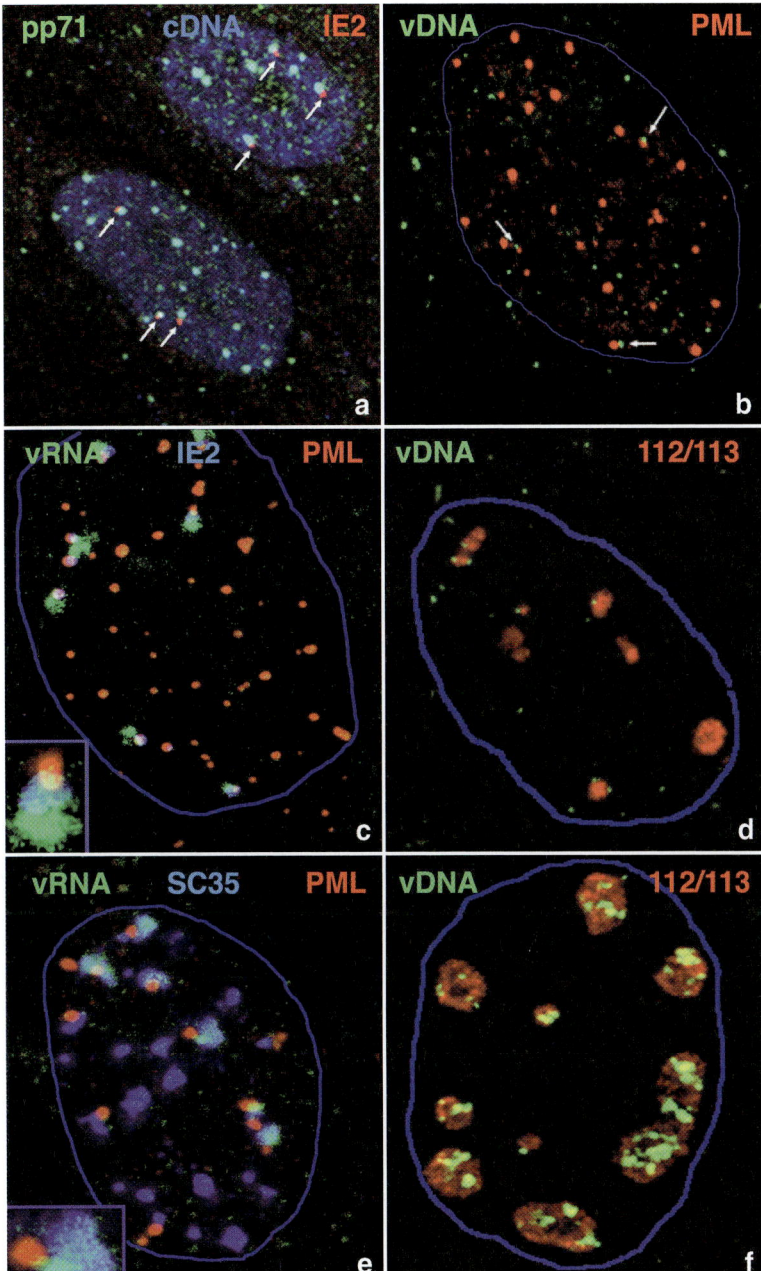

Fig. 1 a HCMV infected HFF (3 h p.i.). *Small green dots* represent pp71 antibodies bound to viral particles. *Larger green dots* in the (*blue*) nucleus represent pp71 antibodies bound to Daxx, which is associated with PML at ND10. The three *red domains* located beside the pp71 stained ND10 denote the localization of IE2 aggregates. **b** Same as **a**, but stained with anti-PML antibodies in *red* indicating ND10 and by in situ hybridization with HCMV DNA labeling viral genomes (*green*). *Arrows* point to viral genomes juxtaposed to ND10. The *blue line* was introduced to show the approximate boundary of the nucleus. **c** Same as **a**, but triple-labeled for the major immediate

The opposite technical problem is the much larger appearance of a structure with dimensions below the wavelength of light due to the dispersion of light from a high-intensity source. In this instance, a structure can appear much larger than it really is. One such type of structure is ND10 (nuclear domain 10), most often identified by immunofluorescence using antibody to the promyelocytic leukemia protein (PML; Fig. 1b, c, e). ND10 figure prominently in the early stages after DNA virus infections and appear to function like nuclear depots (Ishov and Maul 1996; Maul 1998). Most ND10 appear to be substantially larger than 300 nm when examined by fluorescent microscopy but smaller than 300 nm when examined by electron microscopy (Maul et al. 1995). This disparity is important to consider when interpreting images that show physical association between viral genomes and ND10. In human fibroblasts examined at 3 h postinfection (p.i.) (Fig. 1b), HCMV genomes appear throughout the nucleus as diffraction-spot-sized signals. Only a few of these signals localize beside ND10; none localize in these domains. Localization of genomes next to ND10 could be due to random events, particularly when one excludes the large volume of the nucleolus and the apparent location of various DNA viruses in only the interchromosomal space, that is, not within the chromosomal territories but the space occupied where ND10 are positioned (Bell et al. 2001).

Shortly after infection, IE transcripts appear to emanate only from a few ND10 (Fig. 1c and e) (Ishov et al. 1997). The conclusion drawn from these images was that most of the viral genomes that had reached the nucleus were incompetent to transcribe and only those that reached ND10 found a congenial space where transcription was possible. However, most of the major ND10-associated proteins, such as PML and Sp100, are upregulated by interferon (Maul 1998). More ominous, all of these proteins, including the PML-interacting Daxx, are transcriptional repressors involved in the formation of heterochromatin (Seeler et al. 1998; Xu et al. 2001; Ishov et al. 2004). ND10, therefore, appear more like sites for nuclear defense. Also, CMV and other DNA viruses possess genes whose products can eliminate or disperse these nuclear domains, as first identified for herpes simplex virus (Maul et al. 1993). For CMV, the dispersing protein is IE1, which first accumulates at ND10 for some time before ND10 are dispersed (Ishov et al. 1997; Ahn et al. 1998). Surprisingly, the highest concentration of the major immediate early viral transactivator and major immediate early promoter (MIEP) repressor, IE2, is at the site of the highest concentration of viral IE transcripts (Fig. 1c). These transcripts seem to pass from the IE2-covered site into the splicing-factor-containing domain (Fig. 1e). The number of transcripts is much larger than expected, suggesting either a very high transcription rate from a single transcription unit or, as we suspect presently, an inhibition of splicing and thus accumulation of unspliced transcripts. The general arrangement of ND10, the region of accumulated transcripts, and the IE2-containing site were described as the immediate transcript environment which resolves approximately 6 h p.i (Ishov et al. 1997). At first glance, these observations indicate that the virus dispersed a site inhibitory to its replicative success. The immediate early transcript environment also includes the UL112/113 gene products (Fig. 1d). The UL112/113 gene product is part of the expanding prereplication domain and viral genomes are present at the outside of this domain. Prereplication domains apparently become the replication compartments many hours later when they appear hollowed out (Fig. 1f). More than a decade after the first observation that viruses transcribe predominantly at ND10, it remains unclear what advantage the virus gains by dispersing ND10-associated proteins. Is it possible that the questions asked are not relevant and that transcription at ND10, and ultimately replication where ND10 had been present, is simply a consequence of various layers of interactions, none of them simply representing either advantages for the virus or defense mechanisms of the host. Clearly though, those domains focus our mind on the activities that take place in these microenvironments.

Because of the limited resolving power of light microscopy, we still do not know the precise location of transcribing viral genomes relative to ND10 beyond 300 nm resolution, a huge molecular gap. This is important, because it could mean the virus has become part of ND10, where its interaction with a solid interface would have the same mechanistic relevance as recently argued for genes with matrix elements (Kumar et al. 2007). The viral genome may be localizing at the ND10 interface within the interchromosomal space, i.e., randomly localizing in the limited interchromosomal space where ND10 also resides. However, the observations that viral transcription occurs at ND10, that IE1 disperses ND10-associated proteins, and that interferon induces ND10-associated proteins suggest that the association of viral transcription and ND10 is causal rather than casual. To understand this relationship better, we may have to reconstruct the evolutionary balance achieved by a multitude of interactions, each modifying others.

Are ND10 Really the Start Sites of CMV Transcription?

The static images obtained from fixed material 3 h p.i., certainly suggest as much. However, the lateral infection sequence employed for HSV1 (infection from a neighboring cell in the same culture flask plane) indicates that the ND10-associated proteins leave their initial segregated state and move to the viral genomes (Everett and Murray 2005). According to this scenario, new aggregates of ND10-associated proteins form on the virus genome and even prevent the virus genome from moving into the center of the nucleus. Our early observations for HSV1 and HCMV then come from the infection and nuclear pore penetration on the large upper surface of the nuclei, thus requiring a short migration downward to preexisting ND10 (or a short migration of ND10-associated proteins to the virus). We have shown that foreign DNA/foreign protein complexes of bacterial or viral origin attract ND10-associated proteins whether introduced into cells by infection or transfection (Tang et al. 2000, 2003). Reiterated HPV11 origins of replication plus the origin binding protein E2 or integrated bacterial reiterated operon sequences plus GFP-labeled LacI repressor protein attracted ND10-associated proteins (Tang et al. 2001). Foreign DNA alone did not. One interesting finding was that foreign DNA/foreign protein complexes accumulate a limited amount of ND10-associated proteins, i.e., smaller incorporated reiterations had smaller dot size. The dot size did not increase with interferon upregulation of PML or Sp100, indicating these sites were not ND10 nucleating sites. Rather they are new sites of DNA or chromatin that bound PML or Sp100 and entered the visual range because of the reiteration of binding sites. Since most HCMV viral genomes have no visually recognizable PML attached even after entering the nucleus (see Fig. 1b), except for the few particles at ND10 that transcribe, this virus should be re-evaluated for association with ND10. The key experiment would be to determine whether the transcribing viruses are at ND10 and whether such apparent ND10-associated aggregates had moved or were newly formed. Because transcribing genomes cannot presently be labeled in a time-resolved manner, we may use immediate early transcript environment formation as a surrogate, by tracking the accumulation of GFP-tagged IE2 in cells containing Cherry-tagged ND10.

Structural and Functional Aspects of IE1

IE1 is the viral protein that disperses ND10 (Korioth et al. 1996; Ahn and Hayward 1997; Ishov et al. 1997; Wilkinson et al. 1998). In contrast to the immediate early protein ICP0 of HSV-1, it does not do so by the proteosome-dependent destruction of PML and Daxx. It was suggested that IE1 disperses ND10 by binding to constituents and their removal from ND10 over time, because IE1 and PML interact (Ahn et al. 1998). Neither the mechanism of interaction nor the means of dispersal are known. One preliminary finding on MCMV IE1 casts doubt on the idea that stochastic removal of PML by IE1 is the sole mechanism. Removing a 7-amino acid sequence from one of the helices in the N-terminal region of MCMV IE1 eliminates the dispersive function but not the binding function (Q. Tang and G. Maul, unpublished data). We may have to search for a mechanism that includes a function other than simple binding of PML, for dispersion of ND10. We also may have to compare IE1 of HCMV and MCMV; such a comparison could help illuminate the different domains necessary for certain functions.

IE1 has several functional properties that have been used to probe its structure through mutational analysis. It augments viral and host gene transcription, disperses ND10 and binds repressor proteins. These functions are driven either by the indirect augmentation of transcription, possibly by alleviating repression of IE1 binding to p107 (Poma et al. 1996) or HDAC (Tang and Maul 2003; Nevels et al. 2004b), or by the direct augmentation of transcription through IE1 binding of transcription factors (Lukac et al. 1997). Determining how these proteins bind to IE1, and to which interface, is important for developing effective interference strategies. Because IE1 plays an important role during low-particle infections (assumed to be the normal infection mode), a strategy aimed at IE1 inactivation might be successful in blocking the HDAC-binding capacity of IE1 and allowing the host cell to silence competent viral genomes.

Deletion analysis has revealed that the HDAC-binding site in MCMV IE1 is between amino acid residues 100 and 310. A peptide comprised of these amino acids retains HDAC-binding capability and the potential to augment transcription from the MIEP. A deletion in a similar helical region of HCMV IE1 also eliminates the augmenting effect (Stenberg et al. 1990) and abolishes IE1-mediated dispersal of ND10 (Ishov et al. 1997; Lee et al. 2004). More detailed deletion analysis revealed that removing amino acid sequences surrounding the HDAC binding domain eliminates the ability of IE1 to disperse ND10 (Q. Tang and G. Maul, unpublished data). Structurally, this suggests that IE1 possesses a bipartite ND10-binding domain that differs from the HDAC-binding domain. A bipartite p107-binding domain has also been reported for HCMV IE1 (Poma et al. 1996).

HCMV IE1 and MCMV IE1 share only 12% amino acid homology, mostly in the highly acidic C-terminal region. Moreover, HCMV IE1 is 20% shorter than MCMV IE1. Therefore these two proteins appear to have very little in common. However, they share the same genetic structure, as well as the ability to disperse ND10 and augment viral transcription. We need additional comparative and functional analyses, particularly if the mouse system is to be used as an experimental small animal model. Although the primary structures of HCMV IE1 and MCMV IE1 differ significantly, their secondary structures are surprisingly similar. One unresolved aspect of the apparent structural similarity between HCMV IE1 and MCMV IE1 is the position of the small ubiquitin modifier (SUMO). In HCMV IE1, SUMO is at aa 450 (Xu et al. 2001), and in MCMV IE1 we find the SUMO consensus sequence at aa 223. These rather large covalent modifications could have a strong differential influence on the 3D structure of IE1, and thus could influence the functional properties of the two different IE1s. Since SUMO modifies very small amounts of protein at any given time (Johnson 2004), these SUMO subsets may have additional functions. In HCMV IE1, deletion of the SUMO modification site reduces the levels of IE2 transcript and their translation products (Nevels et al. 2004a) and PML desumoylation (Lee et al. 2004). In MCMV IE1, the putative SUMO modification site is within the HDAC binding site (G. Maul, unpublished observations); its precise functions, however, remain unknown.

Though IE1 is not essential for productive infection, it appears to be very important for replicative success. Isolating the respective functions of this molecule and assigning them to its different interfaces may provide a rational basis for the search for small interfering molecules. Such molecules may induce an IE1 minus phenotype, substantially lowering productive infection.

IE1 Counteracts the Host Cell's Silencing Mechanisms

The major immediate early transcript is differentially spliced to produce a number of proteins. The two major and best-investigated proteins, IE1 and IE2, have in common exons 2 and 3 but differ in the larger exon 4 (IE1) and exon 5 (HCMV IE2 and its MCMV homolog, IE3). These proteins act synergistically to activate early viral protein expression, but antagonistically to autoregulate the MIEP (Cherrington and Mocarski 1989; Pizzorno and Hayward 1990; Stenberg et al. 1990; Cherrington et al. 1991; Liu et al. 1991). Both activate or augment viral and host gene transcription (Hagemeier et al. 1992). Most transactivators become part of the basal transcription machinery when they bind to DNA in the promoter region. Apparently, IE1 and IE2 are no exceptions. IE2 binds to specific sequences on early promoters (Cherrington et al. 1991; Meier and Stinski 1997). Some, but not all, transcription factors (TBP, TFIIB) accumulate in the domain adjacent to ND10 together with IE2, where there is just one viral genome (Ishov et al. 1997). IE2 is therefore the likely accumulator of pretranscription complexes, although it might create this enriched microenvironment in some association with the UL112/113 gene products.

IE1's mechanisms of action are not understood as well as those of IE2. Apparently, IE1 is not essential to produce viral progeny, but is necessary for the more natural mode of infection involving low levels of particles. Fibroblasts infected with an IE1-deletion mutant of HCMV require a much larger number of mutant viral particles to achieve the same degree of replicative success as that of wild type viruses, indicating the necessity of multiple genomes of the IE1 mutant. Mocarski and colleagues also found that viral transactivators, such as tegument proteins can compensate for IE1 at high multiplicities of infection (moi) (Mocarski et al. 1996). At low moi, infected cells produce no replication compartments, despite the nearly equal amount of IE2 synthesis (Greaves and Mocarski 1998). This does not support the idea that IE1 is a necessary component of the transcription machinery as IE2 is. The temporal localization of IE1 in specific nuclear compartments, along with the potential interactions of IE1 with nuclear proteins in these compartments, points to additional functions of IE1 that are in line with the often noted augmentation of transcription observed in transfection experiments.

Intuitively, one might suppose that the nuclear site with the highest concentration of a protein is where it functions. IE1 should therefore function in all ND10, and IE2 should function beside a few ND10. However, because not all ND10 have transcribing viral genomes, it follows that IE1 would not act on viral genomes. IE1 does not act on host genes as no host genes have been found in ND10. IE1 must therefore have functions other than transactivation with the basal transcription machinery.

Identifying proteins that interact with IE1 is one way to determine its other functions. IE1 colocalization with ND10 proteins has been used to identify IE1 interaction partners. The interaction between IE1 and PML in HCMV (Ahn et al. 1998) has also been confirmed in MCMV (Tang and Maul 2003). In immunoprecipitation analyses, both of the ND10-associated proteins, PML and Daxx, co-immunoprecipitate with MCMV IE1, suggesting that all three proteins form a complex. At present, no functional assay exists for examining the interaction between IE1 and Daxx. However, MCMV replicates more successfully in Daxx$^{-/-}$ cells (Tang and Maul 2006). However, analysis of the influence of the ND10-associated proteins on overall replicative success has just begun in cells where these proteins and another ND10-associated protein, such as Sp100, have been eliminated or strongly downregulated by siRNA.

Other functions of IE1 are observed. IE1 apparently functions as an antagonist to type1 interferon signal transduction (Paulus et al. 2006). Where this antagonist interferes is of interest since it may be at several levels of interaction. It might be the downstream effect of dispersing PML and thus releasing and possibly changing the Sp100 isotype composition. Circumstantial evidence comes from work that shows that elimination of PML and thus release of all ND10-associated proteins strongly reduces the detrimental Sp100 isotypes (Everett et al. 2006; Negorev et al. 2006).

Effect of ND10-Associated Proteins on CMV

PML

PML is the matrix protein of ND10. Without PML, specifically the SUMO modified form, the aggregation of various ND10-associated proteins does not take place (Ishov et al. 1999). Numerous proteins have been shown to accumulate at ND10, mostly when overexpressed, and many reviews suggest a plethora of supposed functions of these protein accumulations (Negorev and Maul 2001). There seems to be no nuclear function that has not been fingered as dependent, influenced or modulated by these structures, except perhaps splicing. This often indiscriminate assignment of function, based on mostly spurious evidence of colocalization after overexpression, has muddied the field considerably. However, a clearing and clear effect on the replicative cycle of HCMV has recently been provided by Stamminger's group, who showed that the depletion of PML through shRNA substantially increases replicative success. This was all the more convincing because it could be reversed by the reconstitution of a single PML isotype (Tavalai et al. 2006). The finding that depletion of PML can enhance the number of cells replicating HCMV and plaque formation by a factor of 4 shows that IE1 cannot completely overcome the repressive effect, unlike ICP0 of HSV-1 (Everett et al. 2006). However, the much higher (20 times) enhancement of replicative success of the IE1 deletion mutant also shows that IE1 has a suppressive effect on the PML-based inhibition of immediate early transcriptional events and replicative success.

PML may repress the initiation of immediate early transcription, or its progression, or both. PML may repress transcription by interfering directly with Daxx-mediated NFkB binding on the MIEP, or by indirect means such as retaining repressive factors (Daxx, ATRX, Sp100). IE1 in turn may enhance transcriptional activity by lowering free PML, as IE1's concentration generally exceeds that of PML. Observations on HCMV and MCMV show that PML, Daxx and Sp100 initially accumulate in HCMV IE2 or MCMV IE3 domains and the UL112/113 outlined prereplication domains. From there, they disperse at later stages when IE1 is present. In the absence of IE1 in either virus, these ND10-associated proteins remain in the prereplication domains, and later in the replication domains (Tavalai et al. 2006). These observations may merely represent minor shifts in the availability of free repressive proteins, but could also signify inactivation by segregation. The repressors in these domains may also affect the reduction of replicative success. We do not even know whether PML and IE1 interact directly, except that evidence from Daxx$^{-/-}$ cells seems to exclude Daxx as an adapter (Tang and Maul 2003). The mechanism, or at least the molecular domains of PML essential for PML-induced inhibition, can now be investigated by using the PML-depleted cells and manipulating the individual PML isotype, specifically PML IV. This includes the manipulation of its ND10-forming capability. Such investigations may determine whether the HCMV replication sequence is dependent on the ability of PML to multimerize and separately to form complexes by attracting Daxx, and through Daxx, ATRX (Ishov et al. 2004).

Daxx

Daxx interacts at its C-terminal end with sumoylated PML (Ishov et al. 2004), and the SUMO-interacting motif (SIM) at the very end of the molecule is the necessary domain for interaction with may other repressors. Single amino acid changes in the I I V L sequence of Daxx abrogate ND10 association and functional repression of the glucocorticoid receptors (Lin et al. 2006). Daxx therefore likely functions as an adaptor protein or corepressor. The repressive effects of Daxx on HCMV have been documented by siRNA suppression of hDaxx (Cantrell and Bresnahan 2005, 2006; Saffert and Kalejta 2006) and for MCMV, by inference, using mouse Daxx$^{-/-}$ cells (Tang and Maul 2006). The viral counter-defenses seem to center on the tegument protein pp71, which was identified as the first Daxx-interacting protein. Pp71 and Daxx interact in the N-terminal half of the Daxx molecule (Ishov et al. 1999, 2002; Hofmann et al. 2002; Marshall et al. 2002) away from the PML-interacting C-terminal end (Ishov et al. 1999). pp71 is enriched in all ND10 after infection (Ishov et al. 2002). We assume the viral genome deposition at ND10 to come about by interaction of the viral DNA binding to pp71, which in turn is bound to Daxx and thus deposited at highly increased frequency to the high concentration of PML at ND10 (Ishov et al. 1999, 2002; Hofmann et al. 2002; Marshall et al. 2002). If Daxx is a repressor and binds pp71 to produce an inactive transactivator complex, the viral genome should be suppressed at ND10, a nuclear defense. Here IE1 may counter this defense by its binding of Daxx. The balance between these separate interactions should help reveal the choreography of sequential, temporal and spatial interactions that set the stage for the progress or suppression of the lytic cycle.

Our experiments also show that MCMV IE1 appears to interact with PML or Daxx independently. Because Daxx interacts with the histone deacetylases (HDAC) (Li et al. 2000), we tested the possibility that IE1 binds indirectly to HDAC. Identifying an IE1-HDAC interaction, however, may have been fortuitous, since HDAC does not normally localize to ND10, but does so in the presence of IE1. HDAC may be recruited to ND10 by IE1 during the early stages of infection, when

IE1 accumulates at ND10. The IE1-dependent segregation of HDAC to ND10 may not be significant, because there is little segregated, relative to the amount present in the nucleus. On the other hand, the large amounts of IE1 expressed, especially after ND10 dispersal, may be more significant, because IE1 might flood the nucleus sufficiently to reduce free HDAC, relieving the HDAC-associated suppression of chromatinized viral genomes. This is consistent with results from HDAC activity assays, which show that IE1 binding to HDAC inhibits HDAC deacetylation (Tang and Maul 2003). IE1, therefore, may not exert its primary effects at sites where it is most concentrated; rather, it appears that IE1 functions throughout the nucleus as an HDAC scavenger (at least HDAC 1 and 2 or their complexes), and possibly, as a scavenger for other host proteins.

ATRX

ATRX (alpha thalassemia-mental retardation, X linked) protein has not yet been associated with CMV biology. However, ATRX was the first cellular protein found to interact with Daxx at an N-terminal region and localize to ND10 by the adapter function of Daxx (Ishov et al. 2004). ATRX is a member of the SWI/SNF family of helicases or ATPases with chromatin remodeling activity, and it associates with HP1 (Picketts et al. 1996, 1998; Gibbons et al. 1997) and with the SET domain of chromatin modifying proteins (Cardoso et al. 1998). ATRX and Daxx are only removed from ND10 for a short time during the S/G2 interphase, suggesting reestablishment of the epigenetic properties of newly replicated heterochromatin (Ishov et al. 2004). Conditional genetic ablation of ATRX in mice has similar effects on developing brain structures, as has congenital HCMV infection (Berube et al. 2005). It is tempting to speculate that Daxx/ATRX is removed by IE1 during stochastic reactivation events of latent HCMV in the developing embryo. If so, HCMV silencing and development of latency during congenital infection may not be a totally benign cellular defense. We are now investigating the possibility that permissive cells can silence competent virus, and anticipate that ATRX is involved in the suppression of CMV genomes.

Sp100

Sp100 is a constitutive ND10-associated protein that has been shown to affect HSV1 immediate early protein expression by affecting the promoter of these proteins (Taylor et al. 2000; Wilcox et al. 2005; Isaac et al. 2006; Negorev et al. 2006). Sp100A, the dominant isotype found at ND10, produces a mild activation. All other isoforms that have apparent DNA- or chromatin-binding domains (Sp100B-SAND domain; Sp100C-PhD and Bromo domain; Sp100HMG-HMG domain) are repressive, with the Sp100B having the strongest effect. Preliminary experiments, testing the effect of Sp100 isotypes on the MIEP driven-luciferase reporter assay, show that these Sp100 isotypes likely affect CMV replicative properties, but the isotypes that show inhibition are quite different from those affecting HSV1. The properties of Sp100, such as insolubility, extremely low abundance, and cell cycle modifying characteristics do not allow many common approaches to analyze potential interactions directly. However, the differential effect of its isotypes on HSV1 (Negorev et al. 2006), and their differential interferon upregulation and accumulation into the immediate transcript environment and prereplication sites of HCMV (Tavalai et al. 2006), make these proteins worthy of further investigation.

HDAC

When entering the host nucleus, the viral genome may become chromatinized to reduce its size. This would facilitate diffusion through the nucleus and aid repression by the host's deacetylating agents. Such a silencing mechanism would be an effective host defense. Indeed, evidence for silenced viral genomes has been found by precipitating deacetylated chromatinized MIEP with antibodies against deacetylated histones (Meier 2001; Tang and Maul 2003; Reeves et al. 2006) shortly after infection of permissive cells. Consistent with such a host defense mechanism, as well as with the viral counteracting mechanism (IE1-mediated inhibition of the deacetylation of chromatinized viral DNA), is the finding that the deacetylation inhibitor trichostatin A (TSA) rescues an MCMV IE1 deletion mutant (Tang and Maul 2003) and also the HCMV deletion mutant (Nevels et al. 2004b). Somewhat unexpectedly, TSA substantially enhances the viral productivity of permissive cells infected with wild type virus and significantly increases the number of cells exhibiting signs of productive infection. The latter observation suggests that even permissive cells, in the absence of an immune system or the cytokine-based innate immune response, can suppress viral replication after infection by a competent virus and can limit the initial production of IE1 and IE2 (or IE3, in the case of MCMV) (Tang and Maul 2003). The suppression of many individual viral genomes may take place in the same nucleus where some other genomes are actively transcribing (those at ND10).

Perspectives

The ability of the host cell to completely suppress the initiation of the viral replication cycle without complete inhibition of IE transcription may rely on several factors: (1) the number of viral genomes entering the nucleus; (2) the increased amount of tegument-associated transcription factors internalized as a result of fusing dense bodies, i.e., virus particle lacking a capsid and genome but filled with tegument proteins; (3) the cell cycle stage of the cell; and (4) the amounts of silencing factors, such as HDAC, and repressors, such as Daxx, Sp100 and PML. Those factors are likely to affect the immediate early transcription and specifically the production of the major immediate early proteins. Success at the margin appears to depend on the amount of IE1 produced to overcome such repressive properties of the cell in a cell culture environment. Thus, IE1 plays a critical role at the start of infection and might prove to be a target for antiviral drug development. In order to assess the potential of IE1 as a drug development candidate, we need to determine its influence on viral success in an organismal context.

Additionally, we do not know the mechanism of ND10 dispersion and the effect of the release of ND10-associated proteins. The mechanism may be due to desumofication and as such IE1 may be a desumoylation enzyme. Such a function needs to be directly searched for since it may affect more than ND10-associated proteins presently recognized. The release of ND10-associated proteins by IE1 may mimic interferon upregulation of these proteins (PML and Sp100), i.e., sudden availability of additional and in general detrimental proteins, we know now to be specifically segregated and thus inactivated. Interferon downstream effects of Sp100 as the intermediary protein are preliminarily associated with specific inflammation-inducing proteins. Their occasional induction by abortively reactivating latent virus may account for the atherosclerotic association of HCMV despite never finding replicating virus in plaques.

Low priority has been given to splicing of the major immediate early transcription unit. Exon skipping is involved, and how this is regulated will have a major impact on reactivation. We expect to find cellular defense mechanisms that work at the level of splicing inhibitors. A priori they must exist, since many cells may reactivate to produce only IE1 but not IE2 or the MCMV IE3. The splice enhancers and silencers in the nucleotide sequence will be difficult to identify, but it may be essential to do so, so that when mutant viruses are produced with deletions in exon 4 to 5 for in vivo verification of IE1 or IE2 effects, we do not inadvertently assay a splice phenomenon.

References

Ahn JH, Hayward GS (1997) The major immediate-early proteins IE1 and IE2 of human cytomegalovirus colocalize with and disrupt PML-associated nuclear bodies at very early times in infected permissive cells. J Virol 71:4599–4613

Ahn JH, Brignole EJ 3rd, Hayward GS (1998) Disruption of PML subnuclear domains by the acidic IE1 protein of human cytomegalovirus is mediated through interaction with PML and may modulate a RING finger-dependent cryptic transactivator function of PML. Mol Cell Biol 18:4899–4913

Bell P, Montaner LJ, Maul GG (2001) Accumulation and intranuclear distribution of unintegrated human immunodeficiency virus type 1 DNA. J Virol 75:7683–7691

Berube NG, Mangelsdorf M, Jagla M, Vanderluit J, Garrick D, Gibbons RJ, Higgs DR, Slack RS, Picketts DJ (2005) The chromatin-remodeling protein ATRX is critical for neuronal survival during corticogenesis. J Clin Invest 115:258–267

Cantrell SR, Bresnahan WA (2005) Interaction between the human cytomegalovirus UL82 gene product (pp71) and hDaxx regulates immediate-early gene expression and viral replication. J Virol 79:7792–7802

Cantrell SR, Bresnahan WA (2006) Human cytomegalovirus (HCMV) UL82 gene product (pp71) relieves hDaxx-mediated repression of HCMV replication. J Virol 80:6188–6191

Cardoso C, Timsit S, Villard L, Khrestchatisky M, Fontes M, Colleaux L (1998) Specific interaction between the XNP/ATR-X gene product and the SET domain of the human EZH2 protein. Hum Mol Genet 7:679–684

Cherrington JM, Mocarski ES (1989) Human cytomegalovirus ie1 transactivates the alpha promoter-enhancer via an 18-base-pair repeat element. J Virol 63:1435–1440

Cherrington JM, Khoury EL, Mocarski ES (1991) Human cytomegalovirus ie2 negatively regulates alpha gene expression via a short target sequence near the transcription start site. J Virol 65:887–896

Everett RD, Murray J (2005) ND10 components relocate to sites associated with herpes simplex virus type 1 nucleoprotein complexes during virus infection. J Virol 79:5078–5089

Everett RD, Rechter S, Papior P, Tavalai N, Stamminger T, Orr A (2006) PML contributes to a cellular mechanism of repression of herpes simplex virus type 1 infection that is inactivated by ICP0. J Virol 80:7995–8005

Gibbons RJ, Bachoo S, Picketts DJ, Aftimos S, Asenbauer B, Bergoffen J, Berry SA, Dahl N, Fryer A, Keppler K, Kurosawa K, Levin ML, Masuno M, Neri G, Pierpont ME, Slaney SF, Higgs DR (1997) Mutations in transcriptional regulator ATRX establish the functional significance of a PHD-like domain. Nat Genet 17:146–148

Gibson W, Roizman B (1971) Compartmentalization of spermine and spermidine in the herpes simplex virion. Proc Natl Acad Sci USA 68:2818–2821

Greaves RF, Mocarski ES (1998) Defective growth correlates with reduced accumulation of a viral DNA replication protein after low-multiplicity infection by a human cytomegalovirus ie1 mutant. J Virol 72:366–379

Hagemeier C, Walker SM, Sissons PJ, Sinclair JH (1992) The 72K IE1 and 80K IE2 proteins of human cytomegalovirus independently trans-activate the c-fos, c-myc and hsp70 promoters via basal promoter elements. J Gen Virol 73:2385–2393

Hofmann H, Sindre H, Stamminger T (2002) Functional interaction between the pp71 protein of human cytomegalovirus and the PML-interacting protein human Daxx. J Virol 76:5769–5783

Isaac A, Wilcox KW, Taylor JL (2006) SP100B, a repressor of gene expression preferentially binds to DNA with unmethylated CpGs. J Cell Biochem 98:1106–1122

Ishov AM, Maul GG (1996) The periphery of nuclear domain 10 (ND10) as site of DNA virus deposition. J Cell Biol 134:815–826

Ishov AM, Stenberg RM, Maul GG (1997) Human cytomegalovirus immediate early interaction with host nuclear structures: definition of an immediate transcript environment. J Cell Biol 138:5–16

Ishov AM, Sotnikov AG, Negorev D, Vladimirova OV, Neff N, Kamitani T, Yeh ET, Strauss JF 3rd, Maul GG (1999) PML is critical for ND10 formation and recruits the PML-interacting protein Daxx to this nuclear structure when modified by SUMO-1. J Cell Biol 147:221–234

Ishov AM, Vladimirova OV, Maul GG (2002) Daxx-mediated accumulation of human cytomegalovirus tegument protein pp71 at ND10 facilitates initiation of viral infection at these nuclear domains. J Virol 76:7705–7712

Ishov AM, Vladimirova OV, Maul GG (2004) Heterochromatin and ND10 are cell-cycle regulated and phosphorylation-dependent alternate nuclear sites of the transcription repressor Daxx and SWI/SNF protein ATRX. J Cell Sci 117:3807–3820

Johnson ES (2004) Protein modification by sumo. Annu Rev Biochem 73:355–382

Korioth F, Maul GG, Plachter B, Stamminger T, Frey J (1996) The nuclear domain 10 (ND10) is disrupted by the human cytomegalovirus gene product IE1. Exp Cell Res 229:155–158

Kumar PP, Bischof O, Purbey PK, Notani D, Urlaub H, Dejean A, Galande S (2007) Functional interaction between PML and SATB1 regulates chromatin-loop architecture and transcription of the MHC class I locus. Nat Cell Biol 9:45–56

Lee HR, Kim DJ, Lee JM, Choi CY, Ahn BY, Hayward GS, Ahn JH (2004) Ability of the human cytomegalovirus IE1 protein to modulate sumoylation of PML correlates with its functional activities in transcriptional regulation and infectivity in cultured fibroblast cells. J Virol 78:6527–6542

Li H, Leo C, Zhu J, Wu X, O'Neil J, Park EJ, Chen JD (2000) Sequestration and inhibition of daxx-mediated transcriptional repression by PML. Mol Cell Biol 20:1784–1796

Lin DY, Huang YS, Jeng JC, Kuo HY, Chang CC, Chao TT, Ho CC, Chen YC, Lin TP, Fang HI, Hung CC, Suen CS, Hwang MJ, Chang KS, Maul GG, Shih HM (2006) Role of SUMO-interacting motif in Daxx SUMO modification, subnuclear localization, and repression of sumoylated transcription factors. Mol Cell 24:341–354

Liu B, Hermiston TW, Stinski MF (1991) A cis-acting element in the major immediate-early (IE) promoter of human cytomegalovirus is required for negative regulation by IE2. J Virol 65:897–903

Lukac DM, Harel NY, Tanese N, Alwine JC (1997) TAF-like functions of human cytomegalovirus immediate-early proteins. J Virol 71:7227–7239

Marshall KR, Rowley KV, Rinaldi A, Nicholson IP, Ishov AM, Maul GG, Preston CM (2002) Activity and intracellular localization of the human cytomegalovirus protein pp71. J Gen Virol 83:1601–1612

Maul GG (1998) Nuclear domain 10, the site of DNA virus transcription and replication. Bioessays 20:660–667

Maul GG, Guldner HH, Spivack JG (1993) Modification of discrete nuclear domains induced by herpes simplex virus type 1 immediate early gene 1 product (ICP0). J Gen Virol 74:2679–2690

Maul GG, Yu E, Ishov AM, Epstein AL (1995) Nuclear domain 10 (ND10) associated proteins are also present in nuclear bodies and redistribute to hundreds of nuclear sites after stress. J Cell Biochem 59:498–513

Meier JL (2001) Reactivation of the human cytomegalovirus major immediate-early regulatory region and viral replication in embryonal NTera2 cells: role of trichostatin A, retinoic acid, and deletion of the 21-base-pair repeats and modulator. J Virol 75:1581–1593

Meier JL, Stinski MF (1997) Effect of a modulator deletion on transcription of the human cytomegalovirus major immediate-early genes in infected undifferentiated and differentiated cells. J Virol 71:1246–1255

Mocarski ES, Kemble GW, Lyle JM, Greaves RF (1996) A deletion mutant in the human cytomegalovirus gene encoding IE1(491aa) is replication defective due to a failure in autoregulation. Proc Natl Acad Sci USA 93:11321–11326

Negorev D, Maul GG (2001) Cellular proteins localized at and interacting within ND10/PML nuclear bodies/PODs suggest functions of a nuclear depot. Oncogene 20:7234–7242

Negorev DG, Vladimirova OV, Ivanov A, Rauscher F 3rd, Maul GG (2006) Differential role of Sp100 isoforms in interferon-mediated repression of herpes simplex virus type 1 immediate-early protein expression. J Virol 80:8019–8029

Nevels M, Brune W, Shenk T (2004a) SUMOylation of the human cytomegalovirus 72-kilodalton IE1 protein facilitates expression of the 86-kilodalton IE2 protein and promotes viral replication. J Virol 78:7803–7812

Nevels M, Paulus C, Shenk T (2004b) Human cytomegalovirus immediate-early 1 protein facilitates viral replication by antagonizing histone deacetylation. Proc Natl Acad Sci USA 101:17234–17239

Paulus C, Krauss S, Nevels M (2006) A human cytomegalovirus antagonist of type I IFN-dependent signal transducer and activator of transcription signaling. Proc Natl Acad Sci USA 103:3840–3845

Picketts DJ, Higgs DR, Bachoo S, Blake DJ, Quarrell OW, Gibbons RJ (1996) ATRX encodes a novel member of the SNF2 family of proteins: mutations point to a common mechanism underlying the ATR-X syndrome. Hum Mol Genet 5:1899–1907

Picketts DJ, Tastan AO, Higgs DR, Gibbons RJ (1998) Comparison of the human and murine ATRX gene identifies highly conserved, functionally important domains. Mamm Genome 9:400–403

Pizzorno MC, Hayward GS (1990) The IE2 gene products of human cytomegalovirus specifically down-regulate expression from the major immediate-early promoter through a target sequence located near the cap site. J Virol 64:6154–6165

Poma EE, Kowalik TF, Zhu L, Sinclair JH, Huang ES (1996) The human cytomegalovirus IE1–72 protein interacts with the cellular p107 protein and relieves p107-mediated transcriptional repression of an E2F-responsive promoter. J Virol 70:7867–7877

Reeves M, Murphy J, Greaves R, Fairley J, Brehm A, Sinclair J (2006) Autorepression of the human cytomegalovirus major immediate-early promoter/enhancer at late times of infection is mediated by the recruitment of chromatin remodeling enzymes by IE86. J Virol 80:9998–10009

Saffert RT, Kalejta RF (2006) Inactivating a cellular intrinsic immune defense mediated by Daxx is the mechanism through which the human cytomegalovirus pp71 protein stimulates viral immediate-early gene expression. J Virol 80:3863–3871

Seeler JS, Marchio A, Sitterlin D, Transy C, Dejean A (1998) Interaction of SP100 with HP1 proteins: a link between the promyelocytic leukemia-associated nuclear bodies and the chromatin compartment. Proc Natl Acad Sci USA 95:7316–7321

Stenberg RM, Fortney J, Barlow SW, Magrane BP, Nelson JA, Ghazal P (1990) Promoter-specific trans activation and repression by human cytomegalovirus immediate-early proteins involves common and unique protein domains. J Virol 64:1556–1565

Tang Q, Maul GG (2003) Mouse cytomegalovirus immediate-early protein 1 binds with host cell repressors to relieve suppressive effects on viral transcription and replication during lytic infection. J Virol 77:1357–1367

Tang Q, Maul GG (2006) Mouse cytomegalovirus crosses the species barrier with help from a few human cytomegalovirus proteins. J Virol 80:7510–7521

Tang Q, Bell P, Tegtmeyer P, Maul GG (2000) Replication but not transcription of simian virus 40 DNA is dependent on nuclear domain 10. J Virol 74:9694–9700

Tang Q, Li L, Ishov AM, Revol V, Epstein AL, Maul GG (2003) Determination of minimum herpes simplex virus type 1 components necessary to localize transcriptionally active DNA to ND10. J Virol 77:5821–5828

Tavalai N, Papior P, Rechter S, Leis M, Stamminger T (2006) Evidence for a role of the cellular ND10 protein PML in mediating intrinsic immunity against human cytomegalovirus infections. J Virol 80:8006–8018

Taylor JL, Unverrich D, O'Brien WJ, Wilcox KW (2000) Interferon coordinately inhibits the disruption of PML-positive ND10 and immediate-early gene expression by herpes simplex virus. J Interferon Cytokine Res 20:805–815

Wilcox KW, Sheriff S, Isaac A, Taylor JL (2005) SP100B is a repressor of gene expression. J Cell Biochem 95:352–365

Wilkinson GW, Kelly C, Sinclair JH, Rickards C (1998) Disruption of PML-associated nuclear bodies mediated by the human cytomegalovirus major immediate early gene product. J Gen Virol 79:1233–1245

Xu Y, Ahn JH, Cheng M, apRhys CM, Chiou CJ, Zong J, Matunis MJ, Hayward GS (2001) Proteasome-independent disruption of PML oncogenic domains (PODs), but not covalent modification by SUMO-1, is required for human cytomegalovirus immediate-early protein IE1 to inhibit PML-mediated transcriptional repression. J Virol 75:10683–10695

Functional Roles of the Human Cytomegalovirus Essential IE86 Protein

M. F. Stinski(✉), D. T. Petrik

Contents

Introduction 133
Mapping the Functional Domains of the IE86 Protein 135
Autoregulation of the MIE Promoter 139
Transcription from Viral and Cellular Promoters 140
    Viral Promoters 140
    E2F Promoters 141
    NFκ-B Promoters 142
Cell Cycle Progression 142
Perspectives 145
References 146

Abstract The IE86 protein of human cytomegalovirus (HCMV) is unique among viral and cellular proteins because it negatively autoregulates its own expression, activates the viral early and late promoters, and both activates and inhibits cellular promoters. It promotes cell cycle progression from G_0/G_1 to G_1/S and arrests cell cycle progression at the G_1/S interface or at G_2/M. The IE86 protein is essential because it creates a cellular environment favorable for viral replication. The multiple functions of the IE86 protein during the replication of HCMV are reviewed.

Introduction

Human cytomegalovirus (HCMV) infection in utero is the leading infectious cause of birth defects that cause developmental disabilities. Infection of immunocompromised individuals can result in retinitis, pneumonitis, hepatitis, and gastroenteritis. There is no vaccine for HCMV, and the available antiviral therapies are fraught with

---

M.F. Stinski
Department of Microbiology, Interdisciplinary Program in Molecular Biology,
University of Iowa, Iowa City,
IA 52242, USA
mark-stinski@uiowa.edu

limited efficacy and high rates of adverse affects. Investigations of the essential proteins for HCMV replication may lead to novel antiviral strategies. This chapter will focus on the immediate early two gene (IE2) of HCMV, which encodes the essential multifunctional IE protein designated IE86. The IE2 gene is in a region of the viral DNA referred to as the major immediate early (MIE) gene locus. Transcription from the MIE gene locus does not require de novo viral protein synthesis, and MIE viral transcripts are detected in the presence of an inhibitor of protein synthesis. The MIE gene locus consists of an enhancer-containing promoter upstream of the IE1 (UL123) and IE2 (UL122) genes. The transcriptional binding sites and functions of the MIE enhancer have been recently reviewed (Stinski 1999; Meier and Stinski 2006). The functions of the IE1 gene product are reviewed (see the chapter by G. Maul, this volume). This chapter will review our current understanding of the functions of the IE2 gene product.

The MIE enhancer-containing promoter generates a primary transcript that undergoes differential splicing and polyadenylation to produce multiple mRNA species (Fig. 1). The IE1 and IE2 mRNAs contain the first three exons in common, but IE1 contains exon 4, while IE2 contains exon 5 (Fig. 1). Translation initiates in exon 2; consequently the IE1 and IE2 gene products have the first 85 amino acids in common (Stenberg et al. 1984). Minor isoforms are produced from the IE1 and IE2 genes, as diagrammed in Fig. 1. There is also a late promoter in exon 5 that is activated after viral DNA synthesis. Less is known about the functions of the isomers

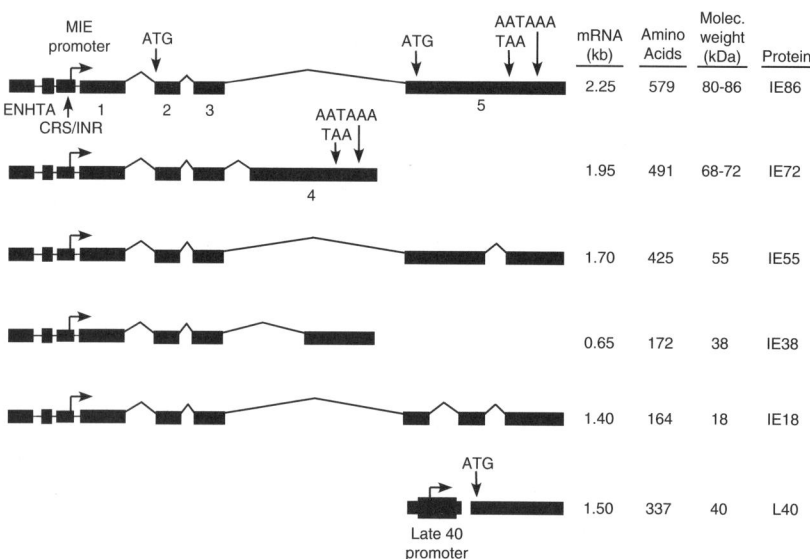

Fig. 1 The mRNAs and viral proteins encoded by the IE1 and IE2 genes of HCMV. Abbreviations: *ENH* enhancer, *TA* TATA box, *crs* cis repression sequence, *INR* initiator, *MIE* major immediate early, *ATG* start codon, *TAA* stop codon, *AATAAA* polyadenylation signal, *kb* kilobases, *kDa* kilodaltons, *IE* immediate early, *L* late

since they are lower in abundance and more difficult to study independently. Even though the isomers are dispensable, they are required for efficient early and late viral gene expression (White et al. 2007). Exon 5 encodes the unique amino acid sequence for the major IE2 gene product designated IE86 (Fig. 1). Recombinant viruses deleted in exon 5 cannot express early or late viral genes; consequently viral replication does not occur (Angulo et al. 2000; Marchini et al. 2001). Therefore, the IE86 protein is essential for viral replication. Both IE72 and IE86 are key regulatory proteins in the switch from latent to lytic infections. However, the molecular mechanisms that trigger and sustain CMV reactivation are related to cellular differentiation and are largely unknown.

After the viral DNA enters the nucleus, subsets of viral genomes are found at nuclear structures referred to as nuclear domain 10 (ND10) or promyelocytic leukemia protein oncogenic domains (PODs). IE72 and IE86 proteins localize at and adjacent to the ND10s, respectively (Ishov et al. 1997). Although IE86 does not directly affect ND10s, it has many other functions, which occur in the nucleus. The functions of the IE86 protein are negative autoregulation of the viral MIE promoter, transactivation of viral and cellular promoters, inactivation of cellular cytokine and chemokine promoters, and control of cell cycle progression. While the IE86 protein is necessary for viral DNA synthesis, initiation of DNA synthesis from oriLyt by an IE86/UL84 protein complex can occur with UL84 alone when a transactivator function in oriLyt is replaced with a constitutive promoter such as SV40 (Xu et al. 2004). The effect of UL84 on the IE86 and lytic viral DNA replication are reviewed (see the chapter by G.S. Pari, this volume). More importantly, IE86 prepares the cell for viral DNA synthesis by activating the expression of early viral genes and cellular genes. The IE86 protein of HCMV is unique among viral and cellular regulatory proteins because it both negatively and positively regulates viral and cellular promoters, and it promotes and arrests cell cycle progression. A better understanding of this essential viral protein may spawn novel strategies for preventing HCMV-induced disease.

Mapping the Functional Domains of the IE86 Protein

The functional domains of the IE86 protein have been studied by in vitro protein-protein binding assays, transient transfection assays, and the construction and isolation of recombinant viruses with either amino acid deletions or substitutions. The use of bacterial artificial chromosomes (BACs) containing the HCMV genome has been useful in confirming and extending the results of these early assays. These studies have indicated that the IE86 protein is a homodimer with critical functional domains located primarily toward the carboxyl end of the viral protein. Structural analysis of the IE86 protein by X-ray crystallography has not been done to date. In addition to tertiary structure of the IE86 protein, posttranslational modifications such as sumoylation and phosphorylation affect the biological activity of the viral protein. The IE86 protein of 579 amino acids is sumoylated at lysine residues 175 and 180 to give an apparent molecular weight of 105 kDa (Hofmann et al. 2000; Ahn et al. 2001). Mutation of lysine residues 175 and 180 negatively affects the efficiency of early viral promoter activation, but sumoylation is not required for viral growth (Lee and Ahn 2004). The IE86 protein is phosphorylated in vitro by the extracellular regulated kinase (ERK) and presumably affected in vivo by both the mitogen activated protein kinase (MAPK) and ERK signal transduction pathways in the infected cell (Harel and Alwine 1998; Heider et al. 2002). The IE86 protein is phosphorylated at multiple serine and threonine residues (Fig. 2a). Mutation of the threonine residues at positions 27 and 233 or the serine residues at positions 144 and 234 positively affects the efficiency of early viral promoter activation (Harel and Alwine 1998). Deletion of the serine residues between 258

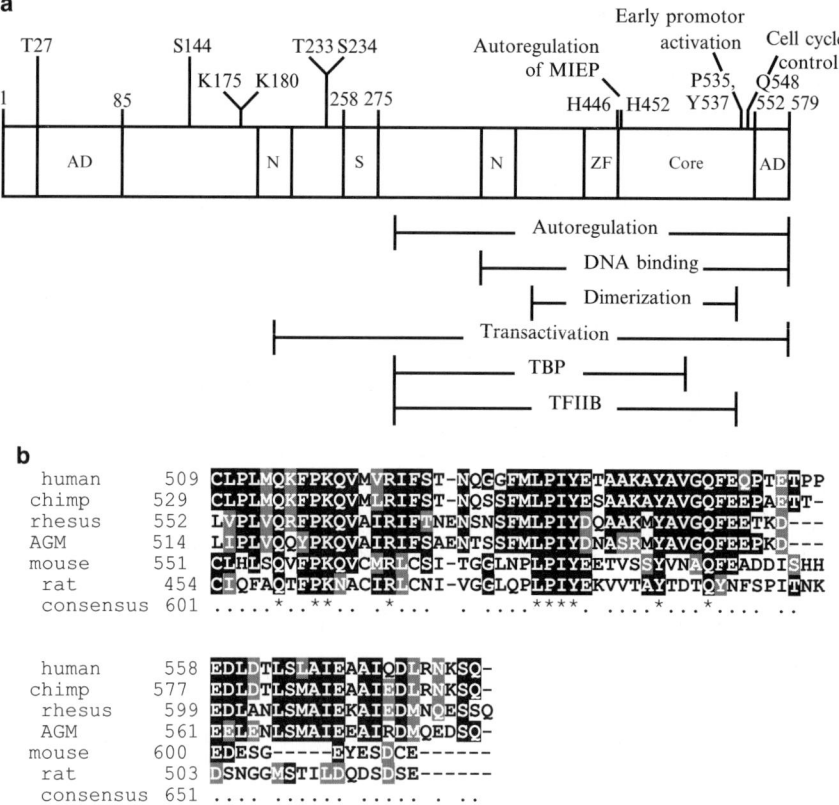

Fig. 2 Functional domains of the HCMV IE86 protein. **a** Abbreviations: *AD* activation domain, *S* serine domain, *K* lysine, *T* threonine, *H* histidine, *P* proline, *Y* tyrosine, *Q* glutamine, *N* nuclear localization signal, *ZF* putative zinc finger. **b** Amino acid sequence alignment of primate and nonprimate IE86 protein homologs. Identical amino acid residues are shaded in black and similar residues in gray. A *star* indicates a consensus sequence, and a *dot* a conserved sequence. A *hyphen* designates a gap in the sequence. MultAlin was used for the alignment of the IE86 residues for Towne strain (AAR31449), chimpanzee (NP612745), rhesus (AAB00488), African green monkey (AAB16881), mouse (AAA74505), and rat (AAB92266)

and 275 positively or negatively affects viral growth depending on the location of the residues (Barrasa et al. 2005). Mutation of residues between 271 and 275 accelerates viral growth and mutation of residues between 258 and 264 or 266 and 269 delays viral growth (Table 1).

The IE86 protein has two nuclear localization signals (N) that can independently target the viral protein to the nucleus (Fig. 2a) (Pizzorno et al. 1991). In the nucleus, the viral protein affects viral promoters through two transcriptional activation domains, one amino and the other carboxyl (Fig. 2a) (Malone et al. 1990; Pizzorno et al. 1991; Yeung et al. 1993; Stenberg 1996). For transcriptional regulatory activity, the IE86 protein functions as a homodimer and dimerizes through the region broadly designated in Fig. 2a (Macias et al. 1996). Within this region, there is a putative zinc finger between amino acids 428 and 452, which may be part of a double zinc finger motif between amino acids 428 and 480 ($CX_5CX_{11}HX_5HXDXCX_{13}HXH$) (Fig. 2a). Double zinc finger motifs are important for interactions with other viral or cellular proteins such as transcription factors that either activate or repress transcription (Bachy et al. 2002; Moreno et al. 2003).

Figure 2a summarizes the regions of posttranslational modification and the broadly mapped functional domains of the IE86 protein. Multiple amino acid deletions were made to determine the functional domains of the IE86 protein and these mutations are summarized in Table 1. Large deletions resulted in nonreplicating genomes and confirmed that the IE86 protein is essential for HCMV replication. Smaller deletions affected the efficiency of either early or late viral gene expression. Frequently more than one viral protein function was affected by these deletions. Mutations that affected DNA binding and negative autoregulation produced high levels of the IE86 protein, which also affected cell viability. Mutations that failed to interact with TBP or TFIIB affected activation of early viral promoters. Mutations that affected phosphorylation of the viral protein affected the rate of viral growth. Delayed viral growth was associated with reduced expression of viral tegument proteins pp65 and pp28 (Sanchez et al. 2002). Many of the mutations made it difficult to assign a particular function to a specific region of the IE86 protein. The region between amino acids 450 and 552 was defined as a core domain because the IE86 regulatory functions of negative autoregulation of the MIE promoter, early promoter transactivation, and cell cycle arrest were all affected by any deletion within this region (Asmar et al. 2004).

Figure 2b shows regions of conserved amino acids in the carboxyl end of the viral protein between primate and nonprimate CMV homologs of the IE86 protein. The carboxy terminus is more conserved than the amino terminus. There are conserved stretches of amino acids suggesting critical structural and functional domains within the protein. Petrik et al. made rationally designed amino acid substitutions in the core region based on sequence conservation (Petrik et al. 2006, 2007); these results are also summarized in Table 1. These mutations separated the transactivation domain from the cell cycle arrest domain and the autoregulation domain from the transactivation domain (Fig. 2a). However, mutations in the putative zinc finger motif affected all functions of the

Table 1 Effect of amino acid deletions or substitutions on the functions of the IE86 protein

Effect on function	Amino acids	Assay	Reference
Dimerization	388–542	DNA binding	Ahn et al. 1998
	463–513		Macias et al. 1996
			Macias and Stinski 1993
Still dimerizes	551–579		Waheed et al. 1998
DNA binding	388–542		Ahn et al. 1998
	346–579		Chiou et al. 1993
	542–579		Jupp et al. 1993b
			Macias and Stinski 1993
Still binds DNA	551–579		Macias et al. 1996
Autoregulation of MIEP	427–435	Reporter	Chiou et al. 1993
	505–511		Hermiston et al. 1990
	H446/H452		Macias and Stinski 1993
			Macias et al. 1996
		Virus	Petrik et al. 2007
Interaction with TBP	290–504	Protein	Caswell et al. 1993
	266–269	Binding	Hagemeier et al. 1992a
	271–275		Jupp et al. 1993a
Interaction with TFIIB	290–542		Caswell et al. 1993
Interaction with CREB	290–410		Lang et al. 1995
Transactivation of promoters	1–98	Reporter	Malone et al. 1990
	25–85		Pizzorno et al. 1991
	169–194		Stenberg et al. 1989
	175–180		Stenberg 1996
	195–579		Yeung et al. 1993
	501–511		
	544–579		
Early gene expression	258–275	Virus	Heider et al. 2002
	266–275		
	E550, E554, E558, D559, D561, E568, D573	Reporter	Yeung et al. 1993
	P535,Y537	Virus	Petrik et al. 2007
Late gene expression	136–290	Virus	Sanchez et al. 2002
	356–359		White et al. 2004
No viral growth	501–511		White et al. 2004
Delayed viral growth	258–275		Barrasa et al. 2005
	266–275		
Accelerated viral growth	271–275	Virus	Barrasa et al. 2005
Interaction with Rb	290–390	Protein Binding	Sommer et al. 1994
Interaction with p53	325–448		Zhang et al. 2006
Interaction with mdm2	326–449		Zhang et al. 2006
Cell cycle arrest	260–279	Reporter	Wiebusch and Hagemeier 1999
	451–579		
	Q548	Virus	Petrik et al. 2006

IE86 protein because this region is necessary for dimerization of the IE86 protein in vivo. The various functions of the IE86 protein and their role in the replication of HCMV are reviewed.

Autoregulation of the MIE Promoter

Negative autoregulation of the MIE promoter by the IE86 protein is almost certainly important to the replication of the virus because recombinant viruses that fail to autoregulate can not be isolated (H. Isomura and M.F. Stinski, unpublished data). The IE86 protein mutated at histidine residues 446 and 452 fails to negatively autoregulate the MIE promoter in in vitro transcription reactions (Macias and Stinski 1993), transient transfection assays (Macias et al. 1996), and recombinant BACs (Petrik et al. 2007). The MIE promoter region of HCMV has a *cis*-acting element that serves as a binding site for the IE86 protein or the late L40 protein (see Fig. 1). Point mutations in histidine residues 446 and 452 abolish DNA binding (Macias and Stinski 1993; Macias et al. 1996). Dimerization of the IE86 protein is required for viral DNA binding (Table 1) (Macias et al. 1996; Waheed et al. 1998). The viral DNA element is located immediately upstream of the transcription start site for IE1/IE2 RNA and is referred to as the *cis*-repression sequence (crs). The crs consists of a pair of dinucleotide CG separated by an A/T-rich region of 10 nucleotides. The CG dinucleotides are critical and the spacing is important for binding by the IE86 protein (Waheed et al. 1998). The crs functions between the TATA box and the transcription start site in either orientation. However, it does not function downstream of the transcription initiation site (+1) (Pizzorno and Hayward 1990; Cherrington et al. 1991; Liu et al. 1991). The IE86 protein binds to the minor groove between −14 and −1 without inhibiting the binding of TBP to the upstream TATA box (Jupp et al. 1993b; Lang and Stamminger 1994).

The first step in negative autoregulation of the MIE promoter by IE86 is blockage of RNA polymerase II occupancy at the transcription start site (Wu et al. 1993; Lee et al. 1996). There is also an initiator-like element between +1 and +7, and cellular protein binding to this element is also affected by the IE86 protein, which in turn negatively affects transcription from the MIE promoter (Macias et al. 1996). The ie3 gene product of murine CMV also negatively autoregulates its MIE promoter (Messerle et al. 1992). Repression of the HCMV MIE promoter is detectable approximately between 6 and 8 h after high multiplicity of infection when the IE86 protein reaches levels to compete for binding to the transcription initiation region (Stamminger et al. 1991; Meier and Stinski 1997).

The second stage of negative autoregulation by the IE86 protein, occurring 24 h postinfection, requires histone deacetylase (HDAC1), which causes deacetylation of histones at the MIE promoter (Reeves et al. 2006). Deacetylated histones are targets for methylation by histone methyltransferases, G9a and Suvar(3-9)H1. Since mutation of the crs abrogates the association of repressive chromatin to the

MIE promoter at late times, it was proposed that IE86 mediated the changes in the chromatin structure (Reeves et al. 2006). During latency, the HCMV MIE promoter is also associated with repressive chromatin (Reeves et al. 2005). The chromatin associated with the MIE promoter during latency and reactivation are reviewed (see the chapter by M. Reeves and J. Sinclair, this volume).

Transcription from Viral and Cellular Promoters

Viral Promoters

A strong heterochromatin structure forms quickly on the HCMV genome in nonpermissive, undifferentiated cells (Reeves et al. 2005; Ioudinkova et al. 2006; Yee et al. 2007). The transcriptional repressors may cause a closed viral chromatin structure (see the chapter by M. Reeves and J. Sinclair, this volume). The MIE promoter is repressed by heterochromatin, and consequently there is no expression of the MIE proteins. The IE86 protein is a master regulator of HCMV transcription required for early viral gene expression. The IE72 protein augments the activity of the IE86 protein by inhibiting histone deacetylase activity (Tang and Maul 2003; Nevels et al. 2004). Early viral gene expression from the viral genome in a latently infected cell, such as HCMV-infected undifferentiated THP-1 cells or murine CMV-infected mice, requires the expression of both of the MIE proteins (Kurz et al. 1999; Kurz and Reddehase 1999; Yee et al. 2007). Nevertheless, this is not sufficient to reactivate virus replication without cellular differentiation. Cellular factors induced or reduced by cellular differentiation appear to be critical for viral DNA replication and production of infectious virus (Murphy et al. 2002; Reeves et al. 2005). Therefore, sporadic expression of the MIE genes may occur, but it is not always sufficient to signal viral reactivation and replication.

The IE86 protein transactivates early viral promoters by interacting with cellular basal transcription machinery and requires a TATA box-containing promoter to transactivate downstream transcription (Lukac et al. 1994). Figure 2a shows the broad regions of the IE86 protein that interact with the basal transcription machinery and Table 1 summarizes the number of mutations in this region. Truncated forms of the IE86 protein interact in vitro with TBP, TFIIB, and TAFII130/TAF4 and the IE86 protein rescues defective TAFII250 (Caswell et al. 1993; Lukac et al. 1994, 1997). Other cellular transcription factors and chromatin remodeling proteins also interact and contribute to the activity of the IE86 protein such as CREB, SP1, Tef-1, Egr-1, p300/CBP, and P/CAF (Lukac et al. 1994; Sommer et al. 1994; Lang et al. 1995; Scully et al. 1995; Schwartz et al. 1996; Yoo et al. 1996; Bryant et al. 2000). Chromatin immunoprecipitation (ChIP) assays demonstrated that TBP is associated with HCMV early promoters, but activation of the viral promoter requires the presence of functional IE86 protein (Petrik et al. 2007). IE86 protein attracts histone acetylases, which are key to activating the early viral promoters (Bryant et al. 2000).

More acetylated chromatin was found on early viral promoters in nonpermissive undifferentiated THP-1 cells when both the IE72 and IE86 proteins were expressed in *trans* prior to infection (Yee et al. 2007).

As shown in Fig. 2a, there is a core domain between amino acids 450 and 552 and when mutated all functions of the IE86 protein are affected (Asmar et al. 2004). The longest stretch of conserved amino acids within the core domain (LPIYE) is shown in Fig. 2b. Mutation of the proline (P) and tyrosine (Y) residues within this conserved stretch of amino acids affected transactivation of early viral promoters without affecting autoregulation of the MIE promoter (Petrik et al. 2007). Mutation of P535 and Y537 to alanines results in a nonreplicating BAC (Table 1). The mutant IE86 protein is found associated with the MIE promoter in ChIP assays, but fails to transactivate early viral genes and is not found associated with viral early promoters in a ChIP assay (Petrik et al. 2007). Therefore, recruitment to the MIE promoter and early viral promoters by the IE86 protein occur through independent mechanisms. The LPIYE conserved sequence may be important for protein-protein interactions. This conserved sequence in the IE86 protein is also found in other cellular proteins that regulate events on a DNA template such as histone methyltransferase, DNA topoisomerase III, and CHROM2. If the LPIYE sequence is mutated in the critical proline and tyrosine residues, there is no activation of the viral early promoters. In contrast, a mutation in the noncritical leucine (L) residue has little to no effect (Petrik et al. 2007). An activation domain of IE86 is carboxyl to the core domain, and it is also necessary to activate transcription of a viral early promoter (Pizzorno et al. 1991; Yeung et al. 1993). This region is rich in acidic amino acids. Mutation of the glutamic and aspartic acids between residues 550 and 573 to valines inactivates the activation domain (Yeung et al. 1993).

E2F Promoters

After HCMV infection, gene microarray assays demonstrated a fourfold or greater increase of at least 124 cellular mRNAs (Zhu et al. 1998). While the IE86 protein can activate cellular promoters through the same mechanisms for viral promoters discussed above (Hagemeier et al. 1992b), it can also activate cellular promoters by inhibiting a repressor of transcription. The expression of genes by the transcription factor E2F is repressed by the cellular tumor suppressor protein pRb. IE86 binds to pRb in vitro and IE86 protein-Rb complexes can be immunoprecipitated from HCMV-infected cells (Hagemeier et al. 1994; Sommer et al. 1994; Fortunato et al. 1997). In the HCMV-infected cell, Rb is converted from the hypophosphorylated to hyperphosphorylated form. An analysis of gene expression using microarrays of cellular genes indicated that HCMV or the IE86 protein strongly activated E2F response genes (Song and Stinski 2002). These are the cellular genes that regulate the enzymes for DNA precursor synthesis, the initiation factors of cellular DNA synthesis, and the movement of the cell cycle. They are expressed prior to the S phase and they include the following: DNA precursor enzymes such as ribonucleotide reductase, thymidylate synthetase, and dihydrofolate reductase, DNA initiation factors such as MCM3 and MCM7, the enzyme DNA polymerase alpha, and the cell cycle control proteins such as cyclin E, cdk-2, and E2F (Salvant et al. 1998; Gribaudo et al. 2000; Song and Stinski 2002). Since HCMV infects terminally differentiated cells of the host that are in the Go phase of the cell cycle, the virus requires a mechanism to overcome cellular quiescence. The IE86 protein-along with the IE72 protein and the viral tegument protein pp71-induce cell cycle progression by inactivating the cellular Rb family of repressor proteins.

NFκ-B Promoters

After HCMV infection, gene microarray assays also demonstrated a fourfold or greater decrease of at least 134 cellular mRNAs (Zhu et al. 1998). Repression of some cellular genes would favor viral replication. In cells infected with an IE2-86 mutant virus lacking amino acids 136-290 (referred to as delta SX) (Sanchez et al. 2002), HMGA2 expression is not significantly reduced, suggesting that the IE86 protein is involved in the regulation of the HMGA2 expression (see the chapter by V. Sanchez and D.H. Spector, this volume). Repression of select cytokine and chemokine expression would favor survival of the virus-infected cell. HCMV infection attenuates interleukin (IL)-1 beta and tumor necrosis factor (TNF)-alpha proinflammatory signaling (Jarvis et al. 2006; Montag et al. 2006). The IE86 protein inhibits the induction of interferon (IFN)-beta, TNF-alpha, RANTES, MIG, MCP-2, and IL-6 (Taylor and Bresnahan 2005, 2006). The effect of IE86 is downstream of the inhibitor of kappa B (IκB) kinase. While IE86 directly effects the binding of NFκ-B to its cognate site on the DNA by an unknown mechanism (Taylor and Bresnahan 2006), IE86 also blocks the activity of NFκ-B when the cellular transcription factor is brought to the promoter via a GAL4 DNA-binding domain (Gealy et al. 2007). Therefore, the inhibition of NFκ-B by IE86 is also subsequent to NFκ-B binding. The regions required to inhibit NFκ-B activity are amino terminal of the IE86 carboxyl activation domain (Gealy et al. 2007). The exact region of the IE86 protein involved in blocking NFκ-B activity requires further investigation.

Cell Cycle Progression

CMVs typically infect terminally differentiated cells in the Go/G1 phase of the cell cycle. The pool of dideoxynucleotide triphosphates and biosynthetic enzymes to make precursors for DNA synthesis are low in these cells. Since CMVs do not encode many of the biosynthetic enzymes for DNA precursor synthesis, the virus must have a mechanism to overcome G1 arrest. The effect of HCMV infection on the cell cycle is reviewed (see the chapter by V. Sanchez and D.H. Spector, this volume). As discussed above, the IE86 protein binds to pRb and activates E2F responsive promoters. Cyclin E/cdk1 expression and activity are increased by HCMV or by the IE86 protein, which causes a feed-forward loop allowing amplification of signals that promote cell cycle progression from the G1 to S phase (Jault et al. 1995; Salvant et al. 1998; McElroy et al. 2000; Sinclair et al. 2000). In a normal cell, p53 induces cell cycle arrest, possibly to allow for DNA repair (Kastan et al. 1992; Kuerbitz et al. 1992) or apoptosis to eliminate cells with damaged genomes (Kuerbitz et al. 1992; Levine 1997; Prives and Hall 1999; Vogelstein et al. 2000). In a $p53^{+/+}$ cell infected with a high multiplicity of HCMV, most of the cells are blocked at G_1/S (Jault et al. 1995; Bresnahan et al. 1996; Lu and Shenk 1996; Morin et al. 1996; Dittmer and Mocarski 1997; Wiebusch and Hagemeier 1999; Casavant et al. 2006). In a p53 mutant cell infected with HCMV, viral replication is delayed and infectious virus production is decreased compared to $p53^{+/+}$ cells (Murphy et al. 2000; Casavant et al. 2006). Wild type p53 expressed in *trans* restores replication efficiency. Although p53 is not an essential cellular protein for HCMV replication, a functional p53 contributes significantly to the progression of infection.

In a $p53^{+/+}$ cell, the IE86 protein binds to p53 and the level of p53 increases in the nucleus (Speir et al. 1994; Bonin and McDougall 1997; Fortunato and Spector 1998). p53 is stabilized by the IE86 protein in an atazia telangiectasis mutated (ATM) kinase-positive cell but not in a ATM kinase-negative cell (Song and Stinski 2004). ATM kinase phosphorylates p53 at serine residue 15, which stabilizes the cellular protein by preventing Mdm2 ubiquitination (Song and Stinski 2004). Mdm2 is a p53-specific E3 ubiquitin ligase that promotes the degradation of p53 (Haupt et al. 1997; Honda et al. 1997; Fang et al. 2000). Mdm2 forms an autoregulatory feedback loop with p53 and allows p53 to control its own level and activity by inducing the expression of Mdm2 (Fang et al. 2000). The IE86 protein interrupts this cellular regulatory control by binding to Mdm2 and facilitating its degradation (Zhang et al. 2006). Therefore, the IE86 protein decreases the level of Mdm2, which increases the level of p53 in the HCMV-infected cell. In addition, phosphorylation at serine residues 15 and 20 of p53 hampers the Mdm2-p53 interaction, which prevents ubiquitination and degradation (Shieh et al. 1997; Dumaz et al. 2001; Louria-Hayon et al. 2003).

There is also a correlation between high levels of IE86 and increased levels of p21 (Shen et al. 2004; Song and Stinski 2004). The early increase in p53 and p21 in the HCMV-infected cell may be related to an early intrinsic cellular defense to virus replication (Garcia et al. 1997). Both IE72 and IE86 induce p21 in a p53-dependent manner (Song and Stinski 2004; Castillo et al. 2005). The p21 promoter is activated by p53, but how IE72 or IE86 proteins increase p21 is currently not understood. The IE86/p53 complex can still bind to the p53 cognate site on DNA (Tsai et al. 1996), but it is uncertain whether IE86 or p53 activates the p21 promoter. Alternatively, p21 could be activated prior to sequestration of p53 in the nucleus with IE86. p21 can inhibit cdk2 and block the activity of cyclin E, arresting the cell cycle progression, but this does not occur in the HCMV-infected cell. It has been reported that IE86 can bind p21 and thereby prevent p21 repression of cyclinE-dependent kinase activity (Sinclair et al. 2000). However, the cyclin E/cdk2 levels increase in both the HCMV-infected cell and the IE86 protein expressing cell and consequently, the virus primes the cell for DNA synthesis (Song and Stinski 2002). In contrast, cyclin A decreases because IE86 protein induces degradation of Mdm2, and Mdm2 normally induces cyclin A expression (Zhang et al. 2006).

The IE86 protein upregulates p21 in a p53$^{+/+}$ cell and not in a p53$^{-/-}$ cell (Song and Stinski 2004). However, p21 is upregulated in a p53-null cell by expressing p53 and IE86 in *trans* (Song and Stinski 2004). In the HCMV-infected cell, the increase in p21 is transient and then p21 levels decrease (Chen et al. 2001). The decrease in p21 may be related to the senescent state of a HCMV-infected cells since senescence causes a decrease in p21 (Stein et al. 1999; Chen et al. 2004). Senescence also occurs in an IE86-expressing HF cell (Noris et al. 2002).

HCMV inhibits cell cycle progression in a p53$^{+/+}$ cell to utilize the cellular DNA precursors for its own DNA synthesis. In a p53$^{+/+}$ cell, IE86 protein activates a quasi G1/S program for cellular gene expression, inhibits cellular DNA synthesis, and consequently co-opts valuable enzymes and macromolecular precursors for viral DNA replication. The inhibition of cellular DNA synthesis appears to be dependent on high multiplicity of infection and on a functional p53 because inhibition does not occur in a p53-null cell unless wild type p53 is introduced in *trans*. Therefore the IE86 protein initially stimulates the cell and then blocks cell cycle progression. The IE86 protein may induce p53 in the permissive HF cell by deregulating E2F activity, activating ATM kinase activity, degrading Mdm2 and consequently stabilizing p53.

The viral IE86 protein and the cellular p53 protein play major roles in control of the cell cycle during HCMV infection. The IE86 protein promotes cell cycle progression in the G_0/G_1 phase. In the presence of the cellular p53 protein, the IE86 protein stops the cell cycle before the S phase. If cells progress into the S phase, then the IE86 protein stops cell cycle progression at the G2/M phase even in the absence of cellular p53 protein (Song and Stinski 2004). The IE86 protein is necessary for cell cycle progression because at least four times more serum-starved p53 mutant glioblastoma U373 or 293T cells are induced into the S phase by wild type IE86 protein compared to a mutant IE86 protein (Murphy et al. 2000). While cell division occurs in the presence of the mutant IE86 protein, it does not occur in the presence of the wild type IE86 protein. In the p53-null Saos-2 cell, wild type IE86 protein blocks cell cycle progression at the G_2/M phase, which correlates with an aberrant increase in cyclin B and cdk1 levels (Song and Stinski 2004). Cells in the S and G_2/M phases suppress transcription of the HCMV MIE genes by an unknown mechanism. When the cell cycle re-enters G_0/G_1, transcription of the MIE genes resumes (Fortunato et al. 2002).

This is important for the replication of the virus because a single amino acid substitution of IE86 at residue 548 from a glutamine to an arginine (Q548R) (Fig. 2b) inactivates the ability of IE86 to arrest the cell cycle at either G1/S or G2/M (Petrik et al. 2006). Under these conditions, the recombinant virus replicates viral DNA slowly, fails to inhibit cellular DNA synthesis or cellular division, and exhibits smaller, nondistinct plaques. The Q548R recombinant virus negatively autoregulates the MIE promoter, activates early viral promoters and cellular

E2F-responsive promoters, but fails to regulate cell cycle progression (Petrik et al. 2006). The virus replicates slowly because viral DNA synthesis is in competition with cellular DNA synthesis. Therefore, arrest of cell cycle progression by the IE86 protein and prevention of cellular DNA synthesis are critical steps for efficient HCMV replication.

The mutation of glutamine at residue 548 to arginine in the Q548R recombinant virus is unique because the IE86 protein fails to regulate the cell cycle at both the G1/S and G2/M phases (Petrik et al. 2006). In stable cell lines, the Q548R mutant IE86 protein complexes with p53 and the cells continue to divide (Bonin and McDougall 1997; Murphy et al. 2000). While the Q548R mutant IE86 protein stimulates cellular E2F gene expression and unscheduled entry into the S phase, it cannot transform cells. The mutant viral protein may not be able to overcome contact inhibition or other cellular processes necessary for cellular transformation. It is likely that the Q548R mutant IE86 protein fails to bind and degrade Mdm2, allowing both cellular DNA synthesis and normal progression into G2/M. The introduction of a positively charged arginine at residue 548 may inhibit IE86 protein interaction with a cellular protein. In contrast, substitution of residue 548 with a neutral alanine residue had little to no effect (Petrik et al. 2006). In addition, the Q548R mutant viral protein may also fail to degrade the anaphase promoting complex/cyclosome (APC/C), which is an E3 ubiquitin ligase that targets regulators of cell division for degradation by the proteasome (Peters 2006). APC/C causes lower levels of cyclin B/cdk1, and cyclin B degradation is necessary to release the cell cycle from G_2/M back into G_1 (Peters 2006). HCMV infection inactivates APC/C and the IE86 protein may be the cause of this inactivation (Wiebusch et al. 2005).

The UL69 gene product is also reported to arrest the cell cycle when overexpressed from a viral vector (Hayashi et al. 2000). Our results with the Q548R recombinant virus indicate that the cell cycle is not arrested when UL69 is present. These results may suggest that both UL69 and a functional IE86 protein are required to efficiently arrest the cell cycle during viral infection. The IE86 protein of HCMV is different from DNA tumor virus early regulatory proteins because it also stops cycle progression at the G1/S interface in $p53^{+/+}$ cells or at G_2/M in p53-null cells. Nevertheless, apoptosis is prevented in the HCMV-infected cell by the viral IE proteins expressed from the UL36, UL37X1, and UL38 genes (see the chapter by A.L. McCormick, this volume). In addition, the IE86 protein itself is reported to inhibit apoptosis under special circumstances (Zhu et al. 1995).

Perspectives

The IE86 protein of HCMV is an essential viral protein that does the following during viral replication:

1. It negatively autoregulates the MIE enhancer-containing promoter, which controls the expression of IE72 and IE86 proteins.

2. It transactivates transcription from viral early and late promoters.
3. It inactivates the cellular Rb repressor, which activates E2F responsive promoters to move the cell cycle from the G_0/G_1 to the G_1/S interface.
4. It prevents expression of cellular cytokines and chemokines expression with antiviral properties.
5. It controls both the G_1/S and G_2/M transition points in the cell cycle.

The regions of the viral protein involved in these functions were broadly mapped, and consequently more than one function was usually affected by the mutation. A core region was identified between amino acid residues 450 and 552 in which mutations in this region affected all identified functions of the IE86 protein. Site-specific mutations in this region demonstrated that the functions of the IE86 protein can be separated and more clearly defined. There are other functional domains and conserved regions within the IE86 protein that need to be defined. The crystal structure of the IE86 protein should be demonstrated to better understand the functional domains. The tertiary structure of the IE86 protein appears critical for function. For example, the histidine residues at 446 and 452 are critical and possibly associated with a double zinc finger domain that determines a three-dimensional structure for the interaction with a variety of cellular and viral proteins. Mutation of the histidine residues at 446 and 452 affects all IE86 protein functions in the context of the viral genome. The proline and tyrosine residues 535 and 537, respectively, which are a part of a conserved five-amino acid stretch in the core domain, affect transcription of early viral promoters. The acidic amino acids between residues 550 and 573 affect early viral promoter activation. The conserved glutamine residue 548 affects control of cell cycle progression. These critical amino acids of the IE86 protein are potential targets for new and unique antivirals. Since the current antivirals for the treatment of HCMV infection are of limited efficacy with high frequencies of adverse effects, it is important to understand the tertiary structure of the carboxyl domain of the IE86 protein to develop new treatments for HCMV infection.

Acknowledgements We thank members of the Stinski lab and Jeffery Meier for critical review of this manuscript. Our work was supported by grant AI-13562 from the National Institutes of Health.

References

Ahn J-H, Chiou C-J, Hayward GS (1998) Evaluation and mapping of the DNA binding and oligomerization domains of the IE2 regulatory protein of human cytomegalovirus using yeast one and two hybrid interaction assays. Gene 210:25-36

Ahn J-H, Xu Y, Jang W-J, Matunis MJ, Hayward GS (2001) Evaluation of interactions of human cytomegalovirus immediate-early IE2 regulatory protein with small ubiquitin-like modifiers and their conjugation enzyme Ubc9. J Virol 75:3859-3872

Angulo A, Ghazal P, Messerle M (2000) The major immediate-early gene ie3 of mouse cytomegalovirus is essential for viral growth. J Virol 74:11129-11136

Asmar J, Wiebusch L, Truss M, Hagemeier C (2004) The putative zinc-finger of the human cytomegalovirus IE286-kilodalton protein is dispensable for DNA-binding and autorepression thereby demarcating a concise core domain in the C-terminus of the protein. J Virol 78:11853-11864

Bachy I, Failli V, Retaux S (2002) A LIM-homeodomain code for development and evolution of forebrain connectivity. Neuroreport 13:A23-A27

Barrasa MI, Harel NY, Alwine JC (2005) The phosphorylation status of the serine-rich region of the human cytomegalovirus 86-kilodalton major immediate-early protein IE2/IEP86 affects temporal viral gene expression. J Virol 79:1428-1437

Bonin LR, McDougall JK (1997) Human cytomegalovirus IE2 86-kilodalton protein binds p53 but does not abrogate G1 checkpoint function. J Virol 71:5861-5870

Bresnahan WA, Boldogh I, Thompson EA, Albrecht T (1996) Human cytomegalovirus inhibits cellular DNA synthesis and arrests productively infected cells in late G1. Virology 224:150-160

Bryant LA, Mixon P, Davidson M, Bannister AJ, Kouzarides T, Sinclair JH (2000) The human cytomegalovirus 86-kilodalton major immediate-early protein interacts physically and functionally with histone acetyltransferase P/CAF. J Virol 74:7230-7237

Casavant NC, Luo MH, Rosenke K, Winegardner T, Zurawska A, Fortunato EA (2006) Potential role for p53 in the permissive life cycle of human cytomegalovirus. J Virol 80:8390-8401

Castillo JP, Frame FM, Rogoff HA, Pickering MT, Yurochko AD, Kowalik TF (2005) Human cytomegalovirus IE1-72 activates ataxia telangiectasia mutated kinase and a p53/p21-mediated growth arrest response. J Virol 79:11467-11475

Caswell R, Hagemeier C, Chiou C-J, Hayward G, Kouzarides T, Sinclair J (1993) The human cytomegalovirus 86K immediate early (IE2) protein requires the basic region of the TATA-box binding protein (TBP) for binding, and interacts with TBP and transcription factor TFIIB via regions of IE2 required for transcriptional regulation. J Gen Virol 74:2691-2698

Chen J-H, Stoeber K, Kingsbury S, Ozanne SE, Williams GH, Hales CN (2004) Loss of proliferative capacity and induction of senescence in oxidatively stressed human fibroblasts. J Biol Chem 279:49439-49446

Chen Z, Knutson E, Kurosky A, Albrecht T (2001) Degradation of p21cip1 in cells productively infected with human cytomegalovirus. J Virol 75:3613-3625

Cherrington JM, Khoury EL, Mocarski ES (1991) Human cytomegalovirus IE2 negatively regulates a gene expression via a short target sequence near the transcription start site. J Virol 65:887-896

Chiou C-J, Zong J, Waheed I, Hayward GS (1993) Identification and mapping of dimerization and DNA-binding domains in the C terminus of the IE2 regulatory protein of human cytomegalovirus. J Virol 67:6201-6214

Dittmer D, Mocarski ES (1997) Human cytomegalovirus infection inhibits G1/S transition. J Virol 71:1629-1634

Dumaz N, Milne DM, Jardine LJ, Meek DW (2001) Critical roles for the serine 20, but not the serine 15, phosphorylation site and for the polyproline domain in regulating p53 turnover. Biochem J 359:459-464

Fang S, Jensen JP, Ludwig RL, Vousden KH, Weissman AM (2000) Mdm2 is a RNIG finger-dependent ubiquitin protein ligase for itself and p53. J Biol Chem 275:8945-8951

Fortunato EA, Sommer MH, Yoder K, Spector DH (1997) Identification of domains within the human cytomegalovirus major immediate-early 86-kilodalton protein and the retinoblastoma protein required for physical and functional interaction with each other. J Virol 71:8176-8185

Fortunato E, Sanchez V, Yen JY, Spector DH (2002) Infection of cells with human cytomegalovirus during S phase results in a blockade to immediate-early gene expression that can be overcome by inhibition of the proteasome. J Virol 76:5369-5379

Fortunato EA, Spector DH (1998) p53 and RPA are sequestered in viral replication centers in the nuclei of cells infected with human cytomegalovirus. J Virol 72:2033-2039

Garcia JF, Piris MA, Lloret E, Orradre JL, Murillo PG, Martinez JC (1997) p53 expression in CMV-infected cells: association with the alternative expression of the p53 transactivated genes p21/WAF1 and MDM2. Histopathology 30:120-125

Gealy C, Humphreys C, Stinski MF, Caswell R (2007) An activation-defective mutant of the HCMV IEp86 protein inhibits NF-kappa B-mediated stimulation of the human IL-6 promoter. J Gen Virol 88:2435-2440

Gribaudo F, Riera L, Lembo D, De Andrea M, Gariglio M, Rudge TL, Johnson LF, Landolfo S (2000) Murine cytomegalovirus stimulates cellular thymidylate synthase gene expression in quiescent cells and requires the enzyme for replication. J Virol 74:4979-4987

Hagemeier C, Walker S, Caswell R, Kouzarides T, Sinclair J (1992a) The human cytomegalovirus 80-kilodalton but not the 72-kilodalton immediate-early protein transactivates heterologous promoters in a TATA box-dependent mechanism and interacts directly with TFIID. J Virol 66:4452-4456

Hagemeier C, Walker SM, Sissons PJ, Sinclair JH (1992b) The 72K IE1 and 80K IE2 proteins of human cytomegalovirus independently trans-activate the c-fos, c-myc and hsp70 promoters via basal promoter elements. J Gen Virol 73:2385-2393

Hagemeier C, Caswell R, Hayhurst G, Sinclair J, Kouzarides T (1994) Functional interaction between the HCMV IE2 transactivator and the retinoblastoma protein. EMBO J 13:2897-2903

Harel NY, Alwine JC (1998) Phosphorylation of the human cytomegalovirus 86-kilodalton immediate-early protein IE2. J Virol 72:5481-5492

Haupt Y, Maya R, Kazaz A, Oren M (1997) Mdm2 promotes the rapid degradation of p53. Nature 387:296-299

Hayashi ML, Blankenship C, Shenk T (2000) Human cytomegalovirus UL69 protein is required for efficient accumulation of infected cells in the G1 phase of the cell cycle. Proc Natl Acad Sci USA 97:2692-2696

Heider JA, Bresnahan WA, Shenk TE (2002) Construction of a rationally designed human cytomegalovirus variant encoding a temperature-sensitive immediate-early 2 protein. Proc Natl Acad Sci USA 99:3141-3146

Hermiston TW, Malone CL, Stinski MF (1990) Human cytomegalovirus immediate-early two protein region involved in negative regulation of the major immediate-early promoter. J Virol 64:3532-3536

Hofmann H, Floss S, Stamminger T (2000) Covalent modification of the transactivator protein IE2-p86 of human cytomegalovirus by conjugation to the ubiquitin-homologous proteins SUMO-1 and hSMT3b. J Virol 74:2510-2524

Honda R, Tanaka H, Yasuda H (1997) Oncoprotein MDM 2 is a ubiquitin ligase E3 for tumor suppressor p53. FEBS Lett 420:25-27

Ioudinkova E, Arcangeletti MC, Rynditch A, De Conto F, Motta F, Covan S, Pinardi F, Razin SV, Chezzi C (2006) Control of human cytomegalovirus gene expression by differential histone modifications during lytic and latent infection of a monocytic cell line. Gene 384:120-128

Ishov AM, Stenberg RM, Maul GG (1997) Human cytomegalovirus immediate early interaction with host nuclear structures: definition of an immediate transcript environment. J Cell Biol 138:5-16

Jarvis MA, Borton JA, Keech AM, Wong J, Britt WJ, Magun BE, Nelson JA (2006) Human cytomegalovirus attenuates interleukin-1beta and tumor necrosis factor alpha proinflammatory signaling by inhibition of NF-kappaB activation. J Virol 80:5588-5598

Jault FM, Jault J-M, Ruchti F, Fortunato EA, Clark C, Corbeil J, Richman D, Spector DH (1995) Cytomegalovirus infection induces high levels of cyclins, phosphorylated RB, and p53, leading to cell cycle arrest. J Virol 69:6697-6704

Jupp R, Flores O, Nelson JA, Ghazal P (1993a) The DNA-binding subunit of human transcription factor IID can interact with the TATA box as a multimer. J Biol Chem 268:16105-16108

Jupp R, Hoffmann S, Depto A, Stenberg RM, Ghazal P, Nelson J (1993b) Direct interaction of the human cytomegalvirus IE86 protein with the cis repression signal does not preclude TBP from binding to the TATA box. J Virol 67:5595-5604

Kastan MB, Zhan Q, el-Deiry WS, Carrier F, Jacks T, Walsh WV, Plunkett BS, Vogelstein B, Fornace AJ Jr (1992) A mammalian cell cycle checkpoint pathway utilizing p53 and GADD45 is defective in ataxia-telangiectasia. Cell 71:587-597

Kuerbitz SJ, Plunkett BS, Walsh WV, Kastan MB (1992) Wild-type p53 is a cell cycle checkpoint determinant following irradiation. Proc Natl Acad Sci USA 89:7491-7495

Kurz SK, Reddehase MJ (1999) Patchwork pattern of transcriptional reactivation in the lungs indicates sequential checkpoints in the transition from murine cytomegalovirus latency to recurrence. J Virol 73:8612-8622

Kurz SK, Rapp M, Steffens H-P, Grzimek NKA, Schmalz S, Reddehase MJ (1999) Focal transcriptional activity of murine cytomegalovirus during latency in the lungs. J Virol 73:482-494

Lang D, Stamminger T (1994) Minor groove contacts are essential for an interaction of the human cytomegalovirus IE2 protein with its DNA target. Nucleic Acids Res 22:3331-3338

Lang D, Gebert S, Arlt H, Stamminger T (1995) Functional interaction between the human cytomegalovirus 86-kilodalton IE2 protein and the cellular transcription factor CREB. J Virol 69:6030-6037

Lee G, Wu J, Luu P, Ghazal P, Flores O (1996) Inhibition of the association of RNA polymerase II with the preinitiation complex by a viral transcriptional repressor. Proc Natl Acad Sci USA 93:2570-2575

Lee H, Ahn J (2004) The sumoylation of HCMV (Towne) IE2 is not required for viral growth in cultured human fibroblast. 29th International Herpesvirus Workshop, Reno, Nevada.

Levine AJ (1997) p53, the cellular gatekeeper for growth and division. Cell 88:323-331

Liu B, Hermiston TW, Stinski MF (1991) A cis-acting element in the major immediate early (IE) promoter of human cytomegalovirus is required for negative regulation by IE2. J Virol 65:897-903

Louria-Hayon I, Grossman T, Sionov RV, Alsheich O, Pandolfi PP, Haupt Y (2003) The promyelocytic leukemia protein protects p53 from Mdm2-mediated inhibition and degradation. J Biol Chem 278:33134-33141

Lu M, Shenk T (1996) Human cytomegalovirus infection inhibits cell cycle progression at multiple points including the transition from G1 to S. J Virol 70:8850-8857

Lukac DM, Manuppello JR, Alwine JC (1994) Transcriptional activation by the human cytomegalovirus immediate-early proteins: requirements for simple promoter structures and interactions with multiple components of the transcription complex. J Virol 68:5184-5193

Lukac DM, Harel NY, Tanese N, Alwine JC (1997) TAF-like functions of human cytomegalovirus immediate-early proteins. J Virol 71:7227-7239

Macias MP, Stinski MF (1993) An in vitro system for human cytomegalovirus immediate early 2 protein (IE2)-mediated site-dependent repression of transcription and direct binding of IE2 to the major immediate early promoter. Proc Natl Acad Sci USA 90:707-711

Macias MP, Huang L, Lashmit PE, Stinski MF (1996) Cellular and viral protein binding to a cytomegalovirus promoter transcription initiation site: effects on transcription. J Virol 70:3628-3635

Malone CL, Vesole DH, Stinski MF (1990) Transactivation of a human cytomegalovirus early promoter by gene products from the immediate-early gene IE2 and augmentation by IE1: mutational analysis of the viral proteins. J Virol 64:1498-1506

Marchini A, Liu H, Ahu H (2001) Human cytomegalovirus with IE-2 (UL122) deleted fails to express early lytic genes. J Virol 75:1870-1878

McElroy AK, Dwarakanath RS, Spector DH (2000) Dysregulation of cyclin E gene expression in human cytomegalovirus-infected cells requires viral early gene expression and is associated with changes in the Rb-related protein p130. J Virol 74:4192-4206

Meier JL, Stinski MF (1997) Effect of a modulator deletion on transcription of the human cytomegalovirus major immediate-early genes in infected undifferentiated and differentiated cells. J Virol 71:1246-1255

Meier JL, Stinski MF (2006) Major immediate-early enhancer and its gene products. In: Reddehase MJ (ed) Cytomegaloviruses molecular and immunology. Caister Academic Press, Norfolk, UK, pp 151-166

Messerle M, Buhler B, Keil GM, Koszinowski UH (1992) Structural organization, expression, and functional characterization of the murine cytomegalovirus immediate-early gene 3. J Virol 66:27-36

Montag C, Wagner J, Gruska I, Hagemeier C (2006) Human cytomegalovirus blocks tumor necrosis factor alpha- and interleukin-1beta-mediated NF-kappaB signaling. J Virol 80:11686-11698

Moreno N, Bachy I, Retaux S, Gonzalez A (2003) Pallial origin of mitral cells in the olfactory bulbs of *Xenopus*. Neuroreport 14:2355-2358

Morin J, Johann S, O'Hara B, Gluzman Y (1996) Exogenous thymidine is preferentially incorporated into human cytomegalovirus DNA in infected human fibroblast. J Virol 70:6402-6404

Murphy EA, Streblow DN, Nelson JA, Stinski MF (2000) The human cytomegalovirus IE86 protein can block cell cycle progression after inducing transition into the S-phase of permissive cells. J Virol 74:7108-7118

Murphy J, Fischle W, Verdin E, Sinclair J (2002) Control of cytomegalovirus lytic gene expression by histone acetylation. EMBO J 21:1112-1120

Nevels M, Brune W, Shenk T (2004) SUMOylation of the human cytomegalovirus 72-kilodalton IE1 protein facilitates expression of the 86-kilodalton IE2 protein and promotes viral replication. J Virol 78:7803-7812

Noris E, Zannetti C, Demurtas A, Sinclair J, De Andrea M, Gariglio M, Landolfo S (2002) Cell cycle arrest by human cytomegalovirus 86-kDa IE2 protein resembles premature senescence. J Virol 76:12135-12148

Peters JM (2006) The anaphase promoting complex/cyclosome: a machine designed to destroy. Nat Rev Mol Cell Biol 7:644-656

Petrik DT, Schmitt KP, Stinski MF (2006) Inhibition of cellular DNA synthesis by the human cytomegalovirus IE86 protein is necessary for efficient virus replication. J Virol 80:3872-3883

Petrik DT, Schmitt KP, Stinski MF (2007) The autoregulatory and transactivating functions of the human cytomegalovirus IE86 protein use independent mechanisms for promoter binding. J Virol 81:5807-5818

Pizzorno MC, Hayward GS (1990) The IE2 gene products of human cytomegalovirus specifically down-regulate expression from the major immediate-early promoter through a target located near the cap site. J Virol 64:6154-6165

Pizzorno MC, Mullen MA, Chang YN, Hayward GS (1991) The functionally active IE2 immediate-early regulatory protein of human cytomegalovirus is an 80-kilodalton polypeptide that contains two distinct activator domains and a duplicated nuclear localization signal. J Virol 65:3839-3852

Prives C, Hall PA (1999) The p53 pathway. J Pathol 187:112-126

Reeves M, Murphy J, Greaves R, Fairley J, Brehm A, Sinclair J (2006) Autorepression of the human cytomegalovirus major immediate-early promoter/enhancer at late times of infection is mediated by the recruitment of chromatin remodeling enzymes by IE86. J Virol 80:9998-10009

Reeves MB, MacAry PA, Lehner PJ, Sissons JG, Sinclair JH (2005) Latency, chromatin remodeling, and reactivation of human cytomegalovirus in the dendritic cells of healthy carriers. Proc Natl Acad Sci USA 102:4140-4145

Salvant BS, Fortunato EA, Spector DH (1998) Cell cycle dysregulation by human cytomegalovirus: influence of the cell cycle phase at the time of infection and effects on cyclin transcription. J Virol 72:3729-3741

Sanchez V, Clark CL, Yen JY, Dwarakanath R, Spector DH (2002) Viable human cytomegalovirus recombinant virus with an internal deletion of the IE2 86 gene affects late stages of viral replication. J Virol 76:2973-2989

Schwartz R, Helmich B, Spector DH (1996) CREB and CREB-binding proteins play an important role in the IE2 86-kilodalton protein-mediated transactivation of the human cytomegalovirus 2.2-kilobase RNA promoter. J Virol 70:6955-6966

Scully AL, Sommer MH, Schwartz R, Spector DH (1995) The human cytomegalovirus IE2 86 kDa protein interacts with an early gene promoter via site-specific DNA binding and protein-protein associations. J Virol 69:6533-6540

Shen YH, Utama B, Wang J, Raveendran M, Senthil D, Waldman WJ, Belcher JD, Vercellotti G, Martin D, Metchelle BM, Wang XL (2004) Human cytomegalovirus causes endothelial injury through the ataxia telangiectasia mutant and p53 DNA damage signaling pathways. Circ Res 94:1-9

Shieh SY, Ikeda M, Taya Y, Prives C (1997) DNA damage-induced phosphorylation of p53 alleviates inhibition by MDM2. Cell 91:325-334

Sinclair J, Baillie J, Bryant L, Caswell R (2000) Human cytomegalovirus mediates cell cycle progression through G(1) into early S phase in terminally differentiated cells. J Gen Virol 6:1553-1565

Sommer MH, Scully AL, Spector DH (1994) Transactivation by the human cytomegalovirus IE2 86-kilodalton protein requires a domain that binds to both the TATA box-binding protein and the retinoblastoma protein. J Virol 68:6223-6231

Song Y-J, Stinski MF (2002) Effect of the human cytomegalovirus IE86 protein on expression of E2F-responsive genes: A DNA microarray analysis. Proc Natl Acad of Sci USA 99:2836-2841

Song Y-J, Stinski MF (2004) Inhibition of cell division by the human cytomegalovirus IE86 protein: role of the p53 pathway or cdk1/cyclin B1. J Virol 79:2597-2603

Speir E, Modali R, Huang E-S, Leon MB, Shawl F, Finkel T, Epstein SE (1994) Potential role of human cytomegalovirus and p53 interaction in coronary restenosis. Science 265:391-394

Stamminger T, Puchtler E, Fleckenstein B (1991) Discordant expression of the immediate-early 1 and 2 gene regions of human cytomegalovirus at early times after infection involves posttranscriptional processing events. J Virol 65:2273-2282

Stein GH, Drullinger LF, Soulard A, Dulic V (1999) Differential roles for cyclin-dependent kinase inhibitors p21 and p16 in the mechanisms of senescence and differentiation in human fibroblasts. Mol Cell Biol 19:2109-2117

Stenberg RM (1996) The human cytomegalovirus major immediate-early gene. Intervirology 39:343-349

Stenberg RM, Thomsen DR, Stinski MF (1984) Structural analysis of the major immediate early gene of human cytomegalovirus. J Virol 49:190-199

Stenberg RM, Depto AS, Fortney J, Nelson J (1989) Regulated expression of early and late RNAs and proteins from the human cytomegalovirus immediate-early gene region. J Virol 63:2699-2708

Stinski MF (1999) Cytomegalovirus promoter for expression in mammalian cells. In: Ferandez JM, Hoeffler JP (eds) Gene expression systems: using nature for the art of expression. Academic Press, San Diego, pp 211-233

Tang Q, Maul GG (2003) Mouse cytomegalovirus immediate-early protein 1 binds with host cell repressors to relieve suppressive effects on viral transcription and replication during lytic infection. J Virol 77:1357-1367

Taylor RT, Bresnahan WA (2005) Human cytomegalovirus immediate-early 2 gene expression blocks virus-induced beta interferon production. J Virol 79:3873-3877

Taylor RT, Bresnahan WA (2006) Human cytomegalovirus immediate-early 2 protein IE86 blocks virus-induced chemokine expression. J Virol 80:920-928

Tsai HL, Kou GH, Chen SC, Wu CW, Lin YS (1996) Human cytomegalovirus immediate-early protein IE2 tethers a transcriptional repression domain to p53. J Biol Chem 271:3534-3540

Vogelstein B, Lane D, Levine AJ (2000) Surfing the p53 network. Nature 408:307-310

Waheed I, Chiou C, Ahn J, Hayward GS (1998) Binding of the human cytomegalovirus 80-kDa immediate-early protein (IE2) to minor groove A/T-rich sequences bounded by CG dinucleotides is regulated by protein oligomerization and phosphorylation. Virology 252:235-257

White EA, Clark CL, Sanchez V, Spector DH (2004) Small internal deletions in the human cytomeaglovirus IE2 gene result in nonviable recombinant viruses with differential defects in viral gene expression. J Virol 78:1817-1830

White EA, Del Rosario CJ, Sanders RL, Spector DH (2007) The IE2 60-kilodalton and 40-kilodalton proteins are dispensable for human cytomegalovirus replication but are required for efficient delayed early and late gene expression and production of infectious virus. J Virol 81:2573-2583

Wiebusch L, Hagemeier C (1999) Human cytomegalovirus 86-kilodalton IE2 protein blocks cell cycle progression in G(1). J Virol 73:9274-9283

Wiebusch L, Bach M, Uecker R, Hagemeier C (2005) Human cytomegalovirus inactivates the G(0)/G(1)-APC/C ubiquitin ligase by Cdh1 dissociation. Cell Cycle 4:1435-1439

Wu J, Jupp R, Stenberg RM, Nelson JA, Ghazal P (1993) Site-specific inhibition of RNA polymerase II preinitiation complex assembly by human cytomegalovirus IE86 protein. J Virol 67:7547-7555

Xu Y, Cei SA, Rodriguez-Huete A, Colletti KS, Pari GS (2004) Human cytomegalovirus DNA replication requires transcriptional activation via an IE2- and UL84-responsive bidirectional promoter element within oriLyt. J Virol 78:11664-11677

Yee L-F, Lin PL, Stinski MF (2007) Ectopic expression of HCMV IE72 and IE86 proteins is sufficient to induce early gene expression but not production of infectious virus in undifferentiated promonocytic THP-1 cells. Virology 363:174-188

Yeung KC, Stoltzfus CM, Stinski MF (1993) Mutations of the cytomegalovirus immediate-early 2 protein defines regions and amino acid motifs important in transactivation of transcription from the HIV-1 LTR promoter. Virology 195:786-792

Yoo YD, Chiou C-J, Choi KS, Yi Y, Michelson S, Kim S, Hayward GS, Kim S-J (1996) The IE2 regulatory protein of human cytomegalovirus induces expression of the human transforming growth factor beta 1 gene through an Egr-1 binding site. J Virol 70:7062-7070

Zhang Z, Evers DL, McCarville JF, Dantonel JC, Huong SM, Huang ES (2006) Evidence that the human cytomegalovirus IE2-86 protein binds mdm2 and facilitates mdm2 degradation. J Virol 80:3833-3843

Zhu H, Shen Y, Shenk T (1995) Human cytomegalovirus IE1 and IE2 proteins block apoptosis. J Virol 69:7960-7970

Zhu H, Cong JP, Mamtora G, Gingeras T, Shenk T (1998) Cellular gene expression altered by human cytomegalovirus: global monitoring with oligonucleotide arrays. Proc Natl Acad Sci USA 95:14470-14475

Nuts and Bolts of Human Cytomegalovirus Lytic DNA Replication

G. S. Pari

Contents

Introduction	154
Essential Region I: IE2-UL84 Responsive Promoter in oriLyt	155
Essential Region II: RNA/DNA Hybrid Structure	158
Viral-Encoded *trans*-Acting Factors Required for Lytic Replication	159
UL84 and IE2	161
Viral and Cellular Encoded UL84 Binding Partners	162
Summary and Perspectives	163
References	163

Abstract HCMV lytic DNA replication is complex and highly regulated. The *cis*-acting lytic origin of DNA replication (oriLyt) contains multiple repeat motifs that comprise two main functional domains. The first is a bidirectional promoter element that is responsive to UL84 and IE2. The second appears to be an RNA/DNA hybrid region that is a substrate for UL84. UL84 is required for oriLyt-dependent DNA replication along with the six core proteins, UL44 (DNA processivity factor), UL54 (DNA polymerase), UL70 (primase), UL105 (helicase), UL102 (primase-associated factor) and UL57 (single-stranded DNA-binding protein). UL84 is an early protein that shuttles from the nucleus to the cytoplasm, binds RNA, suppresses the transcriptional activation function of IE2, has UTPase activity and is proposed to be a member of the DExH/D box family of proteins. UL84 is a key factor that may act in concert with the other core replication proteins to initiate lytic replication by altering the conformation of an RNA stem loop structure within oriLyt. In addition, new data suggests that UL84 interacts with at least one member of the viral replication proteins and several cellular encoded proteins.

G.S. Pari
University of Nevada, Reno, School of Medicine, Reno NV 89557, USA
gpari@med.unr.edu

Introduction

Human cytomegalovirus lytic DNA replication appears to be a complex and highly regulated event. Although immediate early gene expression occurs shortly after infection, the onset of viral DNA synthesis does not take place until approximately 24 h postinfection in cell culture. HCMV DNA replication is thought to involve circularization and concatemer formation (McVoy and Adler 1994). Viral DNA synthesis, which occurs in the nucleus, requires the HCMV core replication machinery and may be regulated by multifunctional immediate early proteins that were originally thought to have a role only in gene regulation. As for the structure of replicating viral DNA, head-to-tail concatemers are evident from examination of the replication products from cloned oriLyt in human fibroblasts. Although a rolling circle model is postulated for lytic DNA replication, there is limited evidence to support this method for HCMV DNA synthesis and actual events may be more complex (McVoy and Adler 1994).

The *cis*-acting lytic origin of replication, oriLyt, is located within the unique long (U_L) region of the genome between AD169 nucleotides 90,500 and 93,930. This region, which is situated between ORFs 57 and 69, was originally determined to contain an origin based on the results of both the transient replication assay, where cloned regions of the HCMV genome were amplified upon transfection and subsequent infection of permissive cells (Anders and Punturieri 1991; Anders et al. 1992; Masse et al. 1992), and a method that took advantage of the chain termination effects of the drug ganciclovir (Hamzeh et al. 1990). Later studies confirmed that the oriLyt identified form both studies was the only functional replicator in the virus genome (Borst and Messerle 2005). Both approaches confirmed that a complex DNA region within the HCMV genome was responsible for the propagation of viral DNA during the lytic phase of viral growth. With the cloned lytic origin in hand, a cotransfection replication assay was developed in an effort to elucidate the viral-encoded *trans*-acting factors that contribute to oriLyt-dependent DNA replication. The transient cotransfection replication assay revealed that eleven loci are required for efficient oriLyt-dependent DNA replication. These genes are: UL36-38, UL44 (pol accessory protein), UL54 (DNA polymerase), UL57 (single-stranded DNA binding protein), UL70 (primase), UL102 (primase-associated factor), UL105 (helicase), UL112/113 (early proteins), IRS1/TRS1 (immediate early RNA-binding), IE1/2 (transactivator) and UL84. Several of these proteins were determined to play ancillary roles in DNA replication (Sarisky and Hayward 1996; Xu et al. 2004b).

Initiation of lytic DNA replication occurs at a single origin of replication and appears to be initially mediated by the viral encoded proteins UL84 and IE2. As our understanding of HCMV lytic DNA matures, it is clear that viral encoded proteins may function in a regulatory as well as enzymatic role such as recognition and unwinding of a distinct structure within oriLyt. What this suggests is that replication proteins themselves may serve to control the onset of DNA synthesis by interacting with, or acting as *trans*-acting factors themselves.

Although the exact mechanism of initiation of HCMV lytic DNA synthesis remains unknown, much of the emphasis has been focused on the *cis*-acting sequences within oriLyt and the protein encoded by the UL84 open reading frame (ORF). UL84 is a unique protein that has no known homolog to any cellular or viral protein. UL84 is a multifunctional protein that is capable of regulating the transcriptional activation mediated by IE2, and evidence indicates that it is a member of the DExH/D-box family of proteins. This review will discuss (a) what is known about the *cis*-acting regions in oriLyt that contribute to DNA synthesis and (b) the properties of the *trans*-acting factors, specifically UL84, required for lytic DNA synthesis and how they relate to the regulation and initiation of HCMV lytic DNA replication.

Essential Region I: IE2-UL84 Responsive Promoter in oriLyt

The HCMV oriLyt region (Fig. 1) was defined according to the smallest fragments that supported amplification in the transient replication assay. The transient replication assay involves the transfection of oriLyt-containing plasmids into human fibroblasts followed by infection with HCMV. Total cellular DNA is harvested and cleaved with two restriction enzymes. The first cleaves input plasmid and cellular DNA. The second, DpnI, cleaves methylated (unreplicated) plasmid DNA. Plasmids propagated in bacterial Dam$^+$ methylase hosts will add a methyl group at an adenosine (which is one base within the DpnI recognition sequence). Once plasmid DNA goes through one round of semiconservative DNA synthesis in mammalian cells, it then becomes unmethylated, since mammalian cells lack an adenosine methylase. Replicated DNA is distinguished from input plasmid DNA based on the sensitivity or resistance to cleavage by DpnI where amplified plasmid DNA migrates more slowly through an agarose gel due to its larger size. Since DpnI has a 4-base pair recognition sequence, input unreplicated DNA migrates toward the bottom of the gel. This powerful assay allowed for the fine mapping of DNA sequences that are essential or contribute to amplification of oriLyt. These initial studies defined the oriLyt region between nts 90,504 and 93,930 on the AD169 genome.

Later studies identified a core domain for oriLyt that contained two essential regions (I and II). These essential regions were originally defined as those sequences that could not be mutated or deleted and still retain a functional oriLyt in transient assays (Zhu et al. 1998). Within essential region I, there are several DNA repeat elements and a highly prymidine-rich DNA sequence referred to as the Y-block (Zhu et al. 1998) as well as two 29-base pair repeat sequences that contain several transcription factor-binding sites. In addition, two IE2 binding sites are also present. The first is a consensus *cis* repression sequence (CRS) that was shown previously to interact with IE2 (Cherrington et al. 1991; Arlt et al. 1994; Tsai et al. 1997; Lashmit et al. 1998; Huang and Chen 2002). This consensus CRS element does not appear to have a functional role in oriLyt promoter activation/repression or DNA

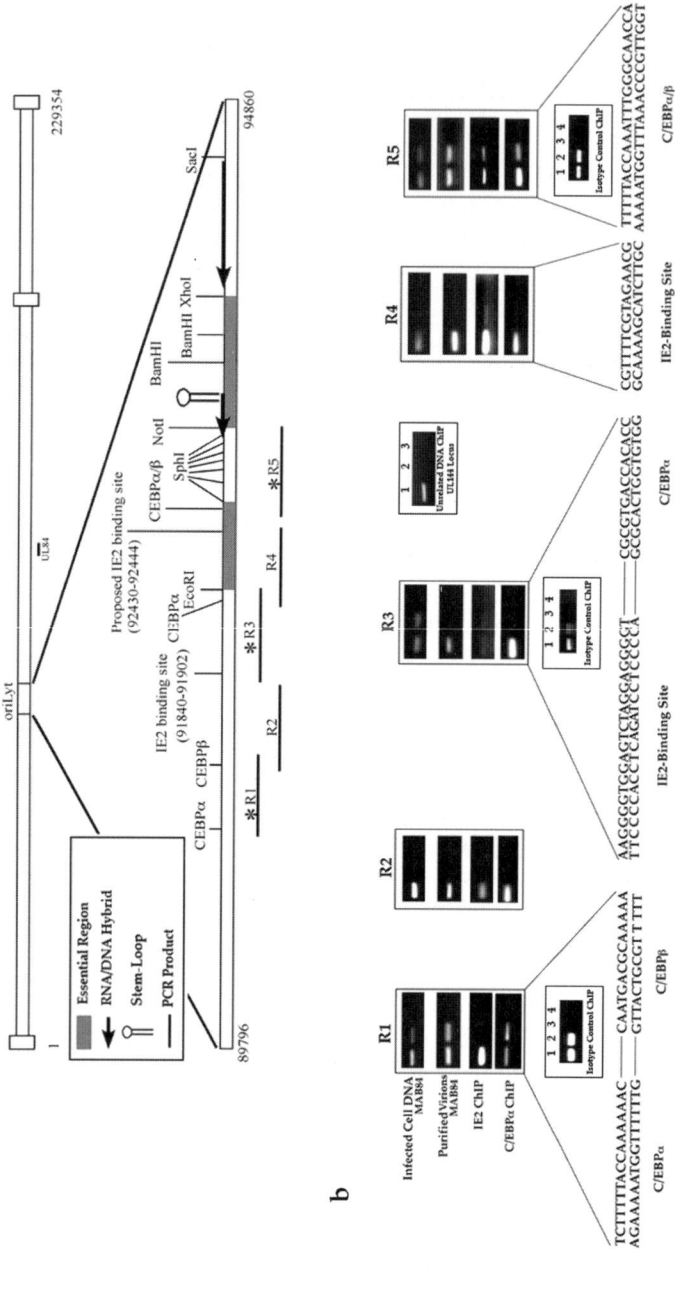

Fig. 1 *Cis*- and *trans*- acting factors interacting with HCMV oriLyt. **a** Schematic of HCMV oriLyt showing the relative position of two essential regions, C/EBPα transcription factor-binding sites, IE2-binding sites, vRNA and the RNA stem loop structure. **b** Chromatin immunoprecipitation (*ChIP*) assay showing the interaction of UL84, IE2 and C/EBPα with oriLyt. Lane one of each set is the PCR product from input DNA, lane two is the PCR product from immunoprecipitation with the anti-UL84 antibody and lane 3 is the PCR product from immunoprecipitations using an isotype matched unrelated antibody

synthesis since it can be deleted without affecting oriLyt amplification. A second IE2-binding site is nonconventional and was discovered using CASTing and SAAB methods (Huang and Chen 2002). This second site appears to be functional in that it was shown to interact with HCMV IE2 (IE86) in vivo using the chromatin immunoprecipitation assay (ChIP) (Xu et al. 2004b). In addition, the DNA sequence containing the nonconventional IE2-binding site was shown to have promoter activity, driving the expression of a luciferase reporter upon HCMV infection, or in the presence of UL84 and IE2 in transient assays in HFF cells (Xu et al. 2004b). The nonconventional IE2-binding site does contain a CRS-like element that is similar to those found in other IE2-responsive promoters, and this DNA motif was shown to be required for efficient promoter activity and oriLyt amplification. In the context of oriLyt, it appears that this promoter element may participate in the initial events surrounding the regulation of DNA synthesis and could serve as the trigger for initiation of lytic DNA synthesis. For example, one such scenario could be that when UL84 protein is produced it interacts with IE2 and facilitates the binding of the protein complex to oriLyt, thereby initiating transcription and signaling the onset of lytic DNA replication.

The presence of an IE2-UL84 responsive promoter suggested that transcription plays an important role in the activation of lytic DNA synthesis. This is not unlike other herpesviruses such as EBV and HHV-8, where promoter elements are part of lytic origins and their presence is essential for efficient oriLyt activity (Hammerschmidt and Sugden 1988; Aiyar et al. 1998; Lin et al. 2003; AuCoin et al. 2004; Wang et al. 2004). The HCMV oriLyt promoter element extends from AD169 nts 91495-92888, and appears to be bidirectional in that transcription driven in either direction can accommodate oriLyt amplification. Transient reporter assays demonstrated that either orientation of the oriLyt promoter region could drive the expression of luciferase. The activity of this promoter is cell type-dependent since transient transfection of luciferase reporter plasmids containing the oriLyt promoter were constitutively active in Vero cells, whereas transcriptional activation in human fibroblasts (HFs) was only achieved by subsequent viral infection or the cotransfection of IE2 and UL84. The constitutively active oriLyt promoter in Vero cells led to the observation that IE2 was not required for oriLyt amplification in the cotransfection replication assay in those cells (Sarisky and Hayward 1996). In HFs however, amplification of oriLyt requires both IE2 and UL84, presumably to activate the oriLyt promoter. Interestingly, when the cotransfection replication assay was performed in HFs constitutively expressing the catalytic subunit of telomerase, these life-expanded cells supported oriLyt amplification without the need for the proteins expressed from the UL36-38 loci. This suggests that in the context of the cotransfection replication assay, ORFs UL36-38 may serve to enhance cell survival.

In an effort to further define the role of transcription in HCMV oriLyt activation, a strong constitutive promoter was inserted in oriLyt in place of the defined oriLyt promoter (Xu et al. 2004b) in a manner similar to that done with the EBV oriLyt (Hammerschmidt and Sugden 1988). The SV40 early promoter was able to functionally substitute for the HCMV oriLyt promoter in transient assays and alleviated the need for IE2, but not UL84, in the replication assay. This strongly suggested that transcription somehow triggers initiation of DNA synthesis. In addition, these experiments identified UL84 as the factor most likely to act as an initiator protein and, since the leftward region of oriLyt is a promoter, the rightward region must contain a substrate for direct or indirect UL84 interaction.

Essential Region II: RNA/DNA Hybrid Structure

In addition to the IE2-UL84 responsive promoter region found in essential region I, the downstream region of oriLyt contains RNA/DNA hybrid structures. These structures were discovered while investigating the presence of alkali-labile regions within the HCMV genome. This RNA/DNA hybrid region stretches from approximately the Not I site (nts 92888) to 300 bp upstream of the BamHI site (93513) (Prichard et al. 1998). Interestingly, this region of the origin is variably reiterated via tandem duplication in the laboratory strain AD169, resulting in about 300 bp of extra sequence in the viral genome. The RNA/DNA hybrids are present within the packaged viral genome and appear to be a stable part of the genome. The RNA/DNA hybrid region contains several G+C-rich repeat sequences, one of which has the capacity to form a stem loop configuration. Since this region can be repeated many times within the viral genome, this stem loop may also be amplified. Although the significance of the RNA/DNA hybrid and the stem loop remains to be seen, it has been postulated that essential region II may be the area of oriLyt that interacts directly with UL84. In vitro RNA binding assays determined that UL84 does interact with a synthetic RNA stem loop oligonucleotide that is the same sequence as that present within the RNA/DNA hybrid region of oriLyt. In addition, UL84 appears to change the conformation of this oligonucleotide in that the UL84-RNA stem loop complex migrates faster in a native gel when more UL84 protein is added. This yields a staircase pattern of binding observed only when RNA stem loop oligonucleotide substrates are used in the binding assay (Colletti et al. 2007).

With respect to UL84 interactions with oriLyt in the infected cell environment, our laboratory used the ChIP assay to demonstrate that UL84 does interact with the RNA/DNA hybrid region of oriLyt in infected cells, as well as in packaged virions. In addition, we demonstrated that UL84 interacts directly or indirectly with regions of oriLyt that contain a CAAT enhancer-binding site (C/EBPα). This interaction suggests that UL84 may cooperate with C/EBPα or use these consensus sequences to facilitate DNA synthesis. This scenario would be consistent with HHV-8 where K-bZIP interacts with C/EBPα and binds to these transcription factor-binding sites within HHV-8 oriLyt (Wang et al. 2003). Another plausible scenario is that UL84 uses the consensus C/EBPα binding sites independent of the C/EBPα transcription factor interacting with oriLyt. Although ChIP results do show C/EBPα binding to this region, we were unable to detect a UL84-C/EBPα protein interaction within infected cells.

Viral-Encoded *trans*-Acting Factors Required for Lytic Replication

Most of the information concerning the *trans*-acting factors involved in HCMV lytic DNA synthesis was elucidated from the cotransfection replication assay. This assay, which is similar to the assay that elucidated the HCMV cloned oriLyt only using a set of plasmids that encode putative replication factors, revealed that eleven loci were required to achieve efficient amplification of cloned oriLyt. The original assay used plasmid clones that expressed the putative replication proteins under the control of each of their native promoters. This meant that those factors that contributed to the efficient expression or activation of replication proteins were also identified. Therefore the complete set of enzymes involved in the mechanics of DNA synthesis, as well as any ancillary factors, were identified using this assay (Pari and Anders 1993; Pari et al. 1993).

Table 1 lists the ORFs identified from the initial cotransfection replication assay performed in human fibroblasts. Six of these genes identified are common to all herpesviruses and are designated the core replication proteins, which comprise a DNA polymerase, polymerase accessory protein, single-stranded DNA binding protein, helicase, primase and primase associated factor. Interestingly, efficient oriLyt-dependent DNA replication can occur when core proteins from one herpesvirus, for example EBV, is used with a different herpesvirus species oriLyt, for example HCMV, in the cotransfection replication assay (Sarisky and Hayward 1996; Xu et al. 2004b). In all cases, the only additional factor that is needed is the initiator protein unique to each herpesvirus. This fact suggests that a set of core enzymes can carry out DNA synthesis on a primed oriLyt substrate independent of the mechanism of initiation. It also suggests that the initial events in DNA synthesis can be performed independent of the enzymes involved in DNA replication, at least in the context of transient assays.

Table 1 Essential ORFs elucidated from the cotransfection-replication assay

Open reading frame	Proposed function
UL54	DNA polymerase
UL44	Pol accessory, binds to UL84
UL105	Helicase
UL70	Primase
UL102	Primase-associated factor
UL57	Single-stranded DNA binding
UL84	UTPase, RNA-binding within oriLyt, regulatory, shuttling protein
IRS1	RNA binding, transactivation
UL112/113	Early proteins, transactivation
UL36–38	Anti-apoptotic
IE2	Transactivation, binds to UL84

Many of the core replication proteins remain poorly characterized at the biochemical level with the exception of UL44 and UL54, which are the best studied of the DNA replication enzymes. UL44 in infected cells is very abundant and found in replication compartments in early and late time of infection. UL44 encodes the polymerase accessory factor and appears to form a C-shaped clamp that interacts with DNA with increased specificity when bound to UL54 (Loregian et al. 2004; Appleton et al. 2006). The C-terminal end of UL54 appears to be the point of interaction with UL44 and a small peptide corresponding to a portion of the UL54 protein efficiently interferes with binding to UL44 in vitro (Loregian et al. 2003). UL44 was shown to interact with UL54 in the cellular environment and this interaction can influence subcellular localization (Alvisi et al. 2006). The crystal structure of UL44 is known and its interaction with UL54 was shown to enhance the affinity of UL44 for DNA (Appleton et al. 2006). UL44 is a substrate for the HCMV-encoded protein kinase UL97 (Krosky et al. 2003) as well as other cellular encoded kinases. This suggests that UL44 activity or interaction with other viral/cellular proteins may be regulated by both viral as well as cellular kinases (Krosky et al. 2003).

UL54 was initially purified from insect cells along with UL44 and found to synthesize DNA in vitro (Ertl and Powell 1992; Ertl et al. 1994). Expression of UL54 is regulated by several different promoter elements depending on the time of infection. In addition, it appears that several regulatory proteins contribute to efficient expression of UL54 in the context of the virus genome (Kerry et al. 1994, 1996).

The helicase-primase complex, composed of UL105, UL70 and UL102, has been studied using recombinant expression systems as well as limited characterization and identification from infected cells (Smith and Pari 1995; Smith et al. 1996). Although this group of proteins comprises the proposed counterparts of the herpes simplex helicase primase proteins, no data has been published demonstrating helicase or primase activity in vitro. Elucidation of the transcription units for UL102 and UL105 revealed that these transcripts are unspliced and approximately 2.7 and 3.4 kb, respectively (Smith and Pari 1995; Smith et al. 1996). The UL105 protein product is a 110-kDa species that is present at early times postinfection. It was demonstrated that all three proteins form a complex and each member of the complex is in contact with every other protein (McMahon and Anders 2002).

HCMV UL57 encodes the single-stranded DNA-binding protein. Initial studies show that UL57 encodes a 140-kDa protein that is confined to the nucleus and associates with replication compartments (Penfold and Mocarski 1997). It is assumed that UL57 acts to facilitate strand separation during DNA synthesis. Interestingly, the HSV-1 counterpart, UL29, was shown to exhibit RNA-binding and aid in the formation of R-loops and participate in helix destabilization (Boehmer 2004).

The proteins encoded by the UL112-113 locus were required in the initial cotransfection replication assay and appear to act as transcriptional enhancers. It was recently demonstrated that the UL112-113 proteins may influence the intranuclear targeting of UL44 to foci involved in DNA replication (Park et al. 2006).

The IRS1/TRS1 locus encode proteins that are part of the HCMV virion. At least one locus is required for viral growth and a deletion of TRS1 results in a decrease in production of infectious virus (Romanowski et al. 1997; Blankenship and Shenk

2002). TRS1 was shown to interact with dsRNA and both IRS1 and TRS1 can block the host protein shutoff response mediated by HSV-1 (Child et al. 2004; Cassady 2005; Hakki and Geballe 2005). The role of IRS1/TRS1 in the initial cotransfection replication assay was assumed to be that of a transcriptional activator based on the findings that TRS1 and IE2 enhanced the expression of UL44 in transient assays (Stasiak and Mocarski 1992; Colberg-Poley 1996).

UL84 and IE2

Since the initial elucidation of the *trans*-acting factors required for oriLyt-dependent DNA replication, the most complex interaction is that of UL84 and IE2. This interaction has become the focus of much of the research with respect to the characterization of a dual assignment for UL84 as a replication and regulatory factor for gene expression. UL84 was first described as the product of a 1,761-bp ORF encoding a 65-kDa protein (He et al. 1992). The mRNA transcript encoding UL84 can be detected as early as 2.5 h postinfection and the protein product is clearly detectable at about 20 h postinfection. Later studies showed that UL84 was found to be associated with IE2 in infected cells (Spector and Tevethia 1994). This association can be found throughout the virus infectious cycle and the only time point where UL84 is not found bound to IE2 is at very early times after infection before the production of UL84. At this time, it is not clear how much of the UL84 or IE2 within infected cells is part of the complex. This is an important point since UL84 has a regulatory effect on IE2-mediated transactivation. Transient assays show that at least some IE2-responsive promoters can be efficiently silenced by the addition of a UL84 expression plasmid (Gebert et al. 1997). The overexpression of UL84 prior to HCMV infection leads to the complete shut down of virus replication (Gebert et al. 1997). This phenomenon is thought to arise from the ability of UL84 to suppress the transactivation function of IE2; however, other mechanisms for inhibition of virus replication are possible.

In the context of the virus genome, UL84 is required for efficient viral replication and regulated gene expression (Xu et al. 2004a). A recombinant HCMV BAC with UL84 deleted results in an aberrant gene expression pattern, especially with respect to late gene transcription. This suggests that UL84 may regulate the effects of IE2, as implicated by the results of the transient reporter assays. In addition, lack of UL84 expression resulted in failure of the formation of replication compartments in infected cells, also consistent with what was observed in transient assays (Sarisky and Hayward 1996; Xu et al. 2004a).

UL84 was shown be a phosphorylated protein and to exhibit UTPase activity in vitro. This activity occurred independent of the presence of DNA or RNA (Colletti et al. 2005). This utilization of UTP suggests the use of an energy-generating system for an as yet unidentified enzymatic activity for UL84. UL84 contains amino acid motifs that are consistent with it belonging to the DExH/D Box family of proteins. This class of proteins has a broad range of activities. DExH/D Box proteins interact with the RNA component of ribonucleoproteins, unwind RNA, shuttle RNA from the nucleus to the cytoplasm, can either up- or downregulate certain promoters, and can directly and actively displace stably bound proteins from RNA (Fairman et al. 2004; Jankowsky and Bowers 2006). There is overwhelming functional evidence that UL84 performs many of the activities similar to those described for DExH/D Box proteins. One of the most interesting features of UL84, mentioned above, also consistent with DExH/D Box protein activity, is that it was shown to shuttle between the nucleus and the cytoplasm (Lischka et al. 2006). This activity serves presumably to increase the accumulation of mRNA in the cytoplasm, although no substrate RNAs were identified. This novel shuttling property of UL84 suggests that UL84 can function in some as yet undefined role in the cytoplasm. This role may involve the export of viral or cellular encoded RNA in order to increase the steady state level of an essential protein.

Viral and Cellular Encoded UL84 Binding Partners

Although it is known that UL84 interacts with IE2, other binding partners have been identified. Our laboratory has performed a proteomics analysis to identify other potential binding partners for UL84 in infected cells. This analysis identified HCMV UL44 and UL83 (pp65) as viral encoded factors that interact with UL84 (Y. Gao et al., 2008). Our analysis also identified CK2 alpha and beta subunits, as well as the cellular protein importin. These results are interesting in that this is the first time that UL84 has been shown to interact with a member of the viral DNA replication machinery as well as a protein (pp65) that is a component of the virion. Additionally, although it was previously reported that UL84 is phosphorylated in infected cells, this result strongly suggests that UL84 is a substrate for casein kinase II (CK2).

Another interesting finding from the proteomics analysis is that UL84 is ubiquitinated in infected cells. This was also discovered in the proteomics analysis of UL84, which revealed two protein spots for UL84 in the 2D gel. Although molecular weights of these two UL84 protein species varied only slightly, their respective pI indicated that one was more acidic than the other. Protein sequence data from one UL84 spot showed that ubiquitin E2-conjugating enzyme was also present. Further analysis using the proteosome inhibitor MG132 demonstrated that UL84 was monoubiquitinated. This apparent mono ubiquitination implicates UL84 as a factor that may influence the degradation of other proteins or may be involved in a multistep regulated degradation pathway.

The interaction of UL84 with UL44 (Pol accessory protein) is very interesting in that the initiator protein for herpes simplex virus type 1 (HSV-1), UL9, also interacts with its Pol accessory protein (Trego and Parris 2003). It is postulated that UL42 increases the ability of UL9 to load on DNA (Trego et al. 2005), a scenario that is also plausible for UL84 with respect to its interaction with UL44. The finding that UL44 interacts with UL84 opens up the possibility that UL44 may have a dual role in HCMV DNA replication. UL44 may act early in the virus life cycle at the initiation of DNA synthesis and also later in its traditional role as a processivity factor along with UL54 at the replication fork.

Summary and Perspectives

HCMV lytic DNA replication is complex and evidence supports a reliance on several viral encoded regulatory proteins for the initiation of DNA synthesis. The preponderance of the evidence points to one key factor, UL84, as a multifunctional protein that has a central role in lytic DNA replication and is involved with both the *cis*- and *trans*-acting replication machinery. UL84 facilitates lytic replication by binding to and regulating the transcriptional activation of IE2 as well as interacting with DNA and RNA motifs within HCMV oriLyt. Indeed our initial understanding of IE2, solely as a transcriptional activator, must be redefined to reflect the fact that, in the cellular environment, UL84 is bound to IE2 and this interaction may influence the activity of both proteins. This IE2-UL84 proteome performs a complex regulatory role controlling viral gene expression as well as defining the timing of initiation of viral lytic DNA replication. Additionally, the interaction of UL84 with other viral and cellular encoded proteins opens up the possibility that replication, regulation of gene expression and RNA transport all act together to orchestrate the onset and timing of viral DNA synthesis.

We are beginning to have a better understanding of HCMV lytic DNA replication; however, the mechanism of initiation of DNA synthesis remains elusive. The questions of how the precise orchestration, assembly and regulation of all the participants necessary to initiate DNA synthesis are still unanswered. It is clear that many factors contribute to the overall mechanism and the function of these factors will be the focus of future studies.

References

Aiyar A, Tyree C, Sugden B (1998) The plasmid replicon of EBV consists of multiple cis-acting elements that facilitate DNA synthesis by the cell and a viral maintenance element. EMBO J 17:6394-6403

Alvisi G, Ripalti A, Ngankeu A, Giannandrea M, Caraffi SG, Dias MM, Jans DA (2006) Human cytomegalovirus DNA polymerase catalytic subunit pUL54 possesses independently acting nuclear localization and ppUL44 binding motifs. Traffic 7:1322-1332

Anders DG, Punturieri SM (1991) Multicomponent origin of cytomegalovirus lytic-phase DNA replication. J Virol 65:931-937

Anders DG, Kacica MA, Pari G, Punturieri SM (1992) Boundaries and structure of human cytomegalovirus oriLyt, a complex origin for lytic-phase DNA replication. J Virol 66:3373-3384

Appleton BA, Brooks J, Loregian A, Filman DJ, Coen DM, Hogle JM (2006) Crystal structure of the cytomegalovirus DNA polymerase subunit UL44 in complex with the C terminus from the catalytic subunit. Differences in structure and function relative to unliganded UL44. J Biol Chem 281:5224-5232

Arlt H, Lang D, Gebert S, Stamminger T (1994) Identification of binding sites for the 86-kilodalton IE2 protein of human cytomegalovirus within an IE2-responsive viral early promoter. J Virol 68:4117-4125

AuCoin DP, Colletti KS, Cei SA, Papouskova I, Tarrant M, Pari GS (2004) Amplification of the Kaposi's sarcoma-associated herpesvirus/human herpesvirus 8 lytic origin of DNA replication is dependent upon a cis-acting AT-rich region and an ORF50 response element and the transacting factors ORF50 (K-Rta) and K8 (K-bZIP). Virology 318:542-555

Blankenship CA, Shenk T (2002) Mutant human cytomegalovirus lacking the immediate-early TRS1 coding region exhibits a late defect. J Virol 76:12290-12299

Boehmer PE (2004) RNA binding and R-loop formation by the herpes simplex virus type-1 single-stranded DNA-binding protein (ICP8). Nucleic Acids Res 32:4576-4584

Borst EM, Messerle M (2005) Analysis of human cytomegalovirus oriLyt sequence requirements in the context of the viral genome. J Virol 79:3615-3626

Cassady KA (2005) Human cytomegalovirus TRS1 and IRS1 gene products block the double-stranded-RNA-activated host protein shutoff response induced by herpes simplex virus type 1 infection. J Virol 79:8707-8715

Cherrington JM, Khoury EL, Mocarski ES (1991) Human cytomegalovirus ie2 negatively regulates alpha gene expression via a short target sequence near the transcription start site. J Virol 65:887-896

Child SJ, Hakki M, De Niro KL, Geballe AP (2004) Evasion of cellular antiviral responses by human cytomegalovirus TRS1 and IRS1. J Virol 78:197-205

Colberg-Poley AM (1996) Functional roles of immediate early proteins encoded by the human cytomegalovirus UL36-38, UL115-119, TRS1/IRS1 and US3 loci. Intervirology 39:350-360

Colletti KS, Xu Y, Yamboliev I, Pari GS (2005) Human cytomegalovirus UL84 is a phosphoprotein that exhibits UTPase activity and is a putative member of the DExD/H box family of proteins. J Biol Chem 280:11955-11960

Colletti KS, Smallenburg KE, Xu Y, Pari GS (2007) Human cytomegalovirus UL84 interacts with an RNA stemloop sequence found within the RNA/DNA hybrid region of oriLyt. J Virol 81:7077-7085

Ertl PE, Powell KL (1992) Physical and functional interaction of human cytomegalovirus and its accessory protein (ICP36). J Virol 66:4126-4133

Ertl PF, Thomas MF, Powell KL (1994) High level expression of DNA polymerases from herpesviruses. J Gen Virol 72:1729-1734

Fairman ME, Maroney PA, Wang W, Bowers HA, Gollnick P, Nilsen TW, Jankowsky E (2004) Protein displacement by DExH/D "RNA helicases" without duplex unwinding. Science 304:730-734

Gao Y, Colletti K, Pari GS (2008) Identification of Human Cytomegalovirus UL84 viral and cellular encoded binding partners using proteomics Analysis. J Virol 82:96–104

Gebert S, Schmolke S, Sorg G, Floss S, Plachter B, Stamminger T (1997) The UL84 protein of human cytomegalovirus acts as a transdominant inhibitor of immediate-early-mediated transactivation that is able to prevent viral replication. J Virol 71:7048-7060

Hakki M, Geballe AP (2005) Double-stranded RNA binding by human cytomegalovirus pTRS1. J Virol 79:7311-7318

Hammerschmidt W, Sugden B (1988) Identification and characterization of oriLyt, a lytic origin of DNA replication of Epstein-Barr virus. Cell 55:427-433

Hamzeh FM, Lietman PS, Gibson W, Hayward GS (1990) Identification of the lytic origin of DNA replication in human cytomegalovirus by a novel approach utilizing ganciclovir-induced chain termination. J Virol 64:6184-6195

He YS, Xu L, Huang E-S (1992) Characterization of human cytomegalovirus UL84 early gene and identification of its putative protein product. J Virol 66:1098-1108

Huang CH, Chen JY (2002) Identification of additional IE2-p86-responsive cis-repressive sequences within the human cytomegalovirus major immediate early gene promoter. J Biomed Sci 9:460-470

Jankowsky E, Bowers H (2006) Remodeling of ribonucleoprotein complexes with DExH/D RNA helicases. Nucleic Acids Res 34:4181-4188

Kerry JA, Priddy MA, Stenberg RM (1994) Identification of sequence elements in the human cytomegalovirus DNA polymerase gene promoter required for activation by viral gene products. J Virol 68:4167-4176

Kerry JA, Priddy MA, Jervey TY, Kohler CP, Staley TL, Vanson CD, Jones TR, Iskenderian AC, Anders DG, Stenberg RM (1996) Multiple regulatory events influence human cytomegalovirus DNA polymerase (UL54) expression during viral infection. J Virol 70:373-382

Krosky PM, Baek MC, Jahng WJ, Barrera I, Harvey RJ, Biron KK, Coen DM, Sethna PB (2003) The human cytomegalovirus UL44 protein is a substrate for the UL97 protein kinase. J Virol 77:7720-7727

Lashmit PE, Stinski MF, Murphy EA, Bullock GC (1998) A cis repression sequence adjacent to the transcription start site of the human cytomegalovirus US3 gene is required to down regulate gene expression at early and late times after infection. J Virol 72:9575-9584

Lin CL, Li H, Wang Y, Zhu FX, Kudchodkar S, Yuan Y (2003) Kaposi's sarcoma-associated herpesvirus lytic origin (ori-Lyt)-dependent DNA replication: identification of the ori-Lyt and association of K8 bZip protein with the origin. J Virol 77:5578-5588

Lischka P, Rauh C, Mueller R, Stamminger T (2006) Human cytomegalovirus UL84 protein contains two nuclear export signals and shuttles between the nucleus and the cytoplasm. J Virol 80:10274-10280

Loregian A, Rigatti R, Murphy M, Schievano E, Palu G, Marsden HS (2003) Inhibition of human cytomegalovirus DNA polymerase by C-terminal peptides from the UL54 subunit. J Virol 77:8336-8344

Loregian A, Appleton BA, Hogle JM, Coen DM (2004) Residues of human cytomegalovirus DNA polymerase catalytic subunit UL54 that are necessary and sufficient for interaction with the accessory protein UL44. J Virol 78:158-167

Masse MJO, Karlin S, Schachtel GA, Mocarski ES (1992) Human cytomegalovirus origin of replication (oriLyt) resides within a highly complex repetitive region. Proc Natl Acad Sci USA 89:5246-5250

McMahon TP, Anders DG (2002) Interactions between human cytomegalovirus helicase-primase proteins. Virus Res 86:39-52

McVoy MA, Adler SP (1994) Human cytomegalovirus DNA replicates after early circularization by concatemer formation, and inversion occurs within the concatemer. J Virol 68:1040-1051

Pari GS, Anders DG (1993) Eleven loci encoding trans-acting factors are required for transient complementation of human cytomegalovirus oriLyt-dependent DNA replication. J Virol 67:6979-6988

Pari GS, Kacica MA, Anders DG (1993) Open reading frames UL44, IRS1/TRS1, and UL36-38 are required for transient complementation of human cytomegalovirus oriLyt-dependent DNA synthesis. J Virol 67:2575-2582

Park MY, Kim YE, Seo MR, Lee JR, Lee CH, Ahn JH (2006) Interactions among four proteins encoded by the human cytomegalovirus UL112-113 region regulate their intranuclear targeting and the recruitment of UL44 to prereplication foci. J Virol 80:2718-2727

Penfold MET, Mocarski ES (1997) Formation of cytomegalovirus DNA replication compartments defined by localization of viral proteins and DNA synthesis. Virology 239:46-61

Prichard MN, Jairath S, Penfold ME, St Jeor S, Bohlman MC, Pari GS (1998) Identification of persistent RNA-DNA hybrid structures within the origin of replication of human cytomegalovirus. J Virol 72:6997-7004

Romanowski MJ, Garrido-Guerrero E, Shenk T (1997) pIRS1 and pTRS1 are present in human cytomegalovirus virions. J Virol 71:5703-5705

Sarisky RT, Hayward GS (1996) Evidence that the UL84 gene product of human cytomegalovirus is essential for promoting oriLyt-dependent DNA replication and formation of replication compartments in cotransfection assays. J Virol 70:7398-7413

Smith JA, Pari GS (1995) Human cytomegalovirus UL102 gene. J Virol 69:1734-1740

Smith JA, Jairath S, Crute JJ, Pari GS (1996) Characterization of the human cytomegalovirus UL105 gene and identification of the putative helicase protein. Virology 220:251-255

Spector DJ, Tevethia MJ (1994) Protein-protein interactions between human cytomegalovirus IE2-5f80aa and pUL84 in lytically infected cells. J Virol 68:7549-7553

Stasiak PC, Mocarski ES (1992) Transactivation of the cytomegalovirus ICP36 gene promoter requires the alpha gene product TRS1 in addition to IE1 and IE2. J Virol 66:1050-1058

Trego KS, Parris DS (2003) Functional interaction between the herpes simplex virus type 1 polymerase processivity factor and origin-binding proteins: enhancement of UL9 helicase activity. J Virol 77:12646-12659

Trego KS, Zhu Y, Parris DS (2005) The herpes simplex virus type 1 DNA polymerase processivity factor, UL42, does not alter the catalytic activity of the UL9 origin-binding protein but facilitates its loading onto DNA. Nucleic Acids Res 33:536-545

Tsai HL, Kou GH, Tang FM, Wu CW, Lin YS (1997) Negative regulation of a heterologous promoter by human cytomegalovirus immediate-early protein IE2. Virology 238:372-379

Wang SE, Wu FY, Fujimuro M, Zong J, Hayward SD, Hayward GS (2003) Role of CCAAT/enhancer-binding protein alpha (C/EBPalpha) in activation of the Kaposi's sarcoma-associated herpesvirus (KSHV) lytic-cycle replication-associated protein (RAP) promoter in cooperation with the KSHV replication and transcription activator (RTA) and RAP. J Virol 77:600-623

Wang Y, Li H, Chan MY, Zhu FX, Lukac DM, Yuan Y (2004) Kaposi's sarcoma-associated herpesvirus ori-Lyt-dependent DNA replication: cis-acting requirements for replication and ori-Lyt-associated RNA transcription. J Virol 78:8615-8629

Xu Y, Cei SA, Huete AR, Pari GS (2004a) Human cytomegalovirus UL84 insertion mutant defective for viral DNA synthesis and growth. J Virol 78:10360-10369

Xu Y, Cei SA, Rodriguez Huete A, Colletti KS, Pari GS (2004b) Human cytomegalovirus DNA replication requires transcriptional activation via an IE2- and UL84-responsive bidirectional promoter element within oriLyt. J Virol 78:11664-11677

Zhu Y, Huang L, Anders DG (1998) Human cytomegalovirus oriLyt sequence requirements. J Virol 72:4989-4996

Interactions of Human Cytomegalovirus Proteins with the Nuclear Transport Machinery

T. Stamminger

Contents

Introduction	168
Nuclear Import and Export Pathways	169
Protein Import and Export	169
Nuclear RNA Export	171
Interaction of the Human Cytomegalovirus Protein pUL69 with the mRNA Export Factor UAP56	173
RNA Export by pUL69	176
Interaction of the HCMV pUL84 with Importin-α Proteins	178
Unconventional Interactions with the Nuclear Transport Machinery: Novel Targets for Antiviral Strategies?	179
References	180

Abstract Accurate cellular localization is crucial for the effective function of most viral macromolecules and nuclear translocation is central to the function of herpesviral proteins that are involved in processes such as transcription and DNA replication. The passage of large molecules between the cytoplasm and nucleus, however, is restricted, and this restriction affords specific mechanisms that control nucleocytoplasmic exchange. In this review, we focus on two cytomegalovirus-encoded proteins, pUL69 and pUL84, that are able to shuttle between the nucleus and the cytoplasm. Both viral proteins use unconventional interactions with components of the cellular transport machinery: pUL69 binds to the mRNA export factor UAP56, and this interaction is crucial for pUL69-mediated nuclear export of unspliced RNA; pUL84 docks to importin-α proteins via an unusually large protein domain that contains functional leucine-rich nuclear export signals, thus serving as a complex bidirectional transport domain. Selective interference with

---

T. Stamminger
Institute for Clinical and Molecular Virology, University Hospital, University of Erlangen-Nürnberg, Schlossgarten 4, 91054, Erlangen, Germany
thomas.stamminger@viro.med.uni-erlangen.de

these unconventional interactions, which disturbs the intracellular trafficking of important viral regulatory proteins, may constitute a novel and attractive principle for antiviral therapy.

Introduction

A defining feature of the eukaryotic cell is its division into nucleoplasm and cytoplasm by a nuclear envelope. This segregation requires specific mechanisms for the continuous transport of large numbers of macromolecules between both compartments. The only conduit between these two compartments are the nuclear pore complexes (NPC), which serve as gatekeepers of the cell nucleus. These are aqueous pores with a diameter of 25–30 nm that allow the free passage of small molecules such as water, ions, nucleotides and small proteins (<40 kDa) but act as an efficient barrier for macromolecules such as proteins, RNA and DNA (Nakielny and Dreyfuss 1999). The mammalian NPC is a large proteinaceous structure of approximately 125 MDa, consisting of 30–35 different proteins termed nucleoporins (Nups) (Reichelt et al. 1990; Cronshaw et al. 2002). Generally, the tightly regulated passage of macromolecules (alternatively called cargos) through the NPC is carried out by a mobile transport machinery that interacts with the NPC and is selective for specific macromolecules (Gorlich and Kutay 1999). The transport receptors by themselves shuttle between the nucleus and the cytoplasm and are able to recognize and bind cargo molecules via specific recognition sequences.

Viruses, in particular those that replicate within the nucleus, have developed various strategies to target the cellular machinery for nucleocytoplasmic exchange (Cullen 2003; Gustin 2003; Fontoura et al. 2005). Several reasons appear to explain the obvious endeavor of viruses to capitalize on these cellular transport pathways. First, viruses have to ensure efficient nuclear import of viral proteins that are necessary within the nucleus to support replication. Second, nuclear export must be optimized, in particular for cargos that differ in structure from typical cellular cargos (e.g., viral RNAs). In this regard, it should be mentioned that research on viruses contributed significantly to the characterization of the major cellular nuclear export pathways (TAP-dependent mRNA export pathway, CRM1/exportin1-dependent pathway) (Cullen 2003). Third, some viruses even block the nucleocytoplasmic transport of cellular molecules. For instance, this has been shown for the vesicular stomatitis virus matrix and the influenza virus NS1 proteins. This is either used as a strategy to compromise host cellular functions or even as a means to impair innate and adaptive immune responses (Faria et al. 2005; Satterly et al. 2007).

Many viral macromolecules have evolved recognition sequences (e.g., nuclear localization signals) that resemble cellular signals and thus allow for conventional interactions with cellular transport receptors. Alternatively, however, several unconventional interactions have also been described that either help a specific viral component to gain access or to modulate the nucleocytoplasmic exchange apparatus of the cell. In this review, we focus on two human cytomegalovirus encoded proteins, which we identified by yeast two-hybrid screening for interactions with components of cellular nucleocytoplasmic transport pathways. Further experiments characterized these interactions as nonconventional but functionally relevant. This is the UL69 protein of HCMV, which interacts with the cellular mRNA export factor UAP56 (Lischka et al. 2006b), and the essential viral regulatory factor pUL84, which interacts via a nonconventional nuclear localization domain with importin-α proteins (Lischka et al. 2003b). The relevance of these interactions for viral replication and the potential implications for antiviral therapeutic approaches will be discussed.

Nuclear Import and Export Pathways

Eukaryotic cells use several nuclear transport pathways, each of which transports a specific range of macromolecules (proteins or various RNA species) either into or out of the nucleus. Three key steps are common to all nuclear trafficking pathways: (1) generation of a complex consisting of cargo and carrier, (2) translocation of this complex through the NPC, and (3) release of the cargo in the target compartment. Most pathways use a homologous family of carrier molecules collectively called β-karyopherins with import carriers called importins and export carriers called exportins (Radu et al. 1995; Gorlich et al. 1994; Stade et al. 1997). The export of mRNA is unique, since it primarily uses the TAP/p15 heterodimer as the main mRNA export receptor, which is not related to β-karyopherins (Gruter et al. 1998).

Protein Import and Export

The best characterized nucleocytoplasmic transport pathway is the classical nuclear import of proteins (Fig. 1). In a first step, importin-α proteins, which act as adaptor molecules, interact with protein cargos via targeting sequences called nuclear localization signals (NLSs). Classical NLSs consist of either one (monopartite) or two (bipartite) short stretches of basic amino acids (for a review see Lange et al. 2007). Molecular recognition of NLSs is crucial for the formation of the import complex and is mediated via specific sites on importin-α that are located within the armadillo repeat region of the protein (Conti et al. 1998; Fontes et al. 2000). Importin-α links the cargo to the β-karyopherin importin-β, which mediates interaction of the trimeric complex with the NPC and translocates the cargo into the nucleus. Within the nucleus, importin-β then binds to RanGTP, inducing a conformational change that results in the release of the cargo. Importin-β complexed with RanGTP is recycled to the cytoplasm, whereas importin-α is exported complexed with the β-karyopherin CAS and RanGTP. Finally, cytoplasmic RanGAP (Ran GTPase-activating protein) stimulates the Ran GTPase, which generates RanGDP and thus releases the importins within the cytoplasm for additional import cycles (reviewed in Stewart 2007).

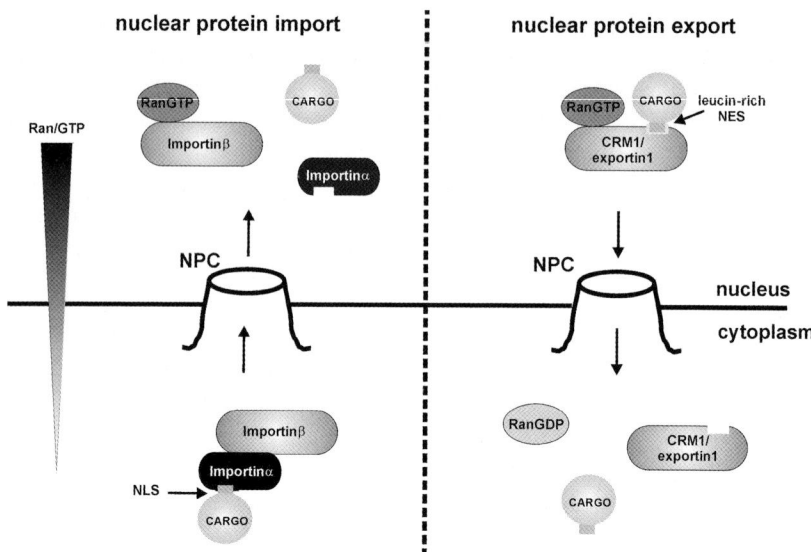

Fig. 1 Classical cellular pathways of nuclear protein import and export. Nuclear protein import: in the cytoplasm, cargo, containing a nuclear localization signal (*NLS*), is bound by the heterodimeric import receptor, importin-α/importin-β; importin-α directly binds to the NLS-containing cargo and importin-β mediates interactions with the nuclear pore complex (*NPC*) during translocation. Within the nucleus, RanGTP binding causes a conformational change of importin-β, resulting in the release of the cargo. Nuclear protein export: in the nucleus, cargo, containing a leucine-rich nuclear export signal (*NES*), interacts with the RanGTP complexed export factor CRM1/exportin1, which directly interacts with components of the NPC. Hydrolysis of RanGTP to RanGDP in the cytoplasm induces the release of the cargo. A gradient of RanGTP across the nuclear envelope, resulting from the activity of the chromatin-associated nucleotide exchange factor RCC1 and the cytoplasmic GTPase-activating protein RanGAP, is considered the major driving force for nuclear protein transport in both directions

Although the classical nuclear import pathway using importin-α as an adaptor is believed to account for the majority of nuclear protein import, several alternative pathways exist. For instance, hnRNPA1 contains a different type of NLS that is rich in aromatic residues and glycine (called M9 sequence), which binds directly to the β-karyopherin transportin1, resulting in docking of the complex at the NPC (Pollard et al. 1996). Also, several viral (e.g., Rex of HTLV-1and Rev of HIV-1) and cellular proteins (e.g., c-fos) are able to bind directly to various members of the importin-β family (importin-β, transportin, importin5, importin7) via arginine-rich sequences without the need for an additional adapter protein (Palmeri and Malim 1999; Henderson and Percipalle 1997; Arnold et al. 2006a, 2006b).

Interestingly, the M9 sequence not only acts as an NLS but also mediates nuclear export, thus constituting a bidirectional signal for nucleocytoplasmic shuttling (Michael et al. 1995). Nuclear export of proteins, however, is in most cases mediated via a signal sequence that is clearly distinct from the NLS. The best characterized nuclear export signal (NES) is the small, hydrophobic, leucine-rich NES, which was identified initially in the HIV-1 Rev protein and the cellular kinase PKI (Fischer et al. 1995; Wen et al. 1995). Functionally related export sequences resembling a leucine-rich NES have been detected since then in many cellular and viral proteins of diverse functions with the capacity to shuttle between the nucleus and the cytoplasm (La Cour et al. 2003). The direct interaction with the importin-β-related export factor CRM1 (exportin1) is essential for the export of proteins containing a leucine-rich NES (Fornerod et al. 1997) (Fig. 1). This interaction can be inhibited specifically by the antibiotic leptomycin B (LMB), resulting in a block of the nuclear export of proteins with a leucine-rich NES (Kudo et al. 1999). CRM1 binds cooperatively to RanGTP and its export cargo, leading to the formation of a trimeric transport complex in the nucleus. After translocation of this complex through the NPC, the cytoplasmic RanGTP-binding protein RanBP1 in concert with RanGAP dissociates the export complex (for a review see Hutten and Kehlenbach 2007).

Nuclear RNA Export

The identification of CRM1 as a protein export factor was initiated by the finding that nuclear export of unspliced HIV-1 RNA depends on binding of the viral protein Rev to CRM1 via a leucine-rich NES (Neville et al. 1997) (Fig. 2). Further studies demonstrated that although CRM1 mediates nuclear export of HIV-1 mRNA, it is not responsible for the export of bulk cellular mRNA (Cullen 2003). Instead, CRM1 acts as a RNA-export receptor for the export of rRNA, Usn RNAs and several specific mRNAs (e.g., c-Fos, Cyclin D1, CD83), which is, however, mediated via RNA-binding adapter proteins that interact with CRM1 (reviewed in Hutten and Kehlenbach 2007) (Fig. 2).

In contrast to complex lentiviruses like HIV-1 encoding Rev-type RNA-binding proteins, incompletely or unspliced RNAs from type D retroviruses are exported due to the presence of a *cis*-acting RNA sequence named constitutive transport element (CTE) (Bray et al. 1994). Investigation of the CTE-mediated export mechanism allowed for the discovery of the major mRNA export receptor, named TAP/NXF1 (in yeast termed Mex67p), which interacts with the CTE element and thus facilitates nuclear export of CTE-containing transcripts (Kang and Cullen 1999) (Fig. 2). Later on, it was demonstrated that TAP interacts with p15, and that this heterodimer is responsible for bulk metazoan mRNA export to the cytoplasm via direct interaction with the nuclear pore (Katahira et al. 1999, 2002). Although TAP is able to bind directly to CTE-containing viral mRNA, additional factors are needed to bridge the interaction between TAP-p15 and metazoan mRNA since metazoan RNA does not contain CTE-like RNA secondary structures (Liker et al. 2000). The production of mature mRNA in eukaryotes involves a complex series of nuclear processing reactions that occur co-transcriptionally and include the addition of the 5' cap, removal

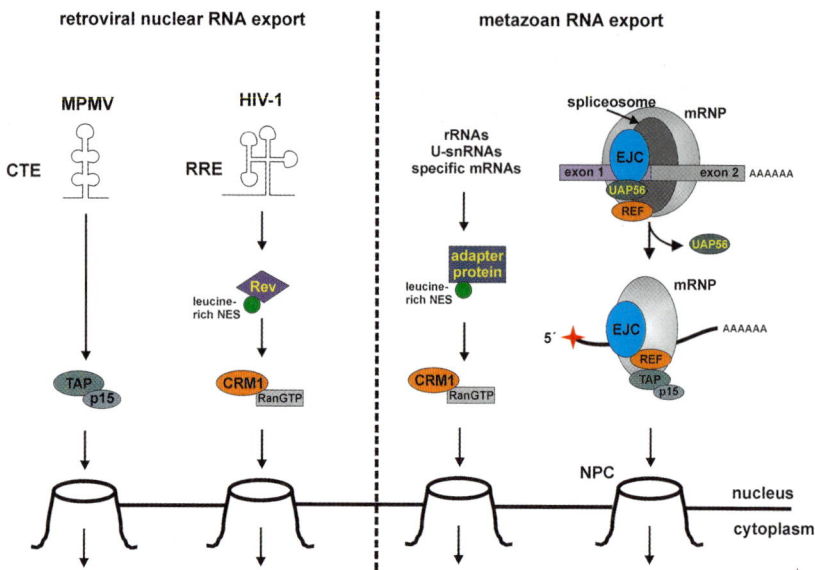

Fig. 2 Nuclear mRNA export pathways. Shown is a schematic overview of several nuclear mRNA export pathways used by retroviruses or by metazoan cells. Retroviral nuclear RNA export: unspliced mRNAs encoded by the simple retrovirus MPMV are exported via direct binding of the constitutive transport element (*CTE*) RNA target to the cellular mRNA export receptor TAP-p15; unspliced RNA encoded by HIV-1 binds via the Rev-responsive element (*RRE*) RNA target to the viral factor Rev. Rev then interacts with the cellular protein CRM1 through its leucine-rich NES, thus mediating nuclear export of unspliced HIV-1 mRNA. Metazoan nuclear RNA export: ribosomal RNAs, small nuclear RNAs (*U snRNAs*) as well as some specific mRNAs are exported from the nucleus via the karyopherin CRM1; however, adaptor proteins binding to CRM1 via a leucine-rich NES are required to mediate these interactions. Bulk cellular mRNA export occurs via the TAP-p15 export receptor: during splicing of vertebrate mRNA, a complex of proteins, the exon junction complex (*EJC*) that contains UAP56 and REF is deposited on spliced mRNAs upstream of exon–exon junctions. REF proteins present in EJCs subsequently recruit the export receptor TAP-p15. Association of TAP-p15 with the mRNPs displaces UAP56

of introns and addition of the 3′ poly(A) tail (Bentley 2002; Maniatis and Reed 2002). The removal of introns results in the deposition of a protein complex, termed exon junction complex (EJC), on the RNA molecule immediately upstream of the splice site (Le Hir et al. 2000). One of the components of the EJC is the DExD/H box protein UAP56 (named Sub2p in *Saccharomyces cerevisiae*), a putative RNA helicase, which is thought to couple mRNA splicing with nuclear export (Reed and Hurt 2002). As a next step, UAP56 recruits the adapter protein Aly/REF (Yra1p in *S. cerevisiae*) to the mRNA, and Aly/REF subsequently interacts with the heterodimeric TAP-p15, leading to the efficient export of the mRNA through the nuclear pore complex (Luo et al. 2001; Strasser and Hurt 2001) (Fig. 2). Interestingly, Sub2p is required for nuclear export of both intron-containing as well as intronless mRNAs, suggesting that the association of the helicase with the mRNA molecule is not necessarily coupled with the splicing reaction (Strasser and Hurt 2001). Indeed, several studies showed that UAP56 recruitment to the mRNA can also occur via co-transcriptional mechanisms (Zenklusen et al. 2002; Kiesler et al. 2002; Strasser et al. 2002). Furthermore, while REF proteins were found to be dispensable for bulk mRNA export, inactivation of UAP56 leads to a nuclear retention of poly(A) mRNA, indicating that this protein plays a central role for mRNA export (Gatfield and Izaurralde 2002; Gatfield et al. 2001; Kapadia et al. 2006).

Interaction of the Human Cytomegalovirus Protein pUL69 with the mRNA Export Factor UAP56

The human cytomegalovirus protein encoded by the open reading frame UL69 belongs to a family of regulatory factors that is conserved among all herpesviruses and includes the proteins ICP27 of herpes simplex type I (HSVI), EB2 of Epstein-Barr virus (EBV), and the ORFs 57 of Kaposi's sarcoma-associated herpesvirus (KSHV) and of Herpesvirus saimiri (HVS) (for reviews see Lischka and Stamminger 2006; Sandri-Goldin 2004; Sandri-Goldin 2001). Although the amino acid identity among these proteins is not very high, ranging from 17% to 36%, they share a region showing a higher conservation of approximately 40% sequence identity. This conserved region can be found at the C-terminus of the α- and γ-herpesvirus proteins, whereas it corresponds to the central part of the β-herpesviral proteins since they have a unique C-terminal domain (Winkler et al. 2000) (see Fig. 3). Recently, we demonstrated that this homology region folds into a globular core domain according to secondary structure predictions and is responsible for a shared

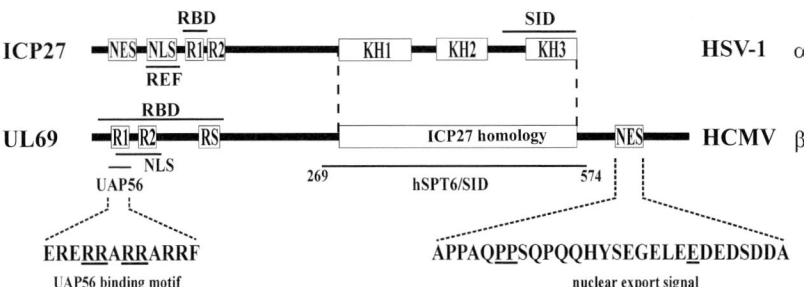

Fig. 3 Domain organization of the HCMV UL69 protein in comparison to the HSV-1 protein ICP27 showing the positions of important functional regions. The sequence of the UAP56-binding motif within pUL69 is depicted, as is the sequence of the leptomycin B-insensitive NES; underlined amino acid residues are critical for the function of the respective motifs. *NES* nuclear export signal, *NLS* nuclear localization signal, *R1, R2, RS* arginine-rich regions, *RBD* RNA-binding domain, *KH1–3* putative KH RNA-binding motifs, *ICP27 homology* domain of pUL69 with high homology to ICP27, *SID* self-interaction domain, *REF* UAP56, *hSPT6* binding sites of the respective cellular factors

property of all members of this protein family, namely the propensity to self-interact and thus to form multimeric protein complexes (Lischka et al. 2007).

A second shared feature of all characterized members of this protein family is a function as posttranscriptional regulators. For instance, the prototype of this protein family, ICP27 of HSVI, has been shown to redistribute small nuclear ribonucleoprotein particles (snRNPs), to inhibit cellular splicing, to bind to intronless viral RNAs, and to shuttle between the nucleus and the cytoplasm, thus acting as a viral mRNA export factor (reviewed by Sandri-Goldin 2004; Sandri-Goldin 2001; Smith et al. 2005). The latter properties are of particular importance for herpesviruses since the majority of viral transcripts are intronless and thus do not interact with the splicing machinery, leading to inefficient nuclear export of viral mRNA. Further investigation of the RNA export mechanism revealed that ICP27 binds viral RNA through an N-terminal RGG box RNA-binding motif (Mears and Rice 1996; Sandri-Goldin 1998). Additionally, ICP27 was shown to interact with the adaptor protein Aly/REF, thereby recruiting intronless viral RNAs to the cellular TAP-p15 mRNA export receptor (Chen et al. 2002; Koffa et al. 2001) (see Fig. 4). An interaction with Aly/REF could also be demonstrated for the γ-herpesviral proteins EB2 of EBV and ORF 57 of KSHV or HVS, suggesting that several ICP27 homologous proteins may use a common mechanism for the nuclear export of viral intronless mRNAs (Hiriart et al. 2003; Malik et al. 2004; Williams et al. 2005).

Initial studies on the HCMV homolog of ICP27, the UL69 protein, revealed several differences between these two regulatory factors:

1. In contrast to the immediate early expression of ICP27, pUL69 could be detected during the early and late phase of the replication cycle and is incorporated as a tegument protein into viral particles (Winkler et al. 1994; Winkler and Stamminger 1996);
2. pUL69 did not repress expression depending on the presence of an intron within a reporter gene but revealed a rather pleiotropic activation of various promoters upstream of the luciferase reporter (Sandri-Goldin and Mendoza 1992; Winkler et al. 1994);
3. No redistribution of snRNPs is induced by pUL69 (Winkler et al. 2000);
4. pUL69 could not complement the growth of an ICP27 deletion mutant of HSVI, further emphasizing the existence of functional differences (Winkler et al. 1994).

In an attempt to unravel the mechanism of pUL69-mediated transactivation, we searched for cellular interaction partners of this viral protein by yeast two hybrid screenings, which revealed at least two proteins with a potential role for mRNA biogenesis and processing. One of the identified cellular proteins corresponded to the transcription elongation factor hSPT6 (Endoh et al. 2004; Kaplan et al. 2000); we were able to demonstrate that this interaction occurs within the conserved homology region of pUL69 and is functionally required for pUL69-mediated transactivation (Winkler et al. 2000). Interestingly, hSpt6 has recently been shown to bind both to the C-terminal domain of RNA polymerase II and to a novel factor

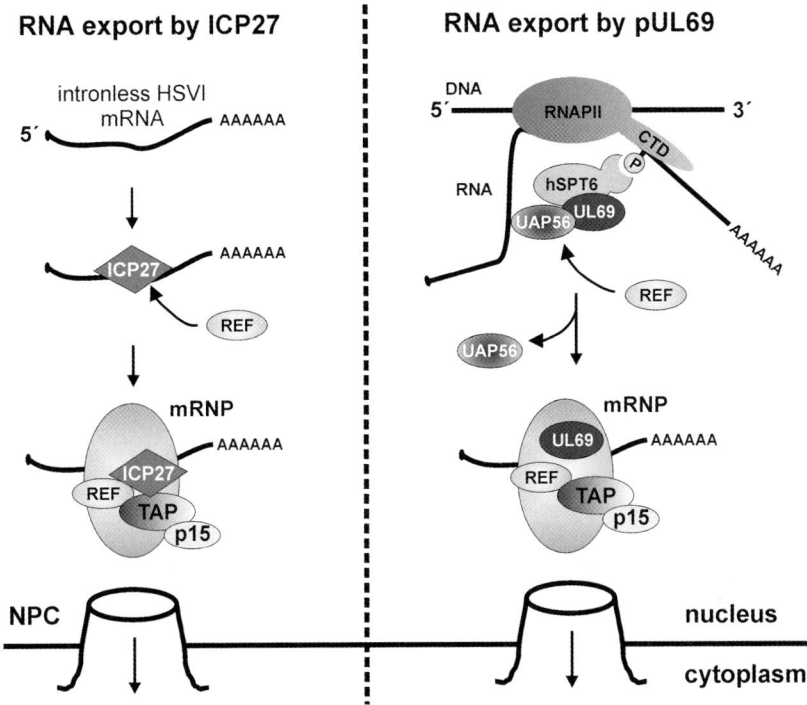

Fig. 4 Models of ICP27- and pUL69-mediated viral mRNA export. ICP27 of HSVI binds directly to viral intronless transcripts. Via its interaction with the adaptor protein REF, the RNA is recruited to the cellular TAP-p15 mRNA export receptor, thus forming an export-competent mRNP. The UL69 protein of HCMV interacts both with the cellular transcription elongation factor hSPT6 and the mRNA export factor UAP56. Thus, pUL69 may facilitate mRNA export by optimizing the co-transcriptional loading of RNA export factors to nascent mRNA

named hIws1, the depletion of which induces nuclear retention of poly(A) RNA. This observation suggests that hSPT6 integrates transcription elongation with downstream mRNA export (Yoh et al. 2007). However, with respect to the documented interaction of ICP27 with Aly/REF, it was even more interesting to identify the putative DExH/D box RNA helicase URH49 as a binding protein of pUL69 by yeast two-hybrid screening. URH49 is highly related to the mRNA export protein UAP56, which is located upstream of Aly/REF in the cellular mRNA export pathway (see Fig. 2) (Pryor et al. 2004). In both yeast two-hybrid and co-immunoprecipitation experiments, we demonstrated that pUL69 interacts not only with URH49, but also with UAP56 (Lischka et al. 2006b). Although details on the biological function of URH49 are unknown so far, initial data support the assumption that both DExH/D box proteins exert largely overlapping functions in the processing and export of mammalian mRNAs (Kapadia et al. 2006; Pryor et al. 2004). In this context, it should be noted that UAP56, alternatively termed BAT1, has been described as a multifunctional cellular protein: in addition to its role in recruiting Aly/REF to mature mRNAs, it plays a well-documented role in pre-mRNA splicing, where it facilitates the association of U2 snRNP with the splice branch point, presumably by dissociating U2AF65 from the polypyrimdine track downstream of the splice branchpoint (Fleckner et al. 1997). Furthermore, several studies propose that the UAP56/BAT1 gene, which is situated in the central region of the major histocompatibility complex, acts as an anti-inflammatory gene, and polymorphism in its promoter region may predispose for specific inflammatory disorders (Allcock et al. 1999, 2001; Ramasawmy et al. 2006). However, it is presently unclear whether this potential antiinflammatory activity of UAP56 is related to its function in mRNA processing and export.

RNA Export by pUL69

The detection of an interaction with URH49 and UAP56 suggested that pUL69, similar to its counterparts in α- and γ-herpesviruses, may modulate nuclear mRNA export. Taking advantage of a widely used functional reporter assay harboring the CAT coding sequence and the Rev responsive element (RRE) of HIV1 inserted into an intron (Hope et al. 1990), both on the level of reporter protein and reporter RNA expression, we showed that pUL69 facilitates the nuclear export of otherwise inefficiently exported, unspliced RNA similar to Rev (Lischka et al. 2006b). Furthermore, we were able to identify a 12-amino-acid sequence motif within the N-terminus of pUL69, which turned out to be crucial for binding to UAP56 (Fig. 3): mutations within this motif abrogated both UAP56 binding and pUL69-mediated nuclear export of unspliced RNA. This indicates that the interaction with UAP56 or URH49 is required for pUL69 to promote cytoplasmic accumulation of unspliced RNA (Lischka et al. 2006b).

However, in contrast to the HIV1 Rev protein, which interacts in a sequence-specific manner with the viral RRE target sequence, this *cis*-acting element was not necessary for pUL69 to facilitate RNA export. Presently, it is unclear whether pUL69 requires a specific *cis*-acting RNA sequence to target viral RNAs for export. Similar to ICP27, pUL69 is able to interact directly with RNA. This is mediated via a complex N-terminal RNA-binding domain consisting of three arginine-rich motifs that overlaps with both the NLS- and the UAP56-binding motif (Fig. 3) (Toth et al. 2006). Surprisingly, however, an RNA-binding-deficient mutant of pUL69, which still interacts with UAP56/URH49, retained its RNA export activity. This suggests that in contrast to its homologs in other herpesviruses, RNA binding is not a prerequisite for pUL69-mediated nuclear RNA export (Toth et al. 2006).

Another surprising finding was that the previously described nuclear export sequence (NES) within pUL69 was clearly distinct from the UAP56 binding motif (Fig. 3) (Lischka et al. 2001, 2006b). Nucleocytoplasmic shuttling is a feature that is common to all ICP27 family members and is also shared by a large number of proteins involved in nuclear mRNA export such as Tap and Aly/REF (Sandri-Goldin 1998; Mears and Rice 1998; Farjot et al. 2000; Goodwin et al. 1999; Bello et al. 1999; Lischka et al. 2001; Bear et al. 1999; Rodrigues et al. 2001). Although it was initially proposed for ICP27 that a leucine-rich region interacting with CRM1/exportin1 acts as a NES (Sandri-Goldin 1998), subsequent studies revealed that a CRM1-independent export mechanism is crucial for the mRNA export activity (Koffa et al. 2001; Chen et al. 2002). This could also be confirmed for other members of the ICP27 family, including pUL69 (Lischka et al. 2001; Hiriart et al. 2003; Williams et al. 2005). Interestingly, CRM1-independent nuclear export of pUL69 is mediated via a novel, transferable nuclear export signal of 28 amino acids that is located within the unique C-terminal domain of the β-herpesviral protein (Lischka et al. 2001) (Fig. 3). Since we detected that UAP56 can also shuttle between the nucleus and the cytoplasm (P. Lischka, M. Thomas, and T. Stamminger, unpublished observations), we initially assumed that the nuclear export of pUL69 may be mediated via docking to UAP56. However, since mutation of the UAP56 binding motif did not abrogate the nucleocytoplasmic shuttling of pUL69, we speculate that the nuclear export pathway accessed by the pUL69 NES still remains to be defined.

In summary, our experiments revealed that interactions of pUL69 with both the transcription elongation factor hSPT6 and the mRNA export factor UAP56 were essential for the mRNA export activity. Since pUL69 binds to these cellular factors via distinct protein domains, the interaction with hSPT6, traveling along with RNA polymerase II, may increase the co-transcriptional loading of UAP56 onto intronless viral transcripts, ultimately leading to the formation of an export-competent mRNP that associates with TAP-p15 (Fig. 4). However, several open questions remain to be answered, in particular regarding the exact functions of pUL69 during viral replication. It is not clear whether pUL69 affects the nuclear export of viral RNAs in general or whether there is specificity for a distinct subgroup of transcripts. Studies using an HCMV mutant virus with a deletion of the UL69 coding region showed that the lack of pUL69 led to a substantially diminished level of several viral late transcripts (Hayashi et al. 2000), suggesting that these RNAs may be targeted by pUL69 for efficient nuclear export. Additionally, the role of RNA binding by pUL69 requires further investigation. Since it was shown that a subset of viral and cellular mRNAs is incorporated into HCMV virions by an as yet unknown mechanism (Bresnahan and Shenk 2000; Greijer et al. 2000), it will be interesting to investigate whether the tegument localization of pUL69 contributes to this. This aspect also emphasizes that pUL69 is certainly a multifunctional protein during viral replication, which is also illustrated by the potential of this protein to induce cell-cycle arrest (Hayashi et al. 2000; Lu and Shenk 1999). The construction and characterization of recombinant viruses expressing pUL69 mutants with a loss of distinct protein functions (e.g., UAP56 binding, RNA binding, nucleocytoplasmic shuttling) will certainly contribute to a further definition pUL69 functions during viral replication.

Interaction of the HCMV pUL84 with Importin-α Proteins

The second cytomegalovirus protein for which we detected a nonconventional interaction with components of nuclear transport pathways is the gene product of the open reading frame UL84. The open reading frame UL84 of HCMV encodes a multifunctional protein with nuclear localization that appears to be absolutely essential for viral replication (Xu et al. 2002, 2004a; Lischka et al. 2003a; He et al. 1992; Yu et al. 2003) (see also the chapter by G. Pari, this volume). Initially, pUL84 was identified as a direct binding partner of the regulatory protein IE2-p86, which is the major transcription-activating protein of HCMV (Spector and Tevethia 1994). Studies concerning the functional consequence of the pUL84–IE2 interaction revealed on the one hand that this interaction downregulates the transactivation activity of IE2 on some early promoters (Gebert et al. 1997). On the other hand, it has been reported that this pUL84–IE2 complex is required for the activation of a bidirectional promoter located within the origin of lytic DNA replication (*ori*-Lyt) (Xu et al. 2004b). Since pUL84 is the only non-core protein required for origin-dependent DNA replication in a transient replication assay (Pari and Anders 1993; Sarisky and Hayward 1996), pUL84 was proposed to act as an initiator protein for viral DNA synthesis of HCMV (Xu et al. 2004b). Initiator proteins of some other herpesviruses were demonstrated to exert an inherent catalytic activity that may unwind a specific region of DNA within *ori*-Lyt, thus allowing the assembly of the DNA replication machinery. In line with this, pUL84 has been shown to display UTPase activity and to exhibit homology to the DExD/H box family of helicases (Colletti et al. 2005).

In a yeast two-hybrid screen that was performed in order to identify cellular binding proteins of pUL84, we were able to select four members of the importin-α protein family as strong interaction partners of this viral protein (Lischka et al. 2003a). Since importin-α proteins function as adapter molecules bridging NLS-containing import cargo proteins to the import receptor importin-β (see Fig. 1), this finding suggested that pUL84 may either access the nuclear import pathway via this interaction or may even be able to modulate this pathway. By performing in vitro import assays using digitonin-permeabilized cells together with purified importin-α and -β proteins we were indeed able to show that pUL84 nuclear import occurs via the well-characterized importin-α/β pathway (Lischka et al. 2003a). Intriguingly, however, the domain of pUL84 interacting with importin-α proteins turned out to be unconventional. While most nuclear proteins dock to importin-α via short, karyophilic amino acid sequences corresponding either to a classical, basic-type NLS (Lange et al. 2007) or to other, short NLS-like sequences (Wang et al. 1997; Wolff et al. 2002), we determined that a long UL84 protein domain comprising 282 amino acids was required for importin-α binding (see Fig. 5). This domain serves as a transferable, importin-α dependent NLS, which was demonstrated by fusing this sequence with a nonkaryophilic protein, resulting in its nuclear translocation (Lischka et al. 2003a). Since we observed that further N- or C-terminal as well as internal deletions abrogated the nuclear translocation as well as the dimerization/multimerization capacity of this domain, we propose that, similar to the cellular transcription factor STAT1 (Fagerlund et al. 2002), a complex overall structure that

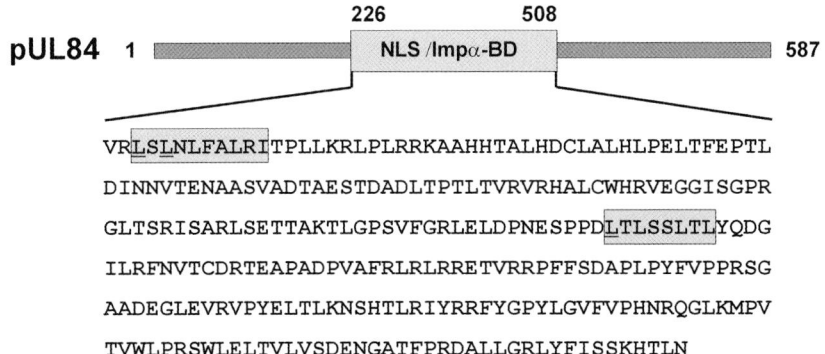

Fig. 5 Schematic representation of the pUL84 NLS/importin-α binding domain. The localization of the two autonomous leucine-rich nuclear export signals within the NLS domain is shown on *gray background*. Leucine residues with a critical function for the nuclear export activity are *underlined*

may depend on protein dimerization generates the functional pUL84 NLS (Lischka et al. 2003a).

Interestingly, sequence inspection of the UL84–importin-α interaction domain revealed the presence of two small leucine-rich regions that exactly match the consensus sequence of a classical nuclear export signal, suggesting that the pUL84 NLS domain may also be able to mediate nuclear export, thus serving as a complex bidirectional transport domain (Fig. 5). Further experimentation revealed that both leucine-rich regions are able to function as autonomous nuclear export signals and are required for CRM-1 dependent nucleocytoplasmic shuttling of pUL84 (Lischka et al. 2006a). This suggests that pUL84, in addition to its role within the nucleus as an initiator protein of origin-dependent viral DNA replication, may carry out an unexpected function within the cytoplasm that has yet to be defined. However, given the recent description of a sequence-specific RNA-binding activity of pUL84 (Colletti et al. 2007) as well as its homology to DExD/H box RNA helicases (Colletti et al. 2005), it is tempting to speculate that pUL84, similar to the UL69 protein, may be able to enhance the accumulation of specific viral transcripts within the cytoplasm of infected cells.

Unconventional Interactions with the Nuclear Transport Machinery: Novel Targets for Antiviral Strategies?

Recent reports emphasize that drug action and delivery can take advantage of cellular compartmentation instead of simply blocking enzymatic active sites, thus suggesting that the interference with nucleocytoplasmic shuttling may be a promising target for novel drug development (Gasiorowski and Dean 2003). A well-studied example where such a strategy has already been used for therapeutic intervention is the inhibition of the nuclear import of transcription factor NF-AT. Before activation, NF-AT is localized to the cytoplasm in a highly phosphorylated state. An influx of Ca^{2+} activates the phosphatase calcineurin, which dephosphorylates NF-AT. This unmasks NLS sequences, allowing the transcription factor to translocate into the nucleus where it activates cytokine genes such as IL2 and IL4 (Shibasaki et al. 1996). This is blocked by the widely used immunosuppressive agents cyclosporin A and FK506, which interact with calcineurin to regulate its activity (Liu et al. 1991). The disadvantage of this approach is the elicitation of side effects resulting from the fact that the inhibition of calcineurin activity affects several other signaling cascades as well. In this respect, the exploitation of unconventional interactions between viral molecules and components of the nuclear transport machinery may represent a more promising approach, since these interactions should be sufficiently different from cellular interactions to ensure a high specificity of the intervention. In line with this, several studies report on small molecule inhibitors that are able to specifically interfere with the nuclear translocation of the HIV preintegration complex correlating with a potent anti-HIV activity of the respective drugs in primary human cells (Al-Abed et al. 2002; Glushakova et al. 2000; Haffar et al. 1998, 2005; Haffar and Bukrinsky 2005). Thus, the detailed characterization of interactions of HCMV proteins fulfilling essential functions for viral replication with components of the nuclear transport machinery may not only contribute to our understanding of molecular mechanisms but may also be useful to develop novel drugs interfering with HCMV replication. Consequently, we are presently investigating whether peptide aptamers targeting the pUL84 NLS domain are able to inhibit the interaction with importin-α proteins and could thus be used to interfere with the nuclear localization of this essential viral regulatory protein, leading to a block of HCMV DNA replication.

Acknowledgements Work presented in this review was supported by the Deutsche Forschungsgemeinschaft (SFB473), the GRK1071, the IZKF Erlangen, and the Wilhelm Sander Stiftung.

References

Al-Abed Y, Dubrovsky L, Ruzsicska B, Seepersaud M, Bukrinsky M (2002) Inhibition of HIV-1 nuclear import via Schiff base formation with arylene bis(methylketone) compounds. Bioorg Med Chem Lett 12:3117–3119

Allcock RJ, Price P, Gaudieri S, Leelayuwat C, Witt CS, Dawkins RL (1999) Characterisation of the human central MHC gene, BAT1: genomic structure and expression. Exp Clin Immunogenet 16:98–106

Allcock RJ, Williams JH, Price P (2001) The central MHC gene, BAT1, may encode a protein that down-regulates cytokine production. Genes Cells 6:487–494

Arnold M, Nath A, Hauber J, Kehlenbach RH (2006a) Multiple importins function as nuclear transport receptors for the Rev protein of human immunodeficiency virus type 1. J Biol Chem 281:20883–20890

Arnold M, Nath A, Wohlwend D, Kehlenbach RH (2006b) Transportin is a major nuclear import receptor for c-Fos: a novel mode of cargo interaction. J Biol Chem 281:5492–5499

Bear J, Tan W, Zolotukhin AS, Tabernero C, Hudson EA, Felber BK (1999) Identification of novel import and export signals of human TAP, the protein that binds to the constitutive transport element of the type D retrovirus mRNAs. Mol Cell Biol 19:6306–6317

Bello LJ, Davison AJ, Glenn MA, Whitehouse A, Rethmeier N, Schulz TF, Barklie CJ (1999) The human herpesvirus-8 ORF 57 gene and its properties. J Gen Virol 80:3207–3215

Bentley D (2002) The mRNA assembly line: transcription and processing machines in the same factory. Curr Opin Cell Biol 14:336–342

Bray M, Prasad S, Dubay JW, Hunter E, Jeang KT, Rekosh D, Hammarskjold ML (1994) A small element from the Mason-Pfizer monkey virus genome makes human immunodeficiency virus type 1 expression and replication Rev-independent. Proc Natl Acad Sci USA 91:1256–1260

Bresnahan WA, Shenk T (2000) A subset of viral transcripts packaged within human cytomegalovirus particles. Science 288:2373–2376

Chen IH, Sciabica KS, Sandri-Goldin RM (2002) ICP27 interacts with the RNA export factor Aly/REF to direct herpes simplex virus type 1 intronless mRNAs to the TAP export pathway. J Virol 76:12877–12889

Colletti KS, Xu Y, Yamboliev I, Pari GS (2005) Human cytomegalovirus UL84 is a phosphoprotein that exhibits UTPase activity and is a putative member of the DExD/H box family of proteins. J Biol Chem 280:11955–11960

Colletti KS, Smallenburg KE, Xu Y, Pari GS (2007) Human cytomegalovirus UL84 interacts with an RNA stemloop sequence found within the RNA/DNA hybrid region of oriLyt. J Virol 81:7777–7085

Conti E, Uy M, Leighton L, Blobel G, Kuriyan J (1998) Crystallographic analysis of the recognition of a nuclear localization signal by the nuclear import factor karyopherin alpha. Cell 94:193–204

Cronshaw JM, Krutchinsky AN, Zhang W, Chait BT, Matunis MJ (2002) Proteomic analysis of the mammalian nuclear pore complex. J Cell Biol 158:915–927

Cullen BR (2003) Nuclear mRNA export: insights from virology. Trends Biochem Sci 28:419–424

Endoh M, Zhu W, Hasegawa J, Watanabe H, Kim DK, Aida M, Inukai N, Narita T, Yamada T, Furuya A, Sato H, Yamaguchi Y, Mandal SS, Reinberg D, Wada T, Handa H (2004) Human Spt6 stimulates transcription elongation by RNA polymerase II in vitro. Mol Cell Biol 24:3324–3336

Fagerlund R, Melen K, Kinnunen L, Julkunen I (2002) Arginine/lysine-rich nuclear localization signals mediate interactions between dimeric STATs and importin alpha 5. J Biol Chem 277:30072–30078

Faria PA, Chakraborty P, Levay A, Barber GN, Ezelle HJ, Enninga J, Arana C, van Deursen J, Fontoura BM (2005) VSV disrupts the Rae1/mrnp41 mRNA nuclear export pathway. Mol Cell 17:93–102

Farjot G, Buisson M, Duc DM, Gazzolo L, Sergeant A, Mikaelian I (2000) Epstein-Barr virus EB2 protein exports unspliced RNA via a Crm-1-independent pathway. J Virol 74:6068–6076

Fischer U, Huber J, Boelens WC, Mattaj IW, Luhrmann R (1995) The HIV-1 Rev activation domain is a nuclear export signal that accesses an export pathway used by specific cellular RNAs. Cell 82:475–483

Fleckner J, Zhang M, Valcarcel J, Green MR (1997) U2AF65 recruits a novel human DEAD box protein required for the U2 snRNP-branchpoint interaction. Genes Dev 11:1864–1872

Fontes MR, Teh T, Kobe B (2000) Structural basis of recognition of monopartite and bipartite nuclear localization sequences by mammalian importin-alpha. J Mol Biol 297:1183–1194

Fontoura BM, Faria PA, Nussenzveig DR (2005) Viral interactions with the nuclear transport machinery: discovering and disrupting pathways. IUBMB Life 57:65–72

Fornerod M, Ohno M, Yoshida M, Mattaj IW (1997) CRM1 is an export receptor for leucine-rich nuclear export signals. Cell 90:1051–1060

Gasiorowski JZ, Dean DA (2003) Mechanisms of nuclear transport and interventions. Adv Drug Deliv Rev 55:703–716

Gatfield D, Izaurralde E (2002) REF1/Aly and the additional exon junction complex proteins are dispensable for nuclear mRNA export. J Cell Biol 159:579–588

Gatfield D, Le HH, Schmitt C, Braun IC, Kocher T, Wilm M, Izaurralde E (2001) The DExH/D box protein HEL/UAP56 is essential for mRNA nuclear export in Drosophila. Curr Biol 11:1716–1721

Gebert S, Schmolke S, Sorg G, Floss S, Plachter B, Stamminger T (1997) The UL84 protein of human cytomegalovirus acts as a transdominant inhibitor of immediate-early-mediated transactivation that is able to prevent viral replication. J Virol 71:7048–7060

Glushakova S, Dubrovsky L, Grivel J, Haffar O, Bukrinsky M (2000) Small molecule inhibitor of HIV-1 nuclear import suppresses HIV-1 replication in human lymphoid tissue ex vivo: a potential addition to current anti-HIV drug repertoire. Antiviral Res 47:89–95

Goodwin DJ, Hall KT, Stevenson AJ, Markham AF, Whitehouse A (1999) The open reading frame 57 gene product of herpesvirus saimiri shuttles between the nucleus and cytoplasm and is involved in viral RNA nuclear export. J Virol 73:10519–10524

Gorlich D, Kutay U (1999) Transport between the cell nucleus and the cytoplasm. Annu Rev Cell Dev Biol 15:607–660

Gorlich D, Prehn S, Laskey RA, Hartmann E (1994) Isolation of a protein that is essential for the first step of nuclear protein import. Cell 79:767–778

Greijer AE, Dekkers CA, Middeldorp JM (2000) Human cytomegalovirus virions differentially incorporate viral and host cell RNA during the assembly process. J Virol 74:9078–9082

Gruter P, Tabernero C, von Kobbe C, Schmitt C, Saavedra C, Bachi A, Wilm M, Felber BK, Izaurralde E (1998) TAP, the human homolog of Mex67p, mediates CTE-dependent RNA export from the nucleus. Mol Cell 1:649–659

Gustin KE (2003) Inhibition of nucleo-cytoplasmic trafficking by RNA viruses: targeting the nuclear pore complex. Virus Res 95:35–44

Haffar O, Bukrinsky M (2005) Nuclear translocation as a novel target for anti-HIV drugs. Expert Rev Anti Infect Ther 3:41–50

Haffar O, Dubrovsky L, Lowe R, Berro R, Kashanchi F, Godden J, Vanpouille C, Bajorath J, Bukrinsky M (2005) Oxadiazols: a new class of rationally designed anti-human immunodeficiency virus compounds targeting the nuclear localization signal of the viral matrix protein. J Virol 79:13028–13036

Haffar OK, Smithgall MD, Popov S, Ulrich P, Bruce AG, Nadler SG, Cerami A, Bukrinsky MI (1998) CNI-H0294, a nuclear importation inhibitor of the human immunodeficiency virus type 1 genome, abrogates virus replication in infected activated peripheral blood mononuclear cells. Antimicrob Agents Chemother 42:1133–1138

Hayashi ML, Blankenship C, Shenk T (2000) Human cytomegalovirus UL69 protein is required for efficient accumulation of infected cells in the G1 phase of the cell cycle. Proc Natl Acad Sci USA 97:2692–2696

He YS, Xu L, Huang ES (1992) Characterization of human cytomegalovirus UL84 early gene and identification of its putative protein product. J Virol 66:1098–1108

Henderson BR, Percipalle P (1997) Interactions between HIV Rev and nuclear import and export factors: the Rev nuclear localisation signal mediates specific binding to human importin-beta. J Mol Biol 274:693–707

Hiriart E, Farjot G, Gruffat H, Nguyen MV, Sergeant A, Manet E (2003) A novel nuclear export signal and a REF interaction domain both promote mRNA export by the Epstein-Barr virus EB2 protein. J Biol Chem 278:335–342

Hope TJ, Huang XJ, McDonald D, Parslow TG (1990) Steroid-receptor fusion of the human immunodeficiency virus type 1 Rev transactivator: mapping cryptic functions of the arginine-rich motif. Proc Natl Acad Sci USA 87:7787–7791

Hutten S, Kehlenbach RH (2007) CRM1-mediated nuclear export: to the pore and beyond. Trends Cell Biol 17:193–201

Kang Y, Cullen BR (1999) The human Tap protein is a nuclear mRNA export factor that contains novel RNA-binding and nucleocytoplasmic transport sequences. Genes Dev 13:1126–1139

Kapadia F, Pryor A, Chang TH, Johnson LF (2006) Nuclear localization of poly(A)+ mRNA following siRNA reduction of expression of the mammalian RNA helicases UAP56 and URH49. Gene 384:37–44

Kaplan CD, Morris JR, Wu C, Winston F (2000) Spt5 and spt6 are associated with active transcription and have characteristics of general elongation factors in *D. melanogaster*. Genes Dev 14:2623–2634

Katahira J, Strasser K, Podtelejnikov A, Mann M, Jung JU, Hurt E (1999) The Mex67p-mediated nuclear mRNA export pathway is conserved from yeast to human. EMBO J 18:2593–2609

Katahira J, Straesser K, Saiwaki T, Yoneda Y, Hurt E (2002) Complex formation between Tap and p15 affects binding to FG-repeat nucleoporins and nucleocytoplasmic shuttling. J Biol Chem 277:9242–9246

Kiesler E, Miralles F, Visa N (2002) HEL/UAP56 binds cotranscriptionally to the Balbiani ring pre-mRNA in an intron-independent manner and accompanies the BR mRNP to the nuclear pore. Curr Biol 12:859–862

Koffa MD, Clements JB, Izaurralde E, Wadd S, Wilson SA, Mattaj IW, Kuersten S (2001) Herpes simplex virus ICP27 protein provides viral mRNAs with access to the cellular mRNA export pathway. EMBO J 20:5769–5778

Kudo N, Matsumori N, Taoka H, Fujiwara D, Schreiner EP, Wolff B, Yoshida M, Horinouchi S (1999) Leptomycin B inactivates CRM1/exportin 1 by covalent modification at a cysteine residue in the central conserved region. Proc Natl Acad Sci U S A 96:9112–9117

La Cour T, Gupta R, Rapacki K, Skriver K, Poulsen FM, Brunak S (2003) NESbase version 1.0: a database of nuclear export signals. Nucleic Acids Res 31:393–396

Lange A, Mills RE, Lange CJ, Stewart M, Devine SE, Corbett AH (2007) Classical nuclear localization signals: definition, function, and interaction with importin alpha. J Biol Chem 282:5101–5105

Le Hir H, Izaurralde E, Maquat LE, Moore MJ (2000) The spliceosome deposits multiple proteins 20–24 nucleotides upstream of mRNA exon-exon junctions. EMBO J 19:6860–6869

Liker E, Fernandez E, Izaurralde E, Conti E (2000) The structure of the mRNA export factor TAP reveals a cis arrangement of a non-canonical RNP domain and an LRR domain. EMBO J 19:5587–5598

Lischka P, Stamminger T (2006) Regulation of viral mRNA export from the nucleus. In: Reddehase M (ed) Cytomegaloviruses: molecular biology and immunology. Norfolk, UK, Caister Academic Press

Lischka P, Rosorius O, Trommer E, Stamminger T (2001) A novel transferable nuclear export signal mediates CRM1-independent nucleocytoplasmic shuttling of the human cytomegalovirus transactivator protein pUL69. EMBO J 20:7271–7283

Lischka P, Sorg G, Kann M, Winkler M, Stamminger T (2003a) A nonconventional nuclear localization signal within the UL84 Protein of human cytomegalovirus mediates nuclear import via the importin alpha/beta pathway. J Virol 77:3734–3748

Lischka P, Rauh C, Mueller R, Stamminger T (2006a) Human cytomegalovirus UL84 protein contains two nuclear export signals and shuttles between the nucleus and the cytoplasm. J Virol 80:10274–10280

Lischka P, Toth Z, Thomas M, Mueller R, Stamminger T (2006b) The UL69 transactivator protein of human cytomegalovirus interacts with DEXD/H-Box RNA helicase UAP56 to promote cytoplasmic accumulation of unspliced RNA. Mol Cell Biol 26:1631–1643

Lischka P, Thomas M, Toth Z, Mueller R, Stamminger T (2007) Multimerization of human cytomegalovirus regulatory protein UL69 via a domain that is conserved within its herpesvirus homologues. J Gen Virol 88:405–410

Liu J, Farmer JD Jr, Lane WS, Friedman J, Weissman I, Schreiber SL (1991) Calcineurin is a common target of cyclophilin-cyclosporin A and FKBP-FK506 complexes. Cell 66:807–815

Lu M, Shenk T (1999) Human cytomegalovirus UL69 protein induces cells to accumulate in G1 phase of the cell cycle. J Virol 73:676–683

Luo ML, Zhou Z, Magni K, Christoforides C, Rappsilber J, Mann M, Reed R (2001) Pre-mRNA splicing and mRNA export linked by direct interactions between UAP56 and Aly. Nature 413:644–647

Malik P, Blackbourn DJ, Clements JB (2004) The evolutionarily conserved Kaposi's sarcoma-associated herpesvirus ORF57 protein interacts with REF protein and acts as an RNA export factor. J Biol Chem 279:33001–33011

Maniatis T, Reed R (2002) An extensive network of coupling among gene expression machines. Nature 416:499–506

Mears WE, Rice SA (1996) The RGG box motif of the herpes simplex virus ICP27 protein mediates an RNA-binding activity and determines in vivo methylation. J Virol 70:7445–7453

Mears WE, Rice SA (1998) The herpes simplex virus immediate-early protein ICP27 shuttles between nucleus and cytoplasm. Virology 242:128–137

Michael WM, Choi M, Dreyfuss G (1995) A nuclear export signal in hnRNP A1: a signal-mediated, temperature-dependent nuclear protein export pathway. Cell 83:415–422

Nakielny S, Dreyfuss G (1999) Transport of proteins and RNAs in and out of the nucleus. Cell 99:677–690

Neville M, Stutz F, Lee L, Davis LI, Rosbash M (1997) The importin-beta family member Crm1p bridges the interaction between Rev and the nuclear pore complex during nuclear export. Curr Biol 7:767–775

Palmeri D, Malim MH (1999) Importin beta can mediate the nuclear import of an arginine-rich nuclear localization signal in the absence of importin alpha. Mol Cell Biol 19:1218–1225

Pari GS, Anders DG (1993) Eleven loci encoding trans-acting factors are required for transient complementation of human cytomegalovirus oriLyt-dependent DNA replication. J Virol 67:6979–6988

Pollard VW, Michael WM, Nakielny S, Siomi MC, Wang F, Dreyfuss G (1996) A novel receptor-mediated nuclear protein import pathway. Cell 86:985–994

Pryor A, Tung L, Yang Z, Kapadia F, Chang TH, Johnson LF (2004) Growth-regulated expression and G0-specific turnover of the mRNA that encodes URH49, a mammalian DExH/D box protein that is highly related to the mRNA export protein UAP56. Nucleic Acids Res 32:1857–1865

Radu A, Blobel G, Moore MS (1995) Identification of a protein complex that is required for nuclear protein import and mediates docking of import substrate to distinct nucleoporins. Proc Natl Acad Sci USA 92:1769–1773

Ramasawmy R, Cunha-Neto E, Fae KC, Muller NG, Cavalcanti VL, Drigo SA, Ianni B, Mady C, Kalil J, Goldberg AC (2006) BAT1, a putative anti-inflammatory gene, is associated with chronic Chagas cardiomyopathy. J Infect Dis 193:1394–1399

Reichelt R, Holzenburg A, Buhle EL Jr, Jarnik M, Engel A, Aebi U (1990) Correlation between structure and mass distribution of the nuclear pore complex and of distinct pore complex components. J Cell Biol 110:883–894

Rodrigues JP, Rode M, Gatfield D, Blencowe BJ, Carmo-Fonseca M, Izaurralde E (2001) REF proteins mediate the export of spliced and unspliced mRNAs from the nucleus. Proc Natl Acad Sci USA 98:1030–1035

Sandri-Goldin RM (1998) ICP27 mediates HSV RNA export by shuttling through a leucine-rich nuclear export signal and binding viral intronless RNAs through an RGG motif. Genes Dev 12:868–879

Sandri-Goldin RM (2001) Nuclear export of herpes virus RNA. Curr Top Microbiol Immunol 259:2–23

Sandri-Goldin RM (2004) Viral regulation of mRNA export. J Virol 78:4389–4396

Sandri-Goldin RM, Mendoza GE (1992) A herpesvirus regulatory protein appears to act post-transcriptionally by affecting mRNA processing. Genes Dev 6:848–863

Sarisky RT, Hayward GS (1996) Evidence that the UL84 gene product of human cytomegalovirus is essential for promoting oriLyt-dependent DNA replication and formation of replication compartments in cotransfection assays. J Virol 70:7398–7413

Satterly N, Tsai PL, van Deursen J, Nussenzveig DR, Wang Y, Faria PA, Levay A, Levy DE, Fontoura BM (2007) Influenza virus targets the mRNA export machinery and the nuclear pore complex. Proc Natl Acad Sci USA 104:1853–1858

Shibasaki F, Price ER, Milan D, McKeon F (1996) Role of kinases and the phosphatase calcineurin in the nuclear shuttling of transcription factor NF-AT4. Nature 382:370–373

Smith RW, Malik P, Clements JB (2005) The herpes simplex virus ICP27 protein: a multifunctional post-transcriptional regulator of gene expression. Biochem Soc Trans 33:499–501

Spector DJ, Tevethia MJ (1994) Protein-protein interactions between human cytomegalovirus IE2–580aa and pUL84 in lytically infected cells. J Virol 68:7549–7553

Stade K, Ford CS, Guthrie C, Weis K (1997) Exportin 1 (Crm1p) is an essential nuclear export factor. Cell 90:1041–1050

Stewart M (2007) Molecular mechanism of the nuclear protein import cycle. Nat Rev Mol Cell Biol 8:195–208

Strasser K, Hurt E (2001) Splicing factor Sub2p is required for nuclear mRNA export through its interaction with Yra1p. Nature 413:648–652

Strasser K, Masuda S, Mason P, Pfannstiel J, Oppizzi M, Rodriguez-Navarro S, Rondon AG, Aguilera A, Struhl K, Reed R, Hurt E (2002) TREX is a conserved complex coupling transcription with messenger RNA export. Nature 417:304–308

Toth Z, Lischka P, Stamminger T (2006) RNA-binding of the human cytomegalovirus transactivator protein UL69, mediated by arginine-rich motifs, is not required for nuclear export of unspliced RNA. Nucleic Acids Res 34:1237–1249

Wang P, Palese P, O'Neill RE (1997) The NPI-1/NPI-3 (karyopherin alpha) binding site on the influenza a virus nucleoprotein NP is a nonconventional nuclear localization signal. J Virol 71:1850–1856

Wen W, Meinkoth JL, Tsien RY, Taylor SS (1995) Identification of a signal for rapid export of proteins from the nucleus. Cell 82:463–473

Williams BJ, Boyne JR, Goodwin DJ, Roaden L, Hautbergue GM, Wilson SA, Whitehouse A (2005) The prototype gamma-2 herpesvirus nucleocytoplasmic shuttling protein, ORF 57, transports viral RNA through the cellular mRNA export pathway. Biochem J 387:295–308

Winkler M, Rice SA, Stamminger T (1994) UL69 of human cytomegalovirus, an open reading frame with homology to ICP27 of herpes simplex virus, encodes a transactivator of gene expression. J Virol 68:3943–3954

Winkler M, Stamminger T (1996) A specific subform of the human cytomegalovirus transactivator protein pUL69 is contained within the tegument of virus particles. J Virol 70:8984–8987

Winkler M, aus Dem ST, Stamminger T (2000) Functional interaction between pleiotropic transactivator pUL69 of human cytomegalovirus and the human homolog of yeast chromatin regulatory protein SPT6. J Virol 74:8053–8064

Wolff T, Unterstab G, Heins G, Richt JA, Kann M (2002) Characterization of an unusual importin alpha binding motif in the Borna disease virus p10 protein that directs nuclear import. J Biol Chem 277:12151–12157

Xu Y, Colletti KS, Pari GS (2002) Human cytomegalovirus UL84 localizes to the cell nucleus via a nuclear localization signal and is a component of viral replication compartments. J Virol 76:8931–8938

Xu Y, Cei SA, Huete AR, Pari GS (2004a) Human cytomegalovirus UL84 insertion mutant defective for viral DNA synthesis and growth. J Virol 78:10360–10369

Xu Y, Cei SA, Rodriguez HA, Colletti KS, Pari GS (2004b) Human cytomegalovirus DNA replication requires transcriptional activation via an IE2- and UL84-responsive bidirectional promoter element within oriLyt. J Virol 78:11664–11677

Yoh SM, Cho H, Pickle L, Evans RM, Jones KA (2007) The Spt6 SH2 domain binds Ser2-P RNAPII to direct Iws1-dependent mRNA splicing and export. Genes Dev 21:160–174

Yu D, Silva MC, Shenk T (2003) Functional map of human cytomegalovirus AD169 defined by global mutational analysis. Proc Natl Acad Sci USA 100:12396–12401

Zenklusen D, Vinciguerra P, Wyss JC, Stutz F (2002) Stable mRNP formation and export require cotranscriptional recruitment of the mRNA export factors Yra1p and Sub2p by Hpr1p. Mol Cell Biol 22:8241–8253

Structure and Formation of the Cytomegalovirus Virion

W. Gibson

Contents

Introduction 187
Formation of the Nucleocapsid 190
Tegumentation and Envelopment 196
Concluding Thoughts 199
References 199

Abstract Transport and protection of the nuclear-replicating double-stranded DNA genome of herpesviruses is accomplished by the virion and its substructures. Studies of the composition, organization, and formation of these particles have provided insight into the molecular mechanisms of virus assembly, leads for antiviral strategies, and information about cellular processes that are required for, resemble, or antagonize virus replication. This chapter updates earlier reviews on the structure and formation human cytomegalovirus (HCMV) virions (Gibson 1996, 2006; Eickmann et al. 2006), and complements several other reviews on herpesvirus structure and replication presented in this volume (see the chapters by E. Murphy and T. Shenk, Z. Ruzsics and U. Koszinowski, R. Kalejta, and G.S. Pari) and elsewhere (Rixon 1993; Steven and Spear 1997; Brown et al. 2002; Varnum et al. 2004; Liu and Zhou 2007).

Introduction

For purposes of brief introduction, the general characteristics of the CMV virion can be summarized as follows. Typical of the herpesvirus group, the virion of HCMV is approximately 230 nm in diameter and is composed of a nucleocapsid, surrounded by

---

W. Gibson
Department of Pharmacology Molecular Sciences, Johns Hopkins University School of Medicine, Baltimore, MD 21205, USA
wgibson@jhmi.edu

a less structured tegument layer, and bounded by a trilaminate membrane envelope (Fig. 1a). The HCMV genome is composed of a linear, double-stranded DNA molecule (236 kbp in wild type virus), the largest among the human herpesviruses, and over 50% larger than that of herpes simplex virus type 1 (HSV-1) (see the chapters by E. Murphy and T. Shenk, and G.S. Pari, this volume). The capsid is isosahedral and about the same diameter as that of HSV (~110 nm, depending on preparation). Accommodating a larger DNA in a similar diameter capsid may be achieved by eliminating the maturational protease (pUL80a) from the interior of CMV capsids (Chan et al. 2002; Loveland et al. 2007). The capsid is composed of four integral protein species (for HCMV, pUL46, pUL80.5, pUL85, pUL104) that are organized into 162 capsomeres (150 hexamers plus 12 pentamers) and 320 triplexes located between the capsomeres. By analogy with HSV-1, one of the pentamer positions is

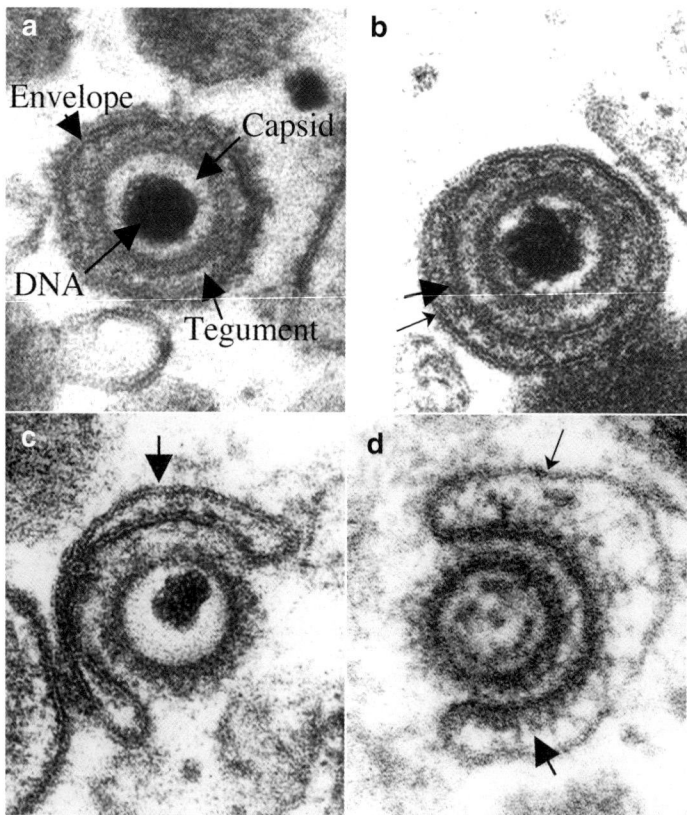

Fig. 1 Particles in cytoplasm of CMV-infected cells. Shown here are electron micrographs of **a** virion with DNA, capsid, tegument, and envelope indicated by arrows, **b** virion within small vesicle or tubule indicated by thinner arrow, **c** tegumented C-capsid, with coarse fibrillar material especially evident on right-hand side, budding into a vesicle or tubule (*arrow*) to become virion, and **d** tegumented B-capsid budding into large tubule or vesicle (*top arrow*) to become NIEP; showing thickening of vesicle membrane where apposed to particle (*bottom arrow*)

shared with or occupied by a portal complex through which DNA enters and leaves the capsid. The tegument region is approximately 50 nm thick and includes seven relatively abundant virus-encoded protein species (see the chapter by R. Kalejta, this volume), at least five of which are phosphorylated. The virion envelope is estimated to be 10 nm thick and contains at least ten abundant protein species. Both the tegument and envelope contain additional less abundant virus-encoded and host-cell proteins, as well as phospholipids, polyamines, and small RNAs.

It is worth noting that determinations of particle composition ultimately depend on the nature of the starting material, which can be influenced by its source and method of preparation. Comparisons of virus particles from different origins, recovered by different methods, and analyzed by different and increasingly sensitive procedures, are focusing even more attention on the challenge of establishing which constituents are integral and present in all particles.

In addition to virions, five other types of virus particles have been recovered from CMV-infected cells and characterized. Three are intracellular and nonenveloped. A- and B-capsids are from nuclei prepared by treating infected cells with NP-40 and have counterparts among the other herpesviruses. A-capsids are shells composed of the four integral protein species and have the simplest structure. B-capsids contain all of the A-capsid proteins and several additional internal scaffolding species (UL80 proteins, Fig. 2). C-capsids are recovered from the cytoplasmic fraction of infected cells treated with NP-40 (Gibson and Roizman 1972; Gibson 1981) and are composed of the DNA genome within an A-capsid shell having some tightly adherent tegument proteins (e.g., pUL32, pUL47, pUL48). The other

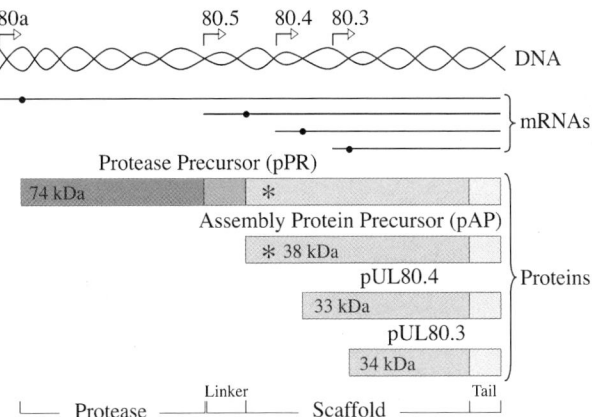

Fig. 2 Nested organization of HCMV UL80 genes and proteins. Shown here is a schematic representation of the nested UL80 genes (*top*, DNA helix; *arrows* indicate separate promoters for each); their 3′ co-terminal mRNAs (*dots* indicate AUG codons starting translation); and the in-frame, carboxy-co-terminal proteins translated from each mRNA (size, name, and abbreviation indicated). *Shading* illustrates portions of each protein shared by the others and the location of the linker and tail portions of the scaffolding domain; *asterisk* indicates location of amino-conserved domain. (Adapted from Gibson 2006)

two types of particles are enveloped and recovered from the culture medium of CMV-infected cells; neither contains DNA. Noninfectious enveloped particles (NIEPs) are enveloped B-capsids that closely resemble virions in structure and composition, but retain the internal B-capsid scaffolding proteins. Dense bodies (DBs), which differ from NIEPs and virions by their larger and more heterogeneous size (~250-600 nm) and by the absence of all nucleocapsid constituents, are solid spheroidal aggregates of a single predominant tegument protein species (i.e., pUL83), surrounded by an envelope so far undistinguished from that of the virion. More detailed descriptions of these particles are presented in earlier reports and reviews (Irmiere and Gibson 1983, 1985; Gibson and Irmiere 1984; Gibson 1996).

Formation of the Nucleocapsid

Capsid assembly is coordinated by the assembly protein precursor (pAP, pUL80.5, 38 kDa) and the genetically related protease precursor (pPR, pUL80a, 74 kDa), both of which are ultimately eliminated from the maturing particle. These proteins are encoded by 3'-coterminal in-frame genes, with the consequence that the carboxyl approximately 60% of pPR is identical to pAP (Fig. 2). Two smaller proteins encoded by the same set of genes (pUL80.4 and pUL80.3) have unknown functions that are dispensable for growth in cell culture (N. Nguyen and W. Gibson, unpublished data) from mutant viruses having one or both translational start methionines replaced with isoleucines. Key amino acid sequences within these proteins that enable their function are illustrated in Fig. 3.

A working model of CMV capsid formation is illustrated in Fig. 4. The earliest steps in the assembly process begin in the cytoplasm. One pathway leads to protocapsomers and is initiated when the amino conserved domain (ACD) promotes pAP

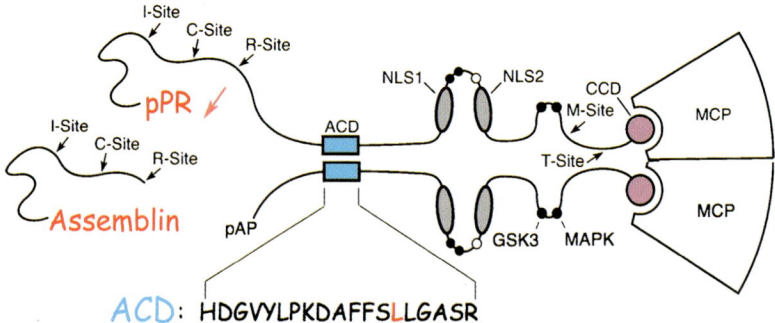

Fig. 3 Landmarks on the assembly protein and protease precursors. The assembly protein precursor (*pAP*) has the same amino acid sequence as the carboxyl half of the protease precursor (*pPR*) and includes the following sequences of interest: the amino conserved domain (*ACD*), which promotes self-interaction of pAP and pPR; the carboxyl conserved domain (*CCD*), which promotes

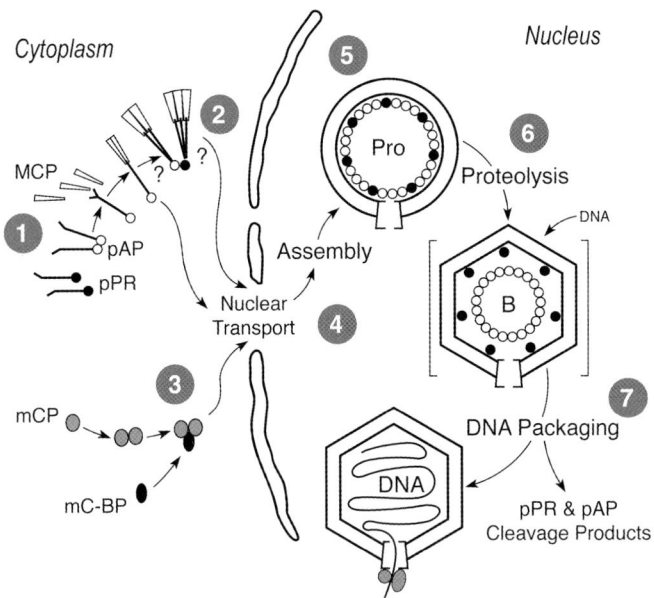

Fig. 4 Capsid assembly model. Shown here are interactions between the major capsid protein (*MCP*, pUL85, *narrow trapezoids*), assembly protein precursor (*pAP*, pUL80.5 *lines with empty circles*), and protease precursor (*pPR*, pUL80a, *lines with filled circles*), and some of the putative complexes they form (*1*). The largest represents a complete capsomer precursor (protocapsomer) (*2*), but there is no direct evidence that its cytoplasmic assembly reaches completion. The minor capsid protein (*mCP*, pUL85, 35 kDa, *light ovals*) and mCP-binding protein (*mCBP*, pUL46, 33 kDa, *darker oval*) also interact with each other in the cytoplasm to form heterotrimers, called triplexes (*3*). The two types of oligomers are translocated into the nucleus (*4*) and coalesce to form the procapsid (*Pro*), incorporating the portal-protein complex (pUL104, 78 kDa, *broken trapezoid at bottom of capsid*) (*5*). The terminase subunits are indicated by *shaded ellipses* below the portal complex. Activation of pPR (*6*) results in cleavage and elimination of the internal scaffolding proteins (pPR and pAP) from the capsid, before or during the process of DNA packaging (*7*). Brackets around the B-capsid (*B*) indicate uncertainty about the nature of putative intermediate(s) between procapsids and DNA-containing nucleocapsids. (Modified from Gibson 2006)

Fig. 3 (continued) interaction of pAP and pPR with MCP; nuclear localization signals 1 and 2 (*NLS1, NLS2*); casein kinase II phosphorylation site (*black dots* between NLS1 and NLS2) of undetermined significance; mitogen-activated protein kinase (*MAPK*) and glycogen synthase kinase 3 (*GSK-3*) sites whose phosphorylation antagonizes pAP self-interaction (Casaday et al. 2004); and the five pPR self-cleavage sites: the maturational site (*M site*, VNA↓S), which severs the linkage of pPR and pAP to MCP; the release site (*R site*, YVKA↓S), which separates the proteolytic domain (assemblin) of pPR from the scaffolding portion (carboxyl end); the internal site (*I site*, VEA↓A), which converts assemblin from an active single-chain form to a two-chain form that retains activity, the cryptic site (*C site*, VDA↓S) that interrupts the assemblin dimer interface, and the tail site (*T site*, VLA↓A) detected upon refolding denatured pPR (Brignole and Gibson 2007). Also shown is the amino acid sequence of the ACD and location of the critical Leu47 (*red*; Leu382 in context of pPR sequence) within it. (Modified from Gibson 2006)

self-interaction, which in turn potentiates or stabilizes pAP binding through its carboxyl conserved domain (CCD) to the major capsid protein (MCP, pUL86) (Wood et al. 1997; see Figs. 3 and 4, step 1). This interaction enables MCP, which lacks its own nuclear localization sequence (NLS), to be translocated into the nucleus (Fig. 4, step 4) as part of the pAP-MCP complex via two NLSs present in pAP (Wood et al. 1997; Plafker and Gibson 1998). Because its carboxyl end is identical to pAP (Figs. 2, 3), the proteinase precursor can interact with itself and pAP through its ACD, and with MCP through its CCD. Mimicking these pAP interactions ensures incorporation of pPR into the capsid cavity, where its enzymatic function is required. The composition and variety of the complexes formed (e.g., pAP-MCP; pAP-pPR-MCP; pPR-MCP) and the extent to which they progress in the cytoplasm toward completely preassembled protocapsomers (Fig. 4, step 2) is unknown.

The biological relevance of ACD-mediated pAP self-interaction was established by using mutant viruses, which were found to replicate slowly, assemble nucleocapsids inefficiently, and yield roughly 20-fold less virus than wild type (Loveland et al. 2007). Mutant viruses have also been used to verify the biological requirement for pAP NLS, by showing that loss of both is lethal, loss of either one alone slows nuclear translocation of MCP, and loss of NLS2 inhibits virus replication more profoundly than loss of NLS1 (Nguyen et al. 2007). Restriction of NLS2 to the β-herpesvirus pAP homologs and its comparatively greater impact on virus replication suggests it may have a different or additional group-specific function.

Through a similar subunit assembly process, the triplex proteins associate in the cytoplasm before translocation into the nucleus (Fig. 4, step 3) (Baxter and Gibson 1997; Spencer et al. 1998). Like MCP, the minor capsid protein (mCP, pUL85) does not enter the nucleus when expressed alone in transfected cells, even though its size is small enough to be allowed entry by diffusion. The similar-size mCBP, in contrast, does enter the nucleus on its own and when mCP and the mCP-binding protein (mCBP, pUL46) are expressed together, they co-localize to the nucleus (Baxter and Gibson 1997). This is attributed to the rapid formation of approximately 70-kDa mCP homodimers that require interaction with an NLS-bearing mCBP for nuclear translocation. These sequential cytoplasmic interactions of pAP and pPR with each other and with MCP, and of mCP with itself and with mCBP, initiate and direct the assembly process and consequently have the potential to help modulate it. Cytoplasmic preassembly may also increase the fidelity and efficiency of procapsid formation by ensuring delivery into the nucleus of correctly organized capsid substructures.

The HCMV portal protein (pUL104, 78 kDa) may also interact with pAP, as it does in HSV (Singer et al. 2005). However, little is known about when and where its self-interaction and interaction with pAP occur. Unlike MCP, the portal protein contains its own NLS (Patel and MacLean 1995; Patel et al. 1996) and would not seem to require interaction with pAP as an NLS-bearing nuclear-translocation escort.

Following translocation into the nucleus, the pAP-MCP, pPR-MCP, and pAP-pPR-MCP complexes or protocapsomers interact more extensively with one another and associate with the triplexes and portal protein complex to form procapsids (Fig. 4, step 5). These unstable particles (Newcomb et al. 1999; Rixon and McNab 1999), first evidenced in HSV (Newcomb et al. 1999; Rixon and

McNab 1999; Newcomb et al. 2000) and constituted or closely approximated in vitro (Newcomb et al. 1999, 2000), are less angular than late-stage capsids and contain scaffolding proteins but no viral DNA. An involvement of the pAP amino-conserved domain during this nuclear stage of capsid formation was discovered with the L47A mutant virus, which showed a dramatically altered distribution of pAP within the nucleus (Fig. 5) and gave rise to overall fewer and aberrant capsids (Loveland et al. 2007).

Absence of detected procapsid formation in the cytoplasm, where all of the necessary proteins are present, may be explained by the comparatively higher protein concentrations in the nucleus or by the presence of specific nuclear initiating or enhancing factors. Alternatively or additionally, there may be assembly-enhancing changes in the complexes that signal or result from nuclear translocation. In HSV, where it has been possible to examine capsid assembly in isolation from other viral proteins, the homologs of MCP, pAP, mCP, and mCBP (i.e., HSV VP5, preVP22a, VP23, VP19c) are all necessary and collectively sufficient to assemble the capsid shell (Tatman et al. 1994; Thomsen et al. 1994). Similar attempts to make CMV capsids from their recombinantly cloned and expressed counterpart genes have not yet succeeded and it is unresolved whether this is due to technical factors (e.g., CMV protein expression levels too low) or perhaps to differences in the minimal compliment of proteins required.

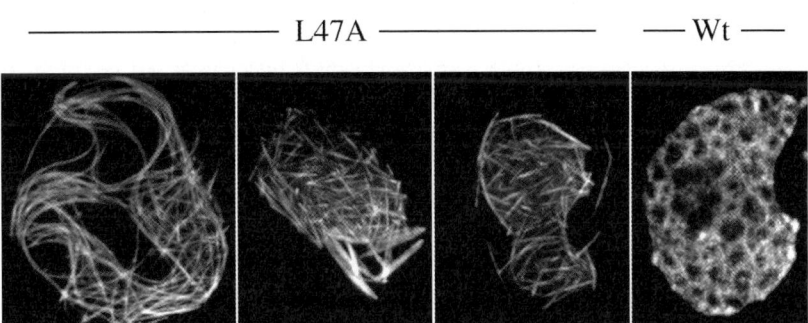

Fig. 5 Distribution of UL80 proteins in nuclei of HCMV-infected cells. An HCMV-AD169 bacmid was mutated to block self-interaction of the UL80 proteins (i.e., point mutation L47A in pAP sequence) (Loveland et al. 2007). Both the HCMV-AD169 bacmid and the parental wild-type bacmid were also modified to incorporate a tetra-cysteine tag (CCPGCC) into the UL80 proteins. Virus derived from each bacmid was used to infect human foreskin fibroblasts, which were stained with the biarsenical dye FlAsH when strong cytopathic effects were observed. Shown here are images of the stained, living cells taken by confocal fluorescence microscopy. The FlAsH-stained UL80 proteins are organized in tubular and rod-shaped structures in nuclei of cells infected with the L47A mutant (*first three panels from the left*), but in an entirely different pattern reminiscent of the intranuclear inclusions typically observed in nuclei of cells infected with wild type virus (*right-hand panel*). (Modified from Loveland et al. 2007)

Taking procapsids as the end point of capsid assembly, two major changes occur during its maturation: elimination of the scaffolding proteins and incorporation of viral DNA. The two processes appear to be coupled. Elimination of the internal scaffolding proteins is effected by proteolysis catalyzed by pPR (Fig. 4, step 6). Although able to cleave itself at five sites, pPR activity appears to be modulated during infection such that its autoproteolysis is initiated following procapsid formation. It has recently been determined using purified HCMV pPR that the active enzyme is a trimer or tetramer whose primary subunit interaction is through the amino-conserved domain of its scaffolding region (Brignole and Gibson 2007). This quaternary structure is very different from that of purified assemblin, whose monomer activates by dimerizing through sequences located in its carboxyl end (Chen et al. 1996; Cole 1996; Darke et al. 1996; Margosiak et al. 1996; Qiu et al. 1996; Shieh et al. 1996; Tong et al. 1996). Active pPR cleaves the R site to release the well-characterized proteolytic domain, assemblin, and cleaves the M site to sever the tail sequence linking itself and pAP to MCP in the capsid shell (Figs. 2, 3).

Although having comparable overall enzymatic rates (e.g., k_{cat}/K_M), pPR and assemblin are distinguished in ways thought to reflect mechanistic differences (Brignole and Gibson 2007). First, whereas imidazole can chemically rescue the enzymatic activity of assemblin substituted with Gly at its catalytically critical His63, it fails to restore activity to the same mutation in pPR (McCartney et al. 2005), indicating a difference in the catalytic sites of the two forms of the enzyme. Second, the sequences driving assemblin dimerization have comparatively little effect on pPR oligomerization, yet this interaction of assemblin is thought to induce catalytic-site changes required for its activity (Tong et al. 1996; Buisson et al. 2002). And third, there is evidence that the enzymatic rates of pPR and assemblin may be comparable because of offsetting differences in their k_{cat} (higher for pPR) and K_M, again suggesting catalytic-site differences between pPR and assemblin that may be important to regulating activity.

Proteolysis results in essentially all M and R sites being cleaved. HCMV pPR makes three additional cleavages. Two are at the internal (I) and cryptic (C) sites within assemblin and reduce production of infectious virus when blocked (Chan et al. 2002; Loveland et al. 2005). These cleavages reduce the size and interactions of the scaffolding proteins, facilitating their elimination from the capsid in preparation for DNA packaging (Fig. 4, step 7). Unlike HSV, which retains assemblin in its mature virion, all remnants of HCMV pAP and pPR, including assemblin, are eliminated from the capsid. Absence of counterpart I and C sites in the HSV assemblin homolog, and persistence of HSV assemblin in the mature virion, may reflect a requirement for additional space within the CMV capsid to accommodate its 51% longer DNA genome (Chan et al. 2002; Loveland et al. 2005). The significance of a fifth cleavage site just discovered in the carboxyl tail (T site) of purified pPR is undetermined. Electrostatic repulsion by the incoming viral DNA has been suggested to play a role in displacing the internal scaffolding proteins (McClelland et al. 2002), and specific pAP phosphorylations that weaken scaffolding protein self-interactions may by important to this process (Casaday et al. 2004; Gibson 2006).

Maturational proteolysis converts at least some procapsids to B-capsids (Fig. 6b), which differ by having an angular appearance (instead of round), and containing cleaved forms of the internal scaffolding proteins (instead of precursors). B-capsids are depicted in Fig. 4 as intermediates in the assembly pathway, but it has been difficult to demonstrate their maturation to DNA-containing nucleocapsids (O'Callaghan and Randall 1976; Ladin et al. 1982; Lee et al. 1988; Sherman and Bachenheimer 1988; Church and Wilson 1997). One explanation is that only a small percentage of the relatively large B-capsid pool incorporates DNA, making their loss from the pool difficult to detect. Moreover, once DNA packaging begins, the particles involved may become compositionally heterogeneous (e.g., a decreasing amount of scaffolding proteins and an increasing amount of DNA) and, consequently, escape detection by methods routinely used to recover and characterize capsids (e.g., sedimentation and equilibrium centrifugation). An alternate and plausible interpretation is that B-capsids are formed when early steps in the procapsid maturation process fail, such as timely cleavage and elimination of the scaffolding proteins or successful initiation of DNA packaging.

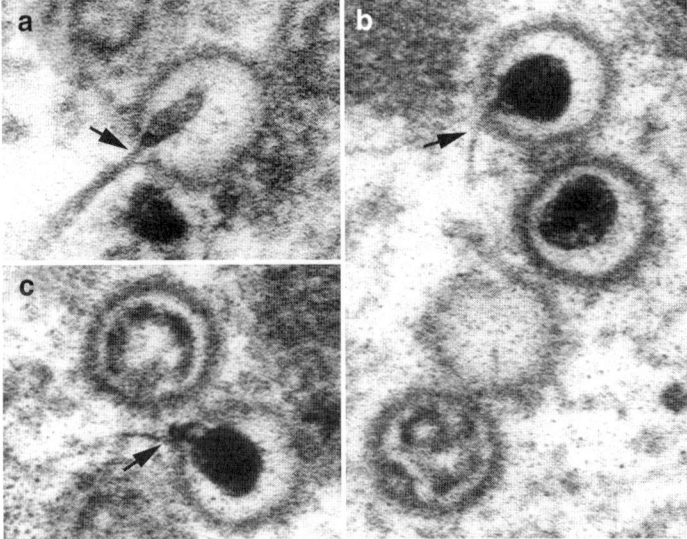

Fig. 6 Capsids in nucleus of CMV-infected cells. These selected particles from electron micrographs of CMV-infected human foreskin fibroblasts show features consistent with the DNA packaging scheme illustrated in Fig. 4. **a** Capsid at top appears to be at comparatively early stage of DNA packaging. The elongated putative nascent core is smaller and less electron dense than those at the top of panel b and bottom of panel c, and is off center to the extent of appearing to touch the inner wall of the capsid and to be continuous with more filamentous material extending through the capsid (presumably via portal) into the nucleoplasm (*arrow*). **b** Capsid at lower left appears to contain scaffolding proteins absent in the particle just above it. Capsid at top contains DNA core with filamentous extension through the capsid shell and into the nucleoplasm (*arrow*). **c** Lower capsid appears to contain scaffolding proteins and upper capsid is interpreted to be nearly finished packaging its DNA, the end of which may extend out to the left of the capsid (*arrow*)

By analogy with the bacteriophage DNA packaging system, once pAP and pPR are eliminated, DNA is incorporated with the help of the terminase-portal complex (Fig. 4, step 7). Examples of intranuclear capsids apparently involved in this process are shown in Fig. 6. The portal protein forms a 12-subunit homo-oligomeric ring at a single vertex of the capsid through which the viral DNA can enter and leave (Newcomb et al. 2001; Trus et al. 2004). At 29-Å resolution the HSV portal complex resembles that of bacteriophage (Trus et al. 2004). Its CMV homolog, pUL104, also appears restricted to a single capsid vertex (Dittmer and Bogner 2005). The herpesvirus DNA cleavage/packaging enzyme (terminase) is composed of two subunits (Poon and Roizman 1993; Baines et al. 1994). The larger (HCMV pUL56, 96 kDa) has properties consistent with it being a counterpart of the large subunit of bacteriophage terminase (Bogner et al. 1993, 1995, 1998; Holzenburg and Bogner 2002; Scheffczik et al. 2002): (a) it associates with a smaller subunit (pUL89, 77 kDa), (b) it binds double-stranded viral DNA, and (c) it binds to the capsid (White et al. 2003). Retention of DNA in the capsid is believed to be stabilized by a protein (HCMV pUL7, 71 kDa) whose HSV homolog (pUL25, 60 kDa) has a mutant phenotype that fails to stably package DNA (McNab et al. 1998; Ogasawara et al. 2001; Sheaffer et al. 2001). This protein is considered a possible counterpart of the bacteriophage cap protein, but may exert its effect by binding at multiple sites on the capsid surface (Newcomb et al. 2006).

Tegumentation and Envelopment

The composition of the HCMV tegument and envelope, and the relationship of their acquisition to virus egress from the nucleus and cell, have recently been reviewed (Eickmann et al. 2006). The cartoon shown in Fig. 7 serves to summarize some of the steps involved. A general consensus of data supports an envelopment/de-envelopment mechanism for nuclear egress of herpesviruses, followed by final envelopment through cytoplasmic membranes (Severi et al. 1988; Gibson 1993; Enquist et al. 1998; Mettenleiter 2002; Leuzinger et al. 2005; Campadelli-Fiume and Roizman 2006). As represented in Fig. 7, step 1, primary envelopment of the capsid occurs at the inner nuclear membrane, enabling the particle to pass into the perinuclear space.

Primary envelopment requires two herpesvirus group-conserved proteins. Their HCMV homologs are pUL50 and pUL53; the respective HSV homologs are pUL34 and pUL31 (Klupp et al. 2000; Roller et al. 2000; Reynolds et al. 2001; Fuchs et al. 2002; Reynolds et al. 2002). HSV pUL34 localizes to both the inner and outer leaflets of the nuclear membrane and is a substrate for the virion-associated US3 protein kinase (Purves et al. 1992), which enhances its membrane localization (Purves et al. 1992; Klupp et al. 2000, 2001; Roller et al. 2000; Reynolds et al. 2001, 2002). Although CMV does not encode a US3 homolog, the murine CMV homolog of pUL34 (MCMV M50/p35) interacts with cellular protein kinase C and carries it to the nuclear membrane where it is proposed to phosphorylate the

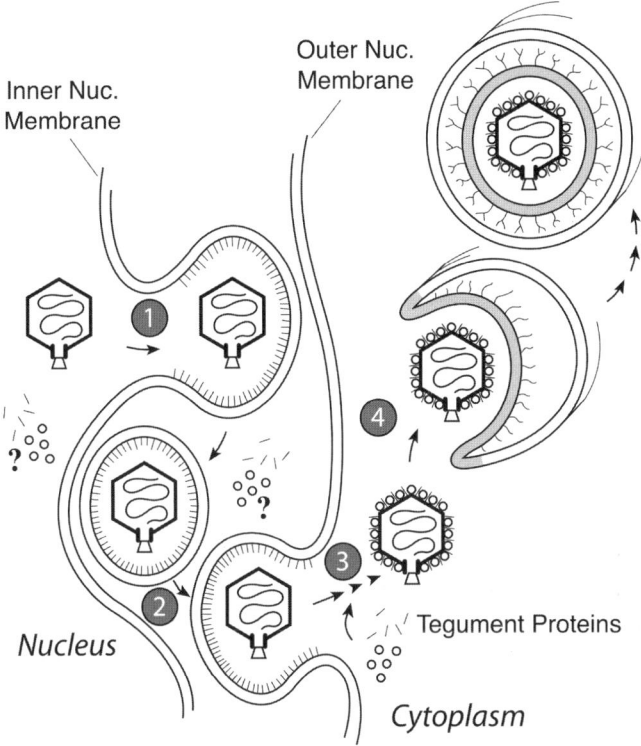

Fig. 7 Egress of nucleocapsid. Shown here is a drawing illustrating some features of the herpesvirus egress pathway. *1* Intranuclear capsids bud through the inner nuclear membrane by a primary envelopment process requiring the homologs of CMV pUL50 and pUL53 (represented by *clustered short lines* on inner and outer nuclear membranes). *2* Resulting enveloped particle in perinuclear space represents a translocation intermediate. *3* De-envelopment at the outer nuclear membrane releases the nucleocapsid into the cytoplasm where it acquires final complement of tegument proteins (*small lines and circles*; also depicted in nucleus and perinuclear space to indicate uncertainty about site(s) of addition). *4* Fully tegumented capsid buds into cytoplasmic vesicles or tubules, through which it completes egress from the cell. The presence of tegumented B-capsids undergoing secondary envelopment in the cytoplasm (see Fig. 1d), indicates this mechanism is not selective for DNA-containing particles

nuclear lamina proteins underlying the inner nuclear membrane, weakening their interaction and dissolving the barrier they pose to capsid egress (Muranyi et al. 2002). Interaction of the membrane-associated member of the nuclear-egress pair (e.g., CMV pUL50) with its nuclear phosphoprotein partner (e.g., CMV pUL53) may then drive primary envelopment (Mettenleiter 2002; Muranyi et al. 2002; Bjerke et al. 2003).

Immature particles in the perinuclear space are then proposed to bud through the outer leaflet of the nuclear membrane, loosing their translocation membrane and becoming nonenveloped cytoplasmic particles (Fig. 7, step 2). Although this general pathway appears to be shared by all herpes viruses, there is less consensus about where the tegument proteins are added.

With regard to the predominant tegument proteins of CMV, three sets of observations are compatible with their proposed addition outside the nucleus (Fig. 7, step 3). First, electron microscopy shows that cytoplasmic capsids have a thick fibrillar coating (deduced to be tegument proteins) that is entirely absent from nuclear capsids (Figs. 1C, 1D, 6; Fig. 1 in Gibson 1993). Second, SDS-PAGE analyses show that capsids recovered from the cytoplasm of infected cells contain the predominant tegument proteins, whereas those from the nucleus do not (Gibson 1981). And third, immunofluorescence studies show accumulations of three abundant tegument proteins (pUL83, pUL32, pUL99) in assembly compartments juxtaposed to the nucleus, but outside of it (Scholl et al. 1988; Sanchez et al. 2000a, 2000b) (reviewed in Eickmann et al. 2006). None of these observations, however, rule out the possibility that some or all of the same tegument proteins bind to capsids within the nucleus or perinuclear space (Hensel et al. 1995; Nii et al. 1998) and are rapidly translocated with the capsid into the cytoplasm where they accumulate to levels more readily detected. Compatible with this possibility, recent studies show that specific mutations in the tegument proteins pUL36 (VP1/2) of HSV, or pUL32 (basic phosphoprotein/pp150) of HCMV, result in their accumulation within the nucleus (O'Hare and Abaitua 2006; J. Wang and W. Gibson, unpublished data from studies using mutant viruses encoding CysCysProGlyCysCys-tagged pUL32 detected in live, infected cells with the biarsenical dye FlAsH).

Once in the cytoplasm and completely tegumented, the capsids bud into cytoplasmic vesicles or tubules to acquire their final envelope (e.g., Fig. 7, step 4). This process required pUL99; in its absence, tegumented capsids accumulate in the cytoplasm (Silva et al., 2003). Several changes appear to accompany this process; including compression or tightening of the tegument and thickening of the membrane with apparent elaboration on its luminal surface. These changes occur where the membrane and capsid are in close proximity (e.g., Fig. 1d), suggesting conformational or compositional changes in both layers as envelopment proceeds. Late maturational events such as phosphorylation by the virion-associated kinase(s) (Roby and Gibson 1986; Nogalski et al. 2007), carbohydrate processing, and possible redistribution of envelope and tegument constituents, are likely to occur as the particle completes its egress and is modified to increase its efficiency as an entry vessel for delivering the viral DNA to the next cell.

The recently discovered ubiquitin-specific cysteine protease (DUB) activity, present and functional in virions at the amino end of the high-molecular-weight tegument protein pUL48 of HCMV and pUL36 of HSV (Kattenhorn et al. 2005; Schlieker et al. 2005; Wang et al. 2006), may be involved in these late events, or have a role at the outset of infection, or both. Viruses mutated in the catalytically critical Cys41 or His163 residues of the HCMV DUB replicate at a reduced efficiency relative to wild type virus, indicating an important even if not absolutely essential role for this new virion-associated enzyme (Wang et al. 2006).

Concluding Thoughts

Cytomegalovirus devotes much of its coding potential to functions required to produce infectious progeny. Delineating the molecular composition of the mature virus particle, the organization of its components, and the processes required to enable its formation will provide new insights into biological assembly mechanisms and critical virus-host interactions, and new leads for antiviral strategies.

Advances in the sensitivity of detecting and quantifying constituents of the virion have led to a more detailed inventory of its composition. However, much remains to be discovered about how those components associate into substructures, what processes drive their associations and dissociations, and what functions are served by the resulting changes.

Ripe for further investigation are several new and less-well understood aspects of how the virus particle is formed and delivers its DNA cargo to the host cell nucleus:

1. What targets the infecting nucleocapsid to the nuclear pore and what triggers DNA release at that destination?
2. What triggers maturational cleavage by the capsid protease, pUL80a, and what drives elimination of the scaffolding elements from the interior of the nascent capsid?
3. What is the function of the CMV-specific basic phosphoprotein (pUL32) that is tightly associated with the outer surface of the capsid and might it help stabilize the capsid to pressure from within exerted by the comparatively large CMV DNA?
4. How is the recently discovered ubiquitin-specific cysteine protease activity of pUL48 involved in virus replication?
5. What are the mechanistic details of nuclear egress by the nucleocapsid?
6. What are the critical interactions between capsid and tegument components, tegument constituents with themselves, and tegument constituents with envelope?

Answers to these and related questions will speed progress toward identifying new targets and strategies for antiviral drugs, and broaden our understanding of this medically important class of viruses.

Acknowledgements Electron micrographs were prepared by Donna Woods. Joe Dieter and Amy Loveland helped me with the illustrations. Contributions of many colleagues to published and unpublished work from our lab are gratefully acknowledged and were aided by Public Health Service research grants AI13718 and AI032957.

References

Baines J, Poon A, Rovnak J, Roizman B (1994) The herpes virus 1 UL15 gene encodes two proteins and is required for cleavage of genomic viral DNA. J Virol 68:8118-8124

Baxter MK, Gibson W (1997) The putative human cytomegalovirus triplex proteins, minor capsid protein (mCP) and mCP-binding protein (mC-BP), form a heterotrimeric complex that localizes to the cell nucleus in the absence of other viral proteins. In: 22nd International Herpesvirus Workshop, La Jolla, CA

Bjerke SL, Cowan JM, Kerr JK, Reynolds AE, Baines JD, Roller RJ (2003) Effects of charged cluster mutations on the function of herpes simplex virus type 1 UL34 protein. J Virol 77:7601-7610

Bogner E, Reschke M, Reis B, Richter A, Mockenhaupt T, Radsak K (1993) Identification of the gene product encoded by ORF UL56 of the human cytomegalovirus. Virology 196:290-293

Bogner E, Radsak K, Stinski MF (1995) The HCMV UL56 gene product is a DNA binding protein. 5th International Cytomegalovirus Conference Abstract P051, p 85

Bogner E, Radsak K, Stinski MF (1998) The gene product of human cytomegalovirus open reading frame UL56 binds the pac motif and has specific nuclease activity. J Virol 72:2259-2264

Brignole EJ, Gibson W (2007) Enzymatic activities of human cytomegalovirus maturational protease assemblin and its precursor (pPR, pUL80a) are comparable: maximal activity of pPR requires self-interaction through its scaffolding domain. J Virol 81:4091-4103

Brown JC, McVoy MA, Homa FL (2002) Packaging DNA into herpesvirus capsids. In: Bogner AHaE (ed) Structure-function relationships of human pathogenic viruses. Kluwer Academic, New York, pp 111-153

Buisson M, Hernandez JF, Lascoux D, Schoehn G, Forest E, Arlaud G, Seigneurin JM, Ruigrok RW, Burmeister WP (2002) The crystal structure of the Epstein-Barr virus protease shows rearrangement of the processed C terminus. J Mol Biol 324:89-103

Campadelli-Fiume G, Roizman B (2006) The egress of herpesviruses from cells: the unanswered questions. J Virol 80:6716-6717; author replies 6717-6719

Casaday RJ, Bailey JR, Kalb SR, Brignole EJ, Loveland AN, Cotter RJ, Gibson W (2004) Assembly protein precursor (pUL80.5 Homolog) of simian cytomegalovirus is phosphorylated at a glycogen synthase kinase 3 site and its downstream "priming" site: phosphorylation affects interactions of protein with itself and with major capsid protein. J Virol 78:13501-13511

Chan CK, Brignole EJ, Gibson W (2002) Cytomegalovirus assemblin (pUL80a): cleavage at internal site not essential for virus growth; proteinase absent from virions. J Virol 76:8667-8674

Chen P, Tsuge H, Almassy RJ, Gribskov CL, Katoh S, Vanderpool DL, Margosiak SA, Pinko C, Matthews DA, Kan C-C (1996) Structure of the human cytomegalovirus protease catalytic domain reveals a novel serine protease fold and catalytic triad. Cell 86:835-843

Church GA, Wilson DW (1997) Study of herpes simplex virus maturation during a synchronous wave of assembly. J Virol 71:3603-3612

Cole JL (1996) Characterization of human cytomegalovirus protease dimerization by analytical centrifugation. Biochemistry 35:15601-15610

Darke PL, Cole JL, Waxman L, Hall DL, Sardana MK, Kuo LC (1996) Active human cytomegalovirus protease is a dimer. J Biol Chem 271:7445-7449

Dittmer A, Bogner E (2005) Analysis of the quaternary structure of the putative HCMV portal protein PUL104. Biochemistry 44:759-765

Eickmann M, Gicklhorn D, Radsak K (2006) Glycoprotein trafficking in virion morphogenesis. In: Reddehase MJ (ed) Cytomegaloviruses molecular biology and immunology. Caister Academic Press, Norfolk, UK, pp 245-264

Enquist LW, Husak PJ, Banfield BW, Smith GA (1998) Infection and spread of alphaherpesviruses in the nervous system. Adv Virus Res 51:237-247

Fuchs W, Granzow H, Klupp BG, Kopp M, Mettenleiter TC (2002) The UL48 tegument protein of pseudorabies virus is critical for intracytoplasmic assembly of infectious virions. J Virol 76:6729-6742

Gibson W (1981) Structural and nonstructural proteins of strain Colburn cytomegalovirus. Virology 111:516-537

Gibson W (1993) Molecular biology of human cytomegalovirus. Springer-Verlag, Berlin New York Heidelberg

Gibson W (1996) Structure and assembly of the virion. Intervirology 39:389-400

Gibson W (2006) Assembly and maturation of the capsid. In: Reddehase MJ (ed) Cytomegaloviruses: molecular biology and immunology. Caister Academic Press, pp 231-244

Gibson W, Irmiere A (1984) Selection of particles and proteins for use as human cytomegalovirus subunit vaccines. Birth Defects 20:305-324

Gibson W, Roizman B (1972) Proteins specified by herpes simplex virus. VIII. characterization and composition of multiple capsid forms of subtypes 1 and 2. J Virol 10:1044-1052

Hensel G, Meyer H, Gartner S, Brand G, Kern HF (1995) Nuclear localization of the human cytomegalovirus tegument protein pp150 (ppUL32). J Gen Virol 76:1591-1601

Holzenburg A, Bogner E (2002) From concatemeric DNA into unit-length genomes - a miracle or clever genes? In: Holzenburg A, Bogner E (eds) Structure-function relationships of human pathogenic viruses, vol 1. Kluwer Academic, New York, pp 155-173

Irmiere A, Gibson W (1983) Isolation and characterization of a noninfectious virion-like particle released from cells infected with human strains of cytomegalovirus. Virology 130:118-133

Irmiere A, Gibson W (1985) Isolation of human cytomegalovirus intranuclear capsids, characterization of their protein constituents, and demonstration that the B-capsid assembly protein is also abundant in noninfectious enveloped particles. J Virol 56:277-283

Kattenhorn LM, Korbel GA, Kessler BM, Spooner E, Ploegh HL (2005) A deubiquitinating enzyme encoded by HSV-1 belongs to a family of cysteine proteases that is conserved across the family Herpesviridae. Mol Cell 19:547-557

Klupp BG, Granzow H, Mettenleiter TC (2000) Primary envelopment of pseudorabies virus at the nuclear membrane requires the UL34 gene product. 25th International Herpesvirus Workshop: Abstract 7.04

Klupp BG, Granzow H, Mettenleiter TC (2001) Effect of the pseudorabies virus US3 protein on nuclear membrane localization of the UL34 protein and virus egress from the nucleus. J Gen Virol 82:2363-2371

Ladin BF, Ihara S, Hampl H, Ben-Porat T (1982) Pathway of assembly of herpesvirus capsids: an analysis using DNA+ temperature-sensitive mutants of pseudorabies virus. Virology 116:544-561

Lee JY, Irmiere A, Gibson W (1988) Primate cytomegalovirus assembly: evidence that DNA packaging occurs subsequent to B capsid assembly. Virology 167:87-96

Leuzinger H, Ziegler U, Schraner EM, Fraefel C, Glauser DL, Heid I, Ackermann M, Mueller M, Wild P (2005) Herpes simplex virus 1 envelopment follows two diverse pathways. J Virol 79:13047-13059

Liu F, Zhou ZH (2007) Comparative virion structures of human herpesviruses. In: Arvin A, Campadelli-Fiume G, Mocarski E, Moore PS, Roizman B, Whitley R, Yamanishi K (eds) Human herpesviruses : biology, therapy, and immunoprophylaxis. Cambridge University Press, Cambridge, pp 27-43

Loveland AN, Chan CK, Brignole EJ, Gibson W (2005) Cleavage of human cytomegalovirus protease pUL80a at internal and cryptic sites is not essential but enhances infectivity. J Virol 79:12961-12968

Loveland AN, Nguyen NL, Brignole EJ, Gibson W (2007) The amino-conserved domain of human cytomegalovirus UL80a proteins is required for key interactions during early stages of capsid formation and virus production. J Virol 81:620-628

Margosiak SA, Vanderpool DL, Sisson W, Pinko C, Kan CC (1996) Dimerization of the human cytomegalovirus protease - kinetic and biochemical characterization of the catalytic homodimer. Biochemistry 35:5300-5307

McCartney SA, Brignole EJ, Kolegraff KN, Loveland AN, Ussin LM, Gibson W (2005) Chemical rescue of I-site cleavage in living cells and in vitro discriminates between the cytomegalovirus protease, assemblin, and its precursor, pUL80a. J Biol Chem 280:33206-33212

McClelland DA, Aitken JD, Bhella D, McNab D, Mitchell J, Kelly SM, Price NC, Rixon FJ (2002) pH reduction as a trigger for dissociation of herpes simplex virus type 1 scaffolds. J Virol 76:7407-7417

McNab AR, Desai P, Person S, Roof LL, Thomsen DR, Newcomb WW, Brown JC, Homa FL (1998) The product of the herpes simplex virus type 1 UL25 gene is required for encapsidation but not for cleavage of replicated viral DNA. J Virol 72:1060-1070

Mettenleiter TC (2002) Herpesvirus assembly and egress. J Virol 76:1537-1547

Muranyi W, Haas J, Wagner M, Krohne G, Koszinowski UH (2002) Cytomegalovirus recruitment of cellular kinases to dissolve the nuclear lamina. Science 297:854-857

Newcomb WW, Homa FL, Thomsen DR, Trus BL, Cheng N, Steven A, Booy F, Brown JC (1999) Assembly of the herpes simplex virus procapsid from purified components and identification of small complexes containing the major capsid and scaffolding proteins. J Virol 73:4239-4250

Newcomb WW, Juhas RM, Thomsen DR, Homa FL, Burch AD, Weller SK, Brown JC (2001) The UL6 gene product forms the portal for entry of DNA into the herpes simplex virus capsid. J Virol 75:10923-10932

Newcomb WW, Trus BL, Cheng N, Steven AC, Sheaffer AK, Tenney DJ, Weller SK, Brown JC (2000) Isolation of herpes simplex virus procapsids from cells infected with a protease-deficient mutant virus. J Virol 74:1663-1673

Newcomb WW, Homa FL, Brown JC (2006) Herpes simplex virus capsid structure: DNA packaging protein UL25 is located on the external surface of the capsid near the vertices. J Virol 80:6286-6294

Nguyen NL, Loveland AN, Gibson W (2008) Nuclear localization sequences in cytomegalovirus capsid assembly proteins (UL80 proteins) are required for virus production: inactivating NLS1, NLS2, or both affects replication to strikingly different extents. J Virol, in press

Nii S, Uno F, Yoshida M, Akatsuka K (1998) Structure and assembly of human beta herpesviruses. Nippon Rinsho 56:22-28

Nogalski MT, Podduturi JP, Demeritt IB, Milford LE, Yurochko AD (2007) The human cytomegalovirus virion possesses an activated casein kinase ii that allows for the rapid phosphorylation of the inhibitor of NF-{kappa}B, I{kappa}B{alpha}. J Virol 81:5305-5314

O'Callaghan DJ, Randall CC (1976) Molecular anatomy of herpesviruses: recent studies. Prog Med Virol 22:152-210

Ogasawara M, Suzutani T, Yoshida I, Azuma M (2001) Role of the UL25 gene product in packaging DNA into the herpes simplex virus capsid: Location of UL25 product in the capsid an demonstration that it binds DNA. J Virol 75:1427-1436

O'Hare P, Abaitua F (2006) Study of herpes simplex virus UL36 gene defect in the temperature sensitive virus TSB7. In: 31st International Herpesvirus Workshop, Seattle, Washington

Patel AH, MacLean JB (1995) The product of the UL6 gene of herpes simplex virus type 1 is associated with virus capsids. 206:465-478

Patel AH, Rixon FJ, Cunningham C, Davison AJ (1996) Isolation and characterization of herpes simplex virus type 1 mutants defective in the UL6 gene. 217:111-123

Plafker SM, Gibson W (1998) Cytomegalovirus assembly protein precursor and proteinase precursor contain two nuclear localization signals that mediate their own nuclear translocation and that of the major capsid protein. J Virol 72:7722-7732

Poon A, Roizman B (1993) Characterization of a temperature-sensitive mutant of the UL15 open reading frame of herpes simplex virus1. J Virol 67:4497-4503

Purves FC, Spector D, Roizman B (1992) UL34, the target of the herpes simplex virus U(S)3 protein kinase, is a membrane protein which in its unphosphorylated state associates with novel phosphoproteins. J Virol 66:4295-4303

Qiu X, Culp JS, DiLella AG, Hellmig B, Hoog SS, Janson DA, Smith WW, Abdel-Meguid SS (1996) Unique fold and active site in cytomegalovirus protease. Nature 383:275-279

Reynolds AE, Ryckman BJ, Baines JD, Zhou Y, Liang L, Roller RJ (2001) U(L)31 and U(L)34 proteins of herpes simplex virus type 1 form a complex that accumulates at the nuclear rim and is required for envelopment of nucleocapsids. J Virol 75:8803-8817

Reynolds AE, Wills EG, Roller RJ, Ryckman BJ, Baines JD (2002) Ultrastructural localization of the herpes simplex virus type 1 UL31, UL34, and US3 proteins suggests specific roles in primary envelopment and egress of nucleocapsids. J Virol 76:8939-8952

Rixon FJ (1993) Structure and assembly of herpesviruses. Sem Virol 4:135-144

Rixon FJ, McNab D (1999) Packaging-competent capsids of a herpes simplex virus temperature-sensitive mutant have properties similar to those of in vitro-assembled procapsids. J Virol 73:5714-5721

Roby C, Gibson W (1986) Characterization of phosphoproteins and protein kinase activity of virions, noninfectious enveloped particles, and dense bodies of human cytomegalovirus. J Virol 59:714-727

Roller RJ, Zhou Y, Schnetzer R, Ferguson J, DeSalvo D (2000) Herpes simplex virus type 1 U(L)34 gene product is required for viral envelopment. J Virol 74:117-129

Sanchez V, Greis KD, Sztul E, Britt WJ (2000a) Accumulation of virion tegument and envelope proteins in a stable cytoplasmic compartment during human cytomegalovirus replication: characterization of a potential site of virus assembly. J Virol 74:975-986

Sanchez V, Sztul E, Britt WJ (2000b) Human cytomegalovirus pp28 (UL99) localizes to a cytoplasmic compartment which overlaps the endoplasmic reticulum-Golgi-intermediate compartment. J Virol 74:3842-3851

Scheffczik H, Savva CG, Holzenburg A, Kolesnikova L, Bogner E (2002) The terminase subunits pUL56 and pUL89 of human cytomegalovirus are DNA-metabolizing proteins with toroidal structure. Nucleic Acids Res 30:1695-1703

Schlieker C, Korbel GA, Kattenhorn LM, Ploegh HL (2005) A deubiquitinating activity is conserved in the large tegument protein of the herpesviridae. J Virol 79:15582-15585

Scholl B-C, Von Hintzenstern J, Borisch B, Traupe B, Broker M, Jahn G (1988) Prokaryotic expression of immunogenic polypeptides of the large phosphoprotein (pp150) of human cytomegalovirus. J Gen Virol 69:1195-1204

Severi B, Landini M-P, Govoni E (1988) Human cytomegalovirus morphogenesis: an ultrastructural study of late cytoplasmic phases. Arch Virol 98:51-64

Sheaffer AK, Newcomb WW, Gao M, Yu D, Weller SK, Brown JC, Tenney DJ (2001) Herpes simplex virus DNA cleavage and packaging proteins associate with the procapsid prior to its maturation. J Virol 75:687-698

Sherman G, Bachenheimer SL (1988) Characterization of intranuclear capsids made by ts morphogenic mutants of HSV-1. Virology 163:471-480

Shieh H-S, Kurumbail RG, Stevens AM, Stegeman RA, Sturman EJ, Pak JY, Wittwer AJ, Palmier MO, Wiegand RC, Holwerda BC, Stallings WC (1996) Three-dimensional structure of human cytomegalovirus protease. Nature 383:279-282

Silva MC, Yu A-C, Enquist L, Shenk T (2003) Human cytomegalovirus UL99-encoded pp28 is required for the cytoplasmic envelopment of tegument-associated capsids. J Virol 77:10594-10605

Singer GP, Newcomb WW, Thomsen DR, Homa FL, Brown JC (2005) Identification of a region in the herpes simplex virus scaffolding protein required for interaction with the portal. J Virol 79:132-139

Spencer JV, Newcomb WW, Thomsen DR, Homa FL, Brown JC (1998) Assembly of the herpes simplex virus capsid: preformed triplexes bind to the nascent capsid. J Virol 72:3944-3951

Steven AC, Spear PG (1997) Herpesvirus capsid assembly and envelopment. In: Chiu W, Burnett RM, Garcea RL (eds) Structural biology of viruses. Oxford University Press, New York, pp 312-351

Tatman JD, Preston VG, Nicholson P, Elliott RM, Rixon FJ (1994) Assembly of herpes simplex virus type 1 capsids using a panel of recombinant baculoviruses. J Gen. Virol 75:1101-1113

Thomsen DR, Roof LL, Homa FL (1994) Assembly of herpes simplex virus (HSV) intermediate capsids in insect cells infected with recombinant baculoviruses expressing HSV capsid proteins. J Virol 68:2442-2457

Tong L, Qian C, Massariol MJ, Bonneau PR, Cordingly MG, Lagacé L (1996) A new serine-protease fold revealed by the crystal structure of human cytomegalovirus protease. Nature 383:272-275

Trus BL, Cheng N, Newcomb WW, Homa FL, Brown JC, Steven AC (2004) Structure and polymorphism of the UL6 portal protein of herpes simplex virus type 1. J Virol 78:12668-12671

Varnum SM, Streblow DN, Monroe ME, Smith P, Auberry KJ, Pasa-Tolic L, Wang D, Camp DG 2nd, Rodland K, Wiley S, Britt W, Shenk T, Smith RD, Nelson JA (2004) Identification of proteins in human cytomegalovirus (HCMV) particles: the HCMV proteome. J Virol 78:10960-10966

Wang J, Loveland AN, Kattenhorn LM, Ploegh HL, Gibson W (2006) High-molecular-weight protein (pUL48) of human cytomegalovirus is a competent deubiquitinating protease: mutant viruses altered in its active-site cysteine or histidine are viable. J Virol 80:6003-6012

White CA, Stow ND, Patel AH, Hughes M, Preston VG (2003) Herpes simplex virus type 1 portal protein UL6 interacts with the putative terminase subunits UL15 and UL28. J Virol 77:6351-6358

Wood LJ, Baxter MK, Plafker SM, Gibson W (1997) Human cytomegalovirus capsid assembly protein precursor (pUL80.5) interacts with itself and with the major capsid protein (pUL86) through two different domains. J Virol 71:179-190

Human Cytomegalovirus Modulation of Signal Transduction

A. D. Yurochko

Contents

Introduction	206
Signaling Overview	207
Receptor/Ligand-Mediated Signaling: Viral Glycoproteins	208
Captured Cellular Enzymes	210
Tegument Protein-Mediated Signaling	211
Other Viral Gene Products That Modulate Signaling	212
Biological Rationale for Modulation of Host Cell Signaling	212
Role of Signaling in Pathogenesis	215
Final Thoughts	215
References	216

Abstract An upregulation of cellular signaling pathways is observed in multiple cell types upon human cytomegalovirus (HCMV) infection, suggesting that a global feature of HCMV infection is the activation of the host cell. HCMV initiates and maintains cellular signaling through a multitiered process that is dependent on a series of events: (1) the viral glycoprotein ligand interacts with its cognate receptor, (2) cellular enzymes and viral tegument proteins present in the incoming virion are released and (3) a variety of viral gene products are expressed. Viral-mediated cellular modification has differential outcomes depending on the cell type infected. In permissive cell types, such as diploid fibroblasts, the upregulation of cellular signaling pathways following infection can initiate the viral gene cascade and promote the efficient transcription of multiple viral gene classes. In other cell types, such as endothelial cells and monocytes/macrophages, the upregulation of cellular pathways initiates functional host changes that allow viral spread to multiple organ systems. Together, the modification of signaling processes appears to be part of a thematic

---

A.D. Yurochko
Department of Microbiology and Immunology, Center for Molecular and Tumor Virology,
Feist-Weiller Cancer Center, Louisiana State University Health Sciences Center,
1501 Kings Highway Shreveport, LA 71130-3932, USA
ayuroc@lsuhsc.edu

strategy deployed by the virus to direct the required functional changes in target cells that ultimately promote viral survival and persistence in the host.

Introduction

HCMV is a species-specific β-herpesvirus found in more than 60% of the human population (Mocarski et al. 2007). HCMV causes severe disease in immunocompromised individuals, where it is a major opportunistic pathogen in AIDS and organ transplant patients, in congenitally infected neonates, and in cancer patients undergoing chemotherapy (see the chapter by W. Britt, this volume). In the immunocompetent host, HCMV causes mononucleosis (see the chapter by W. Britt, this volume) and is associated with chronic human diseases such as atherosclerosis and restenosis (Melnick et al. 1993; Speir et al. 1994; Waldman et al. 1997; Streblow et al. 1999, 2001a) and some forms of cancer (Shen et al. 1993; Cobbs et al. 2002; Soderberg-Naucler 2006).

A hallmark of HCMV infection is a broad cellular tropism in vivo that results in the infection of most host organ tissues (Myerson et al. 1984; Sinzger and Jahn 1996; Mocarski et al. 2007; see the chapter by C. Sinzger et al., this volume). HCMV pathogenesis is a direct result of the infection of host organs and the resulting overt organ disease (Sinzger and Jahn 1996; Mocarski et al. 2007). From an evolutionary standpoint, the ability to infect multiple organs provides the virus access to multiple portals of viral exit and, consequently, allows viral shedding in most human body fluids (Mocarski et al. 2007). Broad cellular tropism necessitates that the virus possess a strategy to productively infect a diverse array of cell types that have unique biochemical features. Regardless of the diversity of cells found in the human host, all cell types utilize cellular signaling pathways as a means of cellular communication and appropriate response to their environment (Cooper and Hausman 2007). Thus, cellular signaling from a general standpoint is a common thread among multiple cell types that, if exploited correctly, would allow HCMV to transcend the differences among cell types. Mechanistically, the exploitation of cellular signaling by the virus provides at least one biological explanation for HCMV's broad tropism in vivo. Certainly viral attachment to an infected cell surface is also a determinant of tropism (see the chapter by C. Sinzger et al., this volume), but because this chapter focuses on the viral modulation of cellular signaling, we will only discuss how cellular signaling can be exploited by the virus to promote persistence and survival in a variety of host cell types. Nevertheless, because we (Yurochko et al. 1995, 1997a; Yurochko and Huang 1999; Bentz and Yurochko, unpublished data) and others (Keay et al. 1995; Boyle et al. 1999; Simmen et al. 2001; Compton et al. 2003; Wang et al. 2003; Boehme et al. 2004; Feire et al. 2004; Wang et al. 2005; Boehme et al. 2006) have strong evidence that viral ligand-mediated signaling is stimulated by the same viral glycoproteins responsible for viral attachment, fusion, and entry (Britt and Mach 1996), it is likely that these two seemingly diverse mechanisms are intimately linked and together provide key control points for the infection of the host. We propose that cellular signaling is a biological aspect exploited by HCMV during infection (from viral entry to post-entry events) to manipulate a variety of cell types. It is the goal of this chapter to provide an overview of the diverse mechanisms HCMV employs to modulate cellular signaling pathways, as well as a discussion of the likely biological rationale for why the virus may have evolved a strategy to dysregulate host cell signaling pathways following infection.

Signaling Overview

HCMV infection results in a wide range of cellular changes including changes in calcium flux and lipid metabolism, activation of kinase signaling cascades (such as calcium/calmodulin-dependent protein kinases, multiple cell cycle-regulated kinases, the epidermal growth factor receptor (EGFR), the IκB kinase (IKK) cascade, the mammalian target of rapamycin pathway, various members of the mitogen activated protein kinase (MAPK) pathway, the phosphatidylinositol 3-kinase (PI(3)K) pathway, and the src family of kinases), cytoskeletal changes, activation of cellular transcription factors (such as AP-1, ATF/CREB, E2F, NFκ-B, Sp1), the induction of proto-oncogenes and other cellular immediate-early (IE) response genes (reviewed in Albrecht et al. 1990, 1993; Evers et al. 2004; DeMeritt and Yurochko 2006). Signaling-induced changes in infected cells can loosely be grouped into two tiers (Table 1): the first tier represents changes that occur prior to the initiation of viral gene expression and, thus, are mediated by the virion itself; and the second tier represents those changes that occur temporally after the production of viral gene products and, thus, are mediated by proteins from the different temporal gene classes. The virion itself is a potent signaling player as the viral envelope glycoproteins initiate rapid cellular responses upon binding to cognate receptors (reviewed in Evers et al. 2004; DeMeritt and Yurochko 2006).

Table 1 Summary of viral-associated signaling[a]

Modulator	Rapid effects[b]	Delayed effects[b]	Function
Viral glycoproteins	X		Receptor/ligand-mediated signaling
Captured cellular enzymes	X	?	Activation of signaling pathways
Tegument proteins[c]	X	?	Activation of signaling pathways/cell cycle regulation
Other viral gene products[d]	–	X	Activation of signaling pathways/cell cycle regulation

[a] Individual gene products are discussed in the text
[b] Signaling induced upon HCMV infection can loosely be grouped into the products that regulate rapid responses (beginning within minutes of infection) and are caused by modulators associated with the virion vs those products that regulate effects later in infection (or delayed compared to the rapid effects) and are caused by viral gene products de novo synthesized following infection
[c] Tegument proteins or tegument-associated virion proteins are included together
[d] Other viral gene products in this table represent those gene products that are synthesized de novo in the infected cell and are not attributed to virion mediated signaling

Glycoprotein-mediated signaling is not the only tool in the virion arsenal, as the virion has evolved a mechanism to capture cellular signal modifying enzymes (Michelson et al. 1996; Gallina et al. 1999; Nogalski et al. 2007), which can be dumped directly into the cytosol following viral entry into host cells. In addition, like all herpesviruses, HCMV has a large number of tegument proteins that modulate cellular signaling (Mocarski et al. 2007). Lastly, viral gene products synthesized following infection can also manipulate host cellular responses (some examples include IE proteins that alter the cell cycle and regulate apoptosis (reviewed in Castillo and Kowalik 2004; Andoniou and Degli-Esposti 2006) or those viral gene products that mimic cellular cytokine/chemokine signaling receptors including US28 (reviewed in Streblow et al. 2001b; Stropes and Miller 2004; van Cleef et al. 2006), a viral G protein-coupled receptor (GPCR), and UL144 (Benedict et al. 1999; Poole et al. 2006), a tumor necrosis factor-like receptor. Together, it is evident that HCMV possesses an array of signal-modifying capabilities that are deployed over a temporal range during the infection process. The likely outcome of this viral-mediated signaling is currently under debate. We suggest the viral-mediated cellular modification is required for multiple critical steps in the viral infection cycle and that the viral-directed signaling can have different outcomes in different cell types. In fibroblasts, for example, the initial signaling seen following receptor/ligand engagement is reported to promote viral entry (Wang et al. 2003; Feire et al. 2004; Wang et al. 2005) and then productive infection by promoting efficient gene transcription (Caposio et al. 2004; DeMeritt et al. 2004, 2006; DeMeritt and Yurochko 2006). In other cell types such as endothelial cells (Bentz et al. 2006) and monocytes (Smith et al. 2004b, 2007), viral-mediated signaling can stimulate the functional changes in these cells required for hematogenous dissemination of the virus. Below we provide a more detailed overview of these different viral-directed steps controlling signaling.

Receptor/Ligand-Mediated Signaling: Viral Glycoproteins

Envelope glycoproteins play an essential role in viral attachment and entry (Britt and Mach 1996; Mocarski et al. 2007; see the chapter by M.K. Isaacson et al., this volume). From a signaling standpoint, these molecules are logical players in the rapid manipulation of the host cell because they are the first viral molecules to contact a target cell. Although HCMV encodes a number of envelope glycoproteins (Britt and Mach 1996; Mocarski et al. 2007), glycoprotein B (gB/UL55; Britt and Mach 1996) and glycoprotein H (gH/UL75 and its associated partners gL/UL115, gO/UL74, and the UL131-UL128 loci; Britt and Mach 1996; Hahn et al. 2004; Wang and Shenk 2005a, b; Patrone et al. 2007) are the glycoproteins documented to be bona fide signaling molecules (Keay et al. 1995; Yurochko et al. 1997a; Boyle et al. 1999; Yurochko and Huang 1999; Simmen et al. 2001; Compton et al. 2003; Wang et al. 2003, 2005; Boehme et al. 2004, 2006; Feire et al. 2004). The gH complex was originally shown to stimulate calcium flux (Keay et al. 1995), while we have demonstrated that both gB and gH stimulate the activation of the cellular transcription factors, NFκ-B and Sp1 (Yurochko et al. 1997a; Yurochko and Huang 1999). Other studies confirmed and expanded these results (Boyle et al. 1999; Simmen et al. 2001; Wang et al. 2003, 2005; Boehme et al. 2004, 2006) and together determined that HCMV fires cellular signal transduction pathways via the actions of the major viral glycoproteins, gB and gH. Viral glycoprotein-mediated signaling occurs in multiple cell types (fibroblasts, monocytes, endothelial cells, etc.), suggesting that the capacity to induce cellular signaling is part of a central theme in the viral infection strategy.

The recent identification of several cellular receptors for HCMV attachment/entry that are found on multiple cell types supports this proposal: HCMV glycoproteins were recently shown to interact with the epidermal growth factor receptor (EGFR; Wang et al. 2003, 2005), integrins ($\alpha_2\beta_1$, $\alpha_6\beta_1$, $\alpha_V\beta_3$; Feire et al. 2004; Wang et al. 2005), and toll-like receptor 2 (TLR2; Compton et al. 2003; Boehme et al. 2006). From a signaling standpoint, the engagement of these receptors by the virus makes sense, as each receptor is biochemically integrated with the signaling machinery. EGFR dimerizes upon ligand binding and then directs downstream signaling events via the action of its intrinsic tyrosine kinase (Wang et al. 2003, 2005). Integrins do not possess intrinsic kinase activity; however, upon their engagement they interact with members of the Src family of tyrosine kinases to modulate downstream signaling events (Wang et al. 2003, 2005). Finally, like all TLRs, TLR2 is part of a signaling network involving a cascade of players (Compton et al. 2003; Boehme et al. 2006).

Mechanistically, it has been documented that gB and gH are responsible for the engagement of the various cellular receptors (EGFR, the integrins, and TLR2) and that, through this receptor/ligand interaction, they rapidly activate signal transduction pathways (Wang et al. 2003, 2005; Boehme et al. 2006). Wang et al. have reported that gB interacts with EGFR and gH interacts with cellular integrins (Wang et al. 2003, 2005), demonstrating that individual receptor/ligand events are controlled by different viral gene products. gB and gH can also interact with TLR2 (Boehme et al. 2006), while gB may additionally interact with cellular integrins (Feire et al. 2004). All three receptors appear to be present on most cell types, suggesting an evolutionarily conserved mechanism may exist for viral binding and receptor engagement during infection of multiple cell types. This possibility is supported by work showing that EGFR and/or integrins are central determinants of signaling and/or attachment/entry in fibroblasts (Wang et al. 2003, 2005), cytotrophoblasts (Maidji E et al. 2007), endothelial cells (Bentz and Yurochko 2008) and monocytes (Yurochko et al. 1992; Chan et al., unpublished data). Nevertheless, the role these receptors play remains controversial, as it was recently reported that EGFR was not required for attachment and signaling on some fibroblast, epithelial and endothelial cell lines (Isaacson et al. 2007). Thus, it remains unclear if all three receptors are utilized on all cell types infected or if different combinations are utilized depending on the cell type. Overall, these findings suggest the following general model (discussed in more detail below): gB and gH binding to cellular receptors initiates the activation of multiple downstream players including the focal

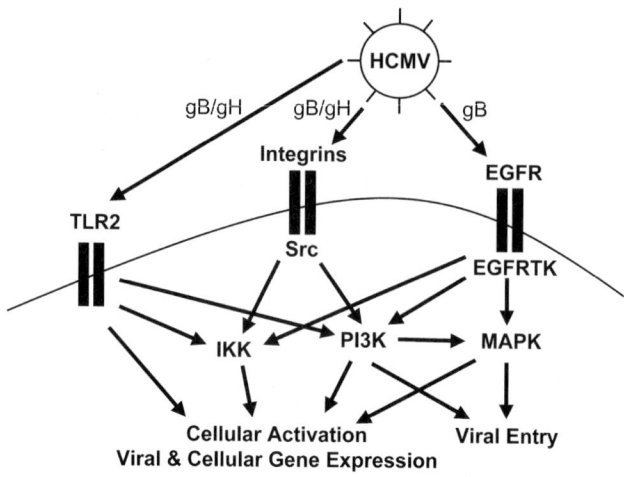

Fig. 1 HCMV binding to cognate receptors initiates signaling cascades. Binding of the envelope glycoproteins, gB and gH, to the cellular receptors, EGFR, integrins and TLR2 begin the outside-in signaling process observed in cells following infection. These known HCMV receptors are integrated with cellular signal transduction pathways; thus viral ligand engagement is the stimulus to fire downstream signaling processes. The initial receptor/ligand-directed signaling modulates a number of pathways, of which a few examples are shown in the drawing. The consequences of this outside-in signaling modulated by the viral glycoproteins include viral entry, cellular activation and transcriptional regulation of cellular and viral genes

adhesion kinase (FAK), the IKK cascade, the MAPK pathway, and the PI(3)K pathway to promote both viral entry and cellular changes such as the activation of NFκ-B and other transcription factors required for the transactivation of key cellular and/or viral genes (Fig. 1).

Captured Cellular Enzymes

The virion has long been known to harbor enzymatic activity (Mar et al. 1981), although the nature of this signaling potential has been unresolved. The signaling potential present in the virion imparts the virus with another mechanism to rapidly mediate distinct cellular changes following infection. Two distinct signaling capabilities are present in the virion: (1) HCMV captures cellular enzymes that directly modify the host cell signaling capabilities following viral fusion (discussed in this section) and (2) tegument proteins found in the mature virion can directly modulate host cell biochemical pathways (discussed in the next section).

The virion contains at least four distinct functional enzyme activities of host cell origin (Michelson et al. 1996; Gallina et al. 1999; Nogalski et al. 2007). A recent mass spectrometry analysis of the HCMV proteome revealed that additional cellular modulators may exist in the virion (Varnum et al. 2004). Michelson et al. first showed that HCMV virions contain serine/threonine protein phosphatase activity due to the cellular protein phosphatases PP1 and PP2A (Michelson et al. 1996). This work provided key evidence that HCMV captures cellular enzymes capable of manipulating phosphorylation. Kinases are also present in the HCMV virion. Gallina et al. showed that HCMV possess serine/threonine kinase activity due to the cellular kinase, (polo-like kinase 1 (Plk1; Gallina et al. 1999)). Plk1 was shown to interact with the major tegument protein, UL83/pp65, identifying a mechanism in which cellular products could be captured by the virus during maturation through a specific interaction with viral tegument proteins. We identified a second serine/threonine kinase, casein kinase II (CKII), that is also incorporated into the mature virion (Nogalski et al. 2007). The virion CKII possesses potent IκB kinase activity and promotes the efficient transactivation of the major IE promoter (MIEP). Why would the virus have evolved a mechanism to capture cellular enzymes? Reversible phosphorylation via the reciprocal action of kinases and phosphatases is an effective and rapid mechanism for modulating cellular function (Arena et al. 2005); thus this biochemical process is an attractive target for a virus that needs to rapidly modulate the host cell for viral infection, survival and persistence. The release of captured enzymes may allow an increase in the local concentration of those enzymes in the viral microenvironment (Nogalski et al. 2007). It is also possible the virion-associated enzymes have a different subcellular localization and thus potentially different targets (Gallina et al. 1999). Additionally, because the virus infects multiple cell types with different biological characteristics, the evolution of multiple mechanisms to drive the rapid activation of the cell may ensure sufficient and appropriate activation of each cell type following infection.

Tegument Protein-Mediated Signaling

HCMV possesses a number of tegument proteins that are able to modulate the host cell, although many tegument proteins do not have identified functions (Mocarski et al. 2007). Because another chapter will cover tegument proteins in detail (see the chapter by R. Kalejta, this volume), the signaling potential of select tegument proteins will only briefly be summarized. UL83, the major tegument protein, has been shown to block the antiviral response through the inhibition of the cellular transcription factors NFκ-B and interferon regulatory factor 1 (Browne and Shenk 2003). Other tegument proteins including UL82 (Schierling et al. 2004; Cantrell and Bresnahan 2006a; Saffert and Kalejta 2006), UL35 (Schierling et al. 2004), US24 (Feng et al. 2006) and UL26 (Stamminger et al. 2002; Munger et al. 2006) can also influence the early events involved with MIEP transactivation and IE gene expression. Tegument proteins also alter the cell cycle (reviewed in Kalejta and Shenk 2002; Kalejta 2004; Mocarski et al. 2007). For example, UL82 promotes cell cycle progression through the degradation of Rb family members (Kalejta et al. 2003; Kalejta and Shenk 2003a, 2003b), while UL69 blocks cell cycle progression by arresting cells in the G_1 phase of the cell cycle (Lu and Shenk 1999). Functionally, UL82 has also been shown to interact with the cellular protein hDaxx resulting in IE gene transcription and viral replication (Cantrell and Bresnahan 2006a, 2006b; Saffert and Kalejta 2006; Hwang and Kalejta 2007; see the chapter by R. Kalejta, this volume).

Other Viral Gene Products That Modulate Signaling

Once viral gene transcription begins, HCMV increases its repertoire of signaling molecules. For example, the major IE genes (IE1-72/UL123 and IE2/UL122) have been shown to interact with a multitude of transcription factors to increase transcription of required viral and cellular genes (reviewed in DeMeritt and Yurochko 2006; Mocarski et al. 2007), as well as interact with cell cycle regulators such as p53, pRB, p107 and others to modulate the cell cycle (reviewed in Kalejta and Shenk 2002; Castillo and Kowalik 2004). IE1-72 has also been reported to contain intrinsic kinase activity and to activate cells through the targeted phosphorylation of members of the E2F family of transcription factors (Pajovic et al. 1997). In addition, IE1-72 and IE2-86 (Zhu et al. 1995) along with the other IE genes, UL36 (viral inhibitor of caspase activation; Skaletskaya et al. 2001; McCormick et al. 2003) and UL37×1 (viral mitochondrial inhibitor of apoptosis; Goldmacher et al. 1999; McCormick et al. 2003; Reboredo et al. 2004), can modulate various survival pathways and provide protection from apoptosis (for additional information see Andoniou and Degli-Esposti 2006). HCMV also encodes other proteins with distinct signaling capabilities such as UL97, a viral kinase that plays a critical role during viral infection through its ability to phosphorylate cellular and viral substrates (Prichard et al. 2005); four putative GPCRs (US27, US28, UL33 and UL78) that have been shown to bind chemokines, activate G proteins in a manner similar to traditional GPCRs, mediate calcium flux, activate various kinases (MAPKs, Src, and FAK) and modulate smooth muscle cell migration (reviewed in Streblow et al. 2001b; Stropes and Miller 2004; van Cleef et al. 2006); and a TNF-like receptor, UL144 that activates NFκ-B through a TRAF6-dependent signaling cascade (Poole et al. 2006).

Biological Rationale for Modulation of Host Cell Signaling

There is little doubt that HCMV binding and/or infection of multiple cell types induces a sequence of signaling events (more detail provided in DeMeritt and Yurochko 2006), of which key points have been discussed briefly above. The question that remains is why the virus has evolved an elaborate strategy involving a multitiered approach to activate host target cells? The available evidence suggests the viral-induced signaling serves to promote multiple steps required for an efficient infection cycle. In human diploid fibroblasts, gB and gH stimulate signal transduction pathways required for viral entry (Wang et al. 2003, 2005; Feire et al. 2004), demonstrating that rapid signaling serves initially to stimulate entry. The same pathways required for this essential first step in the infection process (the activation of the EGFR kinase and Src via binding to EGFR and the integrins, respectively) also rapidly induce transcription factors such as NFκ-B. In our model, this induction is required for efficient transactivation of the MIEP and the production of viral IE gene products (DeMeritt et al. 2004), as well as the later viral gene classes (DeMeritt et al. 2006). It is likely that this facet of the viral biology, the activation of required host cell factors (transcription factors, cell cycle regulators, etc.) through the targeted specific activation of signal transduction pathways, is repeated for other specific pathways documented to be activated during infection of target cells. For example, additional transcription factors such as Sp1 are also induced following viral binding to promote the transactivation of the MIEP (Isomura et al. 2005; Yurochko et al. 1997a, 1997b). Because other signaling players such as the virion-associated CKII (Nogalski et al. 2007) and various tegument proteins (Romanowski et al. 1997; Stamminger et al. 2002; Schierling et al. 2004; Cantrell and Bresnahan 2006a; Feng et al. 2006; Munger et al. 2006; Saffert and Kalejta 2006) also promote the efficient expression of the IE gene products, it appears that multiple signaling pathways, although biochemically distinct, coordinate their efforts to focus on a single goal for the virus such as the upregulation of the MIEP and the initiation of the viral gene cascade. Other steps in the viral infection cycle are also essential to the infection process; thus it is likely that additional viral-mediated signaling pathways converge on a common molecular outcome to benefit the virus. An example is the role various tegument proteins and IE gene products play in ensuring that the required cellular replicative enzymes are available for viral replication (Castillo and Kowalik 2004).

Different cell types have distinct signaling capabilities, and even the same signal transduction pathway can have divergent downstream consequences in different cell types. Thus, we hypothesize that the viral regulation of signaling pathways will have different outcomes in cells such as endothelial cells and monocytes, which are critical cells for in vivo infection. We recently provided evidence for a unique two-pronged strategy for hematogenous dissemination involving endothelial cells and monocytes: (1) HCMV directly infects vascular endothelial cells (see references within Bentz et al. 2006; Mocarski et al. 2007; C. Sinzger et al., this volume), which in turn promotes naïve monocyte transendothelial migration and viral transfer to these migrating monocytes (Bentz et al. 2006), and (2) HCMV directly infects peripheral blood monocytes in order to promote their transendothelial migration (Smith et al. 2004a). Following transendothelial migration, both pools of infected monocytes differentiate into pro-inflammatory macrophages permissive for the replication of the original input virus, even though the original undifferentiated monocyte was not permissive for viral replication at the time of infection. The virus initiates these functional changes in endothelial cells and monocytes through the binding of viral glycoproteins to EGFR and cellular integrins and the resulting modulation of downstream signaling cascades such as the PI(3)K and NFκ-B pathways (Smith et al. 2004b; Bentz et al. 2006; Bentz and Yurochko 2008; Chan et al., unpublished data; Smith et al. 2007). Thus, these signal transduction pathways do not initially drive viral gene expression in these cell types, but instead induce cellular changes required for motility and firm adhesion to endothelial cells and transendothelial migration, suggesting that the biological rationale for the activation of these pathways is to modulate functional changes in cells of the vasculature that favor viral spread to and persistence within host organs. The role EGFR and integrins play in entry and attachment of endothelial cells and monocytes is not clear, although we have data that rapid signaling occurs through these receptors in both cell types (Bentz and Yurochko 2008; Chan et al., unpublished data), similar to that seen in fibroblasts (Wang et al. 2003, 2005; Feire et al. 2004), suggesting that these receptors are globally relevant to infection of multiple cell types. Overall, we propose that viral-induced signaling creates distinct cell-type-specific signaling signatures such that viral infection proceeds appropriately in each cell type (Fig. 2).

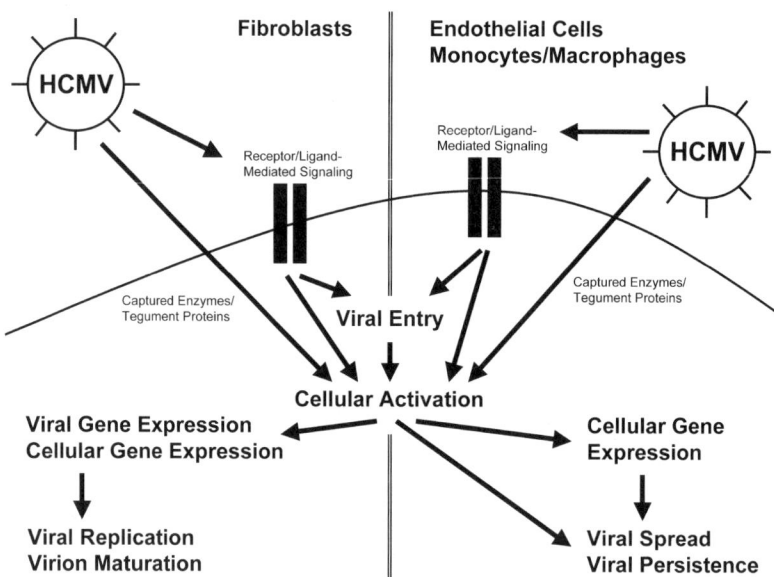

Fig. 2 Potential biological outcome of the viral-mediated signaling. Although unresolved, it is likely that the initially receptor/viral-ligand-mediated signaling promotes viral entry into target cells, regardless of cell type. This same receptor/ligand-mediated signaling also activates multiple biochemical pathways in target cells; both common pathways and cell-type-specific pathways are activated. The other potential mechanisms discussed in this review such as the cellular enzymes and tegument proteins that come in with the virion, as well as various synthesized viral gene products, also play a critical role in cellular modification. The net outcome of the viral-mediated signaling appears to vary depending on the cell type: for example, as represented in this drawing, productive infection is promoted in fibroblasts, while long-term persistence and survival of the virus is promoted in endothelial cells and monocytes/macrophages. Note: monocytes are not

Role of Signaling in Pathogenesis

Aberrant signaling and transcription factor regulation is associated with a multitude of diseases that including birth defects, cancer, and chronic inflammatory diseases such as cardiovascular disease (Kim et al. 2006). Cell cycle abnormalities are equally associated with diseases such as cancer and cardiovascular disease (Castillo and Kowalik 2004; Bentz and Yurochko 2008). Because these same diseases are associated with or caused by HCMV infection, modulation of multiple signaling transduction pathways, although beneficial to the virus, may be a molecular mechanism tying HCMV infection to the onset or severity of viral-mediated disease (reviewed in Evers et al. 2004; DeMeritt and Yurochko 2006; Soderberg-Naucler 2006). Certainly more work is needed to understand the possible direct role that viral-mediated cellular activation has on the infected host. It is also likely that these viral-manipulated cellular pathways required for viral pathogenesis may serve as new therapeutic targets for antiviral agents.

Final Thoughts

Together, it appears that HCMV has evolved a strategy for viral infection, survival, and persistence within the host that involves a complex biochemical manipulation of the host. Because of the possibility of severe effects on the host of unchecked signaling, HCMV as an evolutionarily ancient virus may also have evolved a strategy to mitigate the pathological consequences of this signaling strategy. For example, a recent report shows that HCMV through the UL83/pp65 tegument protein downregulates NFκ-B activity (Browne and Shenk 2003). Although this report runs counter to the data showing that NFκ-B activity is required for viral gene expression (Caposio et al. 2004, 2007; DeMeritt et al. 2004, 2006; Nogalski et al. 2007), if one considers that the virus must walk a fine line when activating a cell between those changes required for viral infection and the activation of cellular antiviral/host defense pathways and/or pathogenic consequences, these divergent results may represent two sides of the same coin. Perhaps this is why other reports have shown that NFκ-B activation negatively regulates or at least does not upregulate MIEP activity (Benedict et al. 2004; Isomura et al. 2004; Eickhoff and Cotten 2005; Gustems et al. 2006) and that for example the viral gene product, IE2p86, can act as a negative regulator of some NFκ-B-dependent cellular promoters (Taylor and Bresnahan 2006a, 2006b; Gealy et al. 2007). Using this example as a model, we argue that

Fig. 2 (continued) productive for viral replication following primary infection, but in response to the viral-mediated signaling, as represented in the drawing, they differentiate into macrophages that support viral replication (Smith et al. 2004a), thus both monocytes and their differentiated counterparts, macrophages, are critical for viral spread and persistence

HCMV needs to activate threshold levels of NFκ-B to initiate gene transcription (cellular and/or viral), but because high levels of this host factor are detrimental to the virus (generation of antiviral responses) and the host (pathogenic consequences), the virus has a mechanism to balance and moderate this transcription factor, or in a more general sense cellular signaling pathways; the virus thus walks a fine line by activating the factors necessary to allow productive infection and life-long persistence within the host with only minimal pathological consequences.

Acknowledgements A.D.Y. is supported by grants from the National Institutes of Health (AI56077 and 1-P20-RR018724). The author wishes to thank S. Adams and Drs. R.S. Scott and S.M. Karst for careful reading of the review of the manuscript and Dr. E.S. Huang for inspiration and support.

References

Albrecht T, Boldogh I, Fons M, AbuBakar S, Deng CZ (1990) Cell activation signals and the pathogenesis of human cytomegalovirus. Intervirology 31:68-75
Albrecht T, Boldogh I, Fons MP, Valyi-Nagy T (1993) Activation of proto-oncogenes and cell activation signals in the initiation and progression of human cytomegalovirus infection. In: Becker Y, Darai G, Huang ES (eds) Molecular aspects of human cytomegalovirus diseases. Springer-Verlag, Berlin, New York, Heidelberg, pp 384-411
Andoniou CE, Degli-Esposti MA (2006) Insights into the mechanisms of CMV-mediated interference with cellular apoptosis. Immunol Cell Biol 84:99-106
Arena S, Benvenuti S, Bardelli A (2005) Genetic analysis of the kinome and phosphatome in cancer. Cell Mol Life Sci 62:2092-2099
Benedict CA, Butrovich KD, Lurain NS, Corbeil J, Rooney I, Schneider P, Tschopp J, Ware CF (1999) Cutting edge: a novel viral TNF receptor superfamily member in virulent strains of human cytomegalovirus. J Immunol 162:6967-6970
Benedict CA, Angulo A, Patterson G, Ha S, Huang H, Messerle M, Ware CF, Ghazal P (2004) Neutrality of the canonical NFκ-B-dependent pathway for human and murine cytomegalovirus transcription and replication in vitro. J Virol 78:741-750
Bentz GL, Jarquin-Pardo M, Chan G, Smith MS, Sinzger C, Yurochko AD (2006) Human cytomegalovirus (HCMV) infection of endothelial cells promotes naïve monocyte extravasation and transfer of productive virus to enhance the hematogenous dissemination of HCMV. J Virol 80:11539-11555
Bentz GL, Yurochko AD (2008) HCMV infection of endothelial cells induces an angiogenic response through viral binding to the epidermal growth factor receptor and the β1 and β3 integrins. Proc Natl Acad Sci U S A (in press)
Boehme KW, Singh J, Perry ST, Compton T (2004) Human cytomegalovirus elicits a coordinated cellular antiviral response via envelope glycoprotein B. J Virol 78:1202-1211
Boehme KW, Guerrero M, Compton T (2006) Human cytomegalovirus envelope glycoproteins B and H are necessary for TLR2 activation in permissive cells. J Immunol 177:7094-7102
Boyle KA, Pietropaolo RL, Compton T (1999) Engagement of the cellular receptor for glycoprotein B of human cytomegalovirus activates the interferon-responsive pathway. Mol Cell Biol 19:3607-3613
Britt WJ, Mach M (1996) Human cytomegalovirus glycoproteins. Intervirology 39:401-412
Browne EP, Shenk T (2003) Human cytomegalovirus UL83-coded pp65 virion protein inhibits antiviral gene expression in infected cells. Proc Natl Acad Sci U S A 100:11439-11444
Cantrell SR, Bresnahan WA (2006a) Human cytomegalovirus (HCMV) UL82 gene product (pp71) relieves hDaxx-mediated repression of HCMV replication. J Virol 80:6188-6191

Cantrell SR, Bresnahan WA (2006b) Interaction between the human cytomegalovirus UL82 gene product (pp71) and hDaxx regulates immediate-early gene expression and viral replication. J Virol 79:7792-7802

Caposio P, Dreano M, Garotta G, Gribaudo G, Landolfo S (2004) Human cytomegalovirus stimulates cellular IKK2 activity and requires the enzyme for productive replication. J Virol 78:3190-3195

Caposio P, Luganini A, Hahn G, Landolfo S, Gribaudo G (2007) Activation of the virus-induced IKK/NFκ-B signalling axis is critical for the replication of human cytomegalovirus in quiescent cells. Cell Microbiol 9: 2040-2054

Castillo JP, Kowalik TF (2004) HCMV infection: modulating the cell cycle and cell death. Int Rev Immunol 23:113-139

Cobbs CS, Harkins L, Samanta M, Gillespie GY, Bharara S, King PH, Nabors LB, Cobbs CG, Britt WJ (2002) Human cytomegalovirus infection and expression in human malignant glioma. Cancer Res 62:3347-3350

Compton T, Kurt-Jones EA, Boehme KW, Belko J, Latz E, Golenbock DT, Finberg RW (2003) Human cytomegalovirus activates inflammatory cytokine responses via CD14 and Toll-like receptor 2. J Virol 77:4588-4596

Cooper GM, Hausman RE (2007) Cell Signaling. In: Press A (ed) The cell: a molecular approach. Sinauer Associates, Inc., Sunderland, MA, pp 599-648

DeMeritt IB, Yurochko AD (eds) (2006) The role of cellular transcription factors in the immediate-early stages of β-herpesvirus replication. Transworld Research Network, Kerla, India

DeMeritt IB, Milford LE, Yurochko AD (2004) Activation of the NFκ-B pathway in human cytomegalovirus-infected cells is necessary for efficient transactivation of the major immediate-early promoter. J Virol 78:4498-4507

DeMeritt IB, Podduturi JP, Tilley AM, Nogalski M, Yurochko AD (2006) Prolonged activation of NFκ-B by human cytomegalovirus promotes efficient viral replication and late gene expression. Virology 346:15-31

Eickhoff JE, Cotten M (2005) NFκ-B activation can mediate inhibition of human cytomegalovirus replication. J Gen Virol 86:285-295

Evers DL, Wang X, Huang ES (2004) Cellular stress and signal transduction responses to human cytomegalovirus infection. Microbes Infect 6:1084-1093

Feire AL, Koss H, Compton T (2004) Cellular integrins function as entry receptors for human cytomegalovirus via a highly conserved disintegrin-like domain. Proc Natl Acad Sci U S A 101:15470-15475

Feng X, Schröer J, Yu D, Shenk T (2006) Human cytomegalovirus pUS24 is a virion protein that functions very early in the replication cycle. J Virol 80:8371-8378

Gallina A, Simoncini L, Garbelli S, Percivalle E, Pedrali-Noy G, Lee KS, Erikson RL, Plachter B, Gerna G, Milanesi G (1999) Polo-like kinase 1 as a target for human cytomegalovirus pp65 lower matrix protein. J Virol 73:1468-1478

Gealy C, Humphreys C, Dickinson V, Stinski M, Caswell R (2007) An activation-defective mutant of the human cytomegalovirus IE2p86 protein inhibits NFκ-B-mediated stimulation of the human interleukin-6 promoter. J Gen Virol 88:2435-2440

Goldmacher VS, Bartle LM, Skaletskaya A, Dionne CA, Kedersha NL, Vater CA, Han JW, Lutz RJ, Watanabe S, McFarland ED, Kieff ED, Mocarski ES, Chittenden T (1999) A cytomegalovirus-encoded mitochondria-localized inhibitor of apoptosis structurally unrelated to Bcl-2. Proc Natl Acad Sci U S A 96:12536-12541

Gustems M, Borst E, Benedict CA, Perez C, Messerle M, Ghazal P, Angulo A (2006) Regulation of the transcription and replication cycle of human cytomegalovirus is insensitive to genetic elimination of the cognate NFκ-B Binding Sites in the Enhancer. J Virol 80:9899-9904

Hahn G, Revello MG, Patrone M, Percivalle E, Campanini G, Sarasini A, Wagner M, Gallina A, Milanesi G, Koszinowski U, Baldanti F, Gerna G (2004) Human cytomegalovirus UL131-128 genes are indispensable for virus growth in endothelial cells and virus transfer to leukocytes. J Virol 78:10023-10033

Hwang J, Kalejta RF (2007) Proteasome-dependent, ubiquitin-independent degradation of Daxx by the viral pp71 protein in human cytomegalovirus-infected cells. Virology 667:334-338

Isaacson MK, Feire AL, Compton T (2007) The epidermal growth factor receptor is not required for human cytomegalovirus entry or signaling. J Virol 81:6241-6247

Isomura H, Tsurumi T, Stinski MF (2004) Role of the proximal enhancer of the major immediate-early promoter in human cytomegalovirus replication. J Virol 78:12788-12799

Isomura H, Stinski MF, Kudoh A, Daikoku T, Shirata N, Tsurumi T (2005) Two Sp1/Sp3 binding sites in the major immediate-early proximal enhancer of human cytomegalovirus have a significant role in viral replication. J Virol 79:9597-9607

Kalejta RF (2004) Human cytomegalovirus pp71: a new viral tool to probe the mechanisms of cell cycle progression and oncogenesis controlled by the retinoblastoma family of tumor suppressors. J Cell Biochem 93:37-45

Kalejta RF, Shenk T (2002) Manipulation of the cell cycle by human cytomegalovirus. Front Biosci 7:295-306

Kalejta RF, Shenk T (2003a) Proteasome-dependent, ubiquitin-independent degradation of the Rb family of tumor suppressors by the human cytomegalovirus pp71 protein. Proc Natl Acad Sci U S A 100:3263-3268

Kalejta RF, Shenk T (2003b) The human cytomegalovirus UL82 gene product (pp71) accelerates progression through the G1 phase of the cell cycle. J Virol 77:3451-3459

Kalejta RF, Bechtel JT, Shenk T (2003) Human cytomegalovirus pp71 stimulates cell cycle progression by inducing the proteasome-dependent degradation of the retinoblastoma family of tumor suppressors. Mol Cell Biol 23:1887-1895

Keay S, Baldwin B, Smith MW, Wasserman SS, Goldman WF (1995) Increases in $[Ca^{2+}]_i$ mediated by the 92.5-kDa putative cell membrane receptor for HCMV gp86. Am J Physiol 269: C11-C21

Kim HJ, Hawke N, Baldwin AS (2006) NFκ-B and IKK as therapeutic targets in cancer. Cell Death Differ 13:738-747

Lu M, Shenk T (1999) Human cytomegalovirus UL69 protein induces cells to accumulate in G1 phase of the cell cycle. J Virol 73:676-683

Maidji E, Genbacev O, Chang HT, Pereira L (2007) Developmental regulation of human cytomegalovirus receptors in cytotrophoblasts correlates with distinct replication sites in the placenta. J Virol 81:4701-4712

Mar EC, Patel PC, Huang ES (1981) Human cytomegalovirus-associated DNA polymerase and protein kinase activities. J Gen Virol 57:149-156

McCormick AL, Skaletskaya A, Barry PA, Mocarski ES, Goldmacher VS (2003) Differential function and expression of the viral inhibitor of caspase 8-induced apoptosis (vICA) and the viral mitochondria-localized inhibitor of apoptosis (vMIA) cell death suppressors conserved in primate and rodent cytomegaloviruses. Virology 316:221-233

Melnick JL, Adam E, DeBakey ME (1993) Human cytomegalovirus and atherogenesis. In: Becker Y, Darai G, Huang ES (eds) Molecular aspects of human cytomegalovirus diseases. Springer-Verlag, Berlin, pp 80-91

Michelson S, Turowski P, Picard L, Goris J, Landini MP, Topilko A, Hemmings B, Bessia C, Garcia A, Virelizier JL (1996) Human cytomegalovirus carries serine/threonine protein phosphatases PP1 and a host-cell derived PP2A. J Virol 70:1415-1423

Mocarski ES Jr, Shenk T, Pass RF (2007) Cytomegaloviruses. In: Knipe DM, Howley PM (eds) Fields virology, vol 2. Lippincott Williams & Wilkins, Philadelphia, pp 2701-2772

Munger J, Yu D, Shenk T (2006) UL26-deficient human cytomegalovirus produces virions with hypophosphorylated pp28 tegument protein that is unstable within newly infected cells. J Virol 80:3541-3548

Myerson D, Hackman RC, Nelson JA, Ward DC, McDougall JK (1984) Widespread presence of histologically occult cytomegalovirus. Hum Pathol 15:430-439

Nogalski MT, Podduturi JP, DeMeritt IB, Milford LE, Yurochko AD (2007) The HCMV virion possesses an activated CKII that allows for the rapid phosphorylation of the inhibitor of NFκ-B, IκBα. J Virol 81:5305-5314

Pajovic S, Wong EL, Black AR, Azizkhan JC (1997) Identification of a viral kinase that phosphorylates specific E2Fs and pocket proteins. Mol Cell Biol 17:6459-6464

Patrone M, Secchi M, Bonaparte E, Milanesi G, Gallina A (2007) Cytomegalovirus UL131-128 products promote gB conformational transition and gB-gH interaction during entry in endothelial cells. J Virol E 81:11479-11488

Poole E, King CA, Sinclair JH, Alcami A (2006) The UL144 gene product of human cytomegalovirus activates NFκ-B via a TRAF6-dependent mechanism. EMBO J 25:4390-4399

Prichard MN, Britt WJ, Daily SL, Hartline CB, Kern ER (2005) Human cytomegalovirus UL97 kinase is required for the normal intranuclear distribution of pp65 and virion morphogenesis. J Virol 79:15494-15502

Reboredo M, Greaves RF, Hahn G (2004) Human cytomegalovirus proteins encoded by UL37 exon 1 protect infected fibroblasts against virus-induced apoptosis and are required for efficient virus replication. J Gen Virol 85:3555-3567

Romanowski MJ, Garrido-Guerrero E, Shenk T (1997) pIRS1 and pTRS1 are present in human cytomegalovirus virions. J Virol 71:5703-5705

Saffert RT, Kalejta RF (2006) Inactivating a cellular intrinsic immune defense mediated by Daxx is the mechanism through which the human cytomegalovirus pp71 protein stimulates viral immediate-early gene expression. J Virol 80:3863-3871

Schierling K, Stamminger T, Mertens T, Winkler M (2004) Human cytomegalovirus tegument proteins ppUL82 (pp71) and ppUL35 interact and cooperatively activate the major immediate-early enhancer. J Virol 78:9512-9523

Shen CY, Ho MS, Chang SF, Yen MS, Ng HT, Huang E-S, Wu CW (1993) High rate of concurrent genital infections with human cytomegalovirus and human papillomaviruses in cervical cancer patients. J Infect Dis 168:449-452

Simmen KA, Singh J, Luukkonen BG, Lopper M, Bittner A, Miller NE, Jackson MR, Compton T, Fruh K (2001) Global modulation of cellular transcription by human cytomegalovirus is initiated by viral glycoprotein B. Proc Natl Acad Sci U S A 98:7140-7145

Sinzger C, Jahn G (1996) Human cytomegalovirus cell tropism and pathogenesis. Intervirology 39:302-319

Skaletskaya A, Bartle LM, Chittenden T, McCormick AL, Mocarski ES, Goldmacher VS (2001) A cytomegalovirus-encoded inhibitor of apoptosis that suppresses caspase-8 activation. Proc Natl Acad Sci U S A 98:7829-7834

Smith MS, Bentz GL, Alexander JS, Yurochko AD (2004a) Human cytomegalovirus induces monocyte differentiation and migration as a strategy for dissemination and persistence. J Virol 78:4444-4453

Smith MS, Bentz GL, Smith PM, Bivins ER, Yurochko AD (2004b) HCMV activates PI(3)K in monocytes and promotes monocyte motility and transendothelial migration in a PI(3)K-dependent manner. J Leukoc Biol 76:65-76

Smith MS, Bivins-Smith ER, Tilley AM, Bentz GL, Chan G, Minard J, Yurochko AD (2007) Roles of PI(3)K and NFκ-B in HCMV-mediated monocyte diapedesis and adhesion: strategy for viral persistence. J Virol 81:7683-7694

Soderberg-Naucler C (2006) Does cytomegalovirus play a causative role in the development of various inflammatory diseases and cancer? J Intern Med 259:219-246

Speir E, Modali R, Huang ES, Leon MB, Shawl F, Finkel T, Epstein SE (1994) Potential role of human cytomegalovirus and p53 interaction in coronary restenosis. Science 265:391-394

Stamminger T, Gstaiger M, Weinzierl K, Lorz K, Winkler M, Schaffner W (2002) Open reading frame UL26 of human cytomegalovirus encodes a novel tegument protein that contains a strong transcriptional activation domain. J Virol 76:4836-4847

Streblow DN, Soderberg-Naucler C, Vieira J, Smith P, Wakabayashi E, Ruchti F, Mattison K, Altschuler Y, Nelson JA (1999) The human cytomegalovirus chemokine receptor US28 mediates vascular smooth muscle cell migration. Cell 99:511-520

Streblow DN, Orloff SL, Nelson JA (2001a) Do pathogens accelerate atherosclerosis? J Nutr 131:2798-2804

Streblow DN, Orloff SL, Nelson JA (2001b) The HCMV chemokine receptor US28 is a potential target in vascular disease. Curr Drug Targets Infect Disord 1:151-158

Stropes MP, Miller WE (2004) Signaling and regulation of G-protein coupled receptors encoded by cytomegaloviruses. Biochem Cell Biol 82:636-642

Taylor RT, Bresnahan WA (2006a) Human cytomegalovirus IE86 attenuates virus- and tumor necrosis factor alpha-induced NFκB-dependent gene expression. J Virol 80:10763-10771

Taylor RT, Bresnahan WA (2006b) Human cytomegalovirus immediate-early 2 protein IE86 blocks virus-induced chemokine expression. J Virol 80:920-928

van Cleef KW, Smit MJ, Bruggeman CA, Vink C (2006) Cytomegalovirus-encoded homologs of G protein-coupled receptors and chemokines. J Clin Virol 35:343-348

Varnum SM, Streblow DN, Monroe ME, Smith P, Auberry KJ, Pasa-Tolic L, Wang D, Camp DGI, Rodland K, Wiley S, Britt W, Shenk T, Smith RD, Nelson JA (2004) Identification of proteins in human cytomegalovirus (HCMV) particles: the HCMV proteome. J Virol 78:10960-10966

Waldman WJ, Adams PW, Knight DA, Sedmak DD (1997) CMV as an exacerbating agent in transplant vascular sclerosis: potential immune-mediated mechanisms modelled in vitro. Transplant Proc 29:1545-1546

Wang D, Shenk T (2005a) Human cytomegalovirus UL131 open reading frame is required for epithelial cell tropism. J Virol 79:10330-10338

Wang D, Shenk T (2005b) Human cytomegalovirus virion protein complex required for epithelial and endothelial cell tropism. Proc Natl Acad Sci U S A 102:18153-18158

Wang X, Huong SM, Chiu ML, Raab-Traub N, Huang E-S (2003) Epidermal growth factor receptor is a cellular receptor for human cytomegalovirus. Nature 424:456-461

Wang X, Huang DY, Huong SM, Huang E-S (2005) Integrin $\alpha_v\beta_3$ is a coreceptor for human cytomegalovirus. Nat Med 11:515-521

Yurochko AD, Huang E-S (1999) HCMV binding to human monocytes induces immuno-regulatory gene expression. J Immunol 162:4806-4816

Yurochko AD, Liu DY, Eierman D, Haskill S (1992) Integrins as a primary signal transduction molecule regulating monocyte immediate-early gene induction. Proc Natl Acad Sci USA 89:9034-9038

Yurochko AD, Kowalik TF, Huong S-M, Huang E-S (1995) HCMV upregulates NFκ-B activity by transactivating the NFκ-B p105/p50- and p65-promoters. J Virol 69:5391-5400

Yurochko AD, Hwang E-S, Rasmussen L, Keay S, Pereira L, Huang E-S (1997a) The human cytomegalovirus UL55 (gB) and UL75 (gH) glycoprotein ligands initiate the rapid activation of Sp1 and NFκ-B during infection. J Virol 71:5051-5059

Yurochko AD, Mayo MW, Poma EE, Baldwin AS Jr, Huang E-S (1997b) Induction of the transcription factor Sp1 during human cytomegalovirus infection mediates upregulation of the p65 and p105/p50 NF-κB promoters. J Virol 71:4638-4648

Zhu H, Shen Y, Shenk T (1995) Human cytomegalovirus IE1 and IE2 proteins block apoptosis. J Virol 69:7960-7970

Chemokines and Chemokine Receptors Encoded by Cytomegaloviruses

P. S. Beisser, H. Lavreysen, C. A. Bruggeman, C. Vink(✉)

Contents

Introduction 222
Evolution of CMV vCK Genes 223
   The Role of vCKs During CMV Infection 226
Chemokine Receptors Encoded by CMVs 227
   Evolution of CMV vGPCR Genes 227
   Modulation of Intracellular Signaling by CMV vGPCRs 229
   CMV vGPCR Gene Deletion Mutants 234
   Localization of CMV vGPCRs 235
Perspectives 237
References 237

Abstract CMVs carry several genes that are homologous to genes of the host organism. These include genes homologous to those encoding chemokines (CKs) and G protein-coupled receptors (GPCRs). It is generally assumed that these CMV genes were hijacked from the host genome during the long co-evolution of virus and host. In light of the important function of the CK and GPCR families in the normal physiology of the host, it has previously been hypothesized that the CMV homologs of these proteins, CMV vCKs and vGPCRs, may also have a significant impact on this physiology, such that lifelong maintenance and/or replication of the virus within the infected host is guaranteed. In addition, several of these homologs were reported to have a major impact in the pathogenesis of infection. In this review, the current state of knowledge on the CMV vCKs and vGPCRs will be discussed.

Abbreviations AC: Adenylyl cyclase; cAMP: Cyclic adenosyl monophosphate; CCMV: Chimpanzee cytomegalovirus; CCR[n] (e.g. CCR5): CC chemokine receptor [n]; CMV: Cytomegalovirus; CRE: Cyclic adenosyl monophosphate responsive

---

C. Vink
Laboratory of Pediatrics, Erasmus Medical Center Rotterdam, PO Box 2040,
3000 CA, Rotterdam, The Netherlands
c.vink@erasmusmc.nl

element; CREB: Cyclic adenosyl monophosphate responsive element-binding factor; ECK-3:England strain rat cytomegalovirus chemokine 3; ELC: Epstein-Barr virus-induced chemokine receptor ligand chemokine; ERK[n] (e.g., ERK1): Extracellular signal-regulated protein kinase [n]; FAK: Focal adhesion kinase; GpCMV: Guinea pig cytomegalovirus; GPCR: G protein-coupled receptor; HCMV: Human cytomegalovirus; HHV-6: Human herpesvirus 6; HHV-7: Human herpesvirus 7; HIV: Human immunodeficiency virus; InsP: Inositol phosphate; KSHV: Kaposi's sarcoma-associated herpesvirus; MCK-2: Murine cytomegalovirus chemokine 2; MCMV: Murine cytomegalovirus; MCP-1: Monocyte chemoattractant protein 1; MDC: Macrophage-derived chemokine; MIP-1α: Macrophage inflammatory protein 1α; NFκ-B: Nuclear factor κB; ORF: Open reading frame; p38/MAPK: 38-kD Mitogen-activated protein kinase; PLC: Phospholipase C; PTK: Protein tyrosine kinase; Rac: Rat sarcoma homolog A-related C3 botulinum toxin substrate; RANTES: Regulated upon activation normal T cell expressed, and secreted; RCK-[n] (e.g. RCK-3): Maastricht strain rat cytomegalovirus chemokine [n]; RCMV: Rat cytomegalovirus; RhCMV: Rhesus macaque cytomegalovirus; RhoA: Rat sarcoma homolog A; SLC: Secondary lymphoid tissue chemokine; SRE: Serum-responsive element; TARC: Thymus and activation-regulated chemokine; vCK: Viral chemokine; vCXC-[n] (e.g. vCXC-1): Human cytomegalovirus CXC chemokine [n]; vGPCR: Viral G protein-coupled receptor; vMIP-II: Kaposi's sarcoma-associated herpesvirus macrophage inflammatory protein II; VSV: Vesicular stomatitis virus

Introduction

Cytomegaloviruses (CMVs) are species-specific betaherpesviruses that establish life-long persistence in their hosts. Their genomes, the largest among herpesviruses, are estimated to contain between 165 (Davison et al. 2003) and 252 (Murphy et al. 2003b; see the chapter by E. Murphy and T. Shenk, this volume) potential open reading frames (ORFs) encompassing up to 241,087 bp of double-stranded DNA (Davison et al. 2003). Approximately 41 human CMV (HCMV) ORFs belong to a core set of genes essential for viral replication in vitro, such as genes encoding DNA polymerase, capsid, matrix and envelope proteins (Yu et al. 2003). Approximately 88 genes were found to be nonessential for efficient CMV replication in vitro (Yu et al. 2003). Some of these nonessential CMV genes are homologous to genes of the host. Among these homologs are genes that share similarities with genes encoding proteins that are associated with the immune system, such as class I MHC proteins (Wills et al. 2005; Prod'homme et al. 2007), a TCR gamma chain (Beck and Barrell 1992), IL-10 (Wagner et al. 2003) and a TNF receptor (Poole et al. 2006). Apparently these gene homologs have been acquired from the host organism and subsequently modified during approximately 180 My of co-evolution (Davison 2002) in order to enable dissemination and maintain life-long persistence. Interestingly, CMVs possess two distinct groups of genes homologous to those of

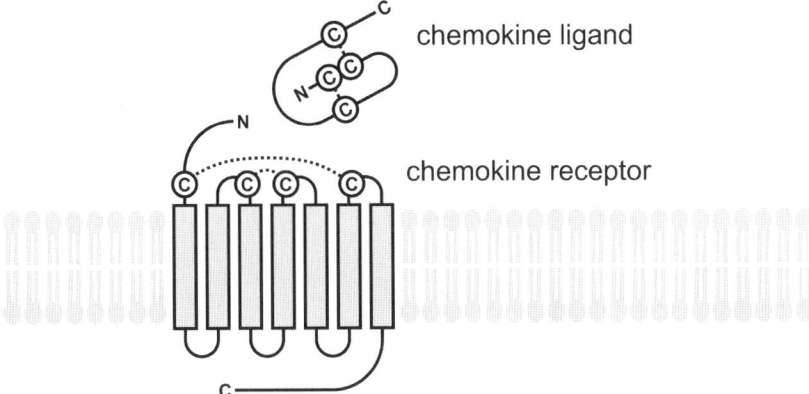

Fig. 1 The 2D peptide structures of a CC chemokine ligand and a chemokine receptor. The *N* and the *C* denote the amino and carboxyl termini, respectively. The encircled *C*s represent conserved cysteine residues. The *dashed lines* indicate conserved disulphide bridges. *Grey boxes* indicate hydrophobic transmembrane alpha helix domains

the host: (a) CC and CXC chemokine genes and (b) G protein-coupled receptors (GPCR) genes, the majority of which resembles chemokine receptor genes. GPCRs form a large family of 7-transmembrane receptors (Fig. 1) that include sensory receptors for sight, smell, and taste as well as receptors for many neurotransmitters, peptide hormones and chemokines. Chemokines comprise a family of immune modulatory cytokine peptides. Currently, four classes of chemokines are known. The classification is based on a conserved structure (Fig. 1) that includes either a single cysteine (C), a CC motif (Fig. 1), a CXC motif or a CX3C motif. Chemokines can be released to initiate inflammatory responses by acting as chemoattractant for infiltrating leukocytes (Glass et al. 2003). They can also stimulate differentiation, maturation and activation of many types of immune-related cells (Glass et al. 2003). Two chemokines, CXCL16 and CX3CL-1, were shown to function as adhesion molecules for leukocytes that are captured from the bloodstream onto the endothelial surface (Haskell et al. 2000; Nakayama et al. 2003). The purpose of this review is to summarize the (putative) functions of the CC and CXC chemokines (vCKs) as well as the chemokine-like GPCRs (vGPCRs) that are encoded by CMV.

Evolution of CMV vCK Genes

To date, three vCK genes have been identified in the HCMV genome (Fig. 2), the CC chemokine-like gene UL128, and the CXC chemokine-like genes UL146 and UL147. To some extent, these genes are conserved among primate CMVs (Table 1). UL128-like genes are also present on the genomes of murine CMV (MCMV) and rat CMV (RCMV), as well as on the genome of human herpesvirus type 6 (HHV-6) (Table 1). The rodent CMV and HHV-6 species lack UL146- and UL147-like

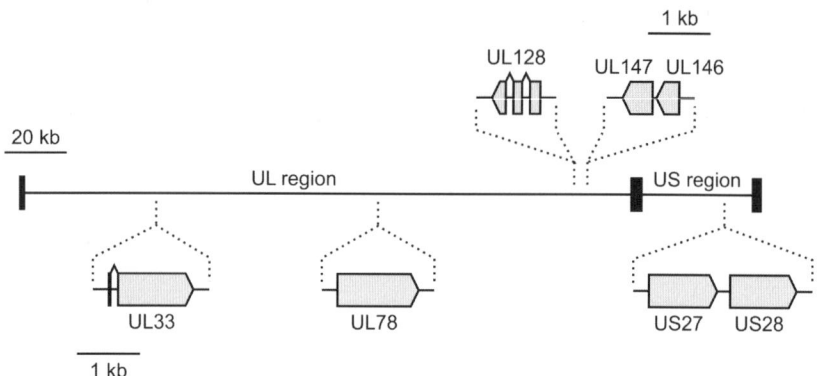

Fig. 2 CMV chemokine and chemokine receptor homolog gene loci. The central line represents the HCMV Merlin strain genome (235,645 kb) derived from GenBank accession NC_006273. The *black boxes* on this line represent repetitive regions. The chemokine homolog gene loci are enlarged above the genomic line, the chemokine receptor homolog gene loci below the genomic line. The *arrowheads* indicate the sizes and directions of the coding content of the genes. The UL128 exon sizes and positions are analogous to those of the HCMV AD169 strain, as indicated in NC_006273. *UL* unique long, *US* unique short

genes. Similar to the UL33- and UL78-like genes, the UL128-like chemokine genes are likely to have been acquired by betaherpesviruses at least 110 My ago. The conservation of this gene family suggests an essential role for these genes in the survival of betaherpesviruses in vivo. This notion is supported by the observation that the UL128 sequences from different clinical HCMV isolates are highly conserved (Baldanti et al. 2006). Paradoxically, the UL128 gene appears to be functional in laboratory strain HCMV AD169, whereas in the clinical, low-passage HCMV strains Toledo and Merlin, as well as in the chimpanzee CMV (CCMV) laboratory strain, the UL128-like genes are disrupted by inversions or frame shift mutations (Table 1). A UL128 counterpart was also identified on the HHV-6 genome, designated U83A. This gene was found to encode a potent CC chemokine (Derwin et al. 2006). Interestingly, the UL130 genes of the primate CMVs also contain chemokine-like sequences, including a CXC domain (Glass et al. 2003). However, the UL130-derived amino acid sequences lack other cysteine residues that are essential for classical chemokine folding (Glass et al. 2003). The MCMV and RCMV genome sequences available to date do not contain UL130-like CXC chemokine sequences. Nevertheless, within the genome of the Maastricht strain of RCMV, a second CC chemokine-like gene, r131, was identified adjacent to r129. These two genes may have originated from duplication of a common ancestor gene (Kaptein et al. 2004; Voigt et al. 2005). In contrast to the UL128-like genes, UL146 and UL147 appear to be restricted to the primate CMVs. Both the HCMV and CCMV UL146 genes encode a CXC vCK (Table 1). Rhesus macaque CMV (RhCMV) does not possess a UL146 homolog (Table 1). Both the HCMV and CCMV genomes contain a UL147 gene, whereas a UL147-like gene is present in

Table 1 Cytomegalovirus chemokine-like genes

CMV Species	Gene (product name)	GenBank accession	References	Comments
HCMV	UL128	NC_006273	Akter et al. 2003, Hahn et al. 2004	Intact in AD169, disrupted in Merlin and Toledo
CCMV	UL128	AF480884	Akter et al. 2003, Hahn et al. 2004	Disrupted
RhCMV	rhUL128	DQ120516	Rivailler et al. 2006	Intact in RhCMV 180.92, not present in RhCMV68.1
RCMV	r129 (RCK–3/ECK–3)	AF232689	Kaptein et al. 2004	
RCMV	r131 (RCK–2)	AF232689	Kaptein et al. 2004	
MCMV	m131–129 (MCK–1; MCK–2, alternative splice product)	U68299	MacDonald et al. 1999	
HHV–6	U83A	NC_001664	Derwin et al. 2006, Catusse et al. 2007	
GpCMV	GPCMV-MIP	AF500307	Haggerty and Schleiss 2002, Penfold et al. 2003a	Unique, MIP–1-like
HCMV	UL146 (vCXC–1)	NC_006273	Penfold et al. 1999	Intact in Towne, Toledo, Merlin, lost in AD169
CCMV	UL146 (vCXC–1)	AF480884	Miller-Kittrell et al. 2007	
CCMV	UL146A	AF480884	Davison et al. 2003	UL146-like, not present in other CMVs
HCMV	UL147 (vCXC–2)	NC_006273	Penfold et al. 1999	Intact in Towne, Toledo, Merlin, lost in AD169
CCMV	UL147	AF480884	Davison et al. 2003	
RhCMV	rh158	AY186194	Rivailler et al. 2006	Intact in 68.1, not present in 180.92
RhCMV	rh156.2	DQ120516	Rivailler et al. 2006	
CCMV	UL156	AF480884	Davison et al. 2003	Splice variant may encode CXC chemokine, not present in other CMVs
CCMV	UL157	AF480884	Davison et al. 2003	UL146-like, not present in other CMVs

only one of two available genomic RhCMV sequences (Table 1). The sequences of UL146 and UL147 derived from numerous clinical HCMV isolates showed an extensive level of variability. This variability was significant among interindividual strains (Hassan-Walker et al. 2004; Stanton et al. 2004). However, the sequence of individual strains remained absolutely stable over time in vitro and in vivo, which indicates that sequence drift is not a mechanism for the observed sequence hypervariability (Lurain et al. 2006). UL146 was found to be the more rapidly evolving paralog (Arav-Boger et al. 2005). Despite the observed hypervariability, no specific UL146 or UL147 genotype was associated with disease outcome in newborns with CMV-associated congenital disorders (Arav-Boger et al. 2006; He et al. 2006). Interestingly, a single CXC vCK-like gene (rh156.2) and two CXC vCK-like genes (UL156 and UL157) have been identified in RhCMV and CCMV, respectively (Table 1). While these genes share significant similarity with UL146-like genes, they do not have counterparts in HCMV or rodent CMVs. This indicates that in contrast to UL128-like genes, the UL146-like and UL147-like genes are rapidly evolving in vivo. Finally, a distinct CC vCK was reported to be encoded by guinea pig cytomegalovirus (GpCMV; Haggerty and Schleiss 2002). Both the genomic localization and the DNA sequence of the MIP gene are unique for GpCMV. No counterparts have been found on the genomes of other herpesviruses (Table 1).

The Role of vCKs During CMV Infection

The HCMV UL128 is functionally clustered within the UL128/UL130/UL131A locus. Three mutant HCMV strains were generated in which any of these genes was disrupted. Each of these strains completely lost its ability to replicate in human umbilical cord endothelial cells, as well as its ability to transfer from one cell to another in cultured polymorphonuclear leukocytes (PMNs) and monocytes (Hahn et al. 2004). Thus, these genes appear to determine endothelial cell tropism as well as cell-to-cell passage in vitro (Hahn et al. 2004). The mechanisms by which the UL128, UL130 and UL131A genes govern these processes have recently been addressed. It was suggested that the proteins encoded by the UL128/UL130/UL131A locus might act as ligands for receptors that convey signals into endothelial cells to facilitate intracellular transport or inactivation of innate intracellular antiviral immunity (Patrone et al. 2005). Moreover, these proteins are a component of the attachment/entry machinery, either by acting as a soluble factor or as a virion component, permitting a viral entry pathway that differs from that used in fibroblasts (Patrone et al. 2005). These proteins were also suggested to be involved in the final stages of virus morphogenesis and maturation at membranes (Hahn et al. 2004), as well as in attraction–adhesion of leukocytes to endothelial cells (Hahn et al. 2004). Finally, UL128, UL130 (Wang and Shenk, 2005) and UL131 (Adler et al. 2006) have been shown to be part of a complex with gH/gL in the virion and to play a direct role in entry into epithelial and endothelial cells. The chemotactic activity of each individual gene product of the UL131–128 locus, as well as the potential cooperation with other viral or cellular gene signaling molecules, remains to be elucidated. The HHV-6 counterpart of UL128, U83A, was found to encode a potent CC chemokine capable of inducing Ca^{2+} mobilization and chemotaxis in T lymphocytes (Derwin et al. 2006). However, this provides little insight in the function of

CMV-encoded UL128-like vCKs, since CMVs and HHV-6 occupy different tissue compartments and cell types in the host.

The MCMV counterpart of UL128, m131, was found to encode a potent CC vCK designated MCP-1, which was capable of inducing Ca^{2+} mobilization and chemotaxis of macrophages and cells expressing human CCR3 (Saederup et al. 1999). Similarly, the unique GpCMV MIP chemokine was found to invoke signaling and chemotaxis in cells expressing human CCR1 (Penfold et al. 2003a). These findings indicate a role for the rodent CMV-encoded CC chemokines in leukocyte attraction. It was hypothesized that leukocytes recruited by vCKs can subsequently be subverted by CMV to serve as vehicles to enable viral dissemination. This claim was supported by deletion mutant experiments in vivo. Mutant MCMV and RCMV strains from which the m129/m131 or r131 locus had been deleted, respectively, exhibited reduced virus levels in salivary gland, liver and spleen tissue during acute infection (Saederup et al. 1999; Kaptein et al. 2004). Moreover, leukocyte infiltration in infected foot pad experiments was significantly lower in mice and rats treated with deletion mutant virus (Saederup et al. 1999; Kaptein et al. 2004). Finally, monocyte-associated viremic peak levels in mice infected with the m129/m131 deletion mutant were dramatically lower than those in mice infected with wild-type MCMV (Saederup et al. 1999).

Both HCMV and CCMV UL146 encode potent vCKs, which were designated vCXC-1. Recombinant vCXC-1 peptides derived from both CMV species were found to be capable of inducing calcium mobilization, chemotaxis, and degranulation via stimulation of CXCR2, as well as inducing integrin upregulation and apoptosis in human neutrophils (Penfold et al. 1999; Miller-Kittrell et al. 2007). To study the function of UL146 and UL147 further, mutant HCMV strains were generated in which both genes were disrupted. Viral passage to PMN was reduced in these strains, but not viral passage in monocytes (Hahn et al. 2004). Thus, the chemotactic factors encoded by HCMV UL146 and UL147 appear to be dispensable for viral growth and dissemination in vitro. In order to determine the role of these genes in vivo, the CCMV model may be the most suitable.

Chemokine Receptors Encoded by CMVs

Evolution of CMV vGPCR Genes

Both vGPCR and vCK genes have been identified within the genomes of beta- and gammaherpesviruses, but not within those of alphaherpesviruses. There is no apparent evolutionary relationship for these genes between the beta- and gammaherpesvirus subfamilies. Yet, within the betaherpesvirus subfamily, both vGPCR and vCK genes appear to be conserved among murine and primate CMV species, as well as the HHV-6 and human herpesvirus type 7 (HHV-7). The HCMV genome contains four vGPCR genes, UL33, UL78, US27 and US28 (Fig. 2) (Murphy et al. 2003a). These genes are conserved among all known primate CMVs (Table 2).

Table 2 Cytomegalovirus chemokine receptor-like genes

CMV Species	Gene	GenBank accession	References
HCMV	UL33	NC_006273	Casarosa et al. 2003
CCMV	UL33	AF480884	Davison et al. 2003
RhCMV	rh56	AY186194	Hansen et al. 2003
MCMV	M33	U68299	Davis-Poynter et al. 1997
RCMV	R33	AF232689	Beisser et al. 1998
GpCMV	GP33	AF355272	Liu and Biegalke 2001
HHV-6	U12	NC_001664	Isegawa et al. 1998
HHV-7	U12	U43400	Nanako et al. 2003
HCMV	UL78	NC_006273	Michel et al. 2005
CCMV	UL78	AF480884	Davison et al. 2003
RhCMV	rh107	AY186194	Hansen et al. 2003
MCMV	M78	U68299	Oliveira et al. 2001
RCMV	R78	AF232689	Beisser et al. 1999
GpCMV	GP78	Unavailable	Stropes and Miller 2004
HHV-6	U51	NC_001664	Milne et al. 2000
HHV-7	U51	U43400	Tadagaki et al. 2005
HCMV	US27	NC_006273	Fraile-Ramos et al. 2002
CCMV	US27	AF480884	Sahagun-Ruiz et al. 2004
HCMV	US28	NC_006273	Gao and Murphy 1994
CCMV	US28	AF480884	Davison et al. 2003
RhCMV	rh214, rh215, rh216, rh218, rh220	AY186194	Sahagun-Ruiz et al. 2004

UL33 and UL78 homologs are also present within the genomes of murine, rat and guinea pig CMV (GpCMV), as well as on the genomes of HHV-6 and -7 (Table 2). These species lack US27- and US28-like genes. HCMV and HHV-6 species were estimated to have diverged approximately 110 My ago (Davison 2002). Since the genomic locations of UL33- and UL78-like genes are highly conserved among all known betaherpesvirus species, it is likely that these genes were acquired by a common ancestor of the betaherpesviruses, rather than independently, after divergence of the different betaherpesvirus species. The notion that the UL33 and UL78 gene families have been maintained over such a long period of time suggests an essential role for these genes in the survival of betaherpesviruses in vivo. The US27- and US28-like genes have only been identified in primate CMV species. It is likely that these genes have emerged somewhere after the branching of a common ancestor of rodents and primates, 100 My ago (Li et al. 1990). Interestingly, RhCMV possesses five consecutive US27/US28-like genes (Penfold et al. 2003b) rather than the two genes (US27 and US28) found in CCMV (Davison et al. 2003) and HCMV (Murphy et al. 2003a). Since the sequences of US27- and US28-like genes in primate CMVs are highly similar and consecutively positioned with their respective genomes, it is likely that these genes have emerged from a single hijacked host GCPR gene-by-gene multiplication.

All CMV vGPCRs, with the exception of those encoded by UL78-like genes, contain the hallmarks of chemokine receptors (Ahuja et al. 1993; Davis-Poynter et al. 1997):

1. An N-linked glycosylation site and several negatively charged amino acid residues located in the extracellular N-terminal region
2. Two cysteine residues, which are likely to form a disulfide bridge, thereby joining the N-terminal region with the third extracellular loop (Fig. 1)
3. Several positively charged amino acid residues within the third intracellular loop
4. Invariant amino acids within the transmembrane regions
5. Several serine and threonine residues in the intracellular C-terminal region

The sequences of US27/US28-like genes have the highest similarity with those of chemokine receptors of the host. The sequences of UL33-like genes are also related to those of chemokine receptors, albeit to a lesser extent. The sequences of UL78-like GPCR genes possess none of the chemokine receptor hallmarks. Interestingly, the HCMV genome contains several putative genes encoding 7-transmembrane proteins: UL100 (the putative structural glycoprotein **M**), US12, US13, US14, US15, US16, US17, US18, US19, US20, and US21 (possibly a multiplication of a US12-like ancestor). These genes share some sequence similarity with genes of well-characterized GPCRs (Rigoutsos et al. 2003). Yet, their predicted amino acid sequences lack the cysteine residues that join the second and third extracellular loop of all known GPCRs and are therefore not considered vGPCRs. Nevertheless, it remains tempting to speculate that these genes have arisen from host GPCR genes.

Modulation of Intracellular Signaling by CMV vGPCRs

Various CMV vGPCRs have been investigated for their putative role in the activation of signal transduction pathways. Attempts have been made to identify ligands that bind to these vGPCRs, as well as to identify the downstream intracellular signaling pathways. The most common signaling factors studied were:

1. Ca^{2+} mobilization
2. Increase of inositol phosphate (InsP) by phospholipase C (PLC)
3. Increase of cAMP by adenylyl cyclase (AC) and subsequent CREB-mediated gene transcription
4. NFκ-B-mediated gene transcription

In some cases, other signaling factors were addressed. The results of these studies are summarized in Table 3. The consequences of modulation of these intermediates by vGPCRs may include either activation or inhibition of immune-related responses, such as cell differentiation and maturation, cell proliferation, cytoskeletal remodeling, cell migration, cytokine release, and cytotoxicity. Apparently, viruses benefit from altering these responses by utilizing their own vGPCRs.

For at least one member of each family included in Table 3, one or more ligands were identified. Interestingly, all of these ligands are CC chemokines, of which RANTES appears to be the most common. Some vGPCRs, most notably those encoded by the HHV-6 U51 and CMV US28, interact with more than one CC chemokine species. The combinations of chemokines that interact with each of these vGPCRs are unique and unrelated to those specific for chemokine receptors of the host. The most direct way of determining ligand-dependent vGPCR signaling, as well as subsequent vGPCR desensitization, is measuring intracellular Ca^{2+} mobilization (Table 3). In addition to Ca^{2+} mobilization, cytoplasmic accumulation of inositol phosphates (InsP) was measured in cells that expressed a member of either of the three betaherpesvirus vGPCR families. Interestingly, in all of these studies, vGPCRs were reported to induce a $G_{q/11}$-mediated increase of InsP in a ligand-independent manner (Table 3). Thus, all vGPCRs were found to be constitutively active, a property that is not shared by chemokine receptors of the host. Constitutive signaling was also demonstrated by subjecting betaherpesvirus vGPCRs to reporter gene assays specific for either cAMP/CREB- or NFκ-B-driven modulation of gene transcription. Most notably, the results of these assays were inconsistent when comparing the activities of members of the UL33 vGPCR family: expression of HCMV UL33 and MCMV M33 resulted in an increase in cAMP-mediated gene transcription, whereas expression of RCMV R33 resulted in a decrease (Table 3). The most important factor responsible for this inconsistency is p38/MAPK, which is activated in cells expressing either HCMV UL33 or MCMV M33, but not in cells expressing RCMV R33 (Table 3). Additionally, expression of M33- and R33-encoded vGPCRs resulted in an increase in NFκ-B-mediated gene transcription, whereas expression of HCMV UL33 had no effect (Table 3). Interestingly, NFκ-B-mediated signaling is also increased in cells expressing HCMV US28. These findings support the hypothesis that the primate UL33-like vGPCRs have lost some signaling properties as similar functions might have become redundantly available upon hijacking US28-like genes. Alternatively, the rodent UL33-like vGPCRs may have gained NFκ-B-stimulating activity following the loss of US28-like genes from their corresponding genomes. Gain and loss of genes as well as activities during evolution of the CMVs has become apparent by comparing HCMV US28 with RhCMV US28-like genes. None of the vGPCRs encoded by the five different US28-like genes within the RhCMV genome possess constitutive signaling activities, nor do they signal in the presence of chemokine ligands that are known to modulate HCMV US28-mediated signaling (Penfold et al. 2003b). Moreover, neither of the InsP-, CREB- and NFκ-B-mediated signaling factors were affected by HCMV US27 expression (Table 3).

A recent study has indicated that the HCMV US28-encoded vGPCR not only acts individually by modulating intracellular signaling, but also constitutes a regulatory switch for signal transduction by other Gi/o-coupled receptors (Bakker et al. 2004). In addition, when either HHV-7 U12 or U51 were expressed together with the host chemokine receptors CCR4 and CCR7, they had a broader ligand specificity than when each of these chemokines were expressed individually (Table 3). This suggested that the gene products of U12 and U51 can interact with

Table 3 Signaling factors associated with CMV cGPCRs

Receptor	Ligand(s)	Ca²⁺ mobilization	PLC/InsP	AC/cAMP/CRE	NF-κB	Other signaling effects
HCMV UL33	No binding with RANTES (Casarosa et al. 2003)	Unknown	Constitutively up via $G_{\alpha q/11}$ (Waldhoer et al. 2002, Casarosa et al. 2003), constitutively up, partially via $G_{\beta\gamma}$ from $G_{i/o}$ (Casarosa et al. 2003)	Overall constitutively up (Waldhoer et al. 2002, Casarosa et al. 2003), constitutively down via $G_{\alpha i/o}$ (Casarosa et al. 2003), constitutively up via G_s and $G_{\beta\gamma}$/Rho/$G_{\alpha s}$ p38 (Waldhoer et al. 2002, Casarosa et al. 2003)	Unaffected (Waldhoer et al. 2002)	$G_{i/o}$ activation mediated by cytoplasmic tail (Casarosa et al. 2003)
MCMV M33	Mouse RANTES (Melnychuk et al. 2005)	Unknown	Constitutively up via G_{q11} (Waldhoer et al. 2002, Sherrill and Miller 2006), GRK2 scavenges $G_{alpaq/11}$, PLCβ stimulation inhibited (Sherrill and Miller 2006)	Constitutively up (Waldhoer et al. 2002, Casarosa et al. 2003) via p38 (Waldhoer et al. 2002), not ERK 1/2/MAPK (Waldhoer et al. 2002)	Constitutively up (Waldhoer et al. 2002)	Rac1 and ERK1/2 stimulation upon mouse RANTES treatment
RCMV R33	No binding with rat RANTES (Gruijthuijsen et al. 2002)	Unknown	Constitutively up via $G_{q/11}$, constitutively up, partially via $G_{i/o}$ (Gruijthuijsen et al. 2002; Casarosa et al. 2003)	Constitutively down via $G_{i/o}$ (Gruijthuijsen et al. 2002, Casarosa et al. 2003)	Constitutively up via $G_{i/o}$ (Gruijthuijsen et al. 2002)	SRE up via $G_{i/o}$, $G_{i/o}$ activation not mediated by cytoplasmic tail (Casarosa et al. 2003)
HHV-6 U12	RANTES, MIP-1α, MIP-1β, MCP-1 (Isegawa et al. 1998)	By RANTES, MIP-1α, MIP-1β, MCP-1, desensitization by all (Isegawa et al. 1998)	Unknown	Unknown	Unknown	

(continued)

Table 3 (continued)

Receptor	Ligand(s)	Ca^{2+} mobilization	PLC/InsP	AC/cAMP/CRE	NF-κB	Other signaling effects
HHV-7 U12	ELC, SLC, TARC, MDC (Tadagaki et al. 2005)	By ELC, SLC, TARC & MDC (Tadagaki et al. 2005)	Unknown	Unknown	Unknown	Broadens ligand specificity for CCR4 and CCR7 (Tadagaki et al. 2007)
HCMV UL78	Unknown	Unknown	Unknown	Unknown	Unknown	
MCMV M78	Unknown	Unknown	Unknown	Unknown	Unknown	
RCMV R78	Unknown	Unknown	Unknown	Unknown	Unknown	
HHV-6 U51	RANTES, eotaxin, MCP-1–3, and -4, KSHV vMIP-II (Milne et al. 2000; Fitzsimons et al. 2006)	By RANTES, not by MCP-1 or eotaxin (Fitzsimons et al. 2006)	Constitutively up via Gq/11, more up by RANTES stimulation, not by MCP-1 or eotaxin (Fitzsimons et al. 2006)	Constitutively down via $G_{q/11}$, up by RANTES, MCP-1 and eotaxin via Gi/o (Fitzsimons et al. 2006)	Unknown	Downregulates RANTES transcription (Milne et al. 2000)
HHV-7 U51	ELC, SLC, TARC, MDC (Tadagaki et al. 2005)	By ELC, SLC, TARC & MDC (Tadagaki et al. 2005)	Unknown	Unknown	Unknown	Alters ligand specificity for CCR4 and CCR7 (Tadagaki et al. 2007)
HCMV US27	Unknown	Unknown	Unaffected (Waldhoer et al. 2002)	Unaffected (Waldhoer et al. 2002)	Unaffected (Waldhoer et al. 2002)	
HCMV US28	RANTES, MCP-1, MIP-1α, MIP-1β,	Via $G_{αi}$ and $G_{α16}$ (Billstrom et al. 1998), by RANTES	Constitutively up via $G_{q/11}$ (Casarosa et al. 2001; Waldhoer et al. 2002), not modulated	Constitutively up (Waldhoer et al. 2002), G_i-independent (Waldhoer et al.	Constitutively up (Casarosa et al. 2001, Waldhoer et al. 2002) via	Various, see text

fractalkine, MCP-3, KSHV vMIP-II (Billstrom et al. 1998; Vieira et al. 1998; Kledal et al. 1999)	(Billstrom et al. 1998; Vieira et al. 1998), MIP-1α (Billstrom et al. 1998), MIP-1β (Vieira et al. 1998), MCP-3 (Billstrom et al. 1998), fractalkine (US patent 20020127544), desensitized for RANTES by MIP-1β (Vieira et al. 1998), MCP-3 (Billstrom et al. 1998)	by RANTES/MCP-1/MCP-3/MIP-1α/MIP-1β (Casarosa et al. 2001), partially down by fractalkine (Casarosa et al. 2001; Waldhoer et al. 2002), RANTES/MCP-1 antagonize fractalkine (Casarosa et al. 2001)	2002), via p38/MAPK (Waldhoer et al. 2002), down by fractalkine (Waldhoer et al. 2002)	$G_{\beta 2\gamma 1}$, down by fractalkine (Casarosa et al. 2001)

CCR4 and CCR7, possibly by receptor dimerization. Dimerization and subsequent changes in receptor kinetics have been demonstrated for CCR2 and CCR5 (Mellado et al. 2001). Similarly, other vGPCRs could modulate intracellular signaling by interacting with host receptors. Such a notion may inspire researchers to reevaluate orphan vGPCRs like those encoded by UL78 and US27, as well as by the rhesus US28-like vGPCR genes, which do not seem to be capable of altering signaling by themselves.

The consequences of GPCR signaling (Table 3) for the survival and replication of CMVs in the host remain in most cases unclear. Only a few of these signaling studies have led to the identification of changes in cellular behavior. The US28-encoded vGPCR was shown to trigger smooth muscle cells to undergo chemokinesis and chemotaxis via the $G_{\alpha 12/13}$/PTK/RhoA/FAK/Src pathway upon stimulation with either RANTES or MCP-1 (Streblow et al. 1999, 2003). Thus, US28 could play a role in the spread of CMV through solid tissue by using infected cells as vehicles. HCMV US28-mediated intracellular signaling could cause infected smooth muscle cells to migrate to inflammatory sites such as atherosclerotic plaques. This, together with the finding that a US28-specific antibodies can cross-react with heat-shock protein 60 (Bason et al. 2003), implies that US28-signaling in CMV-infected cells supports the progress of atherosclerosis. Interestingly, in addition to cell migration, US28-mediated signaling was found to trigger two unusual cellular responses: (a) caspase-dependent apoptosis (Pleskoff et al. 2005) and (b) loss of cell contact inhibition and enhanced cell cycle progression, as well as VEGF-mediated enhancement of tumor progression in vivo (Maussang et al. 2007). More investigation is required to determine how these two phenomena benefit CMV infection and how they relate to pathogenesis of CMV infection in humans.

Similar to HCMV US28, the MCMV M33-encoded vGPCR was shown to trigger smooth muscle cell migration, which, in the case of M33 was enhanced upon stimulation with murine RANTES (Melnychuk et al. 2005). Nevertheless, the direct consequences of altered signaling in cells expressing non-RANTES-binding UL33 and R33 (Table 3) still needs to be elucidated. In addition, numerous other questions regarding the function of putative vGPCRs remain. Most important of these is the question regarding the activities of the proteins encoded by UL78, M78, R78 and US27. To date, none of these proteins has been attributed to any signaling activity.

CMV vGPCR Gene Deletion Mutants

Additional roles for CMV vGPCRs in the host have been determined by studying knockout CMV mutants. The HCMV UL33 gene, as well as MCMV M33 and RCMV R33 are dispensable for viral replication in fibroblasts in vitro (Davis-Poynter et al. 1997; Beisser et al. 1998; Casarosa et al. 2003). Yet, the mortality rate of rats infected with RCMV R33 deletion mutant virus was significantly lower than that of rats infected with wild type virus (Beisser et al. 1998). Additionally, whereas wild type virus can be recovered from salivary glands of both MCMV-infected mice and RCMV-infected rats, their respective M33 and R33 deletion mutant counterparts remained undetectable in these organs throughout infection (Davis-Poynter et al. 1997; Beisser et al. 1998). Other knockout CMVs have been generated that had lower salivary gland replication rate phenotypes (Manning et al. 1992; Xiao et al. 2000; Kaptein et al. 2004), but the M33/R33 deletion mutant studies indicated an absolute requirement of the MCMV M33 and RCMV R33 vGPCR genes for salivary gland tropism in vivo.

Deletion of M78 and R78 from the MCMV or RCMV genome, respectively, resulted in mutant strains that were attenuated both in vitro and in vivo (Beisser et al. 1999; Oliveira and Shenk 2001). In contrast, deletion of UL78 from the HCMV genome resulted in a mutant strain that was neither attenuated in cell culture nor in renal artery explant culture (Michel et al. 2005). Yet, the expression of U51-specific siRNA in HHV-6-infected cells resulted in significantly lower levels of viral replication than the expression of control siRNA in infected cells (Zhen et a. 2005). These results indicate that the vGPCRs encoded by M78, R78 and U51 may fulfill different roles in replication than those encoded by HCMV UL78.

Generation of mutants in which the US28 gene was deleted from the HCMV genome has been instrumental in confirming the US28-specific signaling activities of infected cells. Deletion of US28 resulted in a phenotype in which infected cells were no longer able to bind RANTES and mobilize intracellular Ca^{2+} (Vieira et al. 1998). Infected smooth muscle cells were no longer capable of chemokinesis or chemotaxis (Streblow et al. 1999). This clearly demonstrated a role for US28 in the dissemination of HCMV within the host, by enabling infected cells to navigate through tissue by chemotaxis. Cells infected with US28 deletion mutant virus were no longer capable of sequestering RANTES and MCP-1 (Bodaghi et al. 1998; Randolph-Habecker et al. 2002), allowing more monocytes to be attracted by the medium from cell cultures infected with US28-deleted virus than by medium for cultures infected with wild type HCMV (Randolph-Habecker et al. 2002). Additionally, cells infected with US28 deletion mutant virus showed an increase in IL-8 secretion. These phenotypes indicate that HCMV US28 has an anti-inflammatory effect, which could ensure persistence of the infection.

Localization of CMV vGPCRs

Chemokine receptors are, subsequent to their synthesis, transported to the outer cellular membrane by intrinsic GPCR domains. At the outer membrane, they can interact with both extracellular chemokines and intracellular membrane-bound heterotrimeric G proteins to establish a signaling bridge. Of all CMV vGPCRs detected in either infected or transfected cells, only the vGPCRs encoded by MCMV M33 and RCMV R33 were clearly localized at the outer cellular membrane (Waldhoer et al. 2002; Casarosa et al. 2003). All other CMV vGPCRs were restricted to the perinucleus, as well as to intracellular, endosomic, multivesicular bodies (Oliveira and Shenk 2001; Fraile-Ramos et al. 2002; Waldhoer et al. 2002; Kaptein et al. 2003; Penfold et al. 2003b; Margulies et al. 2006). The HCMV vGPCRs encoded by UL33, US27 and US28 were shown to be subject to multiple internalization mechanisms, via either the beta-arrestin- or clathrin-dependent endocytosis pathways (Fraile-Ramos et al. 2002, 2003; Waldhoer et al. 2002; Miller et al. 2003; Droese et al. 2004, Margulies et al. 2006). Despite the rapid internalization of the vGPCRs encoded by HCMV UL33 and US28, they both show significant constitutive signaling, and, in the case of US28, ligand binding as well as modulation of signaling. Nevertheless, it was reported that the high rate of endocytosis reduces US28-mediated constitutive signaling as well as the signal transduction modulating effects upon stimulation with fractalkine (Waldhoer et al. 2003). Interestingly, for the HHV-6 U51-encoded vGPCR, it was shown that membrane localization required a T cell-specific factor (Menotti et al. 1999). Most CMV vGPCR signaling and internalization assays have been performed in model cell types such as fibroblasts and immortalized kidney epithelial cell lines. Cell types more relevant to CMV infection in vivo, such as endothelial cells and cells of myeloid origin, might likewise provide the correct factor for stable vGPCR membrane expression.

It was suggested that mere expression of HCMV US28 on the leukocyte cell surface might enable adhesion to endothelial cells expressing membrane-bound fractalkine (Haskell et al. 2000). US28 transcription was shown to occur in latently infected monocytic cells in vitro (Beisser et al. 2001), as well as in PBLs in vivo in lung transplant recipients during primary CMV infection (Boomker et al. 2006b). These findings support the hypothesis that the US28-encoded cGPCR can act as an adhesion molecule aiding infected leukocytes to traffic from the bloodstream to solid tissue compartments. Thus, US28 could be a determinant of viral dissemination in the host. Yet, expression of US28-encoded vGPCRs on the outer cellular membrane of HCMV-infected cells in vivo has not yet been reported.

In addition to cell adhesion, US28 expression was also found to enhance viral entry and cell–cell fusion. The US28-encoded vGPCR was shown to interact with envelope proteins such as those from human immunodeficiency virus (HIV) and vesicular stomatitis virus (VSV) particles, enhancing HIV entry by acting as a (co-)receptor and enhancing membrane fusion upon interaction with HIV and VSV surface glycoproteins (Pleskoff et al. 1997, 1998). A similar role was established for HHV-6 U51 (Zhen et al. 2005). Whether such interactions have an impact on the progression of diseases related to HIV- and VZV-like viruses in vivo is still unclear.

Interestingly, most CMV vGPCRs that localize to intracellular multivesicular bodies, rather than the outer cellular membrane, have also been detected in excreted virus particles. These include vGPCRs encoded by HCMV UL33 (Fraile-Ramos et al. 2002), MCMV M78 (Oliveira and Shenk 2001), HCMV US27 (Fraile-Ramos et al. 2002; Waldhoer et al. 2002; Margulies et al. 2006), HCMV US28 (Fraile-Ramos et al. 2002; Waldhoer et al. 2002) and RhCMV RhUS28.5 (Penfold et al. 2003b). It can be hypothesized that vGPCRs, as constituents of the viral envelope, are delivered to the cellular membrane immediately after fusion of envelope and membrane at the initial stage of infection. Upon this rapid delivery, modulation of intracellular processes may take place even before the immediate early stage of CMV infection commences. Interestingly, it was shown that HCMV US28-expressing cells were able to enhance the transcriptional activities regulated by the HCMV major immediate early promoter via the p38/MAPK and NFκ-B pathways (Boomker et al. 2006a). This observation supports a role for virion-borne vGPCR delivery for preimmediate early infection events.

Perspectives

The vCKs and vGPCRs described in this chapter are the result of an ever-continuing arms race between cytomegaloviruses and the host organism. The roles of the UL128-, UL146- and UL147-encoded vCKs will likely remain uncertain until the phenotypes of relevant deletion mutant CMVs are examined in vivo. Since UL33- and UL78-like genes are conserved in rodent CMV species, their contribution to viral dissemination and replication have been well established in vivo. Interestingly, a pantheon of phenomena has been attributed to HCMV US28. Similar to the vCKs, more in vivo work needs to be done to determine how these phenomena benefit CMV growth and survival in the host. CCMV is the most appropriate model virus to address these questions.

Most existing antiviral agents against CMV disease target the viral replication cycle. vCKs and vGPCRs are interesting targets in the development of a new class of antiviral agents, such as synthetic small molecule GPCR ligands and chemokine antagonists and inverse agonists. These agents could inhibit viral reactivation or dissemination rather than replication, thereby increasing the range of available drugs and their effectiveness against CMV disease.

References

Adler B, Scrivano L, Ruzcics Z, Rupp B, Sinzger C, Koszinowski U (2006) Role of human cytomegalovirus UL131A in cell type-specific virus entry and release. J Gen Virol 87:2451–2460

Ahuja SK, Murphy PM (1993) Molecular piracy of mammalian interleukin-8 receptor type B by herpesvirus saimiri. J Biol Chem 268:20691–20694

Akter P, Cunningham C, McSharry BP, Dolan A, Addison C, Dargan DJ, Hassan-Walker AF, Emery VC, Griffiths PD, Wilkinson GW, Davison AJ. (2003) Two novel spliced genes in human cytomegalovirus. J Gen Virol 84:1117–1122

Arav-Boger R, Zong JC, Foster CB (2005) Loss of linkage disequilibrium and accelerated protein divergence in duplicated cytomegalovirus chemokine genes. Virus Genes 31:65–72

Arav-Boger R, Foster CB, Zong JC, Pass RF (2006) Human cytomegalovirus-encoded alpha -chemokines exhibit high sequence variability in congenitally infected newborns. J Infect Dis 193:788–791

Bakker RA, Casarosa P, Timmerman H, Smit MJ, Leurs R (2004) Constitutively active G(q/11)-coupled receptors enable signaling by co-expressed G(i/o)-coupled receptors. J Biol Chem 279:5152–5161

Baldanti F, Paolucci S, Campanini G, Sarasini A, Percivalle E, Revello MG, Gerna G (2006) Human cytomegalovirus UL131A, UL130 and UL128 genes are highly conserved among field isolates. Arch Virol 151:1225–1233

Bason C, Corrocher R, Lunardi C, Puccetti P, Olivieri O, Girelli D, Navone R, Beri R, Millo E, Margonato A, Martinelli N, Puccetti A (2003) Interaction of antibodies against cytomegalovirus with heat-shock protein 60 in pathogenesis of atherosclerosis. Lancet 362:1971–1977

Beck S, Barrell B (1992) An HCMV reading frame which has similarity with both the V and C regions of the TCR gamma chain. DNA Seq 2:33–38

Beisser PS, Vink C, Van Dam JG, Grauls G, Vanherle SJ, Bruggeman CA (1998) The R33 G protein-coupled receptor gene of rat cytomegalovirus plays an essential role in the pathogenesis of viral infection. J Virol 72:2352–2363

Beisser PS, Grauls G, Bruggeman CA, Vink C (1999) Deletion of the R78 G protein-coupled receptor gene from rat cytomegalovirus results in an attenuated, syncytium-inducing mutant strain. J Virol 73:7218–7230

Beisser PS, Laurent L, Virelizier JL, Michelson S (2001) Human cytomegalovirus chemokine receptor gene US28 is transcribed in latently infected THP-1 monocytes. J Virol 75:5949–5957

Billstrom MA, Johnson GL, Avdi NJ, Worthen GS (1998) Intracellular signaling by the chemokine receptor US28 during human cytomegalovirus infection. J Virol 72:5535–5544

Bodaghi B, Jones TR, Zipeto D, Vita C, Sun L, Laurent L, Arenzana-Seisdedos F, Virelizier JL, Michelson S (1998) Chemokine sequestration by viral chemoreceptors as a novel viral escape strategy: withdrawal of chemokines from the environment of cytomegalovirus-infected cells. J Exp Med 188:855–866

Boomker JM, The TH, de Leij LF, Harmsen MC (2006a) The human cytomegalovirus-encoded receptor US28 increases the activity of the major immediate-early promoter/enhancer. Virus Res 118:196–200

Boomker JM, Verschuuren EA, Brinker MG, de Leij LF, The TH, Harmsen MC (2006b) Kinetics of US28 gene expression during active human cytomegalovirus infection in lung-transplant recipients. J Infect Dis 193:1552–1556

Casarosa P, Bakker RA, Verzijl D, Navis M, Timmerman H, Leurs R, Smit MJ (2001) Constitutive signaling of the human cytomegalovirus-encoded chemokine receptor US28. J Biol Chem 276:1133–1137

Casarosa P, Gruijthuijsen YK, Michel D, Beisser PS, Holl J, Fitzsimons CP, Verzijl D, Bruggeman CA, Mertens T, Leurs R, Vink C, Smit MJ (2003) Constitutive signaling of the human cytomegalovirus-encoded receptor UL33 differs from that of its rat cytomegalovirus homolog R33 by promiscuous activation of G proteins of the Gq, Gi, and Gs classes. J Biol Chem 278:50010–50023

Catusse J, Parry CM, Dewin DR, Gompels UA (2007) Inhibition of HIV-1 infection by viral chemokine U83A via high-affinity CCR5 interactions that block human chemokine-induced leukocyte chemotaxis and receptor internalization. Blood 109:3633–3639

Cha TA, Tom E, Kemble GW, Duke GM, Mocarski ES, Spaete RR (1996) Human cytomegalovirus clinical isolates carry at least 19 genes not found in laboratory strains. J Virol 70:78–83

Davis-Poynter NJ, Lynch DM, Vally H, Shellam GR, Rawlinson WD, Barrell BG, Farrell HE (1997) Identification and characterization of a G protein-coupled receptor homolog encoded by murine cytomegalovirus. J Virol 71:1521–159

Davison AJ (2002) Evolution of the herpesviruses. Vet Microbiol 86:69–88

Davison AJ, Dolan A, Akter P, Addison C, Dargan DJ, Alcendor DJ, McGeoch DJ, Hayward GS (2003) The human cytomegalovirus genome revisited: comparison with the chimpanzee cytomegalovirus genome. J Gen Virol 84:17–28

Dewin DR, Catusse J, Gompels UA (2006) Identification and characterization of U83A viral chemokine, a broad and potent beta-chemokine agonist for human CCRs with unique selectivity and inhibition by spliced isoform. J Immunol 176:544–556

Droese J, Mokros T, Hermosilla R, Schülein R, Lipp M, Höpken UE, Rehm A (2004) HCMV-encoded chemokine receptor US28 employs multiple routes for internalization. Biochem Biophys Res Commun 322:42–49

Fitzsimons CP, Gompels UA, Verzijl D, Vischer HF, Mattick C, Leurs R, Smit MJ (2006) Chemokine-directed trafficking of receptor stimulus to different g proteins: selective inducible and constitutive signaling by human herpesvirus 6-encoded chemokine receptor U51. Mol Pharmacol 69:888–898

Fraile-Ramos A, Pelchen-Matthews A, Kledal TN, Browne H, Schwartz TW, Marsh M (2002) Localization of HCMV UL33 and US27 in endocytic compartments and viral membranes. Traffic 3:218–232

Fraile-Ramos A, Kohout TA, Waldhoer M, Marsh M (2003) Endocytosis of the viral chemokine receptor US28 does not require beta-arrestins but is dependent on the clathrin-mediated pathway. Traffic 4:243–253

Gao JL, Murphy PM (1994) Human cytomegalovirus open reading frame US28 encodes a functional beta chemokine receptor. J Biol Chem 269:28539–28542

Glass WG, Rosenberg HF, Murphy PM (2003) Chemokine regulation of inflammation during acute viral infection. Curr Opin Allergy Clin Immunol 3:467–473

Gruijthuijsen YK, Casarosa P, Kaptein SJ, Broers JL, Leurs R, Bruggeman CA, Smit MJ, Vink C (2002) The rat cytomegalovirus R33-encoded G protein-coupled receptor signals in a constitutive fashion. J Virol 76:1328–1338

Haggerty SM, Schleiss MR (2002) A novel CC-chemokine homolog encoded by guinea pig cytomegalovirus. Virus Genes 25:271–279

Hahn G, Revello MG, Patrone M, Percivalle E, Campanini G, Sarasini A, Wagner M, Gallina A, Milanesi G, Koszinowski U, Baldanti F, Gerna G (2004) Human cytomegalovirus UL131-128 genes are indispensable for virus growth in endothelial cells and virus transfer to leukocytes. J Virol 78:10023–10033

Hansen SG, Strelow LI, Franchi DC, Anders DG, Wong SW (2003) Complete sequence and genomic analysis of rhesus cytomegalovirus. J Virol 77:6620–6636

Haskell CA, Cleary MD, Charo IF (2000) Unique role of the chemokine domain of fractalkine in cell capture. Kinetics of receptor dissociation correlate with cell adhesion. J Biol Chem 275:34183–34189

Hassan-Walker AF, Okwuadi S, Lee L, Griffiths PD, Emery VC (2004) Sequence variability of the alpha-chemokine UL146 from clinical strains of human cytomegalovirus. J Med Virol 74:573–579

He R, Ruan Q, Qi Y, Ma YP, Huang YJ, Sun ZR, Ji YH (2006) Sequence variability of human cytomegalovirus UL146 and UL147 genes in low-passage clinical isolates. Intervirology 49:215–223

Isegawa Y, Ping Z, Nakano K, Sugimoto N, Yamanishi K (1998) Human herpesvirus 6 open reading frame U12 encodes a functional beta-chemokine receptor. J Virol 72:6104–6112

Kaptein SJ, Beisser PS, Gruijthuijsen YK, Savelkouls KG, van Cleef KW, Beuken E, Grauls GE, Bruggeman CA, Vink C (2003) The rat cytomegalovirus R78 G protein-coupled receptor gene is required for production of infectious virus in the spleen. J Gen Virol 84:2517–2530

Kaptein SJ, van Cleef KW, Gruijthuijsen YK, Beuken EV, van Buggenhout L, Beisser PS, Stassen FR, Bruggeman CA, Vink C (2004) The r131 gene of rat cytomegalovirus encodes a proinflammatory CC chemokine homolog which is essential for the production of infectious virus in the salivary glands. Virus Genes 29:43–61

Kledal TN, Rosenkilde MM, Schwartz TW (1999) Selective recognition of the membrane-bound CX3C chemokine, fractalkine, by the human cytomegalovirus-encoded broad-spectrum receptor US28. FEBS Lett 441:209–214

Li WH, Gouy M, Sharp PM, O'hUigin C, Yang YW (1990) Molecular phylogeny of Rodentia, Lagomorpha, Primates, Artiodactyla, and Carnivora and molecular clocks. Proc Natl Acad Sci USA 87:6703–6707

Liu Y, Biegalke BJ (2001) Characterization of a cluster of late genes of guinea pig cytomegalovirus. Virus Genes 23:247–256

Lurain NS, Fox AM, Lichy HM, Bhorade SM, Ware CF, Huang DD, Kwan SP, Garrity ER, Chou S (2006) Analysis of the human cytomegalovirus genomic region from UL146 through UL147A reveals sequence hypervariability, genotypic stability, and overlapping transcripts. Virol J 3:4

MacDonald MR, Burney MW, Resnick SB, Virgin HW IV (1999) Spliced mRNA encoding the murine cytomegalovirus chemokine homolog predicts a beta chemokine of novel structure. J Virol 73:3682–3691

Manning WC, Stoddart CA, Lagenaur LA, Abenes GB, Mocarski ES (1992) Cytomegalovirus determinant of replication in salivary glands. J Virol 66:3794–3802

Margulies BJ, Gibson W (2006) The chemokine receptor homologue encoded by US27 of human cytomegalovirus is heavily glycosylated and is present in infected human foreskin fibroblasts and enveloped virus particles. Virus Res 123:57–71

Maussang D, Verzijl D, van Walsum M, Leurs R, Holl J, Pleskoff O, Michel D, van Dongen GA, Smit MJ (2006) Human cytomegalovirus-encoded chemokine receptor US28 promotes tumorigenesis. Proc Natl Acad Sci USA 103:13068–13073

Mellado M, Rodríguez-Frade JM, Vila-Coro AJ, Fernández S, Martín de Ana A, Jones DR, Torán JL, Martínez-A C (2001) Chemokine receptor homo- or heterodimerization activates distinct signaling pathways. EMBO J 20:2497–2507

Melnychuk RM, Smith P, Kreklywich CN, Ruchti F, Vomaske J, Hall L, Loh L, Nelson JA, Orloff SL, Streblow DN (2005) Mouse cytomegalovirus M33 is necessary and sufficient in virus-induced vascular smooth muscle cell migration. J Virol 79:10788–10795

Menotti L, Mirandola P, Locati M, Campadelli-Fiume G (1999) Trafficking to the plasma membrane of the seven-transmembrane protein encoded by human herpesvirus 6 U51 gene involves a cell-specific function present in T lymphocytes. J Virol 73:325–333

Michel D, Milotic I, Wagner M, Vaida B, Holl J, Ansorge R, Mertens T (2005) The human cytomegalovirus UL78 gene is highly conserved among clinical isolates, but is dispensable for replication in fibroblasts and a renal artery organ-culture system. J Gen Virol 86:297–306

Miller WE, Houtz DA, Nelson CD, Kolattukudy PE, Lefkowitz RJ (2003) G-protein-coupled receptor (GPCR) kinase phosphorylation and beta-arrestin recruitment regulate the constitutive signaling activity of the human cytomegalovirus US28 GPCR. J Biol Chem 278:21663–21671

Miller-Kittrell M, Sai J, Penfold M, Richmond A, Sparer TE (2007) Functional characterization of chimpanzee cytomegalovirus chemokine, vCXCL-1(CCMV). Virology 364:454–465

Milne RS, Mattick C, Nicholson L, Devaraj P, Alcami A, Gompels UA (2000) RANTES binding and down-regulation by a novel human herpesvirus-6 beta chemokine receptor. J Immunol 164:2396–2404

Murphy E, Rigoutsos I, Shibuya T, Shenk TE (2003a) Reevaluation of human cytomegalovirus coding potential. Proc Natl Acad Sci USA 100:13585–13590

Murphy E, Yu D, Grimwood J, Schmutz J, Dickson M, Jarvis MA, Hahn G, Nelson JA, Myers RM, Shenk TE (2003b) Coding potential of laboratory and clinical strains of human cytomegalovirus. Proc Natl Acad Sci USA 100:14976–14981

Nakano K, Tadagaki K, Isegawa Y, Aye MM, Zou P, Yamanishi K. (2003) Human herpesvirus 7 open reading frame U12 encodes a functional beta-chemokine receptor. J Virol 77:8108–8115

Nakayama T, Hieshima K, Izawa D, Tatsumi Y, Kanamaru A, Yoshie O (2003) Cutting edge: profile of chemokine receptor expression on human plasma cells accounts for their efficient recruitment to target tissues. J Immunol 170:1136–1140

Oliveira SA, Shenk TE (2001) Murine cytomegalovirus M78 protein, a G protein-coupled receptor homologue, is a constituent of the virion and facilitates accumulation of immediate-early viral mRNA. Proc Natl Acad Sci USA 98:3237–3242

Patrone M, Secchi M, Fiorina L, Ierardi M, Milanesi G, Gallina A (2005) Human cytomegalovirus UL130 protein promotes endothelial cell infection through a producer cell modification of the virion. J Virol 79:8361–8373

Penfold ME, Dairaghi DJ, Duke GM, Saederup N, Mocarski ES, Kemble GW, Schall TJ (1999) Cytomegalovirus encodes a potent alpha chemokine. Proc Natl Acad Sci USA 96:9839–9844

Penfold M, Miao Z, Wang Y, Haggerty S, Schleiss MR (2003a) A macrophage inflammatory protein homolog encoded by guinea pig cytomegalovirus signals via CC chemokine receptor 1. Virology 316:202–212

Penfold ME, Schmidt TL, Dairaghi DJ, Barry PA, Schall TJ (2003b) Characterization of the rhesus cytomegalovirus US28 locus. J Virol 77:10404–10413

Pleskoff O, Tréboute C, Brelot A, Heveker N, Seman M, Alizon M (1997) Identification of a chemokine receptor encoded by human cytomegalovirus as a cofactor for HIV-1 entry. Science 276:1874–1878

Pleskoff O, Tréboute C, Alizon M (1998) The cytomegalovirus-encoded chemokine receptor US28 can enhance cell-cell fusion mediated by different viral proteins. J Virol 72:6389–6397

Pleskoff O, Casarosa P, Verneuil L, Ainoun F, Beisser P, Smit M, Leurs R, Schneider P, Michelson S, Ameisen JC (2005) The human cytomegalovirus-encoded chemokine receptor US28 induces caspase-dependent apoptosis. FEBS J 272:4163–4177

Poole E, King CA, Sinclair JH, Alcami A (2006) The UL144 gene product of human cytomegalovirus activates NF-B via a TRAF6-dependent mechanism. EMBO J 25:4390–4399

Prod'homme V, Griffin C, Aicheler RJ, Wang EC, McSharry BP, Rickards CR, Stanton RJ, Borysiewicz LK, López-Botet M, Wilkinson GW, Tomasec P (2007) The human cytomegalovirus MHC class I homolog UL18 inhibits LIR-1+ but activates LIR-1- NK cells. J Immunol 178:4473–4481

Randolph-Habecker JR, Rahill B, Torok-Storb B, Vieira J, Kolattukudy PE, Rovin BH, Sedmak DD (2002) The expression of the cytomegalovirus chemokine receptor homolog US28 sequesters biologically active CC chemokines and alters IL-8 production. Cytokine 19:37–46

Rigoutsos I, Novotny J, Huynh T, Chin-Bow ST, Parida L, Platt D, Coleman D, Shenk T (2003) In silico pattern-based analysis of the human cytomegalovirus genome. J Virol 77:4326–4344

Rivailler P, Kaur A, Johnson RP, Wang F (2006) Genomic sequence of rhesus cytomegalovirus 180.92: insights into the coding potential of rhesus cytomegalovirus. J Virol 80:4179–4182

Saederup N, Lin YC, Dairaghi DJ, Schall TJ, Mocarski ES (1999) Cytomegalovirus-encoded beta chemokine promotes monocyte-associated viremia in the host. Proc Natl Acad Sci USA 96:10881–10886

Sahagun-Ruiz A, Sierra-Honigmann AM, Krause P, Murphy PM (2004) Simian cytomegalovirus encodes five rapidly evolving chemokine receptor homologues. Virus Genes 28:71–83

Sherrill JD, Miller WE (2006) G protein-coupled receptor (GPCR) kinase 2 regulates agonist-independent Gq/11 signaling from the mouse cytomegalovirus GPCR M33. J Biol Chem 281:39796–39805

Stanton R, Westmoreland D, Fox JD, Davison AJ, Wilkinson GW (2004) Stability of human cytomegalovirus genotypes in persistently infected renal transplant recipients. J Med Virol 75:42–46

Streblow DN, Söderberg-Nauclér C, Vieira J, Smith P, Wakabayashi E, Ruchti F, Mattison K, Altschuler Y, Nelson JA (1999) The human cytomegalovirus chemokine receptor US28 mediates vascular smooth muscle cell migration. Cell 99:511–520

Streblow DN, Vomaske J, Smith P, Melnychuk R, Hall L, Pancheva D, Smit M, Casarosa P, Schlaepfer DD, Nelson JA (2003) Human cytomegalovirus chemokine receptor US28-induced smooth muscle cell migration is mediated by focal adhesion kinase and Src. J Biol Chem 278:50456–50465

Stropes MP, Miller WE (2004) Signaling and regulation of G-protein coupled receptors encoded by cytomegaloviruses. Biochem Cell Biol 82:636–642

Tadagaki K, Nakano K, Yamanishi K (2005) Human herpesvirus 7 open reading frames U12 and U51 encode functional beta-chemokine receptors. J Virol 79:7068–7076

Tadagaki K, Yamanishi K, Mori Y (2007) Reciprocal roles of cellular chemokine receptors and human herpesvirus 7-encoded chemokine receptors, U12 and U51. J Gen Virol 88:1423–1428

Vieira J, Schall TJ, Corey L, Geballe AP (1998) Functional analysis of the human cytomegalovirus US28 gene by insertion mutagenesis with the green fluorescent protein gene. J Virol 72:8158–8165

Voigt S, Sandford GR, Hayward GS, Burns WH (2005) The English strain of rat cytomegalovirus (CMV) contains a novel captured CD200 (vOX2) gene and a spliced CC chemokine upstream from the major immediate-early region: further evidence for a separate evolutionary lineage from that of rat CMV Maastricht. J Gen Virol 86:263–274

Wagner CS, Riise GC, Bergström T, Kärre K, Carbone E, Berg L (2007) Increased expression of leukocyte Ig-like receptor-1 and activating role of UL18 in the response to cytomegalovirus infection. J Immunol 178:3536–3543

Waldhoer M, Kledal TN, Farrell H, Schwartz TW (2002) Murine cytomegalovirus (CMV) M33 and human CMV US28 receptors exhibit similar constitutive signaling activities. J Virol 76:8161–8168

Waldhoer M, Casarosa P, Rosenkilde MM, Smit MJ, Leurs R, Whistler JL, Schwartz TW (2003) The carboxyl terminus of human cytomegalovirus-encoded 7 transmembrane receptor US28 camouflages agonism by mediating constitutive endocytosis. J Biol Chem 278:19473–19482

Wang D, Shenk T (2005) Human cytomegalovirus virion protein complex required for epithelial and endothelial cell tropism. Proc Natl Acad Sci USA 102:18153–18158

Wills MR, Ashiru O, Reeves MB, Okecha G, Trowsdale J, Tomasec P, Wilkinson GW, Sinclair J, Sissons JG (2005) Human cytomegalovirus encodes an MHC class I-like molecule (UL142) that functions to inhibit NK cell lysis. J Immunol 175:7457–7465

Xiao J, Tong T, Zhan X, Haghjoo E, Liu F (2000) In vitro and in vivo characterization of a murine cytomegalovirus with a transposon insertional mutation at open reading frame M43. J Virol 74:9488–9497

Yu D, Silva MC, Shenk T (2003) Functional map of human cytomegalovirus AD169 defined by global mutational analysis. Proc Natl Acad Sci USA 100:12396–12401

Zhen Z, Bradel-Tretheway B, Sumagin S, Bidlack JM, Dewhurst S (2005) The human herpesvirus 6 G protein-coupled receptor homolog U51 positively regulates virus replication and enhances cell-cell fusion in vitro. J Virol 79:11914–11924

Subversion of Cell Cycle Regulatory Pathways

V. Sanchez, D. H. Spector(✉)

Contents

The Host Cell Cycle	243
The Effect of HCMV on the Cell Cycle	246
Cell Cycle Arrest and cdk Dysregulation	246
Checkpoint Control and DNA Damage	246
Inhibition of Cellular DNA Replication	247
Regulation of Cellular RNA Transcription and Protein Stability	247
Importance of Subversion of the Cell Cycle for the Viral Infection	248
Cell Cycle Arrest	248
Checkpoint Control	252
Role of Cyclin-Dependent Kinases in Viral Replication	252
Role of Cyclin-Dependent Kinases in Viral RNA Processing	254
Perspectives	256
References	257

Abstract Human cytomegalovirus (HCMV) has evolved numerous strategies to commandeer the host cell for producing viral progeny. The virus manipulates host cell cycle pathways from the early stages of infection to stimulate viral DNA replication at the expense of cellular DNA synthesis. At the same time, cell cycle checkpoints are by-passed, preventing apoptosis and allowing sufficient time for the assembly of infectious virus.

The Host Cell Cycle

To understand the effects of HCMV on the host cell cycle, it is necessary to first present a brief overview of the phases of the cell cycle and the major regulatory proteins involved (see Fig. 1). The cell cycle is a finely orchestrated sequence of

---

D.H. Spector
Department of Cellular and Molecular Medicine and The Skaggs School of Pharmacy and Pharmaceutical Sciences, University of California San Diego, La Jolla, CA 92093-0712, USA
dspector@ucsd.edu

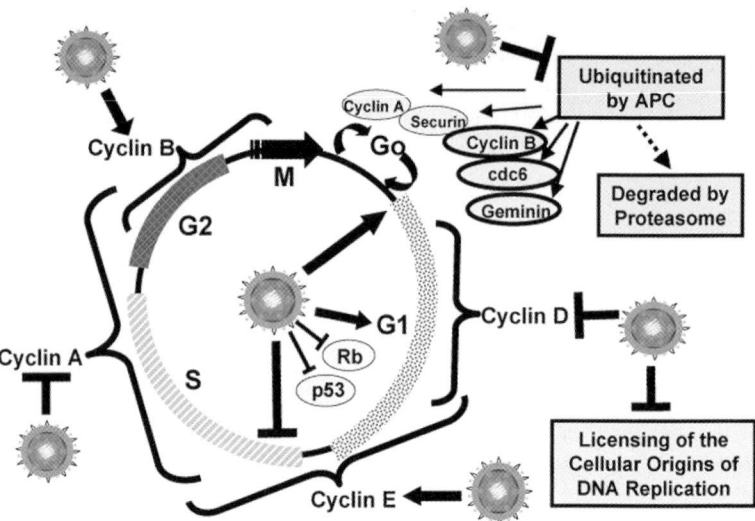

Fig. 1 Effects of HCMV on the cell cycle. The cell cycle is a tightly regulated process through which the cell replicates its DNA and divides into two daughter cells. G_0 is the resting state, and cells are stimulated through growth signals to express cyclin D and enter G_1 phase. During G_1 phase, the anaphase-promoting complex (*APC*), an E3 ubiquitin ligase, remains active, and proteins that are not needed by the cell until later in the cell cycle are ubiquitinated and targeted for degradation by the proteasome. A prereplication complex (licensing) is established at cellular origins of DNA replication and cyclin E is induced. Replication of the cellular DNA in S phase requires cyclin A. G_2 phase follows S phase and marks the transition prior to cell division in M phase. Both cyclin A and cyclin B are required during the G_2/M period. The major guardians of the cell cycle are Rb and p53. Infection of cells during G_0/G_1 phase induces progression through G_1. However, the normal expression of the cyclin-dependent kinases is disrupted, and the cell cycle is blocked before the replication of cellular DNA. HCMV specifically inhibits the expression of cyclin D and cyclin A, but promotes expression of high levels of cyclin E and cyclin B. In infected cells, the activity of the APC is blocked and licensing of the cellular origins of DNA replication is inhibited. p53 is stabilized, but it cannot activate its target promoters, and the inhibitory activity of Rb on the E2F/DP transcription factors is relieved

events by which the cell prepares for division (for review see Sherr 1996, 2000). These events are regulated by heterodimeric kinases that consist of a regulatory cyclin subunit and a catalytic subunit, the cyclin dependent kinase (cdk). Multisubunit E3 ubiquitin ligases that target proteins for degradation by the proteasome also regulate cell cycle progression. A resting cell (in G_0 phase) can be induced to enter the cell cycle by a number of proliferative signals including growth factors and serum stimulation. Initially, there is activation of the expression of the D-type cyclins, which form kinase complexes with cdk4 or 6. The phosphorylation of substrates by the cyclin D/cdk4 and cyclin D/cdk6 kinases releases the quiescent cell from its resting state and permits entry into the G_1 phase of the cell cycle.

The G_1 phase is the period during which multiple transcription factors and genes encoding proteins involved in nucleotide metabolism and DNA replication are induced. In G_1 phase, there is a commitment to DNA synthesis and cell division if the necessary requirements are met. DNA is replicated completely and only once during a single cell cycle (Fujita 1999; Diffley 2001; Lei and Tye 2001). During G_1 phase, prereplication complexes (pre-RC) assemble at the origins of DNA replication. The multisubunit origin recognition complex (ORC) is the first to bind to the DNA and serves as a nucleation point for the recruitment of other factors. Cdc6 and Cdt1 are recruited to the complex and facilitate the loading of the family of six MCM proteins (Maiorano et al. 2000; Nishitani et al. 2000, 2001; Rialland et al. 2002). Cyclin E expression is induced, and the active cdk2/cyclin E kinase promotes the transition into S phase. Cyclin E also facilitates the formation of pre-RC in a kinase-independent fashion through physical interaction with Cdt1 and the MCM proteins (Geng et al. 2003; Ekholm-Reed et al. 2004; Geng et al. 2007).

At the beginning of S phase, cyclin A accumulates and forms an active kinase complex with cdk2. Regulation of cyclin A occurs at both the protein and mRNA levels (Glotzer et al. 1991; Henglein et al. 1994; Desdouets et al. 1995; Schulze et al. 1995; Zwicker et al. 1995; Zwicker and Muller 1997; Bottazzi et al. 2001; Tessari et al. 2003). In S phase, cdk2/cyclin A and cdc7/Dbf4 complexes mediate the firing of DNA origins of replication and promote DNA replication (for review see Nishitani and Lygerou 2002; Machida et al. 2005). At the same time, Cdt1 is released from the replication complex, and its binding to the ORC is prevented by geminin, a protein that accumulates during S, G_2, and M phases. This ensures that DNA replication initiates at each origin only once and prevents polyploidy.

Following the complete replication of the parental chromosomes, cells enter G_2 and proteins involved in mitosis (M phase) begin to accumulate. Cdk1 and cyclin B play a major role at this point. Cyclin B first associates with cdk1 to form an inactive complex. At the G_2/M transition, dephosphorylation of cdk1 by cdc25 phosphatase induces the activation of the kinase, which in turn promotes the onset of mitosis (Millar and Russell 1992). Cdk1/cyclin B and cdk1/cyclin A complexes phosphorylate many substrates that facilitate condensation of the chromosomes, disassembly of the nuclear envelope, and modification of the cell's architecture to ensure that there is an ordered and even segregation of the chromosomes to the daughter cells. Inactivation of the cdk1 complexes occurs during mitosis by degradation of cyclins A and B through the ubiquitin-dependent proteolytic pathway involving the anaphase-promoting complex (APC) E3 ubiquitin ligase and the proteasome (Glotzer et al. 1991). The APC also ubiquitinates geminin and targets it for proteasome degradation as the cells exit mitosis, thereby allowing loading of the pre-RCs onto the chromatin of daughter cells during G_1 phase (McGarry and Kirschner 1998). This degradation of the cyclins and geminin continues until the onset of S phase (Brandeis and Hunt 1996).

Several checkpoints throughout the cell cycle ensure that progression will halt if there is DNA damage or aberrant spindle formation (for review see Lukas et al. 2004), thus protecting the integrity of the genome. The tumor suppressors p53 and the Rb family of pocket proteins (Rb, p107, and p130) are the best studied of these checkpoint sentinels. The Rb proteins, in their hypophosphorylated forms, bind to the E2F family of transcription factors, which then function as transcriptional repressors. Phosphorylation of the Rb proteins in late G_1 dissociates these complexes, allowing the E2F factors to activate transcription of multiple genes, many of which encode proteins required for DNA replication (Dyson 1998). p53 coordinates multiple cellular processes through its activity as a transcriptional activator and repressor in response to stress and growth factors (Vousden and Lu 2002; Slee et al. 2004). Phosphorylation of p53 controls its association with MDM2, which targets p53 for degradation by the proteasome (for review see Lavin and Gueven 2006). In response to DNA damage, nutrient deprivation, and other insults to the cell, p53 levels are stabilized. This can lead to the expression of the cdk inhibitor p21 as well as induction of several pro-apoptotic genes.

The Effect of HCMV on the Cell Cycle

Cell Cycle Arrest and cdk Dysregulation

Cells infected with HCMV in the G_0/G_1 phase of the cycle do not replicate their DNA and arrest in a pseudo-G_1 state that is distinguished by the expression of selected G_1-phase, S-phase and M-phase gene products (Jault et al. 1995; Bresnahan et al. 1996b; Lu and Shenk 1996; Dittmer and Mocarski 1997; Salvant et al. 1998; Wiebusch and Hagemeier 1999; Wiebusch and Hagemeier 2001; Challacombe et al. 2004; Hertel and Mocarski 2004) (see Fig. 1). The G_1/S- and G_2/M-phase cyclins E and B1, respectively, accumulate in infected cells with a concomitant increase in associated kinase activity (Jault et al. 1995; Wiebusch and Hagemeier 2001; Sanchez et al. 2003). In contrast, the steady-state levels of the S-phase cyclin A and G_1-phase cyclin D1 are reduced (Jault et al. 1995; Bresnahan et al. 1996b; Salvant et al. 1998; Wiebusch and Hagemeier 2001). The effects of the virus on cyclins E, A, and D1, are at the level of transcription, while accumulation of cyclin B1 is associated with increased stability of the protein (Salvant et al. 1998; Sanchez et al. 2003).

Checkpoint Control and DNA Damage

The observation that HCMV infection during S phase induces breaks in chromosome 1 has prompted investigations into the effects of infection on DNA damage pathways (Fortunato et al. 2000). Infection of cells with HCMV causes significant changes in proteins involved in G_1 and S phase checkpoint control. For example, the tumor suppressors Rb, p130, and p107, which inhibit the expression of E2F-responsive genes and the activity of cdk2 kinase complexes, are maintained in an inactivate phosphorylated state in infected cells (Jault et al. 1995; McElroy et al. 2000). Interestingly, it has also been reported that HCMV-infected cells express high levels of the p16[Ink4] tumor suppressor (Noris et al. 2002; Zannetti et al. 2006). p16[Ink4] is a marker for senescence and activator of Rb through binding and inactivation of Cdk4 and Cdk6 kinases.

Early cell cycle studies demonstrated the stabilization of p53 in HCMV-infected cells; p53 was sequestered into viral replication centers, thus contributing to its inability to activate the expression of downstream cellular target genes such as p21 (Muganda et al. 1994; Bresnahan et al. 1996a; Fortunato and Spector 1998; Chen and Fang 2001). Later work by Rosenke and co-workers described the binding of p53 to viral promoters and showed that the DNA binding activity of the protein was required for sequestration into the viral replication center (Rosenke et al. 2006).

HCMV infection also activates the ATM and ATR kinases that regulate the double-stranded break and S phase checkpoints, respectively (Shen et al. 2004; Castillo et al. 2005; Gaspar and Shenk 2006; Luo et al. 2007). However, the initiation of a damage response does not appear to be necessary for the infection, as virus replicates normally in cells deficient in ATM (Luo et al. 2007). In addition, a fully functional damage response does not occur due to inefficient relocalization of all of the required proteins (Gaspar and Shenk 2006; Luo et al. 2007). For example, several proteins that are involved in nonhomologous end-joining (NHEJ) response appear to be completely excluded from the viral replication centers (Luo et al. 2007). This may be necessary to prevent the rejoining of ends that are generated during the replication of the virus.

Inhibition of Cellular DNA Replication

In addition to downregulation of cyclin A observed during infection, the replication of cellular DNA is inhibited by several mechanisms during the licensing of the origins. Most importantly, the loading of MCM 2 and 7 (Biswas et al. 2003), as well as MCM 3, 4, 5, and 6 (Wiebusch et al. 2003), onto chromatin is inhibited in infected cells. This appears to be due to the premature accumulation of geminin, which, as discussed above, normally accumulates in S phase to ensure that there is no refiring of origins as the cells proceed through S and G_2/M.

Regulation of Cellular RNA Transcription and Protein Stability

The advent of DNA microarrays provided a means to examine the global effects of HCMV infection on the accumulation of cellular RNAs (Zhu et al. 1997; Browne et al. 2001; Challacombe et al. 2004; Hertel and Mocarski 2004). While increases in the levels of RNA do not necessarily correlate with changes in the steady-state levels of protein, these analyses provide a foundation for defining infection-associated changes in transcription. The general picture that has emerged is that changes in the mRNA levels of key cell cycle proteins is mirrored in the corresponding protein levels, although there are several exceptions.

One example of an exception is where the accumulation of the protein is due to the effects of the infection on protein stability rather than transcript level. For example, cyclin B1 accumulates in infected cells due to stabilization of the protein, with only minimal changes in RNA levels (Salvant et al. 1998; Sanchez et al. 2003). The degradation of cyclin B1 is mediated by the anaphase-promoting complex (APC), an E3 ubiquitin ligase. Interestingly, other substrates of the APC also accumulate in infected cells, including securin, geminin, Aurora B, and cdc6 (Biswas et al. 2003; Wiebusch et al. 2005). The increase in the steady-state levels of these proteins suggests that the APC is inactivated during infection. Data from our lab and others suggest that this dysregulation is due to the dismantling of the APC early in infection (Wiebusch et al. 2005; Tran et al. 2008).

Importance of Subversion of the Cell Cycle for the Viral Infection

Cell Cycle Arrest

A surprising observation regarding the kinetics of HCMV replication was that the initiation of the viral gene expression requires that the cells be in G_0 or G_1 at the time of infection. When cells are infected near or during S phase, many cells are able to pass through S phase and undergo mitosis prior to the synthesis of IE and early gene products (Salvant et al. 1998; Fortunato et al. 2002). The process of cell cycle arrest appears to be important for the early phase of infection, and proteins carried in the virus particle as well as those expressed at immediate early times contribute to this process (see Fig. 2).

Proteins in the Virus Particle

Two virion proteins, pUL69 and pp71 (pUL82), have been shown to modulate cell cycle progression (see the chapter by R. Kalejta, this volume). As components of the virion tegument, they can function as soon as the virus enters the cells. In the case of pUL69, overexpression of this protein stimulates accumulation of cells in G_1 phase of the cell cycle (Lu and Shenk 1999). In addition, cells infected with a virus lacking functional pUL69 do not efficiently undergo cell cycle arrest (Hayashi et al. 2000). This mutant does not replicate to wild type levels, but the growth defect may be attributed to other functions of pUL69.

The deletion of pp71 also creates a virus that is severely impaired for growth. The growth defect can be complemented by expression of the protein in trans (Bresnahan and Shenk 2000; Dunn et al. 2003). pp71 has been shown to interact with the cell growth suppressors Rb, p107, and p130 and to target hypophosphorylated forms of these pocket proteins for degradation by the proteasome (Kalejta et al. 2003). Consistent with this activity, pp71 expression in uninfected cells accelerates their progression through G1 phase, but does not change the overall doubling

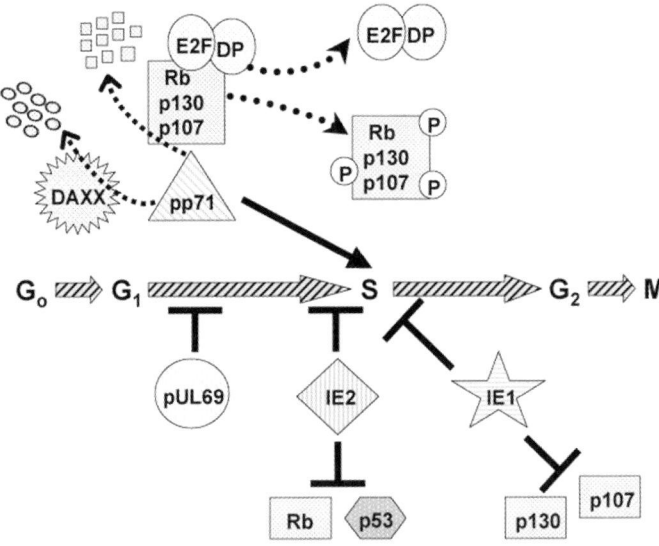

Fig. 2 Modulation of the cell cycle by proteins in the virus particle and immediate early proteins. The tegument proteins pUL69 and pp71 can exert their effects on the cell cycle upon entry of the virus particle into the cell. pUL69 causes accumulation of cells in G_1 phase, while pp71 targets the hypophosphorylated forms of the Rb family of proteins (Rb, p107, p130) and the transcriptional inhibitor Daxx for proteasome-mediated degradation. This degradation of the Rb proteins along with their viral-mediated hyperphosphorylation lead to the release of the E2F/DP transcription factors, which activate many genes involved in DNA replication and promotes S phase. The IE1 protein blocks the activity of p130/p107, and IE2 interferes with at least some functions of Rb and p53. Both IE proteins can prevent cells from passage through S phase

time (Kalejta and Shenk 2003). These results suggest that pp71 delivered to cells as part of the incoming virus particles may stimulate the cell cycle at the beginning of the infection before IE gene expression. The finding that pp71 also interacts with ND10-associated transcription repressor Daxx and promotes its proteasome-mediated degradation suggests an additional function for pp71 in initiating viral transcription at ND10 sites (Hofmann et al. 2002; Ishov et al. 2002; Marshall et al. 2002; Cantrell and Bresnahan 2005; Saffert and Kalejta 2006).

Immediate Early Protein 1

HCMV encodes two IE proteins, IE1-72 and IE2-86, that have been shown to interfere with cell cycle progression in heterologous systems in the absence of infection. Transient expression of IE1-72 in asynchronously cycling cells stimulates their accumulation in the S and G_2/M phases of the cell cycle (Castillo et al. 2000). One possible explanation for this result is that it is due to the interaction of IE1-72 with the pocket protein p107 (Margolis et al. 1995; Poma et al. 1996; Woo et al. 1997; Castano et al. 1998; Hansen et al. 2001; Zhang et al. 2003). IE1-72 alleviates p107-mediated repression of E2F-responsive promoters in transient transfection assays and thus may stimulate S-phase entry. Additionally, IE1-72 can reverse the inhibitory effects of p107 on cdk2/cyclin E kinase activity, which may also facilitate the G_1/S transition. It has been proposed that the formation of an IE1-72/p107 complex mediates these effects (Poma et al. 1996; Johnson et al. 1999; Zhang et al. 2003); however, the finding that IE1-72 can phosphorylate p107, p130, and E2F proteins (Pajovic et al. 1997) raises the possibility that some of the effects of IE1-72 on transcription and the cell cycle result from its reported kinase activity.

Immediate Early Protein 2

Several studies have shown that transient expression of IE2-86 alters cell cycle progression, with a block at the G1/S boundary in a $p53^{+/+}$ cell or after entry into S phase in a p53 mutant cell (Murphy et al. 2000; Wiebusch and Hagemeier 2001; Noris et al. 2002; Wiebusch et al. 2003; Song and Stinski 2005) (see the chapter by M.F. Stinski and D.T. Petrik, this volume). In transient transfection assays, deletion of aa 451–579 abolished the ability of IE2-86 to induce G1 arrest in transient assays in U373 cells (Wiebusch and Hagemeier 1999). Perhaps the most convincing evidence that IE2-86 plays a role in cell cycle arrest is a recent study showing that a mutation of aa 548 of IE2-86 from Q to R results in a growth-impaired virus that does not inhibit cellular DNA synthesis or the cell cycle (Petrik et al. 2006). The observation that this mutant IE2-86 could still autoregulate the MIE promoter and activate viral early genes provides further evidence that efficient viral replication also requires the inhibition of host cell DNA synthesis.

Early observations showing that IE2-86 interacts with several proteins regulating the cell cycle, including Rb and p53, made it reasonable to link some of its functions to cell cycle arrest (Hagemeier et al. 1994; Sommer et al. 1994; Speir et al. 1994; Bonin and McDougall 1997; Fortunato et al. 1997). p53 levels are stabilized in infected cells but the expression of its target gene p21 is repressed (Muganda et al. 1994; Bresnahan et al. 1996b; Fortunato and Spector 1998; Chen et al. 2001). In transient expression and in vitro systems, IE2-86 interacts with the C-terminus of p53, and the binding of p53 to target promoters is inhibited (Speir et al. 1994; Bonin and McDougall 1997; Hsu et al. 2004). IE2-86 expression also inhibits the acetylation of p53 and of histones in proximity to p53-dependent promoters (Hsu et al. 2004), and thus IE2-86 may regulate expression of p53 target genes by multiple mechanisms. These effects on protein acetylation may result from downregulation of p300/CBP histone acetyl transferase (HAT) activity, which was detected in a complex with p53 and IE2-86 (Hsu et al. 2004). It is possible that the inhibition of p300/CBP HAT activity is p53-promoter-specific, as IE2-86 did not suppress histone acetylation globally. The biological relevance of these experiments must be considered with caution, given that none were performed in the context of the viral infection.

In permissive cells, the early phase of the infection is associated with the stimulation of many genes encoding proteins that are required for host cell DNA synthesis and proliferation (Hirai and Watanabe 1976; Estes and Huang 1977; Isom 1979; Boldogh et al. 1991; Wade et al. 1992; Browne et al. 2001). Many of these genes are regulated by the E2F/DP transcription factors, which are inhibited by complex formation with the Rb family of proteins. A role for IE2-86 has been suggested by work showing that there is an increase in the steady state levels of RNA from several E2F-responsive genes in human fibroblasts infected with an adenovirus expressing IE2-86 (Song and Stinski 2002). The key question, however, is what are the underlying mechanisms for the activation of these growth regulatory genes in the context of the infection?

Cyclin E expression is regulated by E2F, and the potential role of IE2-86 in its accumulation during the infection has been the focus of several studies (Bresnahan et al. 1998; McElroy et al. 2000; Wiebusch and Hagemeier 2001; Wiebusch et al. 2003). The majority of the experiments have used transient expression assays to examine the regulation of the cyclin E promoter driving a reporter gene. In one study, it was shown that IE2-86 could bind to sequences in the cyclin E promoter in vitro and could activate expression of a cyclin E promoter-driven reporter construct (Bresnahan et al. 1998). Work by Song and Stinski demonstrated that expression of IE2-86 induces synthesis of endogenous cyclin E mRNA, reinforcing the notion that IE2-86 expression upregulates cyclin E in infected cells (Song and Stinski 2002). In contrast, McElroy et al. reported that early viral gene expression, not IE2-86 expression, was necessary for accumulation of cyclin E protein (McElroy et al. 2000). These conflicting results suggest that cyclin E accumulation in infected cells is controlled by several pathways, and that the increase in cyclin E might be regulated at both the mRNA and protein level.

Besides E2F transcription factors, other proteins regulate the expression of cyclin genes. The finding that the architectural transcription factor HMGA2 (high mobility group AT-hook 2) regulates the transcription of cyclin A (Tessari et al. 2003) prompted our lab to determine whether HCMV affects the expression of this protein. HMGA proteins are referred to as architectural transcription factors because of their ability to organize the assembly of nucleoprotein structures (enhanceosomes), resulting in enhancement or repression of transcription. In the case of cyclin A, HMGA2 activates its expression through derepression of the promoter. Our studies showed that the transcription of the HMGA2 gene is specifically inhibited at the RNA level during the infection (Shlapobersky et al. 2006). To determine whether repression of HMGA2 was important for the HCMV infection, a recombinant virus expressing HMGA2 driven by the MIE promoter was constructed. High multiplicity infection with the HMGA2-expressing virus induced the synthesis of cyclin A mRNA and protein and inhibited virus replication. Furthermore, a role for IE2-86, but not IE1-72, in HMGA2 repression was suggested by additional experiments with HCMV recombinant viruses defective in either IE1-72 or IE2-86. We found that the IE1-72 deletion mutant virus CR208 (Greaves and Mocarski 1998) inhibits HMGA2 transcription. In contrast, in cells infected with an IE2-86 mutant virus lacking aa 136–290 (referred to as delta SX)

(Sanchez et al. 2002), HMGA2 expression is not significantly reduced, suggesting that IE2 is involved in the regulation of the HMGA2 promoter. Cyclin A transcription is also induced in cells infected with delta SX, although this effect is slightly delayed relative to HMGA2 expression. The mechanism involving the downregulation of HMGA2 RNA expression by IE2-86 has yet to be determined. An interesting possibility is that it is related to IE2 interaction with histone deacetylase (HDAC), as HMGA2 is an example of a gene that requires HDAC activity for its expression (Ferguson et al. 2003). Cyclin D1 expression also seems to require HDAC activity (Hu and Colburn 2005) and cyclin D1 and HMGA2 exhibit a similar pattern of expression in the wt and delta SX infected cells (R. Sanders, M. Shlapobersky, and D.H. Spector, unpublished data). Since the levels of IE2-86 are significantly reduced in cells infected with delta SX, there may be less inhibition of HDAC activity in these cells.

Checkpoint Control

Recent studies have focused on the defining the roles of p53 and p16[Ink4] accumulation during infection. As described above, work from Rosenke and colleagues determined that p53 is bound to viral promoters during infection (Rosenke et al. 2006). This group has since shown that active p53 is required for the HCMV infection; p53-null cells infected with HCMV produced 10- to 20-fold less virus than controls (Casavant et al. 2006). The reduction in virus production was attributed to delays in viral DNA replication. Consistent with the binding of p53 to viral promoters, they also observed delays in the expression of early and late proteins in p53 null cells. Reintroduction of wild type but not mutant p53 partially rescued the phenotype, establishing a role for p53 in the viral lytic cycle.

A role for p16[Ink4] during infection was suggested by studies by Zannetti and co-workers, showing that there was a reduction in viral early and late gene expression, DNA synthesis, and production of extracellular virus in cells lacking functional p16 (Zannetti et al. 2006). In this case, p16[Ink4]-null cells and cells containing reduced levels of the protein as a result of treatment with siRNAs were used. The authors proposed that the ras-p16-Rb pathway is activated during the early phases of infection in order to establish an environment favorable for viral replication; however, the hyperphosphorylation of Rb in infected cells and the induction of E2F-responsive genes suggest that other proteins may be targets of p16[Ink4] activity.

Role of Cyclin-Dependent Kinases in Viral Replication

In order to establish the importance of cdk activity during infection, we and others have used cdk inhibitors such as the drug Roscovitine to treat infected cells (Bresnahan et al. 1997; Sanchez et al. 2004). This purine analog has been particularly useful as it has high specificity for cdks 1, 2, 5, 7, and 9 and reversibly inhibits their activity by competing for binding of ATP (De Azevedo et al. 1997). While the inhibitor has provided several important insights, its broad effects on multiple cdks necessitates further experiments to determine the roles of individual cdk complexes during infection. Nevertheless, studies with such chemical inhibitors are a good starting place for understanding the impact of cellular kinase activity on virus replication.

Cdk activity appears to be required primarily at two distinct time intervals in the infection: one during the first 8 h and the second after the onset of viral DNA replication (Bresnahan et al. 1997; Sanchez et al. 2004; Sanchez and Spector 2006). Studies from our lab showed a significant decrease in the expression of select early and late genes and a reduction in viral DNA replication when the drug was added at the time of infection. Notably, processing of IE transcripts was altered in the presence of the drug. However, these effects on IE transcript processing and the inhibition of gene expression and viral DNA replication were not detected if addition of the drug was delayed until 6–8 h postinfection (p.i.). In contrast, a reduction of virus titer was observed even if the drug was not added until 48 h p.i.

The effects on IE gene expression when Roscovitine is added at the beginning of the infection were particularly intriguing and provided the impetus for further studies. IE1-72 and IE2-86 are alternatively spliced transcripts that share their first three exons and differ in their 3' terminal exon, with the 3' exon of IE2-86 being most distal from the start site of transcription. At early phases of the infection, IE1-72 RNA accumulates preferentially compared to IE2-86 RNA; however, as the infection progresses, more IE2-86 transcripts accumulate, while there is only a slight increase in IE1-72 transcripts. If Roscovitine is added at the onset of infection, IE2-86 RNA accumulation is favored at early phases and there is little increase in the level of IE1-72 transcript (Sanchez et al. 2004). In fact, there is a decrease in IE1-72 RNA relative to untreated control samples. Given that the polyadenylation signals for IE1-72 and splicing signals for IE2-86 are juxtaposed, the most likely explanation for these results is that the differential splicing and polyadenylation of the UL122–123 transcript is altered by treatment with the drug. This results in enhanced utilization of the 3' splice acceptor site that generates the IE2-86 transcript and a decrease in the cleavage/polyadenylation of the IE1-72 transcript. Consistent with this hypothesis, the differential processing of the IE UL37 RNAs, which also have signals for polyadenylation and splicing in close proximity, is similarly affected by cdk inhibition.

The requirement for cdks at later times in the infection is directly related to formation of infectious viral particles. In a recent paper, we showed that addition of Roscovitine at 24 h p.i., after the onset of viral DNA synthesis, results in a 1- to 2-log decrease in viral titer (Sanchez and Spector 2006). There is a corresponding decrease in the amount of viral DNA detected in the supernatant from drug-treated cells, indicating that there is a defect in the production or release of extracellular particles. Consistent with this result, we observed changes in the expression, post-translational modification, and localization of virion structural proteins in Roscovitine-treated cells. The levels of the IE2-86 and pp150 (UL32) proteins were significantly reduced, but there was not a corresponding decrease in the mRNAs. The pUL69 matrix protein accumulated in drug-treated cells and the protein was present in a hyperphosphorylated form. pUL69 localized to intranuclear aggregates that did not overlap with viral replication centers. The matrix protein pp65 was also retained in the nucleus (Sanchez et al. 2007). In contrast, the levels of capsid, envelope, and other tegument proteins were only moderately affected in that their expression was slightly delayed. Although preliminary, our recent data with the specific inhibitor of cdk1, RO-3306 (Vassilev et al. 2006), indicate that only some of these effects are due to inhibition of cdk1. We find that the titers are significantly reduced and pUL69 accumulates in a hyperphosphorylated form, but the expression of the IE2-86 and UL32 proteins is not significantly affected (V. Sanchez and D.H. Spector, unpublished data). Thus, there may be a later role for cdk2, cdk7, and cdk9.

Although limited, there have been some attempts to try to decipher the contributions of specific cdks during the infection by expressing dominant-negative forms of the catalytic subunits in infected cells. Work by Bresnahan and colleagues showed a decrease in the levels of some capsid proteins in infected cells transfected with a dominant-negative cdk2, suggesting that the activity of cdk2/cyclin E complexes is essential for viral replication (Bresnahan et al. 1997). In contrast, preliminary work by Hertel et al. suggests that cdk1 activity is dispensable for the infection, given that virus titer was not markedly reduced in cell cultures transduced with a retrovirus expressing a GFP-tagged dominant negative cdk1 (Hertel et al. 2007). The only effect that the dominant negative cdk1 had was to inhibit a pseudomitosis phenotype that they observed in some cells infected with a variant of the HCMV strain AD169 (Hertel and Mocarski 2004). As noted above, however, our preliminary results with a specific inhibitor of cdk1 suggest that this kinase is required for production of infectious virus. Experiments involving expression of siRNAs directed against the individual cdks will likely provide more insight into their individual and combined roles in the infection.

Role of Cyclin-Dependent Kinases in Viral RNA Processing

Our hypothesis for the altered pattern of expression for the IE1-72/IE2-86 and UL37 IE RNAs in the presence of Roscovitine is based on the phosphorylation of the C-terminal domain (CTD) of the large subunit of RNA polymerase II (RNAP II) by cdk7/cyclin H and cdk9/cyclin T. Cdk7/cyclin H is responsible for activating cdks1, 2, and 4 and is also a part of TFIIH, which phosphorylates the carboxyl terminal domain (CTD) of the large subunit of RNA polymerase II within the 52 repeats of the heptapeptide YSPTSPS. Cdk9/cyclin T (P-TEFb) also phosphorylates the CTD. The current consensus is that transcription and RNA processing are integrated events whereby the differential phosphorylation of the CTD repeats at Ser 2 and Ser 5 defines its affinity for various transcription factors, kinases, and RNA processing factors. When the transcription initiation complex is formed on a promoter, the CTD of RNAP II is unphosphorylated (the polymerase is designated RNAP IIa). Cdk7 with cyclin H and MAT1 primarily phosphorylates Ser 5 on the CTD (the polymerase is now designated RNAP IIo), which leads to the recruitment of the RNA capping enzymes. Further phosphorylation of the CTD Ser 2 residues by cdk9/cyclin T occurs upon entry to elongation and is associated with recruitment of the cleavage/polyadenylation and splicing machinery.

Previously, our lab showed that infected cells contain more cdk7, MAT-1, cdk9, and cyclin T1 at 24 h p.i. than the uninfected cells, and the abundance and activity of these proteins increase as the infection proceeds (Tamrakar et al. 2005). All of MAT-1 is complexed with cdk7, although free cdk7 is also present, and most of cdk9 and cyclin T1 are in complex. In accord with the increase in the activity of the cdk9 and cdk7 kinases, an increase in the phosphorylation of the RNAP II CTD, particularly on the Ser 2 and Ser 5 residues of the heptad repeats, was also noted. By immunofluorescence analysis, it was observed that cdk7 and hypophosphorylated RNAP II localize to replication centers. In contrast, cdk9 and ser2-phosphorylated RNAP II are distributed in a punctate pattern throughout the nucleus with some concentration at the periphery of the viral replication centers. Similarly, ser5-phosphorylated RNAP II appears in clusters at the rim of the viral replication centers. These results suggest that at late times in the infection, HCMV may commandeer the RNA polymerase machinery, with viral RNA synthesis initiated within the replication center and active viral transcription occurring at the periphery.

Our lab has also demonstrated that addition of the cdk inhibitor Roscovitine at the time of infection results in decreased CTD phosphorylation in the infected cells and a decrease in the level of the hypophosphorylated RNAP II in both infected and mock-infected cells (Tamrakar et al. 2005). Consistent with our previous results regarding the effect of the cdk inhibitors on the processing and accumulation of the HCMV IE1/IE2 and UL37 IE transcripts, the decrease in CTD phosphorylation does not occur if the drug is added after 8 h p.i. One clue to explain this restricted interval in which cdk activity is required may be found in the differential localization of cdk9 and cdk7 at the beginning of the infection. Upon cell entry, incoming HCMV genomes localize near ND10, where viral IE transcription begins (Ishov and Maul 1996; Ishov et al. 1997) (see the chapter by G. Maul, this volume). Following translation, the IE1-72 and IE2-86 proteins return to nucleus and concentrate near the ND10. IE1-72 mediates the dispersal of ND10 associated proteins and it also disperses, while IE2-86 persists at the site (Kelly et al. 1995; Korioth et al. 1996; Ahn and Hayward 1997; Ishov et al. 1997; Ahn et al. 1998). We find that as early as 4 h p.i., several proteins involved in RNA transcription, including cdk9, cdk7, and Ser2-phosphorylated RNA polymerase II, colocalize with IE2-86 in distinct aggregates (referred to as viral transcriptosomes) adjacent to the ND10 that are undergoing dispersal (Tamrakar et al. 2005) (see Fig. 3). However, if Roscovitine is added at the beginning of the infection, IE2-86 is found in the transcriptosome, but cdk7 and cdk9 are not recruited (Kapasi and Spector 2008). In contrast, both cdks colocalize with IE2-86 in the transcriptosome if Roscovitine is added after 8 h p.i. These results suggest that the formation of a distinct viral transcriptosome at the

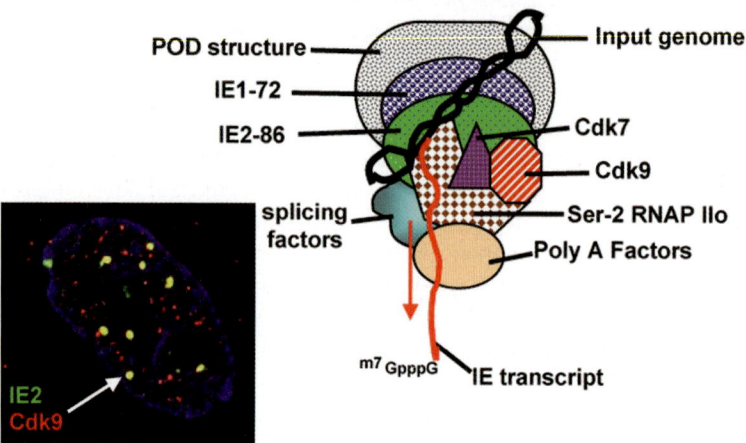

Fig. 3 Model of the viral transcriptosome formed at the beginning of the infection. The input genome that is deposited at the POD structure functions as the template for IE RNA synthesis. Cellular hypophosphorylated RNAP IIa is recruited to the site along with cdk9 and cdk7, which hyperphosphorylate RNAP IIa to the transcriptionally active RNAP IIo that serves as a platform for RNA processing enzymes. The IE transcripts are synthesized and translated into the IE1-72 and IE2-86 proteins, which return to this nuclear body. IE1-72 causes POD dispersal and it also disperses, while IE2-86 remains at the established transcription site, referred to as the transcriptosome. The inset is an infected cell nucleus at 8 h p.i. and shows the accumulation of both cdk9 and IE2 at several transcriptosomes

beginning of the infection serves as a specialized site for recruitment of cellular and viral proteins necessary for viral transcription and replication. The correct phosphorylation of the RNAP II CTD at these sites is essential for accurate processing of the IE transcripts and for transcription of early genes, and it appears that the required level of phosphorylation is established within the first 8 h.

Perspectives

For more than a decade, we have known that HCMV dramatically alters cell cycle regulatory pathways, leading to cell cycle arrest. These alterations begin as soon as the viral particle enters the cell, but sustained effects require early viral gene expression. The molecular mechanisms underlying the viral-mediated effects operate at multiple levels, including altered RNA transcription, changes in the levels and activity of cyclin-dependent kinases as well as other cellular kinases involved in cell cycle control, modulation of protein stability through targeted effects on the ubiquitin-proteasome degradation pathway, and movement of proteins to different cellular locations. A key question regarding all of these effects is which ones are actually important for viral replication? For example, are high levels of cyclin B needed for productive infection or is this simply a side effect of altering the ubiquitin-proteasome pathway to allow accumulation of some other cellular or viral protein that is required? Future experiments involving knockdown of gene expression through the use of siRNAs or induced overexpression with lentiviral vectors should provide some insight into these questions. It is clear, however, that the virus depends on the host cell being in the G_0/G_1 phase to initiate the infection and subverts some G2/M phase activities of the cell for later stages of replication. At this point, only a few viral genes, primarily input virion proteins and IE gene products, have been implicated in these changes. It is critical to identify the other genes, particularly those expressed during the early phase just prior to initiation of viral DNA replication. It also should be recognized that most of the studies to date have been done in fibroblasts, and it will be important to examine the effects of the viral infection on host cell regulatory pathways in other relevant target cells such as endothelial cells and monocytes.

References

Ahn JH, Hayward GS (1997) The major immediate-early proteins IE1 and IE2 of human cytomegalovirus colocalize with and disrupt PML-associated nuclear bodies at very early times in infected permissive cells. J Virol 71:4599–4613

Ahn JH, Brignole ER, Hayward GS (1998) Disruption of PML subnuclear domains by the acidic IE1 protein of human cytomegalovirus is mediated through interaction with PML and may modulate a RING finger-dependent cryptic transactivator function of PML. Mol Cell Biol 18:4899–4913

Biswas N, Sanchez V, Spector DH (2003) Human cytomegalovirus infection leads to accumulation of geminin and inhibition of the licensing of cellular DNA replication. J Virol 77:2369–2376

Boldogh I, AbuBakar S, Deng CZ, Albrecht T (1991) Transcriptional activation of cellular oncogenes *fos*, *jun* and *myc* by human cytomegalovirus. J Virol 65:1568–1571

Bonin LR, McDougall JK (1997) Human cytomegalovirus IE2 86-kilodalton protein binds p53 but does not abrogate G1 checkpoint function. J Virol 71:5831–5870

Bottazzi ME, Buzzai M, Zhu X, Desdouets C, Brechot C, Assoian RK (2001) Distinct effects of mitogens and actin cytoskeleton on CREB and pocket protein phosphorylation control the extent and timing of cyclin A promoter activity. Mol Cell Biol 21:7607–7616

Brandeis M, Hunt T (1996) The proteolysis of mitotic cyclins in mammalian cells persists from the end of mitosis until the onset of S phase. EMBO J 15:5280–5289

Bresnahan WA, Shenk TE (2000) UL82 virion protein activates expression of immediate early viral genes in human cytomegalovirus-infected cells. Proc Natl Acad Sci U S A 97:14506–14511

Bresnahan WA, Boldogh I, Ma T, Albrecht T, Thompson EA (1996a) Cyclin E/CDK2 activity is controlled by different mechanisms in the G0 and G1 phases of the cell cycle. Cell Growth Differ 7:1283–1290

Bresnahan WA, Boldogh I, Thompson EA, Albrecht T (1996b) Human cytomegalovirus inhibits cellular DNA synthesis and arrests productively infected cells in late G1. Virology 224:156–160

Bresnahan WA, Boldogh I, Chi P, Thompson EA, Albrecht T (1997) Inhibition of cellular CDK2 activity blocks human cytomegalovirus replication. Virology 231:239–247

Bresnahan WA, Albrecht T, Thompson EA (1998) The cyclin E promoter is activated by human cytomegalovirus 86-kDa immediate early protein. J Biol Chem 273:22075–22082

Browne EP, Wing B, Coleman D, Shenk T (2001) Altered cellular mRNA levels in human cytomegalovirus-infected fibroblasts: viral block to the accumulation of antiviral mRNAs. J Virol 75:12319–12330

Cantrell SR, Bresnahan WA (2005) Interaction between the human cytomegalovirus UL82 gene product (pp71) and hDaxx regulates immediate-early gene expression and viral replication. J Virol 79:7792–7802

Casavant NC, Luo MH, Rosenke K, Winegardner T, Zurawska A, Fortunato EA (2006) Potential role for p53 in the permissive life cycle of human cytomegalovirus. J Virol 80:8390–8401

Castano E, Kleyner Y, Dynlacht BD (1998) Dual cyclin-binding domains are required for p107 to function as a kinase inhibitor. Mol Cell Biol 18:5380–5391

Castillo JP, Yurochko A, Kowalik TF (2000) Role of human cytomegalovirus immediate-early proteins in cell growth control. J Virol 74:8028–8037

Castillo JP, Frame FM, Rogoff HA, Pickering MT, Yurochko AD, Kowalik TF (2005) Human cytomegalovirus IE1-72 activates ataxia telangiectasia mutated kinase and a p53/p21-mediated growth arrest response. J Virol 79:11467–11475

Challacombe JF, Rechtsteiner A, Gottardo R, Rocha LM, Browne EP, Shenk T, Altherr MR, Brettin TS (2004) Evaluation of the host transcriptional response to human cytomegalovirus infection. Physiol Genomics 18:51–62

Chen J, Fang G (2001) MAD2B is an inhibitor of the anaphase-promoting complex. Genes Devel 15:1765–1770

Chen Z, Knutson E, Kurosky A, Albrecht T (2001) Degradation of p21cip1 in cells productively infected with human cytomegalovirus. J Virol 75:3613–3625

De Azevedo WF, Leclerc S, Meijer L, Havlicek L, Strnad M, Kim SH (1997) Inhibition of cyclin-dependent kinases by purine analogues: crystal structure of human cdk2 complexed with roscovitine. Eur J Biochem 243:518–526

Desdouets C, Matesic G, Molina CA, Foulkes NS, Sassone-Corsi P, Bréchot C, Sobszak-Thepot J (1995) Cell cycle regulation of cyclin A gene expression by the cyclic AMP transcription factors CREB and CREM. Mol Cell Biol 15:3301–3309

Diffley JF (2001) DNA replication: building the perfect switch. Curr Biol 11:R367–R370

Dittmer D, Mocarski ES (1997) Human cytomegalovirus infection inhibits G1/S transition. J Virol 71:1629–1634

Dunn W, Chou C, Li H, Hai R, Patterson D, Stolc V, Zhu H, Liu F (2003) Functional profiling of a human cytomegalovirus genome. Proc Natl Acad Sci U S A 100:14223–14228

Dyson N (1998) The regulation of E2F by pRB-family proteins. Genes Dev 12:2245–2262

Ekholm-Reed S, Mendez J, Tedesco D, Zetterberg A, Stillman B, Reed SI (2004) Deregulation of cyclin E in human cells interferes with prereplication complex assembly. J Cell Biol 165:789–800

Estes JE, Huang E-S (1977) Stimulation of cellular thymidine kinases by human cytomegalovirus. J Virol 24:13–21

Ferguson M, Henry PA, Currie RA (2003) Histone deacetylase inhibition is associated with transcriptional repression of the Hmga2 gene. Nucleic Acids Res 31:3123–3133

Fortunato EA, Sommer MH, Yoder K, Spector DH (1997) Identification of domains within the human cytomegalovirus major immediate-early 86-kilodalton protein and the retinoblastoma protein required for physical and functional interaction with each other. J Virol 71:8176–8185

Fortunato EA, Spector DH (1998) p53 and RPA are sequestered in viral replication centers in the nuclei of cells infected with human cytomegalovirus. J Virol 72:2033–2039

Fortunato EA, Dell'Aquila ML, Spector DH (2000) Specific chromosome 1 breaks induced by human cytomegalovirus. Proc Natl Acad Sci U S A 97:853–858

Fortunato EA, Sanchez V, Yen JY, Spector DH (2002) Infection of cells with human cytomegalovirus during S phase results in a blockade to immediate-early gene expression that can be overcome by inhibition of the proteasome. J Virol 76:5369–5379

Fujita M (1999) Cell cycle regulation of DNA replication initiation proteins in mammalian cells. Front Biosci 4: D816–D823

Gaspar M, Shenk T (2006) Human cytomegalovirus inhibits a DNA damage response by mislocalizing checkpoint proteins. Proc Natl Acad Sci U S A 103:2821–2826

Geng Y, Yu Q, Sicinska E, Das M, Schneider JE, Bhattacharya S, Rideout WM, Bronson RT, Gardner H, Sicinski P (2003) Cyclin E ablation in the mouse. Cell 114:431–443

Geng Y, Lee YM, Welcker M, Swanger J, Zagozdzon A, Winer JD, Roberts JM, Kaldis P, Clurman BE, Sicinski P (2007) Kinase-independent function of cyclin E. Mol Cell 25:127–139

Glotzer M, Murray AW, Kirschner MW (1991) Cyclin is degraded by the ubiquitin pathway. Nature 349:132–138

Greaves RF, Mocarski ES (1998) Defective growth correlates with reduced accumulation of a viral DNA replication protein after low-multiplicity infection by a human cytomegalovirus *ie1* mutant. J Virol 72:366–379

Hagemeier C, Caswell R, Hayhurst G, Sinclair J, Kouzarides T (1994) Functional interaction between the HCMV IE2 transactivator and the retinoblastoma protein. EMBO J 13:2897–2903

Hansen K, Farkas T, Lukas J, Holm K, Roonstrand L, Bartek J (2001) Phosphorylation-dependent and -independent functions of p130 cooperate to evoke a sustained G1 block. EMBO J 20:422–432

Hayashi ML, Blankenship C, Shenk T (2000) Human cytomegalovirus UL69 is required for efficient accumulation of infected cells in the G1 phase of the cell cycle. Proc Natl Acad Sci U S A 97:2692–2696

Henglein B, Chenivesse X, Wang J, Eick D, Bréchot C (1994) Structure and cell cycle-regulated transcription of the human cyclin A gene. Proc Natl Acad Sci U S A 91:5490–5494

Hertel L, Mocarski ES (2004) Global analysis of host cell gene expression late during cytomegalovirus infection reveals extensive dysregulation of cell cycle gene expression and induction of pseudomitosis independent of US28 function. J Virol 78:11988–12011

Hertel L, Chou S, Mocarski ES (2007) Viral and cell cycle-regulated kinases in cytomegalovirus-induced pseudomitosis and replication. PLoS Pathog:e6

Hirai K, Watanabe Y (1976) Induction of α-type DNA polymerases in human cytomegalovirus-infected WI-38 cells. Biochim Biophys Acta 447:328–339

Hofmann H, Sindre H, Stamminger T (2002) Functional interaction between the pp71 protein of human cytomegalovirus and the PML-interacting protein human Daxx. J Virol 76:5769–5783

Hsu C-H, Chang MDT, Tai K-Y, Yang Y-T, Wang P-S, Chen C-J, Wang Y-H, Lee S-C, Wu C-W, Juan L-J (2004) HCMV IE2-mediated inhibition of HAT activity downregulates p53 function. EMBO J 23:2269–2280

Hu J, Colburn NH (2005) Histone deacetylase inhibition down-regulates cyclin D1 transcription by inhibiting nuclear factor-kappaB/p65 DNA binding. Cancer Res 3:100–109

Ishov AM, Maul GG (1996) The periphery of nuclear domain 10 (ND10) as site of DNA virus deposition. J Cell Biol 134:815–826

Ishov AM, Stenberg RM, Maul GG (1997) Human cytomegalovirus immediate early interaction with host nuclear structures: definition of an immediate transcript environment. J Cell Biol 138:5–16

Ishov AM, Vladimirova OV, Maul GG (2002) Daxx-mediated accumulation of human cytomegalovirus tegument protein pp71 at ND10 facilitates initiation of viral infection at these nuclear domains. J Virol 76:7705–7712

Isom HC (1979) Stimulation of ornithine carboxylase by human cytomegalovirus. J Gen Virol 42:265–278

Jault FM, Jault J-M, Ruchti F, Fortunato EA, Clark C, Corbeil J, Richman DD, Spector DH (1995) Cytomegalovirus infection induces high levels of cyclins, phosphorylated RB, and p53, leading to cell cycle arrest. J Virol 69:6697–6704

Johnson RA, Yurochko AD, Poma EE, Zhu L, Huang E-S (1999) Domain mapping of the human cytomegalovirus IE1-72 and cellular p107 protein–protein interaction and the possible functional consequences. J Gen Virol 80:1293–1303

Kalejta RF, Shenk T (2003) The human cytomegalovirus UL82 gene product (pp71) accelerate progression through the G1 phase of the cell cycle. J Virol 77:3451–3459

Kalejta RF, Bechtel JT, Shenk T (2003) Human cytomegalovirus pp71 stimulates cell cycle progression by inducing the proteasome-dependent degradation of the retinoblastoma family of tumor suppressors. Mol Cell Biol 23:1885–1895

Kapasi AJ, Spector DH (2008) Inhibition of the cyclin-dependent kinases at the beginning of the human cytomegalovirus infection specifically alters the levels and localization of the RNA polymerase II carboxyl-terminal domain kinases cdk9 and cdk7 at the viral transcriptosome. J Virol 82:394–407

Kelly C, Driel RV, Wilkinson GW (1995) Disruption of PML-associated nuclear bodies during human cytomegalovirus infection. J Gen Virol 76:2887–2893

Korioth F, Maul GG, Plachter B, Stamminger T, Frey J (1996) The nuclear domain 10 (ND10) is disrupted by the human cytomegalovirus gene product IE1. Exp Cell Res 229:155–158

Lavin MF, Gueven N (2006) The complexity of p53 stabilization and activation. Cell Death Differ 13:941–950

Lei M, Tye BK (2001) Initiating DNA synthesis: from recruiting to activating the MCM complex. J Cell Sci 114:1447–1454

Lu M, Shenk T (1996) Human cytomegalovirus infection inhibits cell cycle progression at multiple points, including the transition from G1 to S. J Virol 70:8850–8857

Lu M, Shenk T (1999) Human cytomegalovirus UL69 protein induces cells to accumulate in G_1 phase of the cell cycle. J Virol 73:676–683

Lukas J, Lukas C, Bartek J (2004) Mammalian cell cycle checkpoints: signalling pathways and their organization in space and time. DNA Repair 3:997–1007

Luo MH, Rosenke K, Czornak K, Fortunato EA (2007) Human cytomegalovirus disrupts both ataxia telangiectasia mutated protein (ATM)- and ATM-Rad3-related kinase-mediated DNA damage responses during lytic infection. J Virol 81:1934–1950

Machida YJ, Hamlin JL, Dutta A (2005) Right place, right time, and only once: replication initiation in metazoans. Cell 123:13–24

Maiorano D, Moreau J, Mechali M (2000) XCDT1 is required for the assembly of pre-replicative complexes in *Xenopus laevis*. Nature 404:622–625

Margolis MJ, Panjovic S, Wong EL, Wade M, Jupp R, Nelson JA, Azizkhan JC (1995) Interaction of the 72-kilodalton human cytomegalovirus IE1 gene product with E2F1 coincides with E2F-dependent activation of dihydrofolate reductase transcription. J Virol 69: 7759–7767

Marshall KR, Rowley KV, Rinaldi A, Nicholson IP, Ishov AM, Maul GG, Preston CM (2002) Activity and intracellular localization of the human cytomegalovirus protein pp71. J Gen Virol 83:1601–1612

McElroy AK, Dwarakanath RS, Spector DH (2000) Dysregulation of cyclin E gene expression in human cytomegalovirus-infected cells requires viral early gene expression and is associated with changes in the Rb-related protein p130. J Virol 74:4192–4206

McGarry TJ, Kirschner MW (1998) Geminin, an inhibitor of DNA replication, is degraded during mitosis. Cell 93:1043–1053

Millar JBA, Russell P (1992) The cdc25 M-phase inducer: an unconventional protein phosphatase. Cell 68:407–410

Muganda P, Mendoza O, Hernandez J, Qian Q (1994) Human cytomegalovirus elevates levels of the cellular protein p53 in infected fibroblasts. J Virol 68:8028–8034

Murphy EA, Streblow DN, Nelson JA, Stinski MF (2000) The human cytomegalovirus IE86 protein can block cell cycle progression after inducing transition into the S phase of the cell cycle. J Virol 74:7108–7118

Nishitani H, Lygerou Z (2002) Control of DNA Replication. Genes Cells 7:523–534

Nishitani H, Lygerou Z, Nishimoto T, Nurse P (2000) The Cdt1 protein is required to license DNA for replication in fission yeast. Nature 404:625–628

Nishitani H, Taraviras S, Lygerou Z, Nishimoto T (2001) The human licensing factor or DNA replication cdt1 accumulates in G1 and is destabilized after initiation of S-phase. J Biol Chem 276:44905–44911

Noris E, Zannetti C, Demurtas A, Sinclair J, De Andrea M, Gariglio M, Landolfo S (2002) Cell cycle arrest by human cytomegalovirus 86-kDa IE2 protein resembles premature senescence. J Virol 76:12135–12148

Pajovic S, Wong EL, Black AR, Azizkhan JC (1997) Identification of a viral kinase that phosphorylates specify E2Fs and pocket proteins. Mol Cell Biol 17:6459–6464

Petrik DT, Schmitt KP, Stinski MF (2006) Inhibition of cellular DNA synthesis by the human cytomegalovirus IE86 protein is necessary for efficient virus replication. J Virol 80:3872–3883

Poma EE, Kowalik TF, Zhu L, Sinclair JH, Huang E-S (1996) The human cytomegalovirus 1E1-72 protein interacts with the cellular p107 protein and relieves p107-mediated transcriptional repression of an E2F-responsive promoter. J Virol 70:7867–7877

Rialland M, Sola F, Santocanale C (2002) Essential role of human CDT1 in DNA replication and chromatin licensing. J Cell Sci 115:1435–1440

Rosenke K, Samuel MA, McDowell ET, Toerne MA, Fortunato EA (2006) An intact sequence-specific DNA-binding domain is required for human cytomegalovirus-mediated sequestration of p53 and may promote in vivo binding to the viral genome during infection. Virology 348:19–34

Saffert RT, Kalejta RF (2006) Inactivating a cellular intrinsic immune defense mediated by Daxx is the mechanism through which the human cytomegalovirus pp71 protein stimulates viral immediate-early gene expression. J Virol 80:3863–3871

Salvant BS, Fortunato EA, Spector DH (1998) Cell cycle dysregulation by human cytomegalovirus: influence of the cell cycle phase at the time of infection and effects on cyclin transcription. J Virol 72:3729–3741

Sanchez V, Spector DH (2006) Cyclin-dependent kinase activity is required for efficient expression and posttranslational modification of human cytomegalovirus proteins and for production of extracellular particles. J Virol 80:5886–5896

Sanchez V, Clark CL, Yen JY, Dwarakanath R, Spector DH (2002) Viable human cytomegalovirus recombinant virus with an internal deletion of the IE2 86 gene affects late stages of viral replication. J Virol 76:2973–2989

Sanchez V, McElroy AK, Spector DH (2003) Mechanisms governing maintenance of cdk1/cyclin B1 kinase activity in cells infected with human cytomegalovirus. J Virol 77:13214–13224

Sanchez V, McElroy AK, Yen J, Tamrakar S, Clark CL, Schwartz RA, Spector DH (2004) Cyclin-dependent kinase activity is required at early times for accurate processing and accumulation of the human cytomegalovirus UL122–123 and UL37 immediate-early transcripts and at later times for virus production. J Virol 78:11219–11232

Sanchez V, Mahr JA, Orazio N, Spector DH (2007) Nuclear export of the human cytomegalovirus tegument protein pp65 requires cyclin-dependent kinase activity and the Crm1 exporter. J Virol 81:11730–11736

Schulze A, Zerfass K, Spitkovsky D, Middendorp S, Berges J, Helin K, Jansen-Dürr P, Henglein B (1995) Cell cycle regulation of the cyclin A gene promoter is mediated by a variant E2F site. Proc Natl Acad Sci U S A 92:11264–11268

Shen YH, Utama B, Wang J, Raveendran M, Senthil D, Waldman WJ, Belcher JD, Vercellotti G, Martin D, Mitchelle BM, Wang XL (2004) Human cytomegalovirus causes endothelial injury through the ataxia telangiectasia mutant and p53 DNA damage signaling pathways. Circ Res 94:1310–1317

Sherr CJ (1996) Cancer cell cycles. Science 274:1672–1677

Sherr CJ (2000) The Pezcoller Lecture: Cancer cell cycles revisited. Cancer Research 60:3689–3695

Shlapobersky M, Sanders R, Clark C, Spector DH (2006) Repression of HMGA2 gene expression by human cytomegalovirus involves the IE2 86-kilodalton protein and is necessary for efficient viral replication and inhibition of cyclin A transcription. J Virol 80:9951–9961

Slee EA, O'Connor DJ, Lu X (2004) To die or not to die: how does p53 decide? Oncogene 23:2809–2818

Sommer MH, Scully AL, Spector DH (1994) Trans-activation by the human cytomegalovirus IE2 86 kDa protein requires a domain that binds to both TBP and RB. J Virol 68:6223–6231

Song Y-J, Stinski MF (2002) Effect of the human cytomegalovirus IE86 protein on expression of E2F responsive genes: a DNA microarray analysis. Proc Natl Acad Sci U S A 99:2836–2841

Song YJ, Stinski MF (2005) Inhibition of cell division by the human cytomegalovirus IE86 protein: role of the p53 pathway or cyclin-dependent kinase 1/cyclin B1. J Virol 79:2597–2603

Speir E, Modali R, Huang E-S, Leon MB, Sahwl F, Finkel T, Epstein SE (1994) Potential role of human cytomegalovirus and p53 interaction in coronary restenosis. Science 265:391–394

Tamrakar S, Kapasi AJ, Spector DH (2005) Human cytomegalovirus infection induces specific hyperphosphorylation of the carboxyl-terminal domain of the large subunit of RNA polymerase II that is associated with changes in the abundance, activity, and localization of cdk9 and cdk7. J Virol 79:15477–15493

Tessari MA, Gostissa M, Altamura S, Sgarra R, Rustighi A, Salvagno C, Caretti G, Imbriano C, Mantovani R, Sal GD, Giancotti V, Manfioletti G (2003) Transcriptional activation of the cyclin A gene by the architectural transcription factor HMGA2. Mol Cell Biol 23:9104–9116

Tran K, Mahr JA, Choi J, Teodoro JG, Green MR, Spector DH (2008) Accumulation of substrates of the anaphase-promoting complex (APC) during human cytomegalovirus infection is associated with the phosphorylation of cdh1 and the dissociation and relocalization of the APC subunits. J Virol 82:529–537

Vassilev LT, Tovar C, Chen S, Knezevic D, Zhao X, Sun H, Heimbrook DC, Chen L (2006) Selective small-molecule inhibitor reveals critical mitotic functions of human CDK1. Proc Natl Acad Sci U S A 103:10660–10665

Vousden KH, Lu X (2002) Live or let die: the cell's response to p53. Nature Rev 2:594–604

Wade M, Kowalik TF, Mudryj M, Huang ES, Azizkhan JC (1992) E2F mediates dihydrofolate reductase promoter activation and multiprotein complex formation in human cytomegalovirus infection. Mol Cell Biol 12:4364–4374

Wiebusch L, Hagemeier C (1999) Human cytomegalovirus 86-kilodalton IE2 protein blocks cell cycle progression in G_1. J Virol 73:9274–9283

Wiebusch L, Hagemeier C (2001) The human cytomegalovirus immediate early 2 protein dissociates cellular DNA synthesis from cyclin dependent kinase activation. EMBO J 20:1086–1098

Wiebusch L, Asmar J, Uecker R, Hagemeier C (2003) Human cytomegalovirus immediate-early protein 2 (IE2)-mediated activation of cyclin E is cell-cycle-independent and forces S-phase entry in IE2-arrested cells. J Gen Virol 84:51–60

Wiebusch L, Bach M, Uecker R, Hagemeier C (2005) Human cytomegalovirus inactivates the G0/G1-APC/C ubiquitin ligase by Cdh1 dissociation. Cell Cycle 4:1435–1439

Woo MS, Sanchez I, Dynlacht BD (1997) p130 and p107 use a conserved domain to inhibit cellular cyclin-dependent kinase activity. Mol Cell Biol 17:3566–3579

Zannetti C, Mondini M, Andrea MD, Caposio P, Hara E, Peters G, Gribaudo G, Gariglio M, Landolfo S (2006) The expression of p16INK4a tumor suppressor is upregulated by human cytomegalovirus infection and required for optimal viral replication. Virology 349:79–86

Zhang Z, Huong S-M, Wang X, Huang DY, Huang E-S (2003) Interactions between human cytomegalovirus IE1-72 and cellular p107: functional domains and mechanisms of up-regulation of cyclin E/cdk2 kinase activity. J Virol 77:12660–12670

Zhu H, Cong JP, Shenk T (1997) Use of differential display analysis to assess the effect of human cytomegalovirus infection on the accumulation of cellular RNAs: induction of interferon-responsive RNAs. Proc Natl Acad Sci U S A 94:13985–13990

Zwicker J, Muller R (1997) Cell-cycle regulation of gene expression by transcriptional repression. Trends in Genetics 13:3–6

Zwicker J, Lucibello FC, Wolfraim LA, Gross C, Truss M, Engeland K, Muller R (1995) Cell cycle regulation of the cyclin A, cdc25C and cdc2 genes is based on a common mechanism of transcriptional repression. EMBO J 15:4514–4522

Modulation of Host Cell Stress Responses by Human Cytomegalovirus

J. C. Alwine

Contents

Introduction 264
Background: PI3K-Akt-TSC-mTOR Signaling..................................... 265
Background: The Complexes of mTOR Kinase and Their Activities................. 266
HCMV and the Activation of the PI3K-Akt-TSC-mTORC1 Pathway................... 268
    Hypoxia 269
    Energy Depletion.. 270
    Amino Acid Depletion.. 270
    Rapamycin... 270
The Effects of HCMV Downstream of Akt 271
HCMV Effects on the mTOR Complexes and Their Substrates..................... 272
HCMV Effects on eIF4E and Mnk-1 273
Conclusions, Questions, Speculations.. 274
References 276

Abstract Human cytomegalovirus (HCMV) induces cellular stress responses during infection due to nutrient depletion, energy depletion, hypoxia and synthetic stress, e.g., endoplasmic reticulum (ER) stress. Cellular stress responses initiate processes that allow the cell to survive the stress; some of these may be beneficial to HCMV replication while others are not. Several studies show that HCMV manipulates stress response signaling in order to maintain beneficial effects while inhibiting detrimental effects. The inhibition of translation is the most common effect of stress responses that would be detrimental to HCMV infection. This chapter will focus on the mechanisms by which cap-dependent translation is maintained during HCMV infection through alterations of the phosphatidylinositol-3′ kinase (PI3K)-Akt-tuberous sclerosis complex (TSC)-mammalian target of rapamycin (mTOR) signaling pathway. The emerging picture is that HCMV affects this pathway in multiple ways, thus

---

J.C. Alwine
Department of Cancer Biology and the Abramson Family Cancer Research Institute,
University of Pennsylvania, 314 Biomedical Research Building, 421 Curie Blvd. Philadelphia, PA 19104-6142, USA
alwine@mail.med.upenn.edu

ensuring that cap-dependent translation is maintained despite the induction of stress responses that would normally inhibit it. Such dramatic alterations of this pathway lead to questions of what other beneficial effects the virus might gain from these changes and how these changes may contribute to HCMV pathogenesis.

Abbreviations 4E-BP: eIF4E binding protein; AICAR: 5-Amino-4-imidazole-carboxamide ribose; Akt: The cellular homolog of the oncoprotein of the AKT8 retrovirus; AMPK: AMP-activated kinase; CaMKKβ: Calcium/calmodulin-dependent protein kinase kinase-β; ER: Endoplasmic reticulum; eIF: Eucaryotic initiation factor; FKBP12: FK506 binding protein; HCMV: Human cytomegalovirus; IR: Insulin receptor; IRS: Insulin receptor substrates; mTOR: Mammalian target of rapamycin; mTORC1: mTOR complex 1; mTORC2: mTOR complex 2; PDK1: Phosphainositide-dependent protein kinase-1; PI3K: Phosphatidylinositol-3′ kinase; PIP2: Phosphatidylinositol-4,5-bisphosphate; PIP3: Phosphatidylinositol-3,4,5-triphosphate; PP2A: Protein phosphatase 2A; PTEN: Phosphatase and tensin homolog; S6K: p70S6 Kinase; TSC: Tuberous sclerosis complex

Introduction

Human cytomegalovirus (HCMV) shares a general life cycle strategy with other mammalian double-stranded DNA viruses that replicate in the nucleus: it must adapt the cellular milieu so the host cell can accommodate the increased demand for nutrients, energy and macromolecular synthesis that accompanies viral infection. For example, successful viral replication requires (1) increased glucose uptake, metabolism and oxygen utilization; (2) abrogation of cellular growth controls; (3) manipulation of the cell cycle to a point that is optimal for virus growth; and (4) inhibition of apoptosis during the productive phase of replication.

These massive changes in the cell's physiology induce cellular stress responses, due to nutrient depletion, energy depletion, hypoxia and synthetic stress, e.g., endoplasmic reticulum (ER) stress. Cellular stress responses are designed to signal the cell when it is in potential trouble and initiate conditions to allow the cell to survive the stress. As a last resort, when the efforts to abate stress fail, apoptosis is induced. Stress responses have many effects on cellular processes; among these some may be beneficial to HCMV replication while others may not. Existing data suggest that HCMV may be able to manipulate stress responses in order to maintain beneficial effects while inhibiting detrimental effects (Isler et al. 2005b; Hakki et al. 2006).

Inhibition of translation is among the most common consequences of cellular stress responses (Kaufman et al. 2002; Arsham et al. 2003; Holcik and Sonenberg 2005; Wouters et al. 2005; Wek et al. 2006). Since translation is an energy-intensive process, its inhibition results in decreased demand for ATP/GTP and decreases the load of proteins entering the ER for processing, consequently relieving ER stress. Translation is well suited to respond to stress, since its inhibition can be accomplished rapidly and reversibly by altering the phosphorylation state of translation regulatory proteins. For example, cap-dependent translation, in which translation initiation depends on recognition of the mRNA's 5'-cap, is controlled through reversible phosphorylation of the eucaryotic initiation factor (eIF) 4E-binding protein (4E-BP) (reviewed in Mamane et al. 2006). This will be discussed in detail below in Sect. 3.

Although inhibition of translation may permit the cell to recover from stress, it would not benefit HCMV replication. The slow replicative cycle characteristic of HCMV requires the virus to maintain the host cell in a metabolically and translationally active state for an extended period; thus HCMV is obliged to abrogate this type of cellular response. A number of studies have shown that HCMV infection induces several mechanisms to overcome the negative effects of stress responses and maintain translation (Child et al. 2004; Kudchodkar et al. 2004; Hakki and Geballe 2005; Isler et al. 2005a, 2005b; Walsh et al. 2005; Hakki et al. 2006; Kudchodkar et al. 2006; Mohr 2006; Kudchodkar et al. 2007). In this chapter, we will concentrate on the mechanisms by which cap-dependent translation is maintained during HCMV infection by modulation of the phosphatidylinositol-3' kinase (PI3K)-Akt-tuberous sclerosis complex (TSC)-mammalian target of rapamycin (mTOR) signaling pathway. The emerging picture is that HCMV-mediated regulation of this pathway is multifaceted, thus ensuring that cap-dependent translation is maintained despite the induction of a variety of cellular stress responses. Such dramatic alterations of this pathway lead one to ask what other beneficial effects the virus might gain from these changes and how these changes may contribute to HCMV pathogenesis.

Background: PI3K-Akt-TSC-mTOR Signaling

Akt (PKB) is the cellular homolog of the oncoprotein of the AKT8 retrovirus (Bellacosa et al. 1991). Members of the mammalian Akt family, Akt1, 2 and 3, are activated by PI3K in response to tropic factors (e.g., insulin and other mitogens); other routes of activation are suspected (Datta et al. 1999; Plas and Thompson 2005; Sarbassov et al. 2005b). In Fig. 1, the binding of insulin to the insulin receptor (IR) is used as an example of an Akt activator. IR activation results in tyrosine phosphorylation of insulin receptor substrates (IRSs), this allows binding of the p85 regulatory subunit of PI3K to IRSs. Consequently, the PI3K catalytic subunit (p110) is activated and phosphorylates phosphatidylinositol (PI)-4,5-bisphosphate (PIP2) to PI-3,4,5-triphosphate (PIP3) on the plasma membrane. Both Akt and phosphoinositide-dependent protein kinase-1 (PDK1) bind PIP3, allowing PDK1 to be positioned to phosphorylate (activate) Akt on threonine 308 (T308).

Activated Akt affects multiple cellular targets that increase metabolism, growth and proliferation while suppressing apoptosis (Summers et al. 1998; Ueki et al. 1998; Cass et al. 1999; Datta et al. 1999; Hill et al. 1999; Plas and Thompson 2005). All of these are beneficial to HCMV lytic growth. Thus, it is not surprising that Akt is activated during HCMV infection (Johnson et al. 2001; Yu and Alwine 2002; Kudchodkar et al. 2006). One of the downstream effects of activated Akt is the activation of mTOR kinase (also known as RAFT1 or FRAP) in mTOR complex 1 (mTORC1, Fig. 1, described in detail in Sect. 3 below). Activation of mTORC1 is critical for the maintenance of cap-dependent translation.

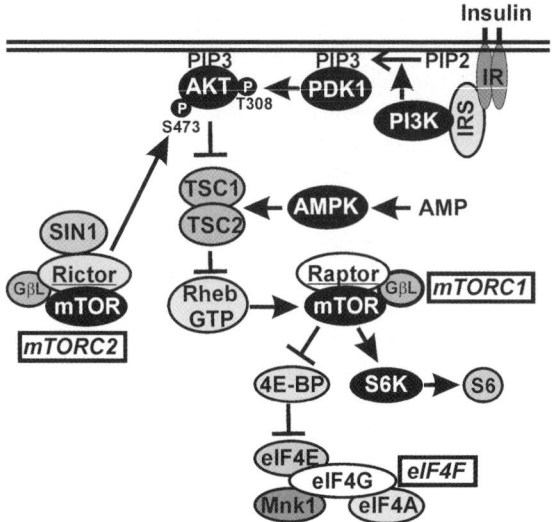

Fig. 1 The PI3K-Akt-mTOR signaling pathway with emphasis on the control of cap-dependant translation via eIF4F. Details are discussed in the text

The link between Akt and mTORC1 is (1) the tuberous sclerosis complex (TSC; reviewed in Luo et al. 2005), made up of TSC1 [hamartin] and TSC2 [tuberin] and (2) Rheb-GTP, a member of the Ras superfamily which binds the N-terminal lobe of the mTOR kinases catalytic domain, allowing mTOR activation (Astrinidis and Henske 2005; Long et al. 2005a, 2005b). Regulation of Rheb-GTP levels is mediated by the GTPase-activating function of the TSC, which stimulates the intrinsic GTPase activity of Rheb, converting it from Rheb-GTP to Rheb-GDP, the inactive form that cannot activate mTORC1. Thus Akt's phosphorylation of the TSC inactivates it, allowing Rheb-GTP levels to remain high in order to activate mTORC1.

Background: The Complexes of mTOR Kinase and Their Activities

mTOR kinase is found in two complexes that differ in their major binding partner (Fig. 1): raptor (*r*egulatory *a*ssociated *p*rotein of *TOR*) in mTORC1 and rictor (*r*apamycin-*i*nsensitive *com*panion of m*TOR*) in mTORC2 (Kim et al. 2002; Sarbassov et al. 2004). Both complexes contain a small protein called GβL that binds to the kinase domain of mTOR kinase and stabilizes the interaction with raptor and rictor (Kim et al. 2003). An additional protein, SIN1, found in mTORC2, maintains the integrity of the complex and regulates activity and substrate specificity (Jacinto et al. 2006; Polak and Hall 2006; Yang et al. 2006a). It is important to note that under normal conditions the two complexes differ in their sensitivity to the drug rapamycin; mTORC1 is sensitive and mTORC2 is insensitive (Sarbassov et al. 2004).

The role of mTORC1 in the control of cap-dependent translation has been extensively studied (Sarbassov et al. 2005a; Reiling and Sabatini 2006). Under conditions of adequate nutrients and oxygen, mTORC1 is activated, permitting phosphorylation of key substrates, p70S6 kinase (S6K) and 4E-BP (Fig. 1). S6K phosphorylation triggers events that promote the formation of translation initiation complexes (reviewed in Mamane et al. 2006); among these is the phosphorylation of ribosomal protein S6, which is often used as a marker for S6K activity. To understand the role of 4E-BP phosphorylation in the control of cap-dependent translation, we must consider its binding partner, eIF4E, and the eIF4F translation initiation complex (Figs. 1 and 2), which binds to the 5'-cap of an mRNA (Fig. 2), the first step in initiation of cap-dependent translation. Functional eIF4F complex consists of the scaffolding protein eIF4G bound to (1) eIF4E, the protein in the complex that directly binds the 5'-cap; (2) Mnk1, a kinase which phosphorylates eIF4E; and (3) eIF4A, an RNA helicase. As indicated in Fig. 2, the functionality of eIF4F depends on eIF4E being bound to eIF4G. However, eIF4E can be removed from eIF4G by binding to 4E-BP, this inhibits cap-dependent translation. It is the role of mTORC1 to control whether or not eIF4E is bound to 4E-BP. Under positive growth conditions, mTORC1 is active and phosphorylates 4E-BP, making 4E-BP unable to bind eIF4E. Thus eIF4E binds eIF4G and completes the eIF4F complex on the 5'-cap. Under these conditions, the polyA binding protein (PABP), bound to the 3'-poly A tail of the mRNA, can also interact with eIF4G (Fig. 2),

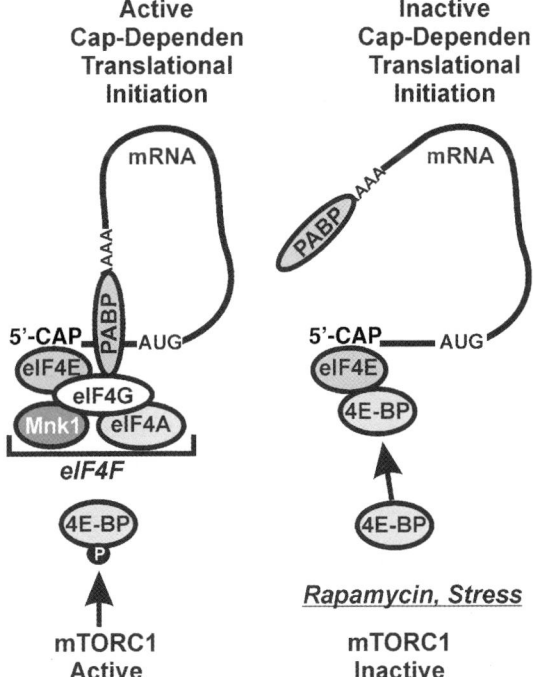

Fig. 2 The control of cap-dependent translation mediated by the activation or inactivation of mTORC1. Details are discusses in the text

setting up conditions in which the 40S ribosome subunit can recognize the mRNA for translation (Mamane et al. 2006). Under negative growth conditions, mTORC1 is inactive and 4E-BP becomes hypophosphorylated. Hypophosphorylated 4E-BP binds to eIF4E, retaining it from eIF4G and the eIF4F complex, thus inhibiting cap-dependent translation.

As discussed below, mTOR kinase activity in mTORC1 is inhibited by many cellular stress responses in order to inhibit phosphorylation of 4E-BP and cap-dependent translation. The drug rapamycin can also do this through direct inhibition of mTOR kinase in mTORC1. It is clear that in order for HCMV to maintain cap-dependent translation, it must find means to maintain hyperphosphorylated 4E-BP. In other words, it needs to circumvent the effects of any cellular stress response that might lead to the inhibition of mTOR kinase activity.

In contrast to the level of understanding of mTORC1, much less is known about the functions of mTORC2. However, RNAi-mediated depletion of rictor in cultured cells demonstrated that mTORC2 plays a role in actin cytoskeleton organization (Jacinto et al. 2004; Sarbassov et al. 2004). At present the only known mTORC2 substrate is serine 473 (S473) of Akt (Sarbassov et al. 2005b; Fig. 1). The role of S473 phosphorylation in Akt activity is controversial, but it has been suggested that it precedes phosphorylation of T308 (the PDK1 site) and may be important for the recognition and phosphorylation of Akt by PDK1 (Sarbassov et al. 2005b) (Fig. 1). This suggests the potential for mTOR and Akt to be involved in an autoregulatory loop. The control of mTORC2 activity is not well understood; however, one study suggests that Rheb-GTP, the activator of mTORC1, does not activate mTORC2 and may inhibit it (Yang et al. 2006b).

HCMV and the Activation of the PI3K-Akt-TSC-mTORC1 Pathway

HCMV infection has dramatic effects on the PI3K-Akt-TSC-mTORC1 pathway. HCMV infection activates Akt through stimulation of T308 phosphorylation via activation of PI3K (Johnson et al. 2001; Yu and Alwine 2002) and stimulation of S473 phosphorylation via activation of mTORC2 (Kudchodkar et al. 2006) (Fig. 1). This occurs by at least two mechanisms. First, transient activation occurs via HCMV attachment to cell receptors which mediate signaling to PI3K (Johnson et al. 2001). However, the identity of this receptor and the means of activating PI3K are under debate (Isaacson et al. 2007; see the chapter by M.K. Isaacson et al., this volume). Second, long-term activation results from the expression of HCMV encoded proteins (Yu and Alwine 2002; Kudchodkar et al. 2006); for example, transfection experiments have shown that expression of the individual major immediate early proteins (either the 72-kDa IE1 or the 86-kDa IE2 proteins) can stimulate phosphorylation of Akt at both sites (Yu and Alwine 2002; Y. Yu and J.C. Alwine, unpublished data).

The HCMV-induced activation of Akt leads to the activation of mTORC1, as indicated by phosphorylation of 4E-BP and S6K, beginning 8-12 h postinfection

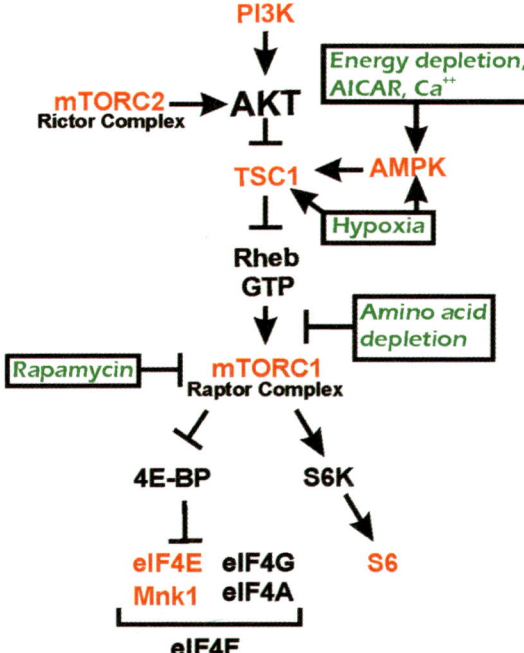

Fig. 3 The effects of stress inducers and HCMV infection on the PI3K-Akt-mTOR pathway, its associated substrates and the components of the eIF4F complex. Stress inducers (*green*) are shown at sites where they affect the PI3K-Akt-mTOR pathway. The points shown in *red* indicate where HCMV infection exerts its effects; Table 1 presents a synopsis of each, see text for details

(Kudchodkar et al. 2004). Although Akt activation is necessary to maintain cap-dependent translation, it is not sufficient. The virus must do more since mTOR kinase activity can be inhibited by many stress responses which exert their effects at points downstream of Akt in the signaling pathway (Fig. 3). Stress responses that do this are induced by the following conditions:

Hypoxia

Hypoxia may occur during the highest rates of metabolic and synthetic activity in HCMV-infected cells. Hypoxia inhibits mTORC1 (Arsham et al. 2002; Brugarolas et al. 2004; Cai et al. 2006) due to the induction of REDD1 (Brugarolas et al. 2004), a protein which activates the TSC (reviewed in van den Beucken et al. 2006). This mechanism is dependent on the induction of hypoxia inducing factor-1 (HIF-1), since the *redd1* gene is a HIF-1 transcriptional target (Schwarzer et al. 2005). A HIF-independent oxygen sensing mechanisms for activation of the TSC involves activation of the AMP-activated kinase (AMPK, Figs. 1 and 3) (Corradetti et al. 2004; Shaw et al. 2004).

Energy Depletion

AMPK activation of the TSC is most often associated with energy depletion (reviewed in Luo et al. 2005; Kimble 2006), which can occur during the highest metabolic and synthetic periods of an HCMV infection. Binding of AMP to AMPK is thought to alter its conformation, permitting subsequent phosphorylation by upstream protein kinases. In contrast, ATP binding prevents phosphorylation and maintains AMPK in its inactive state. Thus, AMPK responds not only to changes in AMP concentration but also to changes in the ratio of AMP to ATP. This means of activating AMPK can be mimicked using the AMP mimetic 5-amino-4-imidazolecarboxamide ribose (AICAR). Under conditions of an increased AMP/ATP ratio or in the presence of AICAR, the tumor suppressor LKB1 phosphorylates AMPK. In addition, the calcium/calmodulin-dependent protein kinase kinase-β (CaMKKβ) also phosphorylates and activates AMPK in response to changes in intracellular calcium concentrations (Hardie 2007).

Both LKB1 and CaMKKβ-mediated mechanisms for activating AMPK could occur during HCMV infection. The increased energy utilization in actively infected cells would increase the AMP:ATP ratio, thus activating AMPK. In addition, it has been shown that the HCMV UL37×1 protein mobilizes calcium from the endoplasmic reticulum into the cytosol (Sharon-Friling et al. 2006); this could potentially activate CaMKKβ.

Amino Acid Depletion

Amino acid depletion inhibits mTORC1 through a mechanism that interferes with the binding of Rheb-GTP to mTOR kinase (reviewed in Avruch et al. 2006). Amino acid depletion, resulting in mTORC1 inhibition, could occur during HCMV infection at times of the greatest viral protein synthesis.

Rapamycin

Direct inhibition of mTOR kinase activity in mTORC1 can be achieved using the drug rapamycin, a macrolide antifungal agent isolated from *Streptomyces hygroscopicus* (Vezina et al. 1975). It is used clinically as an immunosuppressant in organ transplantation, an inhibitor of restenosis of arteries after angioplasty, and an anti-cancer drug (Bjornsti and Houghton 2004). Rapamycin forms a complex with a 12-kDa immunophilin called the FK506 binding protein 12 (FKBP12) and can also bind a 100-amino acid domain (E2015 to Q2114) of mTOR kinase known as the FKBP-rapamycin binding domain (Brown et al. 1994; Sabatini et al. 1994; Chen et al. 1995). The affinity of mTOR's FKBP-rapamycin binding domain for rapamycin in the absence of FKBP12 is modest, but when rapamycin is bound to FKBP12, the affinity is increased by 2,000-fold (Banaszynski et al. 2005). FKBP12-rapamycin bound to the FKBP-rapamycin binding domain is believed to prevent association between mTOR and raptor (Halford et al. 2001; Hara et al. 2002; Kim et al. 2002). The difference in rapamycin sensitivity between mTORC1 and mTORC2 may be due to structural variations that result from the different binding partners in the complexes. In this model, the mTOR-raptor interaction leaves the FKBP-rapamycin binding domain available while the mTOR-rictor interaction occludes it, thus affording rapamycin insensitivity.

These examples of mTORC1 inhibition by stress responses and drugs highlight the need for HCMV to counteract inhibitory mechanisms at multiple points in the PI3K-Akt-TSC-mTOR pathway. That HCMV can do this has been demonstrated in studies showing that HCMV infection can proceed under conditions of hypoxia (Kudchodkar et al. 2004), under conditions where AMPK is activated (Kudchodkar et al. 2007; N.J. Moorman et al., personal communication) and can overcome the inhibitory effects of rapamycin (Kudchodkar et al. 2006). In addition, preliminary data suggest that HCMV can grow, with reduced kinetics, under conditions of glucose and amino acid depletion (our preliminary data). Indeed, a growing body of evidence shows that HCMV infection may affect many aspects of the pathway downstream of Akt. Known points of interaction between the pathway and HCMV are shown in red in Fig. 3. In the following sections we will highlight HCMV's effects on AMPK, TSC, the mTOR complexes and mTOR's downstream effectors.

The Effects of HCMV Downstream of Akt

Experiments using the AMP mimetic AICAR to activate AMPK have been utilized to analyze the effects of activated AMPK during HCMV infection (Kudchodkar et al. 2007; N.J. Moorman et al., personal communication). Addition of AICAR at 4 h p.i. completely inhibits the viral infection due, at least in part, to the AMPK-mediated inhibition of mTORC1 and translation. Under these conditions, the HCMV immediate early proteins are not expressed and the infection cannot be established. However, if immediate early proteins are expressed before the addition of AICAR (e.g., AICAR addition at 12 h p.i.), there is little to no effect on mTORC1 activity, as indicated by the phosphorylation of 4E-BP and S6K. This observation suggests that the expression and action of the immediate early proteins establishes conditions that are resistant to the effects of activated AMPK. This may involve direct effects on AMPK or downstream at the TSC or Rheb-GTP (Fig. 1). Existing data suggest that HCMV targets AMPK and the TSC (Kudchodkar et al. 2007; N.J. Moorman et al., personal communication); no studies have been reported examining Rheb-GTP.

HCMV infection appears to induce a means to limit the level of phosphorylated AMPK in the infected cell, possibly by dephosphorylation (Kudchodkar et al.

2007). However, this mechanism seems to be of marginal effectiveness since it is overwhelmed by the extensive phosphorylation of AMPK caused by the addition of AICAR. However, under conditions of high levels of phosphorylated AMPK, HCMV inhibits its downstream effects by inactivating the TSC. It has been shown that the HCMV protein pUL38 binds to and inactivates the TSC (N.J. Moorman et al., personal communication), thus protecting mTORC1 from inhibition. This function of pUL38, an early protein, may explain why mTORC1-mediated phosphorylation of 4E-BP and S6K became resistant to the effects of AICAR by 12 h p.i. It is important to note that the inhibition of the TSC by pUL38 may not only block the effects of activated AMPK, but also the effects of other stress responses such as hypoxia that function through activating the TSC.

One question that arises is why would HCMV maintain a limited ability to dephosphorylate AMPK as well as inhibit the TSC with pUL38? It would seem that the inhibition of the TSC would be sufficient to protect mTORC1 from inactivation caused by AMPK activity. A putative answer arises from the observation that HCMV strives to maintain beneficial aspects of cellular stress responses while inhibiting disadvantageous ones. The beneficial aspects of having phosphorylated, activated AMPK are that it activates pathways that increase: (1) ATP production, (2) glucose transport and glycolysis, and (3) fatty acid oxidation (Luo et al. 2005). Thus, HCMV's limited ability to dephosphorylate AMPK may provide a situation where some activated AMPK is maintained during infection. In this regard, data suggest that the different effects of AMPK are activated differentially, requiring different levels of phosphorylated AMPK (Jones et al. 2005). For example, the amount of phosphorylated AMPK needed to increase ATP production, glucose transport, glycolysis and fatty acid oxidation is thought to be much less than the amount needed to activate the TSC and inhibit translation. Hence it would be to the virus's advantage to be able to maintain levels of phosphorylated AMPK capable of inducing beneficial effects but below the level needed to activate the TSC. However, under conditions where this mechanism is overwhelmed, and high levels of phosphorylated AMPK accumulate, the pUL38-mediated inhibition of the TSC would still protect mTOR kinase activity and circumvent translation inhibition.

HCMV Effects on the mTOR Complexes and Their Substrates

One might think that the inhibition of the TSC would be sufficient to assure the maintenance of mTORC1 activity and cap-dependent translation in the infected cell. However, HCMV infection also targets the mTOR complexes and functionally alters them in order to maintain 4E-BP phosphorylation and cap-dependent translation. The first indication of such a mechanism was the observation that in the presence of rapamycin, which directly inhibits mTORC1, translation was maintained in infected human fibroblasts (Kudchodkar et al. 2004). Rapamycin causes a 12- to 24-h delay in the first appearance of progeny virions when compared to a normal viral growth curve; however, once viral production initiates, it proceeds with normal kinetics to produce near normal levels of progeny (Kudchodkar et al. 2004). Thus the normal rapamycin sensitivity of mTORC1 had been altered; the delay of 12-24 h in viral growth indicated a period in which the viral infection adapted to rapamycin. Indeed, the establishment of rapamycin resistance of viral growth correlated with the development of rapamycin-insensitive phosphorylation of 4E-BP (Kudchodkar et al. 2004, 2006). Interestingly, the phosphorylation of mTORC1's other substrate, S6K, does not become rapamycin-insensitive (Kudchodkar et al. 2004, 2006). Thus HCMV's effect on mTOR kinase activity is substrate-specific, suggesting that HCMV either (1) targets and alters the mTOR complexes directly for the purpose of facilitating 4E-BP phosphorylation or (2) it induces an mTOR-independent mechanism for 4E-BP phosphorylation.

Present data suggest that HCMV alters the mTOR complexes. Experiments using shRNAs to deplete rictor or raptor have shown that HCMV infection alters the substrate specificity of the rictor-containing complex (Kudchodkar et al. 2006). Under normal conditions in uninfected cells, only the raptor-containing complex uses 4E-BP and S6K as substrates; thus depletion of raptor with shRNA results in a severe inhibition of 4E-BP and S6K phosphorylation while depletion of rictor has little effect (Sarbassov et al. 2006b). However, in HCMV-infected cells similar depletion of raptor or rictor caused only about 50% inhibition of 4E-BP and S6K phosphorylation. This suggests that 4E-BP and S6K can be phosphorylated by either the rictor- or the raptor-containing complex; this was confirmed by other experiments (Kudchodkar et al. 2006). Thus the substrate specificity of the rictor-containing complex is expanded, most likely by structure/function modification, during HCMV infection.

An additional indication of HCMV-induced structural modification of the raptor- and rictor-containing complexes is the observation that the rapamycin sensitivity of 4E-BP phosphorylation is altered in infected cells. The raptor-containing complex, normally sensitive to rapamycin, is insensitive in infected cells; and the rictor-containing complex, normally insensitive for the phosphorylation of Akt, is rapamycin-sensitive with respect to 4E-BP phosphorylation (Kudchodkar et al. 2006). How the viral infection mediates these changes remains to be determined; however, virally induced structural modifications of each complex could account for them. Just as structural modifications could alter the substrate specificity of the rictor-containing complexes, they could also change the accessibility of mTOR's FKBP-rapamycin binding domain such that it becomes accessible in rictor-containing complexes and occluded in raptor-containing complexes, thus altering rapamycin sensitivity.

HCMV Effects on eIF4E and Mnk-1

While the HCMV-induced phosphorylation of 4E-BP may seem sufficient to allow eIF4E to be in the eIF4F complex to promote cap-dependent translation, HCMV maintains an additional mechanism. During HCMV infection, the phosphorylation of eIF4E by Mnk-1 is enhanced to maintain translation; moreover, the levels of eIF4E are increased in infected cells (Walsh et al. 2005). Increasing the amount of eIF4E would be predicted to titrate out hypophosphorylated 4E-BP while still maintaining sufficient free eIF4E to bind eIF4G and preserve the integrity of the eIF4F complex. This is again a situation where HCMV appears to use redundant mechanisms: (1) hyperphosphorylation of 4E-BP to lower its affinity for eIF4E and (2) increased levels of eIF4E to titrate out hypophosphorylated 4E-BP. Such redundancy indicates the importance of maintaining cap-dependent translation during HCMV infection. In addition, the two different mechanisms may be required to maintain cap-dependent translation in different cell types and under different growth conditions.

Conclusions, Questions, Speculations

The points shown in red in Fig. 3 indicate where HCMV infection exerts its effects on the PI3K-Akt-TSC-mTOR pathway, its associated substrates and the components of the eIF4F complex; Table 1 presents a synopsis of each. The striking feature is the number of ways the virus has evolved to manipulate this pathway and thereby manipulate the effects of cellular stress responses. The effects of stress responses that would be inhibitory to viral growth are circumvented, while potentially beneficial effects may be maintained. It is also striking that in more than one case the virus has introduced multiple means to circumvent inhibitory effects of stress responses: for example, the phosphorylation of 4E-BP plus the increased levels of eIF4E and the inhibition of the TSC plus the dephosphorylation of AMPK. The maintenance of redundant mechanisms suggests that they provide the virus with additional capabilities yet to be discovered and understood.

Table 1 Effects of HCMV infection on the PI3K-Akt-TSC-mTOR pathway, its associated substrates and the components of the eIF4F complex

Target	HCMV-mediated mechanism
PI3K	Activated by viral attachment, MIEPs and possible other viral or induced cellular proteins
Akt	Phosphorylated by HCMV activated PI3K-PDK1 and mTORC2
AMPK	Inactivated by dephosphorylation
TSC	Inactivated by HCMV UL38
mTORC1	Activated by HCMV infection
	Altered producing rapamycin insensitivity
mTORC2	Activated by HCMV infection
	Altered producing rapamycin sensitivity and gaining 4E-BP and S6K as substrates
eIF4E	Level increased during HCMV infection
Mnk1	Phosphorylation of eIF4E increased by HCMV infection
S6	Phosphorylated by an HCMV-induced mechanism that does not involve S6K (Kudchodkar et al. 2004)

A number of important and exciting questions remain:

1. How does HCMV infection activate PI3K? Initially, viral attachment to receptors may transiently activate PI3K but later in the infection viral protein synthesis is required. The observation that the major immediate early proteins (IE1 and IE2) can mediate PI3K activation suggests a mechanism involving transcriptional activation of viral or cellular genes, the products of which may activate PI3K. However, IE1 and IE2, being complex proteins, may mediate functions, such as the activation of PI3K, which do not depend on transcriptional mechanisms.
2. How are mTORC1 and mTORC2 altered such that their rapamycin sensitivities are reversed and mTORC2 gains 4E-BP and S6K as substrates? Structural modifications of the complexes have been discussed above. It is clearly possible that a cellular protein, induced or misplaced by the viral infection, or a viral protein, becomes part of the complexes and alters their specificities and functions. Indeed, it is conceivable that mTOR is not the only kinase associated with raptor and rictor in infected cells. The shRNA-mediated depletion of mTOR kinase in HCMV infected cells significantly decreases but does not eliminate 4E-BP and S6K phosphorylation (Kudchodkar et al. 2006), thus another kinase could be active with rictor and raptor in infected cells.
3. What mechanism is HCMV using to trigger mTORC2-mediated phosphorylation of Akt S473? As mentioned above in Sect. 3, the control of mTORC2 activity is not well understood. One study suggests that Rheb-GTP, the activator of mTORC1, may inhibit mTORC2 (Yang et al. 2006b). If this is correct, the inhibition of the TSC by pUL38 would be expected to increase Rheb-GTP levels and potentially inhibit mTORC2. However, this does not happen: the rapamycin insensitive phosphorylation of Akt S473 by mTORC2 is activated during an HCMV infection (Kudchodkar et al. 2006). Hence there is still much to learn about HCMV's effects of mTORC2 and the phosphorylation of Akt.

In the above discussion, we have not considered the potential role of phosphatases in the control of the PI3K-Akt-mTOR pathway. PTEN (phosphatase and tensin homolog) is the phosphatase that counteracts PI3K (Baker 2007); its inhibition during infection could account, in part, for the activation of Akt. In human fibroblasts, PTEN levels do not change during infection (Y. Yu and J.C. Alwine, unpublished observations), but it has not been determined whether its activity is altered. One report suggests that PTEN activity may be increased in primary human aortic endothelial cells infected with HCMV (Shen et al. 2006). This could reduce Akt activation and slow the growth of HCMV in these specialized cells, possibly resulting in specific pathogenesis. Protein phosphatase 2A (PP2A) is believed to be the phosphatase that counteracts mTOR kinase by dephosphorylating 4E-BP and S6K. Indeed, PP2A is a central controlling factor in many cellular processes (Mumby 2007). Little is known about the HCMV's interactions with PP2A, but it is likely to be an HCMV target.

Finally, our discussion has largely used the maintenance of cap-dependent translation as the main reason for HCMV targeting the PI3K-Akt-mTOR pathway. However, this is a narrow view. The many interactions that HCMV has with this pathway (Fig. 3 and Table 1) opens the possibility for vast effects on cellular physiology. The resulting pathogenic effects could implicate HCMV as a subtle, and unexpected, cofactor in many maladies. For example, nearly every one of the targets listed in Table 1 can be an oncoprotein when mutated, inappropriately activated or inappropriately expressed. While not suggesting that HCMV is a frank transforming agent, it is possible that the virus serves as a co-factor with other agents/mutations to promote transformation. The effects of HCMV on oncoproteins such as PI3K, Akt, mTOR, mTOR's effectors and eIF4E could increase the oncogenic potential of a cell, serving as one factor among several which cause transformation, as suggested by the Knudson multi-hit hypothesis (Knudson 1988). The understanding of the means by which HCMV adapts cellular stress response signaling will provide new insight into HCMV pathogenesis.

Acknowledgements J.C.A. is funded by Public Health Service grants R01 CA28379-27 and R01 GM45773-15 from the National Institutes of Health and by the Abramson Family Cancer Research Institute.

References

Arsham AM, Plas DR, Thompson CB, Simon MC (2002) PI3-K/Akt signaling is neither required for hypoxic stabilization of HIF-1 nor sufficient for HIF-1-dependent target gene transcription. J Biol Chem 277:15162-15170

Arsham AM, Howell JJ, Simon MC (2003) A novel hypoxia-inducible factor-independent hypoxic response regulating mammalian target of rapamycin and its targets. J Biol Chem 278:29655-29660

Astrinidis A, Henske EP (2005) Tuberous sclerosis complex: linking growth and energy signaling pathways with human disease. Oncogene 24:7475-7481

Avruch J, Hara K, Lin Y, Liu M, Long X, Ortiz-Vega S, Yonezawa K (2006) Insulin and amino-acid regulation of mTOR signaling and kinase activity through the Rheb GTPase. Oncogene 25:6361-6372

Baker SJ (2007) PTEN enters the nuclear age. Cell 128:25-28

Banaszynski LA, Liu CW, Wandless TJ (2005) Characterization of the FKBP-rapamycin-FRB ternary complex. J Am Chem Soc 127:4715-4721

Bellacosa A, Testa JR, Staal SP, Tsichlis PN (1991) A retroviral oncogene, akt, encoding a serine-threonine kinase containing an SH2-like region. Science 254:274-277

Bjornsti MA, Houghton PJ (2004) The TOR pathway: a target for cancer therapy. Nat Rev Cancer 4:335-348

Brown EJ, Albers MW, Shin TB, Ichikawa K, Keith CT, Lane WS, Schreiber SL (1994) A mammalian protein targeted by G1-arresting rapamycin-receptor complex. Nature 369:756-758

Brugarolas J, Lei K, Hurley RL, Manning BD, Reiling JH, Hafen E, Witters LA, Ellisen LW, Kaelin WG Jr (2004) Regulation of mTOR function in response to hypoxia by REDD1 and the TSC1/TSC2 tumor suppressor complex. Genes Dev 18:2893-2904

Cai SL, Tee AR, Short JD, Bergeron JM, Kim J, Shen J, Guo R, Johnson CL, Kiguchi K, Walker CL (2006) Activity of TSC2 is inhibited by AKT-mediated phosphorylation and membrane partitioning. J Cell Biol 173:279-289

Cass LA, Summers SA, Prendergast GV, Backer JM, Birnbaum MJ, Meinkoth JL (1999) Protein kinase A-dependent and -independent signaling pathways contribute to cyclic AMP-stimulated proliferation. Mol Cell Biol 19:5882-5891

Chen J, Zheng XF, Brown EJ, Schreiber SL (1995) Identification of an 11-kDa FKBP12-rapamycin-binding domain within the 289-kDa FKBP12-rapamycin-associated protein and characterization of a critical serine residue. Proc Natl Acad Sci U S A 92:4947-4951

Child SJ, Hakki M, De Niro KL, Geballe AP (2004) Evasion of cellular antiviral responses by human cytomegalovirus TRS1 and IRS1. J Virol 78:197-205

Corradetti MN, Inoki K, Bardeesy N, DePinho RA, Guan KL (2004) Regulation of the TSC pathway by LKB1: evidence of a molecular link between tuberous sclerosis complex and Peutz-Jeghers syndrome. Genes Dev 18:1533-1538

Datta SR, Brunet A, Greenberg ME (1999) Cellular survival: a play in three Akts. Genes Dev 13:2905-2927

Hakki M, Geballe AP (2005) Double-stranded RNA binding by human cytomegalovirus pTRS1. J Virol 79:7311-7318

Hakki M, Marshall EE, De Niro KL, Geballe AP (2006) Binding and nuclear relocalization of protein kinase R by human cytomegalovirus TRS1. J Virol 80:11817-11826

Halford WP, Kemp CD, Isler JA, Davido DJ, Schaffer PA (2001) ICP0, ICP4, or VP16 expressed from adenovirus vectors induces reactivation of latent herpes simplex virus type 1 in primary cultures of latently infected trigeminal ganglion cells. J Virol 75:6143-6153

Hara K, Maruki Y, Long X, Yoshino K, Oshiro N, Hidayat S, Tokunaga C, Avruch J, Yonezawa K (2002) Raptor, a binding partner of target of rapamycin (TOR), mediates TOR action. Cell 110:177-189

Hardie DG (2007) AMP-activated protein kinase as a drug target. Annu Rev Pharmacol Toxicol 47:185-210

Hill MM, Clark SF, Tucker DF, Birnbaum MJ, James DE, Macaulay SL (1999) A role for protein kinase Bbeta/Akt2 in insulin-stimulated GLUT4 translocation in adipocytes. Mol Cell Biol 19:7771-7781

Holcik M, Sonenberg N (2005) Translational control in stress and apoptosis. Nat Rev Mol Cell Biol 6:318-327

Isaacson MK, Feire AL, Compton T (2007) The epidermal growth factor receptor is not required for human cytomegalovirus entry or signaling. J Virol 81:6241-6247

Isler JA, Maguire TG, Alwine JC (2005a) Production of infectious HCMV virions is inhibited by drugs that disrupt calcium homeostasis in the endoplasmic reticulum. J Virol 79:15338-15397

Isler JA, Skalet AH, Alwine JC (2005b) Human cytomegalovirus infection activates and regulates the unfolded protein response. J Virol 79:6890-6899

Jacinto E, Loewith R, Schmidt A, Lin S, Ruegg MA, Hall A, Hall MN (2004) Mammalian TOR complex 2 controls the actin cytoskeleton and is rapamycin insensitive. Nat Cell Biol 6:1122-1128

Jacinto E, Facchinetti V, Liu D, Soto N, Wei S, Jung SY, Huang Q, Qin J, Su B (2006) SIN1/MIP1 maintains rictor-mTOR complex integrity and regulates Akt phosphorylation and substrate specificity. Cell 127:125-137

Johnson RA, Wang X, Ma XL, Huong SM, Huang ES (2001) Human cytomegalovirus up-regulates the phosphatidylinositol 3-kinase (PI3-K) pathway: inhibition of PI3-K activity inhibits viral replication and virus-induced signaling. J Virol 75:6022-6032

Jones RG, Plas DR, Kubek S, Buzzai M, Mu J, Xu Y, Birnbaum MJ, Thompson CB (2005) AMP-activated protein kinase induces a p53-dependent metabolic checkpoint. Mol Cell 18:283-293

Kaufman RJ, Scheuner D, Schroder M, Shen X, Lee K, Liu CY, Arnold SM (2002) The unfolded protein response in nutrient sensing and differentiation. Nature Rev Mol Cell Biol 3:411-421

Kim DH, Sarbassov DD, Ali SM, King JE, Latek RR, Erdjument-Bromage H, Tempst P, Sabatini DM (2002) mTOR interacts with raptor to form a nutrient-sensitive complex that signals to the cell growth machinery. Cell 110:163-175

Kim DH, Sarbassov DD, Ali SM, Latek RR, Guntur KV, Erdjument-Bromage H, Tempst P, Sabatini DM (2003) GβL, a positive regulator of the rapamycin-sensitive pathway required for the nutrient-sensitive interaction between raptor and mTOR. Mol Cell 11:895-904

Kimble SR (2006) Interaction between the AMP-activated protein kinase and mTOR signaling pathways. Med Sci Sports Exercise 38:1958-1964

Knudson AG (1988) The genetics of childhood cancer. Bull Cancer 73:135-138

Kudchodkar SB, Yu Y, Maguire T, Alwine JC (2004) Human cytomegalovirus infection induces rapamycin insensitive phosphorylation of downstream effectors of mTOR kinase. J Virol 78:11030-11039

Kudchodkar SB, Yu Y, Maguire TG, Alwine JC (2006) Human cytomegalovirus infection alters the substrate specificities and rapamycin sensitivities of raptor- and rictor-containing complexes. Proc Natl Acad Sci U S A 103:14182-14187

Kudchodkar SB, Del Prete GQ, Maguire TG, Alwine JC (2007) AMPK-mediated inhibition of mTOR kinase is circumvented during immediate-early times of human cytomegalovirus infection. J Virol 81:3649-3651

Long X, Lin Y, Ortiz-Vega S, Yonezawa K, Avruch J (2005a) Rheb binds and regulates the mTOR kinase. Curr Biol 15:702-713

Long X, Ortiz-Vega S, Lin Y, Avruch J (2005b) Rheb binding to mammalian target of rapamycin (mTOR) is regulated by amino acid sufficiency. J Biol Chem 280:23433-23436

Luo Z, Saha AK, Xiang X, Ruderman NB (2005) AMPK, the metabolic syndrome and cancer. Trends Pharmacol Sci 26:69-76

Mamane Y, Petroulakis E, LeBacquer O, Sonenberg N (2006) mTOR, translation initiation and cancer. Oncogene 25:6416-6422

Mohr I (2006) Phosphorylation and dephosphorylation events that regulate viral mRNA translation. Virus Res 119:89-99

Mumby M (2007) The 3D structure of protein phosphatase 2A: new insights into a ubiquitous regulator of cell signaling. ACS Chem Biol 2:99-103

Plas DR, Thompson CB (2005) Akt-dependent transformation: there is more to growth than just surviving. Oncogene 24:7435-7442

Polak P, Hall MN (2006) mTORC2 Caught in a SINful Akt. Dev Cell 11:433-434

Reiling JH, Sabatini DM (2006) Stress and mTORture signaling. Oncogene 25:6373-6383

Sabatini DM, Erdjument-Bromage H, Lui M, Tempst P, Snyder SH (1994) RAFT1: a mammalian protein that binds to FKBP12 in a rapamycin-dependent fashion and is homologous to yeast TORs. Cell 78:35-43

Sarbassov DD, Ali SM, Kim DH, Guertin DA, Latek RR, Erdjument-Bromage H, Tempst P, Sabatini DM (2004) Rictor, a novel binding partner of mTOR, defines a rapamycin-insensitive and raptor-independent pathway that regulates the cytoskeleton. Curr Biol 14:1296-1302

Sarbassov DD, Ali SM, Sabatini DM (2005a) Growing roles for the mTOR pathway. Curr Opin Cell Biol 17:596-603

Sarbassov DD, Guertin DA, Ali SM, Sabatini DM (2005b) Phosphorylation and regulation of Akt/PKB by the rictor-mTOR complex. Science 307:1098-1101

Schwarzer R, Tondera D, Arnold W, Giese K, Klippel A, Kaufmann J (2005) REDD1 integrates hypoxia-mediated survival signaling downstream of phosphatidylinositol 3-kinase. Oncogene 24:1138-1149

Sharon-Friling R, Goodhouse J, Colberg-Poley AM, Shenk T (2006) Human cytomegalovirus pUL37×1 induces the release of endoplasmic reticulum calcium stores. Proc Natl Acad Sci U S A 130:19117-19122

Shaw RJ, Bardeesy N, Manning BD, Lopez L, Kosmatka M, DePinho RA, Cantley LC (2004) The LKB1 tumor suppressor negatively regulates mTOR signaling. Cancer Cell 6:91-99

Shen YH, Zhang L, Utama B, Wang J, Gan Y, Wang X, Wang J, Chen L, Vercellotti GM, Coselli JS, Mehta JL, Wang XL (2006) Human cytomegalovirus inhibits Akt-mediated eNOS activation through upregulating PTEN (phosphatase and tensin homolog deleted on chromosome 10). Cardiovasc Res 69:502-511

Summers SA, Garza LA, Zhou H, Birnbaum MJ (1998) Regulation of insulin-stimulated glucose transporter GLUT4 translocation and Akt kinase activity by ceramide. Mol Cell Biol 18:5457-5464

Ueki K, Yamamoto-Honda R, Kaburagi Y, Yamauchi T, Tobe K, Burgering BM, Coffer PJ, Komuro I, Akanuma Y, Yazaki Y, Kadowaki T (1998) Potential role of protein kinase B in insulin-induced glucose transport, glycogen synthesis, and protein synthesis. J Biol Chem 273:5315-5322

van den Beucken T, Koritzinsky M, Wouters BG (2006) Translational control of gene expression during hypoxia. Cancer Biol Therapy 5:749-755

Vezina C, Kudelski A, Sehgal SN (1975) Rapamycin (AY-22,989), a new antifungal antibiotic. I. Taxonomy of the producing streptomycete and isolation of the active principle. J Antibiotics 28:721-726

Walsh D, Perez C, Notary J, Mohr I (2005) Regulation of the translation initiation factor eIF4F by multiple mechanisms in human cytomegalovirus-infected cells. J Virol 79:8057-8064

Wek RC, Jiang HY, Anthony TG (2006) Coping with stress: eIF2 kinases and translational control. Biochem Soc Trans 34:7-11

Wouters BG, van den Beucken T, Magagnin MG, Koritzinsky M, Fels D, Koumenis C (2005) Control of the hypoxic response through regulation of mRNA translation. Sem Cell Dev Biol 16:487-501

Yang Q, Inoki K, Ikenoue T, Guan K-L, Iaccheri L (2006a) Identification of Sin1 as an essential TORC2 component required for complex formation and kinase activity. Genes Dev 20:2820-2832

Yang Q, Inoki K, Kim E, Guan K-L (2006b) TSC1/TSC2 and Rheb have different effects on TORC1 and TORC2 activity. Proc Natl Acad Sci U S A 103:6811-6816

Yu Y, Alwine JC (2002) Human cytomegalovirus major immediate-early proteins and simian virus 40 large T antigen can inhibit apoptosis through activation of the phosphatidylinositide 3'-OH kinase pathway and cellular kinase Akt. J Virol 76:3731-3738

Control of Apoptosis by Human Cytomegalovirus

A. L. McCormick

Contents

Introduction	282
vMIA Controls Mitochondria-Dependent Death	283
vICA Controls Caspase-8	287
IE1$_{491aa}$, IE2$_{579aa}$, and Akt-Dependent Pro-survival Pathways	287
UL38 Decreases Intrinsic Stress	288
M45 Is a Cell Type-Specific Survival Factor	289
m41, Late Infection, and the Golgi Apparatus	290
β2.7 and Mitochondrial Respiratory Complex I	290
Summary and Perspectives	290
References	291

Abstract Caspase-dependent apoptosis has an important role in controlling viruses, and as a result, viruses often encode proteins that target this pathway. Caspase-dependent apoptosis can be activated from within the infected cell as an intrinsic response to replication-associated stresses or through death-inducing signals produced extrinsically by immune cells. Cytomegaloviruses (CMVs) encode a mitochondria-localized inhibitor of apoptosis, vMIA, and a viral inhibitor of caspase activation, vICA, the functional homologs of Bcl-2 related and c-FLIP proteins, respectively. Evidence from viral mutants deleting either vMIA or vICA suggests that each is necessary and sufficient to promote survival of infected cells undergoing caspase-dependent apoptosis. Additional proteins, including pUL38, IE1$_{491aa}$, and IE2$_{579aa}$, can prevent apoptosis induced by various stimuli, while viruses with deletions of UL38, M45, or m41 undergo apoptosis. The viral RNA, β2.7, binds mitochondrial respiratory complex I, maintains ATP production late in infection, and prevents death induced by a mitochondrial poison. Thus, CMV

A.L. McCormick
Department of Microbiology & Immunology, Emory Vaccine Center, Emory University
Atlanta, GA 30322, USA
louise.mccormick@emory.edu

alters cell intrinsic defenses employing apoptosis, and multiple viral gene products together control death-inducing stimuli to promote survival.

Introduction

Apoptosis is an evolutionarily conserved process that removes cells during development and homeostasis and that can limit viral replication (Roulston et al. 1999). Apoptosis results from the hierarchical activation of a family of cysteine proteases, the caspases, that follows extrinsic or intrinsic pro-death signaling (Festjens et al. 2006). Extrinsic signals engage death receptors, a subset of the TNF superfamily, promoting the recruitment of cytoplasmic proteins and activation of initiator caspase-8. This caspase is highly regulated, including by c-FLIP proteins that can prevent proteolytic activation (Barnhart et al. 2003). Intrinsic signals following DNA damage, ER stress, or other stresses alter mitochondria membrane permeability, promote release of cytochrome c and additional pro-death factors, and activate initiator caspase-9 (Festjens et al. 2006). Both extrinsic and intrinsic pathways converge on downstream executioner caspase-3, which targets specific proteins. For most types of cells, extrinsic signals are also amplified by way of mitochondrial alterations and caspase-9 activation (Barnhart et al. 2003). The cellular Bcl-2 family proteins tightly regulate the mitochondria membrane permeability transition (Kuwana and Newmeyer 2003; Green and Kroemer 2004; Sharpe et al. 2004; Antignani and Youle 2006). Proteins in this family include one of four distinct amino acid sequence domains, known as Bcl-2 homology (BH) domains that are important to function (Petros et al. 2004). The balance of pro- and antiapoptotic Bcl-2 proteins determines whether a cell undergoes apoptosis (Kuwana and Newmeyer 2003; Green and Kroemer 2004; Sharpe et al. 2004; Antignani and Youle 2006). The proapoptotic proteins Bax and Bak are directly linked to the release of pro-death factors from mitochondria. While still controversial, one suggested mechanism employs the inherent pore-forming properties of these proteins. The actions of the proapoptotic Bcl-2-related proteins are balanced by antiapoptotic Bcl-2 and Bcl-x_L. Likewise, pro-death signals can be balanced by more global pro-survival signals. Sensors located in various organelles, including the nucleus, endoplasmic reticulum, lysosomes, and the Golgi apparatus can promote death through apoptosis (Ferri and Kroemer 2001); thus, events leading to death can occur from multiple cellular sites. Many viral factors that counteract caspase-dependent apoptosis are homologs of key cellular regulatory proteins, including the Bcl-2 related proteins and the c-FLIP proteins (Irusta et al. 2003; Polster et al. 2004).

CMV genes that impact apoptosis have been identified by three different strategies. In the first, the antiapoptotic designation followed transient expression and increased cell survival in well-defined models of apoptosis. A random search employing this strategy uncovered the viral mitochondria-localized inhibitor of apoptosis (vMIA) encoded by UL37×1 and the viral inhibitor of caspase-8 activation (vICA), encoded by UL36 (Goldmacher et al. 1999; Skaletskaya et al. 2001). Both functions increase infected cell resistance to apoptosis (Skaletskaya et al. 2001; Menard et al. 2003; Reboredo et al. 2004; McCormick et al. 2005). A direct assessment of $IE1_{491aa}$ and $IE2_{579aa}$ lead to observed impacts on pro-survival signaling mediated by the kinase Akt (Lukac and Alwine; Yu and Alwine 2002). The mechanisms and direct impact of these pro-survival activities on viral growth remain unexplored. Viral genetics highlighted the contributions of UL38, M45, and m41 to survival from apoptosis induced by replication (Brune et al. 2001, 2003; Terhune et al. 2007). Although the antiapoptotic mechanism remains unknown, pUL38 is sufficient to increase survival in apoptosis models (Terhune et al. 2007). Lastly, interaction studies revealed the RNA, β2.7, binds mitochondrial complex I and as a result, controls mitochondrial function and cell survival following death induced by respiration poisons. Thus, multiple CMV genes encode pro-survival or antiapoptotic factors. The phenotypes of viral mutants combined with results of exogenous expression analyses, suggest the UL36-38 genomic region is a cell death suppression locus.

vMIA Controls Mitochondria-Dependent Death

The UL37×1 ORF encoding vMIA is included on multiple viral transcripts. (Tenney and Colberg-Poley 1990, 1991a, 1991b; Goldmacher et al. 1999) (Fig. 1). The predominant, unspliced transcript yields the 163 aa vMIA, while splicing to UL37×2 and UL37×3 yields the larger antiapoptotic glycoprotein gpUL37 and $pUL37_M$ (Goldmacher et al. 1999). Additional less abundant spliced transcripts are predicted to encode antiapoptotic proteins as well, but have not yet been tested for function (Adair et al. 2003). vMIA localizes to mitochondria and prevents the release of pro-death

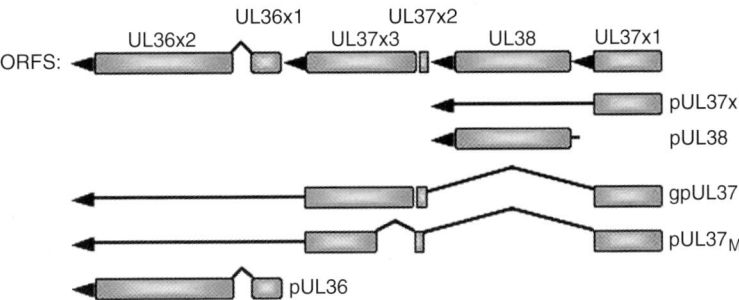

Fig. 1 A map of the HCMV UL36-UL38 cell death suppression locus indicating the relative positions of open reading frames (ORFs) and major transcripts of the region. *Rectangles* represent the ORFs and include an *arrowhead* to denote the direction of transcription. *Arrows* represent the 3′ nontranslated regions. A *raised line* connecting ORFs indicates splicing. Splicing events producing minor transcripts of the UL37 gene (Adair et al. 2003) are not shown

factors similar to Bcl-2 or Bcl-x_L (Goldmacher et al. 1999). To date, vMIA is the most broadly antiapoptotic CMV protein known and analogous to the cellular Bcl-2 proteins, is highly effective against a myriad of stimuli including intrinsic stresses as well as extrinsic, immune-regulated signals (Goldmacher et al. 1999; Belzacq et al. 2001; Vieira et al. 2001; Jan et al. 2002; Roumier et al. 2002; Boya et al. 2003; Andreau et al. 2004; Arnoult et al. 2004; Boya et al. 2005; McCormick et al. 2005). However, vMIA does not encode any BH-domains that characterize the cellular proteins (Goldmacher et al. 1999).

vMIA function requires an amino terminal mitochondrial-targeting domain (aa 2-34) and a carboxyl-terminal antiapoptotic domain (AAD, aa 118-147) (Hayajneh et al. 2001) that together are sufficient for function. The mitochondrial-targeting domain includes an amino-terminal hydrophobic signal followed by highly conserved basic residues, and both are required for mitochondrial trafficking (Mavinakere and Colberg-Poley 2004). Evidence suggests a mitochondrial membrane association with the targeting domain spanning the membrane and the AAD exposed to the cytoplasm (Mavinakere et al. 2006). The carboxyl-terminal AAD includes a predicted amphipathic α-helix motif (aa 126-140) critical to function (Smith and Mocarski 2005). Point mutations predicted to disrupt an α-helical structure alter amphipathicity or place charge on the hydrophobic face of the AAD α-helix, each completely abrogate vMIA function. In contrast, the hydrophilic face of the AAD α-helix tolerates significant substitutions with as many as five or six amino acid substitutions required to disrupt function (Smith and Mocarski 2005).

The growth arrest and DNA damage 45 alpha (GADD45α) protein interacts directly with vMIA in yeast and mammalian cells, fails to bind vMIA mutant proteins, and is essential for vMIA-mediated antiapoptotic activity (Smith and Mocarski 2005). Targeted knockdown of GADD45α, GADD45β, and GADD45γ reduced vMIA activity, and each GADD45 family protein individually enhanced vMIA activity. GADD45α increased both the overall amount of vMIA and that associated with mitochondrial fractions. Thus, the DNA damage response pathway is directly linked to vMIA-mediated cell death suppression. Further, vMIA was shown to bind the antiapoptotic Bcl-2 family protein Bcl-x_L in mammalian cells. Collectively, these data suggest that vMIA acts together with Bcl-x_L and GADD45 to regulate the mitochondrial release of proapoptotic factors (Fig. 2).

In addition to GADD45 proteins, vMIA also binds the proapoptotic Bcl-2 family protein Bax (Arnoult et al. 2004), which has more recently been connected to mitochondrial morphogenesis during life (Karbowski et al. 2006). In most instances, Bax is distributed in the cytoplasm, but Bax oligomerization and relocalization to mitochondria mediates the release of proapoptotic factors from the organelle (Antonsson et al. 2001). In the presence of vMIA, however, oligomerized Bax at mitochondria fails to promote apoptosis, suggesting sequestration as a component of the antiapoptotic mechanism (Arnoult et al. 2004; Poncet et al. 2004). Thus, the vMIA-dependent antiapoptotic mechanism is distinct from that of cellular and viral Bcl-2 proteins that prevent Bax relocalization and oligomerization at mitochondria. Recruitment and sequestration

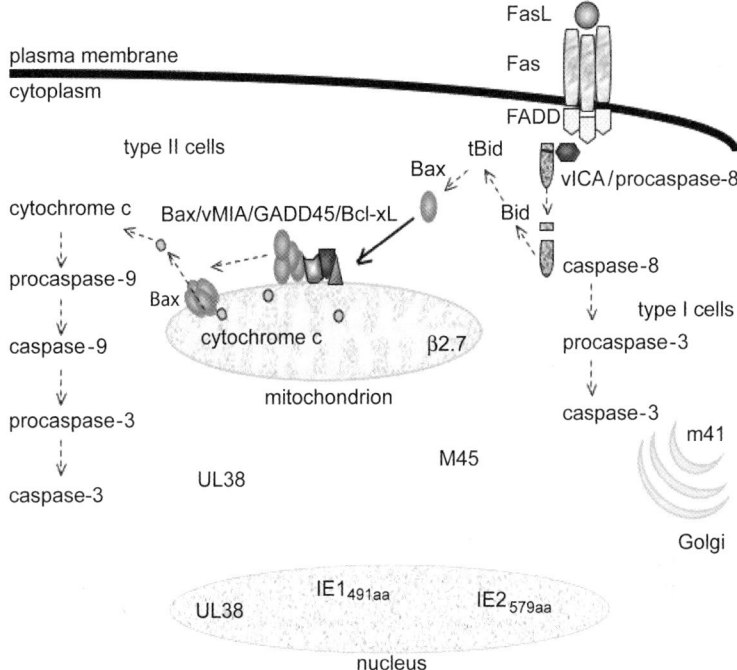

Fig. 2 A representation of the apoptosis pathway and CMV-mediated alterations preventing death, as described in the text. *Dashed arrows* indicate events prevented by the viral proteins, vICA or vMIA, as indicated. The *solid arrow* indicates vMIA-dependent relocalization of Bax to mitochondria. At mitochondria, a complex(es) of proteins including Bax as an oligomer, vMIA, GADD45, and Bcl-x_L prevents the release of mitochondrial protein cytochrome c. vICA binds procaspase-8 and is depicted as a complex that prevents procaspase-8 activation following addition of extrinsic death signals. The mechanisms and/or direct physical interactions that promote survival in the presence of the remaining viral proteins, pUL38, $IE1_{491aa}$, $IE2_{579aa}$, M45, and m41 are unclear, and these are placed according to the anticipated site of localization within the cell. For simplicity, many important regulatory components have not been included

of Bax at mitochondria has also been suggested as the mechanism (Karbowski et al. 2006) for vMIA-dependent disruption of reticular mitochondrial networks (McCormick et al. 2003b); however, more recent evidence of vMIA mutants that disrupt networks but fail to bind Bax (Pauleau et al. 2007) suggests other factors may also be important.

vMIA prevents apoptosis during infection; however, vMIA is not required for replication, and replication in the absence of vMIA does not induce caspase-dependent apoptosis (McCormick et al. 2005). A vMIA deletion mutant made in the laboratory-propagated strain, Towne*var*ATCC, produces yields nearly equivalent to parental virus. In contrast, vMIA is more critical for efficient replication of the laboratory strain AD169*var*ATCC (Reboredo et al. 2004). These strain-dependent variations may suggest that the quantity or quality of intrinsic stresses produced by individual strains varies because evidence suggests vMIA is an important regulator of the viral response to stress (McCormick et al. 2005). The phenotype produced by disruption of vMIA in AD169*var*ATCC is highly variable (Brune et al. 2003; Yu et al. 2003; Sharon-Friling et al. 2006), perhaps due to other factors that prevent vMIA-dependent release of calcium from the endoplasmic reticulum (Sharon-Friling et al. 2006) or increase ATP levels (Poncet et al. 2006), and further analyses are needed to resolve the role of vMIA in that strain.

Although of limited impact on replication in HFs, the Towne*var*ATCC mutant revealed a role for vMIA in regulating caspase-independent death. Caspase-3 activation underlies caspase-dependent apoptosis; however, this protease is not required for other cell death pathways that are considered to be apoptosis-like (Leist and Jaattela 2001; Lockshin and Zakeri 2002; Jaattela 2004). Thus, UL37×1 deletion can promote CMV-induced caspase-3-dependent cell death in the case of AD169*var*ATCC (Reboredo et al. 2004), or a caspase-3-independent cell death in the case of Towne*var*ATCC (McCormick et al. 2005), and vMIA regulates both forms of death during infection (McCormick et al. 2005). From studies so far, the context where caspase-3-independent cell death is a significant obstacle to the virus is unknown.

Chimpanzee CMV, rhesus macaque CMV, and African green monkey CMV each retain a vMIA homolog that could be identified through computer analysis (McCormick et al. 2003a). Each of these proteins share sequence similarity with the mitochondrial-targeting and AAD domains of vMIA. Rhesus macaque CMV vMIA retains similarity only to the amino- and carboxyl-terminal domains of human CMV (HCMV) vMIA and functions as an antiapoptotic protein. It is expected that all primate CMVs encode functional homologs. In contrast, the identification of rodent CMV functional homologs encoded by ORFs, m38.5 and r38.5, required more extensive analyses due to limited sequence homology (McCormick et al. 2003a, 2005; Brocchieri et al. 2005). Initial searches for murine CMV (MCMV) mitochondrial localized proteins with vMIA function were executed in HeLa cells utilizing methods that revealed vMIA (Goldmacher et al. 1999; McCormick et al. 2003a). Increasing the repertoire of stimuli revealed that m38.5 prevents proteasome inhibitor-induced, intrinsic apoptosis but not extrinsic, Fas-mediated apoptosis in HeLa cells (McCormick et al. 2005) or a telomerase-immortalized retinal epithelial cell line of human origin (Jurak and Brune 2006). Thus, MCMV m38.5 encodes an antiapoptotic protein that localizes to mitochondria (McCormick et al. 2005). The rodent CMV ORFs map to positions on the viral genomes analogous to UL37×1 (McCormick et al. 2003a; Brocchieri et al. 2005), indicating that rodent and primate CMVs each encode vMIA and vICA homologs.

Limited sequence similarity and differences in protective function in human cells suggest the human and rodent vMIA homologs retain elements that are specific to function in the appropriate host (McCormick et al. 2005). Identification of additional MCMV proteins localized to mitochondria may also suggest the potential for synergism or even replacement of m38.5 function in specific cells (Tang et al. 2006). Interestingly, vMIA apparently protects from specific apoptotic stimuli in a species-dependent fashion as well. Thus, vMIA fails to prevent mitochondrial damage induced by staurosporine in wild type murine fibroblasts, apparently due to the role of murine Bak in that setting (Arnoult et al. 2004) but prevents staurosporine-induced death in HeLa cells (Andreau et al. 2004). Thus, properties of the antiapoptotic proteins encoded by these viruses reflect the evolutionary divergence of the host. In fact, one aspect of the species barrier that restricts CMVs is reportedly due to functions that for MCMV can be provided by vMIA (Jurak and Brune 2006). Given the genomic organization and studies thus far, it is likely that m38.5 will retain vMIA functions relevant to survival in the host and that all CMVs rely on vMIA function.

vICA Controls Caspase-8

vICA, the UL36 gene product, interferes with caspase-8-dependent apoptosis by binding procaspase-8 and preventing proteolytic activation (Skaletskaya et al. 2001) (Fig. 2). The role of caspase-8 as an initiator protease activated by extrinsic, immune-regulated signals implies vICA is important to survival in the host (Skaletskaya et al. 2001). vICA is highly conserved among mammalian betaherpesviruses both in sequence and function, suggesting a conserved biologic role (McCormick et al. 2003a; Menard et al. 2003). In contrast, passage in tissue culture has promoted adventitious mutations that impact antiapoptotic function (Skaletskaya et al. 2001). Although early work employed a recombinant virus made in AD169*var*ATCC, a laboratory strain that had already acquired mutations in vICA (Patterson and Shenk 1999; Skaletskaya et al. 2001), deletion of the gene from Towne-BAC, a viral strain that retains vICA function, confirmed that both the UL36 gene and vICA function can be altered without impacting replication in cultured fibroblasts (Dunn et al. 2003). MCMV mutants impacting M36 also grow in fibroblasts; however, this gene is required for growth in cultured macrophages (Menard et al. 2003) and in mice (Cicin-Sain et al. 2005). Importantly, infected macrophages elevated caspase-8 activity only in the absence of M36 (Menard et al. 2003). These observations are consistent with the expectation that vICA prevents caspase-8 activation, thereby performing a critical role for survival in the host.

$IE1_{491aa}$, $IE2_{579aa}$, and Akt-Dependent Pro-survival Pathways

Pathways leading to death are balanced by pro-survival pathways, including those regulated by trophic factors that signal through phosphatidylinositide 3′-OH kinase (PI3K)/Akt kinase (Datta et al. 1999). Evidence suggests CMV requires the PI3K/Akt pathway for replication (Johnson et al. 2001). The $IE1_{491aa}$ and $IE2_{579aa}$ are nuclear proteins that regulate transcription and have important roles in viral replication (see the chapter by M.F. Stinski and D.T. Petrik, this volume and Stinski and

Meier 2007; White and Spector 2007). Each has been connected to the PI3K/Akt pro-survival pathway through the following. Initially, antiapoptotic roles for $IE1_{491aa}$ and $IE2_{579aa}$ were suggested from results of transient and stable expression in HeLa cells (Zhu et al. 1995). Here, either protein protects from short exposure (8 h) to TNF or infection by E1B-19-kDa-deficient adenovirus, but not from UV irradiation. Mechanisms are suggested to differ because antiapoptotic activity maps to unique sequences. In comparison with vMIA, neither $IE1_{491aa}$ nor $IE2_{579aa}$ protect HeLa cells undergoing TNF- or Fas-mediated apoptosis when evaluated in a more rigorous assay (24 h) (Goldmacher et al. 1999). Thus, these proteins present challenges for mechanistic studies.

One experimental approach that has suggested the mechanism of $IE1_{491aa}$ and $IE2_{579aa}$ antiapoptotic function employed the temperature-sensitive (*ts*) BHK-21 cell line *ts*13 (Lukac et al. 1997; Lukac and Alwine 1999; Yu and Alwine 2002). At the nonpermissive temperature, a mutation in $TAF_{II}250$ produces transcription alterations in specific genes that results in a block to cell cycle progression and the induction of apoptosis in these hamster cells (Talavera and Basilico 1977; Sekiguchi et al. 1988, 1995). When expressed from a genomic construct, $IE1_{491aa}$ and $IE2_{579aa}$ prevent apoptosis and rescue promoter-specific transcription through independent mechanisms that do not rescue cell cycle defects (Lukac et al. 1997; Lukac and Alwine 1999). Further, IE rescue of transcription is primarily due to $IE2_{579aa}$ (Lukac et al. 1997), while either IE protein rescues apoptosis (Yu and Alwine 2002). Protein domains required for protective function remain unidentified. Although the protective mechanism in this setting relies on PI3K activation of Akt, how $IE1_{491aa}$ and $IE2_{579aa}$ promote this activation is unknown, as is the significance of these results to infection. Nevertheless, these studies suggest hypotheses that may elucidate $IE1_{491aa}$ and $IE2_{579aa}$ contributions to survival of infected cells.

UL38 Decreases Intrinsic Stress

The UL38 ORF maps to an intron of the UL37 gene (Tenney and Colberg-Poley 1990, 1991a, 1991b) (Fig. 1). The UL38 sequence is included on the unspliced vMIA transcript and on a unique transcript with early kinetics. During infection, pUL38 initially localizes to the nucleus but is well distributed between the nucleus and cytoplasm by 24 h (Terhune et al. 2007). Mutagenesis of the UL38 ORF in either Towne-BAC (Dunn et al. 2003) or pAD/Cre (Yu et al. 2002; Terhune et al. 2007) reduces yield by approximately 100-fold during a single round of replication (Terhune et al. 2007).

The premature death induced by the UL38-null mutant virus and rescue by addition of the pan-caspase inhibitor zVAD-fmk prompted further analysis of pUL38 as an antiapoptotic factor (Terhune et al. 2007). Replication of the UL38 deletion mutant is also largely restored by zVAD-fmk. Death initiates very early (24 h) and reaches more than 50% by 72 h, suggesting pUL38 is required very early. Consistently, TUNEL labeling, expected to reflect caspase-3-mediated activation of nucleases, occurs by 48 h. In contrast, detection of active caspase-3 or cleaved substrate (PARP) is apparently variable and occurs consistently only by 96 h. Thus, evaluation of specific steps along the apoptosis pathway may highlight important viral controls. Nevertheless, a deletion mutant of UL38 in pAD/Cre induces an apoptotic death that is repaired by growth on pUL38-expressing cells. pUL38 is sufficient to inhibit death induced by E1B-19-kDA-deficient adenovirus or thapsigargin, but is ineffective against Fas-mediated apoptosis. Thus, pUL38 protects from intrinsic but not extrinsic death signals. Thus far, little is known of the pUL38-dependent antiapoptotic mechanism or protein domains required for function, but these studies will likely be included in future endeavors.

M45 Is a Cell Type-Specific Survival Factor

Betaherpesviruses, including HCMV, encode the UL45 genes that are related by sequence but not function to ribonucleotide reductase (Chee et al. 1990; Patrone et al. 2003; Lembo et al. 2004). Viral mutants that disrupt the MCMV M45 gene induce apoptosis and are growth-restricted in endothelial cells and macrophages but not fibroblasts, bone marrow stromal cells, or hepatocytes (Brune et al. 2001). Although early and late genes are expressed, infection induces apoptosis and infectious progeny are not produced. Assays predictive of apoptosis, including nuclease activity and phosphatidylserine exposure, implicate this pathway; however, a direct link between decreased apoptosis and rescued growth has not been established. Further, M45-dependent survival in apoptosis models or viral replication has not been demonstrated, and it is unclear how the phenotype of the mutant virus relates to apoptosis pathways and M45. However, disruption of M45 produces a nonpathogenic virus (Lembo et al. 2004), and further studies will likely answer these important questions.

Neither replication in endothelial cells or resistance to induced intrinsic stress requires HCMV UL45 (Hahn et al. 2002). Thus, the intrinsic stresses revealed by deletion of MCMV M45 are apparently controlled by other viral factors in HCMV. In contrast, UL45 does increase viral production of the laboratory strain AD169*var*ATCC in fibroblasts following a low multiplicity infection (Patrone et al. 2003). However, decreased yield does not result from increased apoptosis. The resistance of HCMV to extrinsic but not intrinsic apoptosis is also halved, but the contribution of UL45 is unknown and UL45-expressing fibroblasts remain sensitive to Fas-mediated apoptosis. Considerable effort has been required to evaluate the potential of UL45 as a ribonucleotide reductase (Patrone et al. 2003; Lembo et al. 2004). Future studies that define the UL45 function that increases low multiplicity growth and the M45 function that permits replication in endothelial cells will likely clarify the role of this perplexing gene.

m41, Late Infection, and the Golgi Apparatus

The CMV ORFs m41 and r41 are apparently unique to rodent CMVs and, as such, do not share sequence homology with HCMV ORFs (Chee et al. 1990; Rawlinson et al. 1996; Mocarski et al. 1997; Vink et al. 2000; Brocchieri et al. 2005). Expression constructs produce a single protein of 19 kDa while polypeptides of 19 and 21-kDa polypeptides are produced during infection, suggesting splicing to upstream or downstream ORFs (Brune et al. 2003). Recombinants that disrupt the m41 ORF induce apoptosis very late in infection, as suggested both by apoptosis-induced molecular changes, including phosphatidylserine exposure and nuclease-driven chromatin alterations, and increased survival in the presence of caspase inhibitors. A more dramatic impact on replication occurs in endothelial cells where viral yields are reduced 50-fold. Thus far, neither yield nor death has been directly related to pm41, but Golgi localization is likely important to pm41 function.

β2.7 and Mitochondrial Respiratory Complex I

The highly abundant early transcript β2.7 (McDonough and Spector 1983; McDonough et al. 1985) is polysome-associated (Wathen and Stinski 1982); however, sequence analyses suggest a noncoding RNA (McSharry et al. 2003). Although the gene is conserved in both laboratory-adapted viral strains and clinical isolates, the RL4 ORF is not. Northwestern screening suggested β2.7 binds proteins of the nicotinamide adenine dinucleotide-ubiquinone oxidoreductase (mitochondrial respiration complex I) (Reeves et al. 2007). β2.7 also increases survival from the mitochondrial poison rotenone and maintains ATP production late in infection. The importance of continued mitochondrial function during CMV infection has been suggested from several studies that have evaluated mitochondrial DNA synthesis, mitochondrial protein expression profiles, and ATP production (Furukawa et al. 1976; Hertel and Mocarski 2004; Reeves et al. 2007). β2.7 contributes to viral production by maintaining ATP production (Reeves et al. 2007), and in addition, prevents death that follows intrinsic stresses associated with decreased ATP.

Summary and Perspectives

Several CMV gene products that impact apoptosis have already been identified. vMIA and vICA are the most extensively characterized with regard to proposed mechanism and antiapoptotic roles during infection. pUL38, M45, and m41 remain largely uncharacterized with regard to mechanism, and the contributions of $IE1_{491aa}$ and $IE2_{579aa}$ antiapoptotic functions during replication remain to be addressed. Additional genes like β2.7, which increase survival in response to stress, are likely to be identified through efforts that determine viral control of cellular stress responses (see the chapter by A.L. Alwine, this volume) and these genes will also impact viral control of apoptosis. Three types of cell death-apoptosis, necrosis, and autophagy-have been well characterized by morphology, biochemical events, and host responses (Leist and Jaattela 2001; Jaattela 2004; Lockshin and Zakeri 2004; Vandenabeele et al. 2006; Golstein and Kroemer 2007). Given overlapping regulation and expectations from cellular homologs with cross-inhibitory properties, future efforts will undoubtedly reveal as yet unappreciated connections between CMV antiapoptotic proteins and other cell death pathways. In summary, the cell tropism of CMV (Mocarski et al. 2006) likely means the virus must be armed against multiple forms of death and the combination of all suppressors encoded by the virus likely balances the apoptotic threshold in a direction supporting replication.

References

Adair R, Liebisch GW, Colberg-Poley AM (2003) Complex alternative processing of human cytomegalovirus UL37 pre-mRNA. J Gen Virol 84:3353-3358

Andreau K, Castedo M, Perfettini JL, Roumier T, Pichart E, Souquere S, Vivet S, Larochette N, Kroemer G (2004) Preapoptotic chromatin condensation upstream of the mitochondrial checkpoint. J Biol Chem 279:55937-55945

Antignani A, Youle RJ (2006) How do Bax and Bak lead to permeabilization of the outer mitochondrial membrane? Curr Opin Cell Biol 18:685-689

Antonsson B, Montessuit S, Sanchez B, Martinou JC (2001) Bax is present as a high molecular weight oligomer/complex in the mitochondrial membrane of apoptotic cells. J Biol Chem 276:11615-11623

Arnoult D, Bartle LM, Skaletskaya A, Poncet D, Zamzami N, Park PU, Sharpe J, Youle RJ, Goldmacher VS (2004) Cytomegalovirus cell death suppressor vMIA blocks Bax- but not Bak-mediated apoptosis by binding and sequestering Bax at mitochondria. Proc Natl Acad Sci U S A 101:7988-7993

Barnhart BC, Alappat EC, Peter ME (2003) The CD95 type I/type II model. Semin Immunol 15:185-193

Belzacq AS, El Hamel C, Vieira HL, Cohen I, Haouzi D, Metivier D, Marchetti P, Brenner C, Kroemer G (2001) Adenine nucleotide translocator mediates the mitochondrial membrane permeabilization induced by lonidamine, arsenite and CD437. Oncogene 20:7579-7587

Boya P, Gonzalez-Polo RA, Poncet D, Andreau K, Vieira HL, Roumier T, Perfettini JL, Kroemer G (2003) Mitochondrial membrane permeabilization is a critical step of lysosome-initiated apoptosis induced by hydroxychloroquine. Oncogene 22:3927-3936

Boya P, Gonzalez-Polo RA, Casares N, Perfettini JL, Dessen P, Larochette N, Metivier D, Meley D, Souquere S, Yoshimori T, Pierron G, Codogno P, Kroemer G (2005) Inhibition of macroautophagy triggers apoptosis. Mol Cell Biol 25:1025-1040

Brocchieri L, Kledal TN, Karlin S, Mocarski ES (2005) Predicting coding potential from genome sequence: application to betaherpesviruses infecting rats and mice. J Virol 79:7570-7596

Brune W, Menard C, Heesemann J, Koszinowski UH (2001) A ribonucleotide reductase homolog of cytomegalovirus and endothelial cell tropism. Science 291:303-305

Brune W, Nevels M, Shenk T (2003) Murine cytomegalovirus m41 open reading frame encodes a Golgi-localized antiapoptotic protein. J Virol 77:11633-11643

Chee MS, Bankier AT, Beck S, Bohni R, Brown CM, Cerny R, Horsnell T, Hutchison CAI, Kouzarides T, Martignetti JA, Preddie E, Satchwell SC, Tomlinson P, Weston KM, Barrell BG (1990) Analysis of the protein-coding content of the sequence of human cytomegalovirus strain AD169. Curr Top Microbiol Immunol 154:125-170

Cicin-Sain L, Podlech J, Messerle M, Reddehase MJ, Koszinowski UH (2005) Frequent coinfection of cells explains functional in vivo complementation between cytomegalovirus variants in the multiply infected host. J Virol 79:9492-9502

Datta SR, Brunet A, Greenberg ME (1999) Cellular survival: a play in three Akts. Genes Dev 13:2905-2927

Dunn W, Chou C, Li H, Hai R, Patterson D, Stolc V, Zhu H, Liu F (2003) Functional profiling of a human cytomegalovirus genome. Proc Natl Acad Sci U S A 100:14223-14228

Ferri KF, Kroemer G (2001) Organelle-specific initiation of cell death pathways. Nat Cell Biol 3:E255-E263

Festjens N, Cornelis S, Lamkanfi M, Vandenabeele P (2006) Caspase-containing complexes in the regulation of cell death and inflammation. Biol Chem 387:1005-1016

Furukawa T, Sakuma S, Plotkin SA (1976) Human cytomegalovirus infection of WI-38 cells stimulates mitochondrial DNA synthesis. Nature 262:414-416

Goldmacher VS, Bartle LM, Skaletskaya A, Dionne CA, Kedersha NL, Vater CA, Han J, Lutz RJ, Watanabe S, McFarland ED, Kieff ED, Mocarski ES, Chittenden T (1999) A cytomegalovirus-encoded mitochondria-localized inhibitor of apoptosis structurally unrelated to Bcl-2. Proc Natl Acad Sci U S A 96:12536-12541

Golstein P, Kroemer G (2007) Cell death by necrosis: towards a molecular definition. Trends Biochem Sci 32:37-43

Green DR, Kroemer G (2004) The pathophysiology of mitochondrial cell death. Science 305:626-629

Hahn G, Khan H, Baldanti F, Koszinowski UH, Revello MG, Gerna G (2002) The human cytomegalovirus ribonucleotide reductase homolog UL45 Is dispensable for growth in endothelial cells, as determined by a BAC-cloned clinical isolate of human cytomegalovirus with preserved wild-type characteristics. J Virol 76:9551-9555

Hayajneh WA, Colberg-Poley AM, Skaletskaya A, Bartle LM, Lesperance MM, Contopoulos-Ioannidis DG, Kedersha NL, Goldmacher VS (2001) The sequence and antiapoptotic functional domains of the human cytomegalovirus UL37 exon 1 immediate early protein are conserved in multiple primary strains. Virology 279:233-240

Hertel L, Mocarski ES (2004) Global analysis of host cell gene expression late during cytomegalovirus infection reveals extensive dysregulation of cell cycle gene expression and induction of pseudomitosis independent of US28 function. J Virol 78:11988-12011

Irusta PM, Chen YB, Hardwick JM (2003) Viral modulators of cell death provide new links to old pathways. Curr Opin Cell Biol 15:700-705

Jaattela M (2004) Multiple cell death pathways as regulators of tumour initiation and progression. Oncogene 23:2746-2756

Jan G, Belzacq AS, Haouzi D, Rouault A, Metivier D, Kroemer G, Brenner C (2002) Propionibacteria induce apoptosis of colorectal carcinoma cells via short-chain fatty acids acting on mitochondria. Cell Death Differ 9:179-188

Johnson RA, Wang X, Ma XL, Huong SM, Huang ES (2001) Human cytomegalovirus up-regulates the phosphatidylinositol 3-kinase (PI3-K) pathway: inhibition of PI3-K activity inhibits viral replication and virus-induced signaling. J Virol 75:6022-6032

Jurak I, Brune W (2006) Induction of apoptosis limits cytomegalovirus cross-species infection. EMBO J 25:2634-2642

Karbowski M, Norris KL, Cleland MM, Jeong SY, Youle RJ (2006) Role of Bax and Bak in mitochondrial morphogenesis. Nature 443:658-662

Kuwana T, Newmeyer DD (2003) Bcl-2-family proteins and the role of mitochondria in apoptosis. Curr Opin Cell Biol 15:691-699

Leist M, Jaattela M (2001) Four deaths and a funeral: from caspases to alternative mechanisms. Nat Rev Mol Cell Biol 2:589-598

Lembo D, Donalisio M, Hofer A, Cornaglia M, Brune W, Koszinowski U, Thelander L, Landolfo S (2004) The ribonucleotide reductase R1 homolog of murine cytomegalovirus is not a functional enzyme subunit but is required for pathogenesis. J Virol 78:4278-4288

Lockshin RA, Zakeri Z (2002) Caspase-independent cell deaths. Curr Opin Cell Biol 14: 727-733

Lockshin RA, Zakeri Z (2004) Apoptosis, autophagy, and more. Int J Biochem Cell Biol 36: 2405-2419

Lukac DM, Alwine JC (1999) Effects of human cytomegalovirus major immediate-early proteins in controlling the cell cycle and inhibiting apoptosis: studies with ts13 cells. J Virol 73:2825-2831

Lukac DM, Harel NY, Tanese N, Alwine JC (1997) TAF-like functions of human cytomegalovirus immediate-early proteins. J Virol 71:7227-7239

Mavinakere MS, Colberg-Poley AM (2004) Dual targeting of the human cytomegalovirus UL37 exon 1 protein during permissive infection. J Gen Virol 85:323-329

Mavinakere MS, Williamson CD, Goldmacher VS, Colberg-Poley AM (2006) Processing of human cytomegalovirus UL37 mutant glycoproteins in the endoplasmic reticulum lumen prior to mitochondrial importation. J Virol 80:6771-6783

McCormick AL, Skaletskaya A, Barry PA, Mocarski ES, Goldmacher VS (2003a) Differential function and expression of the viral inhibitor of caspase 8-induced apoptosis (vICA) and the viral mitochondria-localized inhibitor of apoptosis (vMIA) cell death suppressors conserved in primate and rodent cytomegaloviruses. Virology 316:221-233

McCormick AL, Smith VL, Chow D, Mocarski ES (2003b) Disruption of mitochondrial networks by the human cytomegalovirus UL37 gene product viral mitochondrion-localized inhibitor of apoptosis. J Virol 77:631-641

McCormick AL, Meiering CD, Smith GB, Mocarski ES (2005) Mitochondrial cell death suppressors carried by human and murine cytomegalovirus confer resistance to proteasome inhibitor-induced apoptosis. J Virol 79:12205-12217

McDonough SH, Spector DH (1983) Transcription in human fibroblasts permissively infected by human cytomegalovirus strain AD169. Virology 125:31-46

McDonough SH, Staprans SI, Spector DH (1985) Analysis of the major transcripts encoded by the long repeat of human cytomegalovirus strain AD169. J Virol 53:711-718

McSharry BP, Tomasec P, Neale ML, Wilkinson GW (2003) The most abundantly transcribed human cytomegalovirus gene (beta 2.7) is non-essential for growth in vitro. J Gen Virol 84:2511-2516

Menard C, Wagner M, Ruzsics Z, Holak K, Brune W, Campbell AE, Koszinowski UH (2003) Role of murine cytomegalovirus US22 gene family members in replication in macrophages. J Virol 77:5557-5570

Mocarski ES, Prichard MN, Tan CS, Brown JM (1997) Reassessing the organization of the UL42-UL43 region of the human cytomegalovirus strain AD169 genome. Virology 239:169-175

Mocarski ES Jr, Shenk T, Pass RF (2006) Cytomegaloviruses. In: Knipe DM, Howley PM, Griffin DE, Lamb RA, Martin MA (eds) Fields virology, 5th edn. Lippincott Williams & Wilkins, Philadelphia, pp 2701-2772

Patrone M, Percivalle E, Secchi M, Fiorina L, Pedrali-Noy G, Zoppe M, Baldanti F, Hahn G, Koszinowski UH, Milanesi G, Gallina A (2003) The human cytomegalovirus UL45 gene product is a late, virion-associated protein and influences virus growth at low multiplicities of infection. J Gen Virol 84:3359-3370

Patterson CE, Shenk T (1999) Human cytomegalovirus UL36 protein is dispensable for viral replication in cultured cells. J Virol 73:7126-7131

Pauleau AL, Larochette N, Giordanetto F, Scholz SR, Poncet D, Zamzami N, Goldmacher VS, Kroemer G (2007) Structure-function analysis of the interaction between Bax and the cytomegalovirus-encoded protein vMIA. Oncogene 26:7067-7080

Petros AM, Olejniczak ET, Fesik SW (2004) Structural biology of the Bcl-2 family of proteins. Biochim Biophys Acta 1644:83-94

Polster BM, Pevsner J, Hardwick JM (2004) Viral Bcl-2 homologs and their role in virus replication and associated diseases. Biochim Biophys Acta 1644:211-227

Poncet D, Larochette N, Pauleau AL, Boya P, Jalil AA, Cartron PF, Vallette F, Schnebelen C, Bartle LM, Skaletskaya A, Boutolleau D, Martinou JC, Goldmacher VS, Kroemer G, Zamzami

N (2004) An anti-apoptotic viral protein that recruits Bax to mitochondria. J Biol Chem 279:22605-22614

Poncet D, Pauleau AL, Szabadkai G, Vozza A, Scholz SR, Le Bras M, Briere JJ, Jalil A, Le Moigne R, Brenner C, Hahn G, Wittig I, Schagger H, Lemaire C, Bianchi K, Souquere S, Pierron G, Rustin P, Goldmacher VS, Rizzuto R, Palmieri F, Kroemer G (2006) Cytopathic effects of the cytomegalovirus-encoded apoptosis inhibitory protein vMIA. J Cell Biol 174:985-996

Rawlinson WD, Farrell HE, Barrell BG (1996) Analysis of the complete DNA sequence of murine cytomegalovirus. J Virol 70:8833-8849

Reboredo M, Greaves RF, Hahn G (2004) Human cytomegalovirus proteins encoded by UL37 exon 1 protect infected fibroblasts against virus-induced apoptosis and are required for efficient virus replication. J Gen Virol 85:3555-3567

Reeves MB, Davies AA, McSharry BP, Wilkinson GW, Sinclair JH (2007) Complex I binding by a virally encoded RNA regulates mitochondria-induced cell death. Science 316:1345-1348

Roulston A, Marcellus RC, Branton PE (1999) Viruses and apoptosis. Annu Rev Microbiol 53:577-628

Roumier T, Vieira HL, Castedo M, Ferri KF, Boya P, Andreau K, Druillennec S, Joza N, Penninger JM, Roques B, Kroemer G (2002) The C-terminal moiety of HIV-1 Vpr induces cell death via a caspase-independent mitochondrial pathway. Cell Death Differ 9:1212-1219

Sekiguchi T, Miyata T, Nishimoto T (1988) Molecular cloning of the cDNA of human X chromosomal gene (CCG1) which complements the temperature-sensitive G1 mutants, tsBN462 and ts13, of the BHK cell line. EMBO J 7:1683-1687

Sekiguchi T, Nakashima T, Hayashida T, Kuraoka A, Hashimoto S, Tsuchida N, Shibata Y, Hunter T, Nishimoto T (1995) Apoptosis is induced in BHK cells by the tsBN462/13 mutation in the CCG1/TAFII250 subunit of the TFIID basal transcription factor. Exp Cell Res 218:490-498

Sharon-Friling R, Goodhouse J, Colberg-Poley AM, Shenk T (2006) Human cytomegalovirus pUL37×1 induces the release of endoplasmic reticulum calcium stores. Proc Natl Acad Sci USA 103:19117-19122

Sharpe JC, Arnoult D, Youle RJ (2004) Control of mitochondrial permeability by Bcl-2 family members. Biochim Biophys Acta 1644:107-113

Skaletskaya A, Bartle LM, Chittenden T, McCormick AL, Mocarski ES, Goldmacher VS (2001) A cytomegalovirus-encoded inhibitor of apoptosis that suppresses caspase-8 activation. Proc Natl Acad Sci USA 98:7829-7834

Smith GB, Mocarski ES (2005) Contribution of GADD45 family members to cell death suppression by cellular Bcl-xL and cytomegalovirus vMIA. J Virol 79:14923-14932

Stinski MF, Meier JL (2007) Immediate-early viral gene regulation and function. In: Arvin AM, Mocarski ES, Moore P, Whitley R, Yamanishi K, Campadelli-Fiume G, Roizman B (eds) Human herpesviruses: biology, therapy and immunoprophylaxis. Cambridge Press, Cambridge, pp 241-263

Talavera A, Basilico C (1977) Temperature sensitive mutants of BHK cells affected in cell cycle progression. J Cell Physiol 92:425-436

Tang Q, Murphy EA, Maul GG (2006) Experimental confirmation of global murine cytomegalovirus open reading frames by transcriptional detection and partial characterization of newly described gene products. J Virol 80:6873-6882

Tenney DJ, Colberg-Poley AM (1990) RNA analysis and isolation of cDNAs derived from the human cytomegalovirus immediate-early region at 0.24 map units. Intervirology 31:203-214

Tenney DJ, Colberg-Poley AM (1991a) Expression of the human cytomegalovirus UL36-38 immediate early region during permissive infection. Virology 182:199-210

Tenney DJ, Colberg-Poley AM (1991b) Human cytomegalovirus UL36-38 and US3 immediate-early genes: temporally regulated expression of nuclear, cytoplasmic, and polysome-associated transcripts during infection. J Virol 65:6724-6734

Terhune S, Torigoi E, Moorman N, Silva M, Qian Z, Shenk T, Yu D (2007) Human cytomegalovirus UL38 protein blocks apoptosis. J Virol 81:3109-3123

Vandenabeele P, Vanden Berghe T, Festjens N (2006) Caspase inhibitors promote alternative cell death pathways. Sci STKE 2006:pe44

Vieira HL, Belzacq AS, Haouzi D, Bernassola F, Cohen I, Jacotot E, Ferri KF, El Hamel C, Bartle LM, Melino G, Brenner C, Goldmacher V, Kroemer G (2001) The adenine nucleotide translocator: a target of nitric oxide, peroxynitrite, and 4-hydroxynonenal. Oncogene 20:4305-4316

Vink C, Beuken E, Bruggeman CA (2000) Complete DNA sequence of the rat cytomegalovirus genome. J Virol 74:7656-7665

Wathen MW, Stinski MF (1982) Temporal patterns of human cytomegalovirus transcription: mapping the viral RNAs synthesized at immediate early, early, and late times after infection. J Virol 41:462-477

White EA, Spector DH (2007) Early viral gene regulation and function. In: Arvin AM, Mocarski ES, Moore P, Whitley R, Yamanishi K, Campadelli-Fiume G, Roizman B (eds) Human herpesviruses: biology, therapy and immunoprophylaxis. Cambridge Press, Cambridge, pp 264-294

Yu Y, Alwine JC (2002) Human cytomegalovirus major immediate-early proteins and simian virus 40 large T antigen can inhibit apoptosis through activation of the phosphatidylinositide 3'-OH kinase pathway and the cellular kinase Akt. J Virol 76:3731-3738

Yu D, Smith GA, Enquist LW, Shenk T (2002) Construction of a self-excisable bacterial artificial chromosome containing the human cytomegalovirus genome and mutagenesis of the diploid TRL/IRL13 gene. J Virol 76:2316-2328

Yu D, Silva MC, Shenk T (2003) Functional map of human cytomegalovirus AD169 defined by global mutational analysis. Proc Natl Acad Sci USA 100:12396-12401

Zhu H, Shen Y, Shenk T (1995) Human cytomegalovirus IE1 and IE2 proteins block apoptosis. J Virol 69:7960-7970

Aspects of Human Cytomegalovirus Latency and Reactivation

M. Reeves, J. Sinclair(✉)

Contents

Introduction 298
Latency, Carriage and Reactivation of HCMV in the Cells of the Myeloid Lineage 298
Models of HCMV Latency Using Experimental Infection 300
Viral Gene Expression Associated with HCMV Latency 301
Key Aspects of HCMV Latency and Reactivation 302
    The Establishment of HCMV Latency.. 302
    The Maintenance of HCMV Latency... 304
    Reactivation of HCMV from Latency 305
Other Sites of HCMV Latency 308
Conclusion 309
References 309

Abstract Primary infection of healthy individuals with human cytomegalovirus (HCMV) is usually asymptomatic and results in the establishment of a lifelong latent infection of the host. Although no overt HCMV disease is observed in healthy carriers, due to effective immune control, severe clinical symptoms associated with HCMV reactivation are observed in immunocompromised transplant patients and HIV sufferers. Work from a number of laboratories has identified the myeloid lineage as one important site for HCMV latency and reactivation and thus has been the subject of extensive study. Attempts to elucidate the mechanisms controlling viral latency have shown that cellular transcription factors and histone proteins influence HCMV gene expression profoundly and that the type of cellular environment virus encounters upon infection may have a critical role in determining a lytic or latent infection and subsequent reactivation from latency. Furthermore, the identification of a number of viral gene products expressed during latent infection suggests a more active role for HCMV during latency. Defining the role of these viral proteins in latently infected cells will be important for our full understanding of HCMV latency and reactivation in vivo.

J. Sinclair
Department of Medicine, University of Cambridge, Addenbrooke's Hospital Cambridge, CB2 2QQ, UK
js@mole.bio.cam.ac.uk

Introduction

The ability of human cytomegalovirus (HCMV), like all herpes viruses, to establish a lifelong persistent infection plays a crucial role in the long-term carriage of this opportunistic pathogen in the human host. It is likely that HCMV persistence, in vivo, involves sites in the host which continually produce low levels of virus. However, it is now clear that it also involves sites which carry the viral genome latently in the absence of any productive infection.

Although HCMV causes few overt symptoms following primary infection of healthy individuals, significant morbidity and mortality is observed in the immunonaïve, immunocompromised and immunosuppressed (Ho 1990; Zaia 1990). Primary infection is an important factor in HCMV-mediated disease (see the chapter by W. Britt, this volume), particularly following congenital infection (Griffiths and Walter 2005). However, reactivation from latency is a major cause of disease in certain transplant patients (both solid organ and bone marrow transplantation) and also in late-stage HIV patients suffering from AIDS (Adler 1983; Rubin 1990; Sissons and Carmichael 2002). Consequently, developing an understanding of the mechanisms that regulate latency and reactivation in vivo is of paramount importance for future clinical intervention.

In order to do this, a number of fundamental questions about the basic biology of HCMV need to be addressed: Firstly, in which cells does the latent virus reside? Secondly, in which cells does the virus reactivate. Thirdly, what regulates this latency and reactivation?

The ability to detect latent HCMV, particularly prior to the development of highly sensitive techniques such as the polymerase chain reaction (PCR), is in contrast to the ease of detecting productive HCMV infection during disease. Acute HCMV infection is manifest in numerous tissues (Rubin 1990; Sissons and Carmichael 2002). Epithelial, endothelial, smooth muscle, stromal, fibroblast and neuronal cells all support lytic HCMV replication in vivo (Sinzger et al. 1995; Plachter et al. 1996; see the chapter by C. Sinzger et al., this volume) and, consequently, HCMV pathology can be seen in a diverse range of organs throughout the body. In contrast, HCMV latency appears to be restricted to subpopulations of cell types.

Latency, Carriage and Reactivation of HCMV in the Cells of the Myeloid Lineage

Some of the first instructive observations regarding HCMV latency came from clinical studies. Although it was extremely difficult to detect infectious virus in the blood of normal healthy individuals, it was evident that blood transfusions from healthy seropositive donors often resulted in the transmission of HCMV to blood donor recipients (Adler 1983). However, the incidence of this transmission was significantly reduced if leukocyte-depleted blood products were used (Yeager et al. 1981; Tolpin et al.

1985; Gilbert et al. 1989), which strongly suggested that viral transmission was cell-based and not mediated by free virus. Consequently, one of the cellular sites of latency was believed to be in the peripheral blood compartment.

We now know that the latent load of HCMV in healthy carriers is around 1 genome-positive cell per 10,000 peripheral blood mononuclear cells (PBMCs) (Slobedman and Mocarski 1999), clearly below the detection limits of Northern, Southern and Western analyses. However, the use of a highly-sensitive PCR approach finally permitted the analysis of HCMV latency in vivo and defined the myeloid lineage as an important site of HCMV latency (Fig. 1). An experimental approach that isolated different fractions of cells from the blood of healthy seropositives showed that carriage of HCMV DNA occurred predominantly in the leukocyte fraction of peripheral blood - particularly in the CD14$^+$ monocyte population (Fig. 1a). HCMV was not found in the lymphocyte fraction or the polymorphonuclear cells (Fig. 1b) (Bevan et al. 1991; Taylor-Wiedeman et al. 1991, 1993; Stanier et al. 1992). Monocytes, however, represent a short-lived continually renewable population of cells that arise from haematopoietic cell precursors (CD34$^+$ cells) present in the bone marrow (Metcalf 1989). These cells were also shown to be HCMV genome-positive, suggesting that the bone marrow represents one latent reservoir of virus (Mendelson et al. 1996; Sindre et al. 1996). Although CD34$^+$ cells are sites of latency for HCMV and are a common precursor of both lymphoid and myeloid cells, the carriage of virus appears to be restricted to cells of the monocyte/myeloid lineage by, as yet, undefined mechanisms (for review see Sinclair and Sissons 2006).

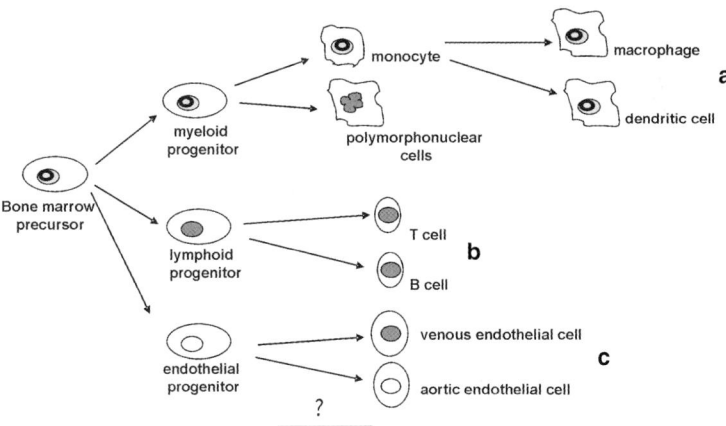

Fig. 1 HCMV latency is established in bone marrow progenitors and is carried in the myeloid lineage. During natural latency, HCMV DNA can be detected in bone marrow progenitor cells that give rise multiple lineages. However, the carriage of viral genomes is detected in the myeloid lineage (**a**) and not the lymphoid lineage (**b**). A third endothelial lineage (**c**) has been proposed, but not proven (**?**), which may also provide another site of HCMV latency in aortic, and not venous, endothelial cells

One major difference between viral latency and reactivation is characteristically defined by the absence of lytic gene transcription during latent carriage of virus. Although transcripts arising from the major IE region of HCMV have been detected during latency (Kondo and Mocarski 1995; Kondo et al. 1996) (see Sect. 4 below), the major IE transcripts IE72 or IE86 are not expressed in naturally latent CD34+ cells or monocytes (Taylor-Wiedeman et al. 1994; Mendelson et al. 1996). Indeed, it is only upon terminal differentiation of these cells to mature macrophage or dendritic cell phenotypes that viral lytic gene expression is observed, which, under certain conditions, can result in complete reactivation and release of infectious virus (Taylor-Wiedeman et al. 1994; Soderberg-Naucler et al. 1997; Soderberg-Naucler et al. 2001; Reeves et al. 2005b). However, attempts to dissect the mechanisms of HCMV latency and reactivation have been hampered by the frequency of seropositive cells in vivo and a lack of a robust tissue culture model which allows a more thorough, large-scale analysis of natural latency.

Models of HCMV Latency Using Experimental Infection

The low frequency of latently infected cells in vivo has resulted in a number of studies of HCMV latency which have been used in experimental infection of cord blood CD34+ cells, fetal liver CD34+ cells (GM-Ps), G-CSF mobilised CD34+ cells and CD34+ cells isolated from bone marrow aspirates (Kondo et al. 1994; Minton et al. 1994; Hahn et al. 1998; Maciejewski and St Jeor 1999; Goodrum et al. 2002; Slobedman et al. 2002; Reeves et al. 2005a). These experimentally infected latent model systems can result in 20%-90% of cells carrying latent virus depending on the cell type, virus strain and multiplicity of infection used. Consequently, such experimental latency model systems result in an increase in the level of latently infected cells in the experimental population and have made it possible to perform more comprehensive analyses which can then be tested in naturally latently infected cells.

Although it is not possible to fully review the wealth of data obtained from these studies here (for reviews see Streblow and Nelson 2003; Bego and St Jeor 2006; Sinclair and Sissons 2006), a number of instructive observations have been made. In general, they suggest that long-term carriage of viral genomes during latency occurs in the absence of any significant viral gene expression and the carriage of latent genomes appears to be specific to certain cell populations which include the precursors of monocytes (CD34/CD33/CD14) and dendritic cells (CD34/CD33/CD1a) (Hahn et al. 1998). Furthermore, reactivation of lytic gene expression requires terminal differentiation of such progenitors to macrophages or dendritic cells (Maciejewski and St Jeor 1999; Reeves et al. 2005a). Thus, there is good agreement between studies on experimental and natural latency and, consequently, these experimental models have been used extensively to address one of the more intriguing aspects of HCMV latency: latent viral gene expression.

Viral Gene Expression Associated with HCMV Latency

Using an experimental latent model system involving in vitro infection of GM-Ps, Kondo et al. first identified cytomegalovirus latency-expressed transcripts (CLTs) from both strands of the MIE region of the genome (Kondo and Mocarski 1995; Kondo et al. 1996). Subsequently, some of the CLTs were identified in the cells of healthy seropositives' bone marrow aspirates and antibodies to these six ORFs were detected in infected individuals (Kondo et al. 1996; Landini et al. 2000). However, their specific role, if any, in latency is unclear as they are expressed during productive infection and deletion of one CLT (pORF94) had no impact on the ability of the virus to establish latency or to reactivate in vitro (White et al. 2000).

Using the same experimentally latent GM-P model, another putative latency-associated transcript has been identified (Jenkins et al. 2004): the UL111a transcript, also expressed during lytic infection (Kotenko et al. 2000). UL111a encodes a viral homologue of interleukin-10 (vIL-10). Unlike its cellular counterpart, vIL-10 encodes only the immunosuppressive functions associated with cellular IL-10 (Spencer et al. 2002) and thus may play a role in avoiding immune surveillance (see the chapter by C. Powers et al., this volume). This provides an attractive mechanism for the increased survival of latently infected cells in vivo. However, detection of the transcript in vivo did not correlate with HCMV serostatus: monocytes from some seronegative individuals were also positive for the v-IL10 transcript. Whether such seronegative donors were DNA-positive or were, perhaps, sero-converting at the time of the analysis was never determined.

More comprehensive analyses of viral gene expression associated with experimental latency have also been carried out using microarrays (Goodrum et al. 2002; Cheung et al. 2006). These detected a large number of viral RNAs, including IE transcripts that were expressed transiently following infection of CD34$^+$ cells or GM-Ps. Consequently, whether all these viral RNAs represent truly latent transcripts requires more in depth analysis. The possibility that some of these viral RNAs may reflect detection of low-level persistent infection in some cells of the experimental latent cultures needs to be completely ruled out. Although the possibility that expression of these virals RNAs are required to establish latent infection, which are progressively switched off during long-term latency, needs consideration (Cheung et al. 2006).

One transcript, however, initially identified by Goodrum et al. with their microarray analysis (Goodrum et al. 2002), has also been shown to be expressed during natural latency in some seropositive monocytes and CD34$^+$ cells (Goodrum et al. 2007). This transcript, encoded by UL138 of the viral genome, may be required for HCMV latency as recombinant viruses lacking UL138 have an impaired ability to establish a latent infection in an experimental model system (Goodrum et al. 2007). However, the exact role of this transcript during latency will require further investigation.

Another recently identified putative latency-associated transcript encoded by HCMV is the LUNA (also known as latency-associated nuclear antigen) transcript (Bego et al. 2005). Identified in a screen of a monocyte cDNA library prepared from RNA isolated from a healthy HCMV-seropositive individual, this transcript is partially anti-sense to the viral UL81 and UL82 genes. Although a protein product is expressed during productive infection, it has been suggested that LUNA could function as an anti-sense RNA during latency, perhaps mediating the inhibition of pp71 expression from the UL82 ORF. As pp71 is a potent transactivator of the MIEP (major immediate-early promoter) (Bresnahan and Shenk 2000), such suppression of the MIEP could help maintain latency.

Key Aspects of HCMV Latency and Reactivation

Although RNAs expressed in experimentally infected latent models do need to be carefully interpreted, these model systems have also been used extensively to attempt to address other key aspects of viral latency, namely latency establishment and maintenance as well as reactivation.

The Establishment of HCMV Latency

One of the critical steps for establishing latency is likely to include the silencing of the viral MIEP: such control of viral major IE gene expression is a credible mechanism by which all subsequent viral lytic gene expression will be regulated (Fig. 2). Thus, what regulates the MIEP? The MIEP appears to be regulated by multiple cellular transcription factors and higher-order chromatin structure during both lytic (Meier and Stinski 1996; Nevels et al. 2004; Ioudinkova et al. 2006; Reeves et al. 2006) and latent infection (Sinclair and Sissons 2006). Promoter transfection assays have identified a number of factors that repress the MIEP: including YY1 (ying yang 1) (Liu et al. 1994), ERF (ets-2 repressor factor) (Bain et al. 2003) and Gfi-1 (growth factor independent-1) (Zweidler-Mckay et al. 1996). These factors are expressed at high levels in nonpermissive cells and, interestingly, ERF and YY1 are known to interact with chromatin-modifying enzymes (Thomas and Seto 1999; Wright et al. 2005). Consistent with this, during both experimental and natural latency, the transcriptionally inactive MIEP is associated with markers of repressed chromatin, such as Heterochromatin protein 1 (HP1), and is responsive to the histone deacetylase inhibitor Trichostatin A (TSA) providing a model for silencing of the MIEP during experimental (Meier 2001; Murphy et al. 2002; Reeves et al. 2005a) and natural latency (Reeves et al. 2005b). Interestingly, during experimental latency in GM-Ps, cellular factors associated with the formation of repressive chromatin (i.e. AML-1b) are known to be upregulated (Slobedman et al. 2004). Therefore, such cellular responses to latent infection, coupled with a cellular environment already high in levels of repressors of the MIEP, may be critical determinants for the establishment and maintenance of latent carriage of viral genomes.

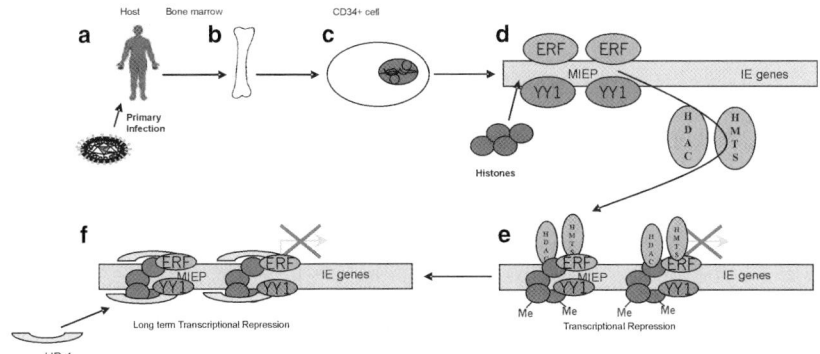

Fig. 2 The establishment of HCMV latency is promoted by chromatin structure. Following infection (**a**), HCMV infects the cells of the bone marrow (**b**) and establishes a latent infection of the CD34$^+$ cells resident therein (**c**). High levels of cellular transcriptional repressors such as ERF and YY1 (**d**) repress the MIEP. As well as transcription factor binding, histone proteins are recruited to the MIEP which become targets for histone deacetylases and histone methyltransferases that are recruited by YY1 and ERF (**e**). These methylated histones become targets for the recruitment of HP-1, which augments repression and the establishment of latency (**f**). Whether any viral products expressed during latency that are important for the repression of the MIEP in this model is, to date, unknown

To date, there is a good consensus that HCMV infects CD34$^+$ haematopoietic stem cells and establishes a latent infection in them. Whilst this is demonstrably true, it has also been suggested that subsets of CD34$^+$ cells may show susceptibility for HCMV productive infection. A study by Goodrum et al. (2004) that analysed differential outcomes of HCMV infection in sorted populations of haematopoietic CD34$^+$ stem cells concluded that infection of one subset of CD34$^+$ cells (CD34$^+$ but CD38$^-$) established the hallmarks of a latent infection (Goodrum et al. 2004), i.e. no detectable virus production but the ability to reactivate upon cellular differentiation. In contrast, other CD34$^+$ cell subpopulations were fully productive for HCMV infection, whilst more mature CD34$^+$ stem cell subpopulations appeared to undergo abortive infection and failed to maintain latent viral genomes. This suggests that the outcome of infection of different CD34$^+$ stem cell subpopulations could depend on the exact phenotype of each stem cell subpopulation. Indeed, there is increasing evidence of early lineage commitment in the haematopoietic stem cell compartment such that a dendritic cell fate, although not irreversible (O'Garra and Trinchieri 2004), is thought to be determined at earlier stages of progenitor cell development (Olweus et al. 1997; Monji et al. 2002). Taken together, these observations could support the hypothesis that HCMV infection of CD34$^+$ stem cells, resulting in latent viral carriage, is restricted to certain subpopulations of CD34$^+$ stem cells. These cells are restricted to specific myeloid cell fates and this mechanism may explain why the carriage of HCMV genomes occurs in some but not all cell types of the myeloid lineage.

Alternatively, infection of CD34⁺ stem cells could, in itself, promote lineage commitment of the latently infected cell to specific myeloid cell types. Although there is no direct evidence for this, differences in the cellular transcriptome of experimentally latently infected GM-Ps compared with uninfected GM-Ps suggest that such changes in cellular gene expression upon latent infection could, hypothetically, promote lineage commitment of these myeloid progenitor cells (Slobedman et al. 2004). Thus, whether viral genome carriage in only certain myeloid cells is a consequence of HCMV initially infecting specific CD34⁺ subpopulations which are already committed to different lineages or is due to the latent infection itself, promoting lineage commitment to specific myeloid cell types, is unclear, but both are plausible.

It is clear that a critical determinant of whether the outcome of an infection is productive or latent is dependent on the regulation of IE gene expression: if repression of the MIEP prevails, latency will probably be established (Sinclair and Sissons 1996). However, whether this also involves expression of other specific viral genes is not clear. As stated previously, experimental latency models have identified a wide range of viral transcripts which appear to be expressed during latency. However, many of these are not exclusive to latent infection (Lunetta and Wiedeman 2000; Bego et al. 2005; Goodrum et al. 2007) and their expression has not been confirmed in natural latency (Beisser et al. 2001; Goodrum et al. 2002; Cheung et al. 2006). Consequently, they may simply represent noise from abortive or productive infection in certain subpopulations of progenitor cell types or they may represent a class of viral gene products required to establish latent infection (see Sect. 4, above).

The Maintenance of HCMV Latency

Latency is established within myeloid cells, a cell type with substantial proliferative capacity (Metcalf 1989). Between two and ten copies (Slobedman and Mocarski 1999) of the HCMV genome are carried in an episomal form in mononuclear cells in the peripheral blood of healthy seropositive individuals (Bolovan-Fritts et al. 1999). How the viral genome is maintained in these dividing progenitor cells is unclear. Contrast this with the gamma herpesvirus Epstein-Barr Virus (EBV), which has a defined latent origin of replication (Yates et al. 1984; Adams 1987) and encodes a number of latently expressed genes. Some of these viral genes have defined roles in viral latent genome maintenance in B cells (Leight and Sugden 2000; Young et al. 2000). HCMV does not have similar genes and does not have a known latent origin of replication. Genome replication does not appear to be due to any low-level persistence in the cells in which HCMV is carried in vivo. Recently, Mocarski et al. reported the identification of a segment of genome with proximity to the IE region of HCMV that appears to be important for carriage of the viral genome in an experimentally infected GM-Ps (Mocarski et al. 2006). Since the carriage of viral genomes in these cells occurs without lytic gene expression (Hahn et al. 1998), and presumably without lytic DNA replication, this region of the viral genome may act as a latent origin of replication or encode a viral gene product essential for maintenance of the viral genome. Characterisation of this region awaits further study.

If latent replication of HCMV genomes does occur, how is segregation of viral genomes to the daughter cells accomplished? This is achieved in EBV and KSHV by using chromatin tethering proteins such as EBNA-1 (Leight and Sugden 2000) and LANA (Cotter and Robertson 1999), respectively. In the absence of an identified homologue in HCMV, how the viral genomes are replicated and segregated during latent carriage is not known.

One alternative, which needs to be considered, is that there is actually no latent replication of HCMV DNA. Instead, $CD34^+$ haematopoietic cells exiting the bone marrow are continually infected from a low level subclinical persistent infection in, for instance, surrounding stromal cells. Once these $CD34^+$ cells enter the bloodstream, they are relatively short-lived and quite quickly differentiate, which results in virus reactivation and virus production from these terminally differentiated myeloid cells. Whether or not any of these models are true will require further study, in perhaps more tractable model systems than we have at present.

The maintenance of HCMV latency is likely to require the continued repression of viral gene expression - particularly IE gene expression - to prevent the virus from re-entering the lytic cycle. In experimentally latent cells, the MIEP is associated with repressed chromatin upon infection (Murphy et al. 2002) and throughout long-term culture as $CD34^+$ cells (Reeves et al. 2005a). Similarly, in naturally latent cells from healthy individuals the MIEP is associated with repressed chromatin (Reeves et al. 2005b). This suggests that virus is kept latent, at least in part, by the recruitment of repressive chromatin factors to the MIEP. However, chromatin is a highly dynamic structure and thus it may be a possibility that yet undefined viral functions have a role in the maintenance of the MIEP in a repressed form, which may also impact on the carriage of the viral genome in a latent state.

Reactivation of HCMV from Latency

Latency is operationally defined as the persistence of the viral genome in the absence of viral lytic gene expression, but importantly, with the capacity of the virus to re-enter its lytic life cycle. The ease and rapidity with which HCMV reactivates in vivo, causing severe disease, is in stark contrast to the ability to reactivate in vitro.

Observations from both experimentally and naturally latent cells suggest that the cellular environment is a key factor in HCMV reactivation: changes in the cellular environment result in the induction of viral lytic gene expression and, hence, virus reactivation (Fig. 3). A number of functions associated with virus infection are known to augment viral IE gene expression. Virus binding on the surface of the cell results in significant changes to the cellular environment by targeting a number of

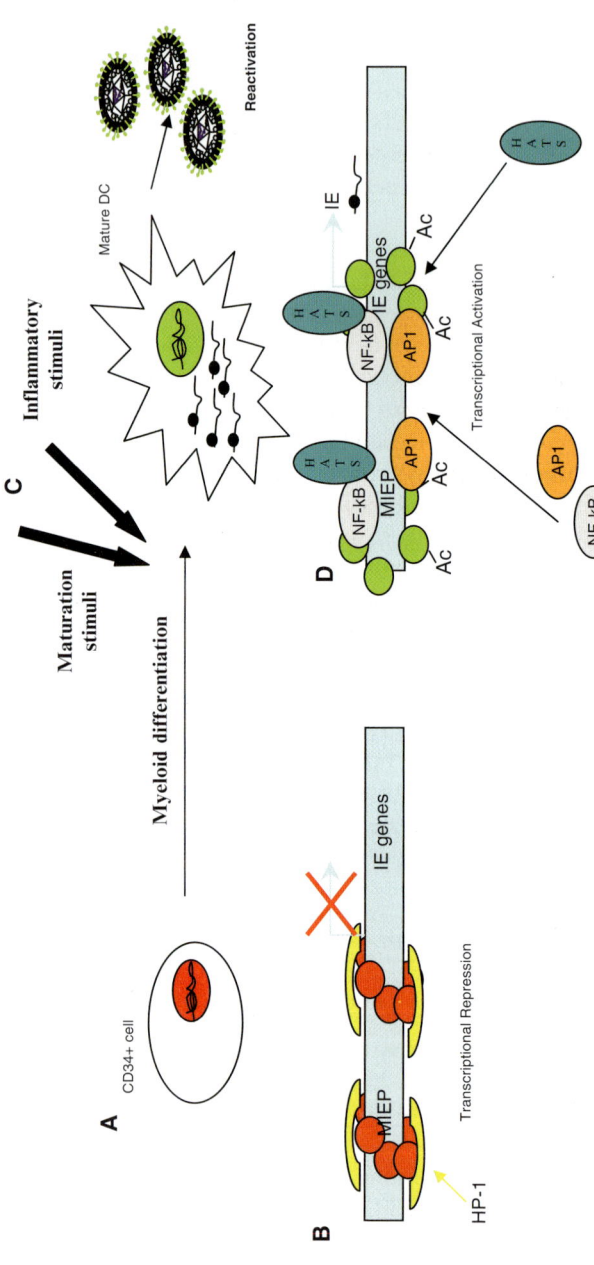

Fig. 3 Reactivation may be promoted by pro-inflammatory signals and is differentiation dependent. CD34+ cells carry the HCMV genome in a latent form (**A**), maintained by a repressive chromatin structure around the MIEP (**B**). Exposure of CD34+ cells to growth factors and/or inflammatory cytokines promotes myeloid differentiation to mature macrophages and dendritic cells (**C**). Differentiation is concomitant with changes in the levels of specific cellular transcription factors and co-factors. These changes promote the reactivation of HCMV immediate-early gene expression, which is consistent with changes in the chromatin structure of the MIEP from a repressed to an active conformation (**D**). Whether this change is totally regulated by cellular factors or requires a viral product remains unknown

signaling pathways (Fortunato et al. 2000; Simmen et al. 2001; Johnson and Hegde 2002; see the chapters by A. Yurochko and M.K. Isaacson et al., this volume). Viral tegument proteins are also delivered to the cell which can target cellular functions (Everett 2006; see the chapter by G. Maul, this volume). Finally, some of these viral tegument proteins delivered by the virion particle are known to transactivate gene expression to promote high levels of viral IE transcription (Liu and Stinski 1992; Bresnahan and Shenk 2000; Schierling et al. 2004; see the chapter by R; Kaletja, this volume). However, none of these are likely to be involved in reactivation from latency as no virions will be present in the apparent absence of these events during latency; the switch from a latent to a reactivating phenotype requires a latency breaking step. Whether this is a virally encoded latent function or is a consequence of changes to the cellular environment is presently under intense investigation in a number of laboratories.

In our laboratory, we have shown that reactivation of viral gene expression and productive infection in natural (Reeves et al. 2005b) or experimental latency (Murphy et al. 2002; Reeves et al. 2005a) is associated with differentiation of $CD34^+$ cells to a DC phenotype. Histone acetylation at the MIEP facilitates an open chromatin conformation which is permissive for MIEP transcription (Reeves et al. 2005b). Consequently, the implication is that normal changes in cellular transcriptional regulators which occur upon terminal differentiation of myeloid cells could be enough to trigger reactivation of virus IE gene expression.

The likelihood that reactivation from latency occurs in the absence of virally encoded transactivators of IE gene expression implies that the viral genome senses reactivation signals from cellular mediators. The first report of reactivation in vitro from myeloid cells involved the stimulation of monocytes with cytokines derived from allogeneically stimulated T cells (Soderberg-Naucler et al. 1997), including TNF-alpha, interferon-gamma, interleukins and GM-CSF (Soderberg-Naucler et al. 2001). Pro-inflammatory factors or induction of myeloid cell differentiation have been responsible for promoting reactivation of viral major IE gene expression. This scenario may have strong clinical relevance, considering the known association of virus reactivation and CMV disease with transplantation (Sissons et al. 2002). Attempts to further characterise the role of the cytokines have, so far, proved inconclusive and await further study.

Besides the basic regulation of viral IE expression, it is also clear that the interplay between the host immune system and reactivating virus has a profound role in HCMV reactivation in vivo (Sissons et al. 2002; Peggs and Mackinnon 2004). Possibly, HCMV reactivation is a sporadic event, occurring infrequently when certain inflammatory conditions are encountered locally in the host. Alternatively, it could be a more common event, occurring whenever latently infected myeloid cells naturally differentiate. In both cases, any reactivation and virus dissemination, which could result in severe disease, is efficiently controlled by a robust immune response. This may be the reason for the unprecedented number of memory T cells that recognise lytic HCMV antigens from healthy carriers in vivo (Riddell et al. 1991; McLaughlin-Taylor et al. 1994; Wills et al. 1996; Sylwester et al. 2005). Both scenarios are possible and it is likely that that the exact mechanism could lay somewhere between these two extremes; either could account for the success of this virus as an opportunistic pathogen.

In contrast, during transplantation and likely accompanying immune suppression, the cytokine storm induced by transplantation could drive virus reactivation which is uncontrollable in the absence of a good cytotoxic T cell (CTL) response. However, studies to address these types of questions are extremely difficult, requiring analyses with naturally, latently infected cells and homologous or nonhomologous CTLs. Given the frequency of HCMV DNA-positive cells, this would be a challenging exercise. It is becoming increasingly clear that conducting these and other analyses pertinent to HCMV latency will require a system for enriching for HCMV-positive cells in cell populations from naturally infected individuals.

Other Sites of HCMV Latency

Whilst most analyses of HCMV latency have focused on the myeloid lineage, it must be acknowledged that there may be other cellular sites of HCMV latency in vivo. Upon virus reactivation, a diverse number of tissues become infected very rapidly. Whilst this could be attributed to mononuclear cell-mediated dissemination of the virus, it could also equally be attributed to other sites of reactivation (Fig. 1c).

$CD34^+$ stem cells are also believed to be the precursors of some endothelial cells (Quirici et al. 2001), and latent carriage and/or reactivation in these cells has been suggested to be involved in atherosclerotic disease (Epstein et al. 1996; see the chapter by D.N. Streblow et al., this volume). Although the data linking atherosclerosis and HCMV is circumstantial and contentious (Danesh et al. 1997), HCMV reactivation in endothelial cells (ECs) could result from changes in cellular gene expression (Maussang et al. 2006; Reinhardt et al. 2006). Countering this interpretation is the observation that latent HCMV genomes have not been detected in ECs isolated from saphenous vein tissue of seropositive donors (Reeves et al. 2004). Although all ECs are likely to arise from the same progenitor, such an analysis does not preclude the possibility that aortic ECs are sites of latent carriage of HCMV (Jarvis and Nelson 2007). Alternatively, the possibility that a selective segregation of viral genomes also occurs between venous and aortic ECs, as appears to be the case in the myeloid linage, cannot be ruled out (Fig. 1c).

Another possibility is that ECs are sites of low-level persistent HCMV infection, in vivo, rather than sites of true viral latency. Some studies have shown that infected ECs are sites of virus production which show little sign of cell lysis (Fish et al. 1995, 1998). However, other studies have suggested that the infection of ECs is fully cytopathic, resulting in complete cell lysis (Kahl et al. 2000). Whether these differences result from different types of ECs used in the studies or virus strain variation is unclear. What is clear, though, is that the difficulty in obtaining aortic tissue from healthy individuals means it is almost impossible to analyse directly whether these cells are latently or persistently infected.

Conclusion

Fully understanding the establishment and maintenance of HCMV latent infection as well as reactivation from latency remains a weighty problem. As molecular techniques continue to progress, it is likely that global analyses of the effects of HCMV infection on the cellular transcriptome and proteome will become even more possible. As our understanding of the functional heterogeneity of populations of CD34+ haematopoietic precursors increases, it may become possible to further clarify the mechanisms which promote viral latency in only specific cell types. Why the maintenance of HCMV genomes occurs only in particular subsets of cells derived from a common ancestor is unknown. As a consequence, what conditions in these cells facilitate virus reactivation is also unknown.

Acknowledgements We thank the many members of the laboratory, past and present, whose work has contributed to the studies presented here. We also apologise to colleagues in the field whose work has not been cited due to space limitations. Work from our laboratory which has contributed to these studies was supported by the United Kingdom Medical Research Council and the Wellcome Trust.

References

Adams A (1987) Replication of latent Epstein-Barr virus genomes in Raji cells. J Virol 61:1743-1746

Adler SP (1983) Transfusion-associated cytomegalovirus infections. Rev Infect Dis 5:977-993

Bain M, Mendelson M, Sinclair J (2003) Ets-2 repressor factor (ERF) mediates repression of the human cytomegalovirus major immediate-early promoter in undifferentiated non-permissive cells. J Gen Virol 84:41-49

Bego MG, St Jeor S (2006) Human cytomegalovirus infection of cells of hematopoietic origin: HCMV-induced immunosuppression, immune evasion, and latency. Exp Hematol 34:555-570

Bego M, Maciejewski J, Khaiboullina S, Pari G, St Jeor S (2005) Characterization of an antisense transcript spanning the UL81-82 locus of human cytomegalovirus. J Virol 79:11022-11034

Beisser PS, Laurent L, Virelizier JL, Michelson S (2001) Human cytomegalovirus chemokine receptor gene US28 is transcribed in latently infected THP-1 monocytes. J Virol 75:5949-5957

Bevan IS, Daw RA, Day PJ, Ala FA, Walker MR (1991) Polymerase chain reaction for detection of human cytomegalovirus infection in a blood donor population. Br J Haematol 78:94-99

Bolovan-Fritts CA, Mocarski ES, Wiedeman JA (1999) Peripheral blood CD14(+) cells from healthy subjects carry a circular conformation of latent cytomegalovirus genome. Blood 93:394-398

Bresnahan WA, Shenk TE (2000) UL82 virion protein activates expression of immediate early viral genes in human cytomegalovirus-infected cells. Proc Natl Acad Sci USA 97:14506-14511

Cheung AK, Abendroth A, Cunningham AL, Slobedman B (2006) Viral gene expression during the establishment of human cytomegalovirus latent infection in myeloid progenitor cells. Blood 108:3691-3699

Cotter MA 2nd, Robertson ES (1999) The latency-associated nuclear antigen tethers the Kaposi's sarcoma-associated herpesvirus genome to host chromosomes in body cavity-based lymphoma cells. Virology 264:254-264

Danesh J, Collins R, Peto R (1997) Chronic infections and coronary heart disease: is there a link? Lancet 350:430-436

Epstein SE, Speir E, Zhou YF, Guetta E, Leon M, Finkel T (1996) The role of infection in restenosis and atherosclerosis: focus on cytomegalovirus. Lancet 348 [Suppl 1]:s13-s17

Everett RD (2006) Interactions between DNA viruses, ND10 and the DNA damage response. Cell Microbiol 8:365-374

Fish KN, Stenglein SG, Ibanez C, Nelson JA (1995) Cytomegalovirus persistence in macrophages and endothelial cells. Scand J Infect Dis Suppl 99:34-40

Fish KN, Soderberg-Naucler C, Mills LK, Stenglein S, Nelson JA (1998) Human cytomegalovirus persistently infects aortic endothelial cells. J Virol 72:5661-5668

Fortunato EA, McElroy AK, Sanchez I, Spector DH (2000) Exploitation of cellular signaling and regulatory pathways by human cytomegalovirus. Trends Microbiol 8:111-119

Gilbert GL, Hayes K, Hudson IL, James J (1989) Prevention of transfusion-acquired cytomegalovirus infection in infants by blood filtration to remove leucocytes. Neonatal Cytomegalovirus Infection Study Group. Lancet 1:1228-1231

Goodrum FD, Jordan CT, High K, Shenk T (2002) Human cytomegalovirus gene expression during infection of primary hematopoietic progenitor cells: a model for latency. Proc Natl Acad Sci USA 99:16255-16260

Goodrum F, Jordan CT, Terhune SS, High K, Shenk T (2004) Differential outcomes of human cytomegalovirus infection in primitive hematopoietic cell subpopulations. Blood 104:687-695

Goodrum F, Reeves M, Sinclair J, High K, Shenk T (2007) Human cytomegalovirus sequences expressed in latently infected individuals promote a latent infection in vitro. Blood 110:937-945

Griffiths PD, Walter S (2005) Cytomegalovirus. Curr Opin Infect Dis 18:241-245

Hahn G, Jores R, Mocarski ES (1998) Cytomegalovirus remains latent in a common precursor of dendritic and myeloid cells. Proc Natl Acad Sci USA 95:3937-3942

Ho M (1990) Epidemiology of cytomegalovirus infections. Rev Infect Dis 12 [Suppl 7]: S701-S710

Ioudinkova E, Arcangeletti MC, Rynditch A, De Conto F, Motta F, Covan S, Pinardi F, Razin SV, Chezzi C (2006) Control of human cytomegalovirus gene expression by differential histone modifications during lytic and latent infection of a monocytic cell line. Gene 384:120-128

Jarvis MA, Nelson JA (2007) Human cytomegalovirus tropism for endothelial cells: not all endothelial cells are created equal. J Virol 81:2095-2101

Jenkins C, Abendroth A, Slobedman B (2004) A novel viral transcript with homology to human interleukin-10 is expressed during latent human cytomegalovirus infection. J Virol 78:1440-1447

Johnson DC, Hegde NR (2002) Inhibition of the MHC class II antigen presentation pathway by human cytomegalovirus. Curr Top Microbiol Immunol 269:101-115

Kahl M, Siegel-Axel D, Stenglein S, Jahn G, Sinzger C (2000) Efficient lytic infection of human arterial endothelial cells by human cytomegalovirus strains. J Virol 74:7628-7635

Kondo K, Mocarski ES (1995) Cytomegalovirus latency and latency-specific transcription in hematopoietic progenitors. Scand J Infect Dis Suppl 99:63-67

Kondo K, Kaneshima H, Mocarski ES (1994) Human cytomegalovirus latent infection of granulocyte-macrophage progenitors. Proc Natl Acad Sci USA 91:11879-11883

Kondo K, Xu J, Mocarski ES (1996) Human cytomegalovirus latent gene expression in granulocyte-macrophage progenitors in culture and in seropositive individuals. Proc Natl Acad Sci USA 93:11137-11142

Kotenko SV, Saccani S, Izotova LS, Mirochnitchenko OV, Pestka S (2000) Human cytomegalovirus harbors its own unique IL-10 homolog (cmvIL-10). Proc Natl Acad Sci USA 97:1695-1700

Landini MP, Lazzarotto T, Xu J, Geballe AP, Mocarski ES (2000) Humoral immune response to proteins of human cytomegalovirus latency-associated transcripts. Biol Blood Marrow Transplant 6:100-108

Leight ER, Sugden B (2000) EBNA-1: a protein pivotal to latent infection by Epstein-Barr virus. Rev Med Virol 10:83-100

Liu B, Stinski MF (1992) Human cytomegalovirus contains a tegument protein that enhances transcription from promoters with upstream ATF and AP-1 cis-acting elements. J Virol 66:4434-4444

Liu R, Baillie J, Sissons JG, Sinclair JH (1994) The transcription factor YY1 binds to negative regulatory elements in the human cytomegalovirus major immediate early enhancer/promoter and mediates repression in non-permissive cells. Nucleic Acids Res 22:2453-2459

Lunetta JM, Wiedeman JA (2000) Latency-associated sense transcripts are expressed during in vitro human cytomegalovirus productive infection. Virology 278:467-476

Maciejewski JP, St Jeor SC (1999) Human cytomegalovirus infection of human hematopoietic progenitor cells. Leuk Lymphoma 33:1-13

Maussang D, Verzijl D, van Walsum M, Leurs R, Holl J, Pleskoff O, Michel D, van Dongen GA, Smit MJ (2006) Human cytomegalovirus-encoded chemokine receptor US28 promotes tumorigenesis. Proc Natl Acad Sci USA 103:13068-13073

McLaughlin-Taylor E, Pande H, Forman SJ, Tanamachi B, Li CR, Zaia JA, Greenberg PD, Riddell SR (1994) Identification of the major late human cytomegalovirus matrix protein pp65 as a target antigen for CD8+ virus-specific cytotoxic T lymphocytes. J Med Virol 43:103-110

Meier JL (2001) Reactivation of the human cytomegalovirus major immediate-early regulatory region and viral replication in embryonal NTera2 cells: role of trichostatin A, retinoic acid, and deletion of the 21-base-pair repeats and modulator. J Virol 75:1581-1593

Meier JL, Stinski MF (1996) Regulation of human cytomegalovirus immediate-early gene expression. Intervirology 39:331-342

Mendelson M, Monard S, Sissons P, Sinclair J (1996) Detection of endogenous human cytomegalovirus in CD34+ bone marrow progenitors. J Gen Virol 77:3099-3102

Metcalf D (1989) The molecular control of cell division, differentiation commitment and maturation in haemopoietic cells. Nature 339:27-30

Minton EJ, Tysoe C, Sinclair JH, Sissons JG (1994) Human cytomegalovirus infection of the monocyte/macrophage lineage in bone marrow. J Virol 68:4017-4021

Mocarski ES, Hahn G, White KL, Xu J, Slobedman B, Hertel L, Aguirre SA, Noda S (2006) Myeloid cell recruitment and function in pathogenesis and latency. In: Reddehase MJ (ed) Cytomegaloviruses: molecular biology and immunology. Caister, Wymondon, UK, pp 465-481

Monji T, Petersons J, Saund NK, Vuckovic S, Hart DN, Auditore-Hargreaves K, Risdon G (2002) Competent dendritic cells derived from CD34+ progenitors express CMRF-44 antigen early in the differentiation pathway. Immunol Cell Biol 80:216-225

Murphy JC, Fischle W, Verdin E, Sinclair JH (2002) Control of cytomegalovirus lytic gene expression by histone acetylation. EMBO J 21:1112-1120

Nevels M, Paulus C, Shenk T (2004) Human cytomegalovirus immediate-early 1 protein facilitates viral replication by antagonizing histone deacetylation. Proc Natl Acad Sci USA 101:17234-17239

O'Garra A, Trinchieri G (2004) Are dendritic cells afraid of commitment? Nat Immunol 5:1206-1208

Olweus J, BitMansour A, Warnke R, Thompson PA, Carballido J, Picker LJ, Lund-Johansen F (1997) Dendritic cell ontogeny: a human dendritic cell lineage of myeloid origin. Proc Natl Acad Sci USA 94:12551-12556

Peggs KS, Mackinnon S (2004) Cytomegalovirus: the role of CMV post-haematopoietic stem cell transplantation. Int J Biochem Cell Biol 36:695-701

Plachter B, Sinzger C, Jahn G (1996) Cell types involved in replication and distribution of human cytomegalovirus. Adv Virus Res 46:195-261

Quirici N, Soligo D, Caneva L, Servida F, Bossolasco P, Deliliers GL (2001) Differentiation and expansion of endothelial cells from human bone marrow CD133(+) cells. Br J Haematol 115:186-194

Reeves MB, Coleman H, Chadderton J, Goddard M, Sissons JG, Sinclair JH (2004) Vascular endothelial and smooth muscle cells are unlikely to be major sites of latency of human cytomegalovirus in vivo. J Gen Virol 85:3337-3341

Reeves MB, Lehner PJ, Sissons JG, Sinclair JH (2005a) An in vitro model for the regulation of human cytomegalovirus latency and reactivation in dendritic cells by chromatin remodelling. J Gen Virol 86:2949-2954

Reeves MB, MacAry PA, Lehner PJ, Sissons JG, Sinclair JH (2005b) Latency, chromatin remodeling, and reactivation of human cytomegalovirus in the dendritic cells of healthy carriers. Proc Natl Acad Sci USA 102:4140-4145

Reeves M, Murphy J, Greaves R, Fairley J, Brehm A, Sinclair J (2006) Autorepression of the human cytomegalovirus major immediate-early promoter/enhancer at late times of infection is mediated by the recruitment of chromatin remodeling enzymes by IE86. J Virol 80:9998-10009

Reinhardt B, Winkler M, Schaarschmidt P, Pretsch R, Zhou S, Vaida B, Schmid-Kotsas A, Michel D, Walther P, Bachem M, Mertens T (2006) Human cytomegalovirus-induced reduction of extracellular matrix proteins in vascular smooth muscle cell cultures: a pathomechanism in vasculopathies? J Gen Virol 87:2849-2858

Riddell SR, Rabin M, Geballe AP, Britt WJ, Greenberg PD (1991) Class I MHC-restricted cytotoxic T lymphocyte recognition of cells infected with human cytomegalovirus does not require endogenous viral gene expression. J Immunol 146:2795-2804

Rubin RH (1990) Impact of cytomegalovirus infection on organ transplant recipients. Rev Infect Dis 12 [Suppl 7]:S754-S766

Schierling K, Stamminger T, Mertens T, Winkler M (2004) Human cytomegalovirus tegument proteins ppUL82 (pp71) and ppUL35 interact and cooperatively activate the major immediate-early enhancer. J Virol 78:9512-9523

Simmen KA, Singh J, Luukkonen BG, Lopper M, Bittner A, Miller NE, Jackson MR, Compton T, Fruh K (2001) Global modulation of cellular transcription by human cytomegalovirus is initiated by viral glycoprotein B. Proc Natl Acad Sci USA 98:7140-7145

Sinclair J, Sissons P (1996) Latent and persistent infections of monocytes and macrophages. Intervirology 39:293-301

Sinclair J, Sissons P (2006) Latency and reactivation of human cytomegalovirus. J Gen Virol 87:1763-1779

Sindre H, Tjoonnfjord GE, Rollag H, Ranneberg-Nilsen T, Veiby OP, Beck S, Degre M, Hestdal K (1996) Human cytomegalovirus suppression of and latency in early hematopoietic progenitor cells. Blood 88:4526-4533

Sinzger C, Grefte A, Plachter B, Gouw AS, The TH, Jahn G (1995) Fibroblasts, epithelial cells, endothelial cells and smooth muscle cells are major targets of human cytomegalovirus infection in lung and gastrointestinal tissues. J Gen Virol 76:741-750

Sissons JG, Carmichael AJ (2002) Clinical aspects and management of cytomegalovirus infection. J Infect 44:78-83

Sissons JG, Bain M, Wills MR (2002) Latency and reactivation of human cytomegalovirus. J Infect 44:73-77

Slobedman B, Mocarski ES (1999) Quantitative analysis of latent human cytomegalovirus. J Virol 73:4806-4812

Slobedman B, Mocarski ES, Arvin AM, Mellins ED, Abendroth A (2002) Latent cytomegalovirus down-regulates major histocompatibility complex class II expression on myeloid progenitors. Blood 100:2867-2873

Slobedman B, Stern JL, Cunningham AL, Abendroth A, Abate DA, Mocarski ES (2004) Impact of human cytomegalovirus latent infection on myeloid progenitor cell gene expression. J Virol 78:4054-4062

Soderberg-Naucler C, Fish KN, Nelson JA (1997) Reactivation of latent human cytomegalovirus by allogeneic stimulation of blood cells from healthy donors. Cell 91:119-126

Soderberg-Naucler C, Streblow DN, Fish KN, Allan-Yorke J, Smith PP, Nelson JA (2001) Reactivation of latent human cytomegalovirus in CD14(+) monocytes is differentiation dependent. J Virol 75:7543-7554

Spencer JV, Lockridge KM, Barry PA, Lin G, Tsang M, Penfold ME, Schall TJ (2002) Potent immunosuppressive activities of cytomegalovirus-encoded interleukin-10. J Virol 76:1285-1292

Stanier P, Kitchen AD, Taylor DL, Tyms AS (1992) Detection of human cytomegalovirus in peripheral mononuclear cells and urine samples using PCR. Mol Cell Probes 6:51-58

Streblow DN, Nelson JA (2003) Models of HCMV latency and reactivation. Trends Microbiol 11:293-295

Sylwester AW, Mitchell BL, Edgar JB, Taormina C, Pelte C, Ruchti F, Sleath PR, Grabstein KH, Hosken NA, Kern F, Nelson JA, Picker LJ (2005) Broadly targeted human cytomegalovirus-specific CD4+ and CD8+ T cells dominate the memory compartments of exposed subjects. J Exp Med 202:673-685

Taylor-Wiedeman J, Sissons JG, Borysiewicz LK, Sinclair JH (1991) Monocytes are a major site of persistence of human cytomegalovirus in peripheral blood mononuclear cells. J Gen Virol 72:2059-2064

Taylor-Wiedeman J, Hayhurst GP, Sissons JG, Sinclair JH (1993) Polymorphonuclear cells are not sites of persistence of human cytomegalovirus in healthy individuals. J Gen Virol 74:265-268

Taylor-Wiedeman J, Sissons P, Sinclair J (1994) Induction of endogenous human cytomegalovirus gene expression after differentiation of monocytes from healthy carriers. J Virol 68:1597-1604

Thomas MJ, Seto E (1999) Unlocking the mechanisms of transcription factor YY1: are chromatin modifying enzymes the key? Gene 236:197-208

Tolpin MD, Stewart JA, Warren D, Mojica BA, Collins MA, Doveikis SA, Cabradilla C Jr, Schauf V, Raju TN, Nelson K (1985) Transfusion transmission of cytomegalovirus confirmed by restriction endonuclease analysis. J Pediatr 107:953-956

White KL, Slobedman B, Mocarski ES (2000) Human cytomegalovirus latency-associated protein pORF94 is dispensable for productive and latent infection. J Virol 74:9333-9337

Wills MR, Carmichael AJ, Mynard K, Jin X, Weekes MP, Plachter B, Sissons JG (1996) The human cytotoxic T-lymphocyte (CTL) response to cytomegalovirus is dominated by structural protein pp65: frequency, specificity, and T-cell receptor usage of pp65-specific CTL. J Virol 70:7569-7579

Wright E, Bain M, Teague L, Murphy J, Sinclair J (2005) Ets-2 repressor factor recruits histone deacetylase to silence human cytomegalovirus immediate-early gene expression in non-permissive cells. J Gen Virol 86:535-544

Yates J, Warren N, Reisman D, Sugden B (1984) A cis-acting element from the Epstein-Barr viral genome that permits stable replication of recombinant plasmids in latently infected cells. Proc Natl Acad Sci USA 81:3806-3810

Yeager AS, Grumet FC, Hafleigh EB, Arvin AM, Bradley JS, Prober CG (1981) Prevention of transfusion-acquired cytomegalovirus infections in newborn infants. J Pediatr 98:281-287

Young LS, Dawson CW, Eliopoulos AG (2000) The expression and function of Epstein-Barr virus encoded latent genes. Mol Pathol 53:238-247

Zaia JA (1990) Epidemiology and pathogenesis of cytomegalovirus disease. Semin Hematol 27:5-10; discussion 28-19

Zweidler-Mckay PA, Grimes HL, Flubacher MM, Tsichlis PN (1996) Gfi-1 encodes a nuclear zinc finger protein that binds DNA and functions as a transcriptional repressor. Mol Cell Biol 16:4024-4034

Murine Model of Cytomegalovirus Latency and Reactivation

M. J. Reddehase(✉), C. O. Simon, C. K. Seckert, N. Lemmermann, N. K. A. Grzimek

Contents

Introduction 316
Definitions and Caveats: The Difference Between Latency and Persistence 317
Latent Viral Genome Load Defining the Risk of Recurrence 320
Bidirectional Gene Pair Architecture of the Regulatory Major Immediate Early Locus 321
Stochastic Desilencing of the Major Immediate Early Locus During Latency 323
Extrinsic Signals Triggering Transcriptional Reactivation and Recurrence 323
Role of Viral Chromatin Remodeling 324
Dynamic Control of Latency at Immunological Checkpoints:
  The Immune Sensing Hypothesis 325
Concluding Thoughts and Perspectives 328
References 328

Abstract Efficient resolution of acute cytopathogenic cytomegalovirus infection through innate and adaptive host immune mechanisms is followed by lifelong maintenance of the viral genome in host tissues in a state of *replicative latency*, which is interrupted by episodes of virus reactivation for transmission. The establishment of latency is the result of aeons of co-evolution of cytomegaloviruses and their respective host species. Genetic adaptation of a particular cytomegalovirus to its specific host is reflected by *private* gene families not found in other members of the cytomegalovirus group, whereas basic functions of the viral replicative cycle are encoded by *public* gene families shared between different cytomegaloviruses or even with herpesviruses in general. Private genes include genes coding for immunoevasins, a group of glycoproteins specifically dedicated to dampen recognition by the host's innate and adaptive immune surveillance to protect the virus against elimination. Recent data in the mouse model of cytomegalovirus latency have indicated that viral replicative latency established in the immunocompetent host is

---

M.J. Reddehase
Institute for Virology, Johannes Gutenberg-University, Obere Zahlbacher Strasse 67,
Hochhaus am Augustusplatz, 55131, Mainz, Germany
Matthias.Reddehase@uni-mainz.de

a dynamic state characterized by episodes of viral gene desilencing and immune sensing of reactivated presentation of antigenic peptides at immunological checkpoints by CD8 T cells. This sensing maintains viral replicative latency by triggering antiviral effector functions that terminate the viral gene expression program before infectious viral progeny are assembled. According to the *immune sensing hypothesis of latency control*, immunological checkpoints are unique for each infected individual in reflection of host MHC (HLA) polymorphism and the proteome(s) of the viral variant(s) harbored in latency.

Introduction

Cytomegalovirus (CMV) disease is an accident for both the virus and its host, and occurs only if the intricate balance between them is disrupted in an immunologically immature or immunocompromised individual. Disease is obviously harmful for the host, but self-limiting replication by lethal infection of the host was apparently also not a successful survival strategy in the evolutionary selection of virus strains, at least not for the members of the herpesvirus family, including the CMVs. The establishment of latent infection after resolution of the acute infection by the host's innate and adaptive immune mechanisms is a hallmark of herpesvirus biology. Resolution of acute infection implies that productively infected cells are recognized by cells of the immune system, resulting in an interruption of the viral productive replication cycle by cytolytic or noncytolytic immune effector mechanisms. Ideally, this occurs at a stage prior to completion of the replicative cycle and to virion exit. On the other hand, establishment of a latent infection implies that the virus escapes elimination through mechanisms of immune evasion, which may be manifold.

CMVs are host-specific and the adaptation to their respective host species appears to be particularly highly evolved, as is suggested by a very significant number of *private* genes not shared between different species of CMVs. In the case of murine CMV (mCMV), the nomenclature introduced by Rawlinson and colleagues (1996), who have greatly expedited the field by sequencing the Smith strain, is instructive, as mCMV's private genes are marked by lowercase letter "m", whereas genes with homology to genes of human CMV (hCMV) are marked by capital letter "M". As a rule of thumb, we expect that private genes are *host adaptation genes*, and, indeed, viral open reading frames (ORFs) encoding proteins involved in the modulation of the murine host's innate and adaptive immune surveillance usually belong to the group of private genes. For most genes of this category, however, the function is still unknown; moreover, most of them appear to be dispensable for virus replication in cell culture (Brune et al. 2006). It is predictable that for many of the private genes a function will be revealed only by in vivo studies, and a great deal of scientific intuition and creativity will be necessary to design the proper experiments. In our view, every viral gene has its role in virus biology, the problem only is to find it! An intriguing example presenting us with riddles is *m128*, also known as mCMV *ie2*, which is part of a bidirectional gene pair constituting the major immediate early (MIE) locus (see Sect. 4), a regulatory locus thought to be of key importance for kick-starting viral gene expression in acute infection and reactivation from latency; nevertheless, no in vivo function could so far be attributed to *m128* (*ie2*) in acute infection or latency (Cardin 1995). As latency and reactivation are events in virus biology that are closely linked to host cell functions such as cell cycle, gene silencing, and nucleotide metabolism, one may propose that currently unknown functions of at least some of the viral private genes will be involved in the establishment, maintenance, and breaking of latency.

By definition, due to the lack of homology, the function of mCMV private genes gives no direct prediction for the function of hCMV private genes. Although this might be considered a weakness of the murine model, there is good reason to anticipate that principles of virus-host interactions in latency and reactivation are analogous between different virus-host pairs. This hope is justified from the previous experience with mCMV private genes *m152* and *m06*, which interfere with the MHC class I pathway of antigen presentation. Specifically, m152/gp40 was the first CMV immunoevasin to be described and has greatly stimulated the search for a similar function in hCMV. As in the co-evolution between mCMV and its host species mouse, co-evolution of hCMV and the human species has indeed led to gene functions interfering with the MHC class I pathway of antigen presentation, namely *US2/US11*, *US3*, and *US6* (for a review, see Reddehase 2002). This example indicates that analogous selective forces, in this case exerted by the respective hosts' immune systems, can lead to convergent results, albeit the solutions may differ in the molecular details. The value of the murine model is that the consequences of mutations in private genes—consequences that we expect to occur only upon interaction with host functions in the context of host tissues and organs—can be tested systematically in vivo.

Complete gene silencing in a viral episome with closed higher-order chromatin-like structure, that is, immune evasion by *hiding*, appears to be a reasonable mechanism for maintenance of the viral genome in the face of fully developed host immunity. Although this is probably true for the vast majority of latent viral genomes in host tissues at any particular time point, recent data have indicated that latency is not static, but is highly dynamic, involving a permanent *immune surveillance* by CD8 T cell-mediated sensing of early stages of transcriptional reactivation.

Definitions and Caveats: The Difference Between Latency and Persistence

There is some semantic confusion in the field regarding the terms *latency* and *persistence*, which are sometimes used as synonyms to describe the situation that an infection is not cleared but persists in the host for an extended period of time.

Without claiming to be right, in this review we would like to reserve the term latency for the maintenance of the viral genome in the absence of productive infection, that is, without assemblage of infectious virions. In contrast, the term persistence is used to describe a situation of enduring or at least long-lasting virus production, which may occur at a low level. In the past there was a debate whether low-level productive infection can ever be excluded to meet the strict definition of latency. Improved assays for detecting infectious virus combined with highly-sensitive RT-PCRs specific for transcripts of essential genes, such as *M122* (*ie3*) and *M55* (*gB*), have provided reasonable evidence allowing the conclusion that the viral genome can indeed be maintained without completion of the productive cycle (Kurz et al. 1997; 1999; Pollock and Virgin 1995). In accordance with this, conditionally replication-incompetent, temperature-sensitive mutants of mCMV were shown to establish and maintain latency, and to reactivate gene expression up to the mutation, at which point the reactivation stopped (Bevan et al. 1996).

Whereas most host organs can be sites of mCMV latency, mCMV persistence occurs at a privileged site, namely the tissue of the salivary glands and, therein, specifically the glandular epithelial cell, which is a polarized cell type specializing in secretion. Virion morphogenesis in this cell type is special in that huge numbers of monocapsid virions gather in secretory vacuoles and are released into the salivary duct, which is an important route for virus host-to-host transmission via saliva. Persistence in glandular epithelial cells occurs despite a vigorous immune response in salivary gland tissue (Cavanaugh et al. 2003), although it eventually ceases, unless the host is depleted of CD4 T cells (Jonjic et al. 1989), indicating that CD4 T cells function as antiviral effector cells and clear salivary gland infection in the long run. Recent work by Humphreys and colleagues (2007) has suggested that mCMV exploits the privileged mucosal environment, which favors the development of interleukin (IL)-10-expressing CD4 T cells and disfavors the development of antiviral interferon (IFN)-γ-expressing effector CD4 T cells. In fact, these authors demonstrated that systemic blockade of IL-10-receptor signaling leads to accumulation of IFN-γ-expressing CD4 T cells in the salivary glands associated with control of salivary gland infection, though it remained open to question whether the virus failed to disseminate to the salivary glands or was controlled there locally.

Viral functions, however, appear to be actively involved, since several viral genes have been implicated in persistent infection of the salivary glands. *M33*, *M43*, and *sgg1* mutants, competent for replication in cell culture as well as in various organs, showed a defect specifically in salivary gland replication (Davis-Poynter et al. 1997; Lagenaur et al. 1994; Xiao et al. 2000). Interestingly, such a specific effect on salivary gland tropism, and thus on mCMV persistence, was also recently attributed to a noncoding 7.2-kb RNA ortholog of the hCMV 5-kb RNA, which is a stable intron (Kulesza and Shenk 2006).

Persistence in the salivary glands can occur, logically, only when the virus disseminates to the salivary glands and infects the glandular epithelial cells. As a consequence, deletion of or mutations in viral genes involved in replicative fitness, in cell-to-cell spread, in long-distance dissemination, or in peripheral immune control, are likely to show also a phenotype in persistence. This has to be considered before viral genes are claimed to be specifically involved in the regulation of persistence. This caveat likewise applies also to genes thought to regulate latency and reactivation. A virus mutant that fails to disseminate to a particular site is trivially also unable to establish latency at that site and to reactivate from that site.

The classical definition of latency is based on an organismal point of view, demanding the absence of productive virus replication from the whole organism. According to this definition, latency is established only after cessation of persistence in the salivary glands. Although such a definition is helpful to define the end of the risk of virus transmission through saliva, which is relevant from an epidemiologist's perspective, linking latency in other organs to the termination of persistence in the very particular mucosal environment of the salivary glands makes little sense for a molecular understanding of latency. As we have discussed in greater detail recently (Simon et al. 2006b), from a molecular viewpoint, virus latency may be established in a particular organ, while virus replication is still going on in another organ; moreover, latency may even be established in a particular cell type within an organ, while virus replication is still going on in another cell type of the very same organ.

It is currently understood that the phenomena of latency and reactivation are closely linked to gene silencing and desilencing, to closed and open viral chromatin structure, respectively (see the chapter by M. Reeves and J. Sinclair, this volume). Sometimes, the term *molecular latency* is used to indicate the proposed latent state of the viral genome. Although we have used this term in previous publications, it may be misleading in that it suggests that all viral genes need to be in a silenced state. As we will discuss in greater detail below, not all viral genomes are transcriptionally silent during latency. CMV genomes encompass many essential genes that all need to be desilenced for completion of the productive cycle. Thus, limited desilencing of only some essential genes or desilencing of nonessential genes leads to transcriptional activity from the viral genome without breaking latency. It is our understanding that latency is not a *transcriptional latency* but is better described as a *replicative latency*. Sensitivity problems have so far precluded a microarray analysis of the latent viral transcriptome; moreover, there exists the fundamental problem that patterns of silenced and desilenced genes may differ between individual viral genomes at any particular time point and may fluctuate with time. As a consequence, even if one would find transcripts from all essential viral genes in an infected tissue representing a statistical ensemble of numerous viral genomes, this would not prove that virus replication is occurring in any single cell. On the other hand, even if the most sensitive RT-PCR fails to detect transcripts from any of the many essential genes, from *M55* (*gB*) as only one example, this is reasonable but not firm evidence for replicative latency. Theoretically at least, a desilencing event may have led to the synthesis of the essential protein, which may remain present in the cell while the corresponding mRNA is degraded and the respective gene silenced again. In the case of the MIE locus genes, however, the corresponding mRNAs are fairly stable with half-lives of more than 24 h and of approximately 6 h determined for IE1 and IE2, respectively (Simon et al. 2007). Thus, absence of MIE locus transcripts, of IE1 mRNA in particular, is a good negative marker for latency. We have recently proposed to use the term *MIE locus latency* to characterize the state of viral genomes that are silenced at this essential regulatory gene locus (Simon et al. 2006b).

Latent Viral Genome Load Defining the Risk of Recurrence

Epidemiologically, CMV infections can occur at any age, but frequently, CMVs are acquired by their respective hosts perinatally or during early childhood. Regardless of the time of virus acquisition, latency is eventually established. A CMV-specific immune status, i.e., the presence of CMV-specific memory CD4 and CD8 T cells as well as of antibodies, testifies for a resolved acute infection in the past, which may have been years or decades ago. Importantly, clinicians use CMV-antibody status to define donor-recipient risk constellations for CMV recurrence and disease in solid organ and hematopoietic stem cell transplantations. The individual infection history of donor and recipient, however, is usually unknown. Although it is clear that presence of latent CMV in either the donor or the recipient, or in both, indicates a risk in a qualitative sense, the incidence of CMV reactivation is difficult to predict.

The mouse model has revealed that infection history matters (Reddehase et al. 1994). Mice infected as neonates within 24 h after birth showed delayed clearance of acute, productive infection in organs and prolonged persistence in the salivary glands, whereas fully immunocompetent mice infected as adults rapidly controlled acute infection in all organs and showed a shortened persistence in the salivary glands. Months later, during replicative latency, this different history of primary infection was reflected by high or low load of viral genomes, respectively, in organs including lungs, spleen, heart, kidney, adrenal glands, and salivary glands. Importantly, viral DNA load is a predictor of the risk of virus recurrence after hematoablative total-body γ-irradiation, with high or low load predicting high or low incidence of recurrence, respectively. Recurrence turned out to be a focal and stochastic event occurring independently in different latently infected organs, thus leading to all possible combinations of virus detection in single or multiple organs. These findings indicate that recurrence is a generally rare event emerging from a minority of the latent viral genomes, and they provide an immediate explanation for the causal relationship between the number of latent virus genomes in tissues and the probability of reactivation. In accordance with such a causal relationship, limitation of acute infection by CD8 T cells in a mouse model of adoptive immunotherapy of CMV infection resulted in a dose-dependent reduction of both latent viral genome load and incidence of virus recurrence (Steffens et al. 1998).

This, again, shows the difficulty in identifying virus genes specifically implicated in the molecular control of latency and reactivation. Mutations of any viral gene that directly or indirectly influences the latent viral genome load can have a phenotype in latency and reactivation. This includes genes directly involved in viral replicative fitness and spread, but also genes involved in the immune control of virus replication and spread either by encoding an antigen or by encoding an immunomodulatory protein. Clearly, this is not what we look upon as being a true *latency gene*.

Bidirectional Gene Pair Architecture of the Regulatory Major Immediate Early Locus

It is widely accepted that the MIE locus of CMVs, which includes an essential enhancer region (Dorsch-Häsler et al. 1985; Ghazal et al. 2003; for a review, see Meier and Stinski 2006), is a key regulatory unit that kick-starts the viral transcriptional program in acute infection as well as in reactivation from latency (see the chapters by G. Maul; M.F. Stinski and D.T. Petrik, this volume). An open viral chromatin structure at the MIE locus appears to be a primary condition for reactivation to be initiated (Bain et al. 2006; see the chapter by M. Reeves and J. Sinclair, this volume). The MIE locus in mCMV is unique in that it is structurally organized as a *bidirectional gene pair*. Bidirectional gene pair architecture is defined as two neighboring genes arranged head-to-head on opposite strands of the DNA and regulated by a shared *cis*-acting regulatory unit. It has long been known that this organization applies to the mCMV MIE locus (Fig. 1) (for a review see Simon et al. 2006b). Interest in this structural feature of the mCMV MIE locus (Chatellard et al. 2007; Simon et al. 2007) results from the recent finding that bidirectional gene pairs are common in the human genome, often conserved among mouse orthologs and thought to provide a unique mechanism of regulation for a significant number of mammalian genes, in particular of genes involved in DNA repair (Adachi and Lieber 2002; Li et al. 2006; Trinklein et al. 2004). That such an architecture is used by mCMV just for the MIE locus, a locus of outstanding regulatory importance, is intriguing and underlines the close host-relatedness of this highly host-adapted virus. In this context, it is of interest to note that hCMV uses a bidirectional promoter element within oriLyt, another locus with a key function in acute infection and reactivation (Xu et al. 2004).

Recent data suggest that the mCMV MIE enhancer region is actually a tandem of two bona fide enhancers $E^{1/3}$ and E^2 (Chatellard et al. 2007) driving the transcription from genes *m123/M122* (*ie1/ie3*) and *m128* (*ie2*), respectively, in opposite directions. Alternative splicing of the IE1/IE3 precursor RNA leads to IE1 mRNA (coding exons 2, 3, and 4) and IE3 mRNA (coding exons 2, 3, and 5); splicing of IE2 precursor RNA leads to IE2 mRNA (coding exon 3).

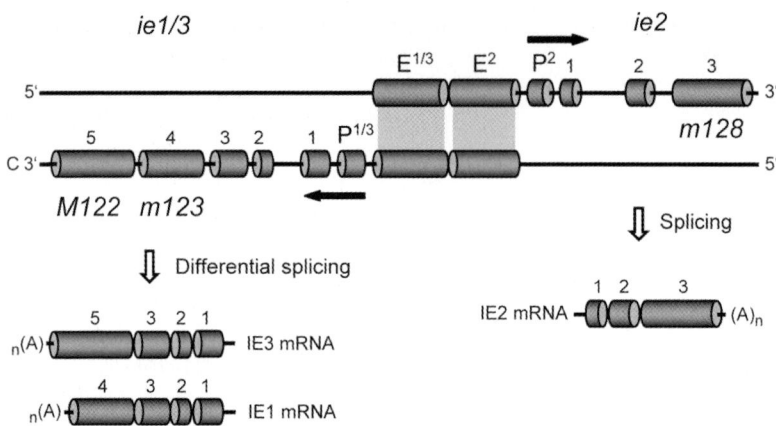

Fig. 1 Bidirectional gene pair architecture of the mCMV MIE locus. *Arrows* indicate the direction of transcription. Numbered cylindrical boxes represent exons. *C*, complementary strand. *M* and *m* indicate mCMV ORFs homologous to hCMV ORFs and mCMV private ORFs, respectively, according to the nomenclature used by Rawlinson et al. (1996). *P* promoter, *E* enhancer

The 76/89-kDa IE1 protein is involved in breaking epigenetic host cell defense by early disruption of nuclear domains (ND)10 (see the chapter by G. Maul, this volume); it co-transactivates the expression of viral early (E)-phase genes and autostimulates its own promoter. Interestingly, it also acts as a transactivator of cellular genes involved in dNTP biosynthesis, such as thymidylate synthase (Gribaudo et al. 2000) and ribonucleotide reductase (Lembo et al. 2000), a property thought to facilitate virus replication in resting cells (for reviews, see Simon et al. 2006b; Tang and Maul 2006). Clearly, efficient provision of dNTPs could possibly be a key parameter in virus reactivation from latently infected cells, which are most likely resting cells, and it will be intriguing to test this idea. The importance of the 88- to 90-kDa IE3 protein for virus reactivation is undoubted, as IE3 is the essential transactivator of viral E gene expression (Angulo et al. 2000). Thus, beyond MIE locus transcription initiation, differential splicing generating IE3 mRNA is a crucial second molecular checkpoint in the transition from mCMV latency to reactivation. Strikingly, no essential function could so far be identified for the 43-kDa IE2 protein (Cardin et al. 1995; Messerle et al. 1991). It remains an open question whether IE2 is really dispensable or whether we just failed to ask the correct questions, to design the proper experiments, and to look at the right place and time. In this context, it is of interest that recent work by Ishiwata et al. (2006) has localized IE2 protein in the brain of neonatally infected mice at a stage of prolonged infection selectively in neurons of the cortex and hippocampus, while the IE3 protein was preferentially expressed in glial cells only at an early stage of the infection. It is a challenge to identify the function of IE2 and to understand why its expression is paired with IE1 and IE3 in bidirectional gene organization, a gene architecture adopted from the mammalian host.

Stochastic Desilencing of the Major Immediate Early Locus During Latency

In a model of latency established in the lungs, highly sensitive RT-PCRs detected IE1 and IE2 transcripts in absence of any infectivity (Grzimek et al. 2001). For a statistical analysis of the frequency of MIE locus transcription, the lungs were subdivided into 18 pieces that were tested individually. This approach resulted in patterns of transcript-positive and transcript-negative lung tissue pieces and allowed calculation of transcription frequencies from the fraction of negative pieces by using the Poisson distribution equation. Such a variegated expression, also referred to as mosaic expression, is reminiscent of transgene expression (Fiering et al. 2000) and indicates stochastic desilencing of an otherwise silenced MIE locus. It is important to emphasize that MIE gene expression is not a constitutive, latency-associated feature of latent mCMV genomes. MIE locus activity in latently infected lungs was found to be approximately 10-20 events per 10^6 viral genomes. Therefore, most viral genomes are in a state of MIE locus latency at any one time, with a minor fraction moving a first step toward reactivation. We propose that all latent viral genomes may be desilenced at the MIE locus some time. Unfortunately, this prediction is difficult to test in a living organism.

Notably, desilencing of the bidirectional gene pair was found to be not coordinated, indicating an independent regulation of the genes that flank the enhancer region (Grzimek et al. 2001). This was true also after extrinsic signaling to the enhancer, which showed that the enhancer does not synchronize transcription from the bidirectional gene pair but rather operates as a switch (Simon et al. 2007). It is currently open to question whether this feature is related to the tandem structure of the mCMV enhancer region. According to our current understanding, isolated IE2 expression should not be able to initiate the replicative cycle. Interestingly, however, in the absence of extrinsic signaling to the enhancer, stochastic desilencing at the *ie1/ie3* promoter did not initiate the replicative cycle either, as IE3 transcripts were absent (Kurz et al. 1999). This identified IE3 splicing as a second molecular checkpoint of latency and explained why replicative latency was maintained despite MIE locus transcription.

Extrinsic Signals Triggering Transcriptional Reactivation and Recurrence

Classical and more recent review articles have dealt in greater detail with extrinsic signals and conditions, such as immune cell depletion, allogeneic transplantation, ischemia/reperfusion injury, polymicrobial sepsis, inflammatory disease states in general, and experimental tissue explantation, which apparently all trigger mCMV transcriptional reactivation and recurrence (Hummel and Abecassis 2002; Jordan 1983; Reddehase et al. 2002). All these conditions are multifactorial and many involve a cytokine storm and deprivation of immune control. Attempts to dissect these multifactorial conditions on a molecular level have recently focused on tumor necrosis factor (TNF)-α, which is known to signal to CMV MIE enhancers through transcription factors NF-κB and AP-1 (Hummel et al. 2001; Prösch et al. 1995) (for enhancer mechanics, see the review by Meier and Stinski 2006).

Although the canonical NF-κB-dependent pathway of MIE enhancer activation appears to be dispensable for virus replication in cultured fibroblasts (Benedict et al. 2004), TNF-α signaling clearly enhanced the prevalence of mCMV MIE gene expression as well as the amount of MIE transcripts per transcriptional event in latently infected lungs within 24 h after intravenous administration of the cytokine (Simon et al. 2005). Notably, this signaling also helped to overcome the splicing checkpoint, which resulted in the expression of the transactivator IE3 mRNA, but transcriptional reactivation did not proceed to expression of the essential gene *M55* (*gB*), thus predicting the existence of further checkpoints beyond MIE locus gene expression. As a consequence, although TNF-α may play an important role in enhancing the initiation of virus reactivation, additional conditions must be met for completion of the productive cycle. Notably, intraperitoneal administration of TNF-α or IL-1β triggered mCMV recurrence in latently infected lungs with a delay of 3 weeks (Cook et al. 2006). Likewise, as a correlate for the sepsis model of mCMV reactivation (Cook et al. 2002), lipopolysaccharide was found to reactivate latent mCMV through a TLR-4-dependent pathway (Cook et al. 2006).

These findings are not in conflict with the direct effects of TNF-α on the MIE enhancer but may rather indicate that TNF-α as well as other proinflammatory mediators can trigger a cascade of events that eventually provide all the conditions for virus recurrence. As mCMV recurrence in the study by Cook et al. (2006) was achieved with near-lethal doses of TNF-α, the question arises of whether the cytokine's effects on the endothelia, causing microvascular injury, may be involved in a pathogenetic process facilitating virus reactivation.

Role of Viral Chromatin Remodeling

There exists reasonable evidence to suggest that the latent hCMV genome is maintained as an episome associated with cellular histones in a nucleosome-like but not nucleosome-identical higher-order chromatin-like structure, and remodeling of viral chromatin is considered to have a critical role in regulating MIE locus activity (for a review see Bain et al. 2006; see the chapter by M. Reeves and J. Sinclair, this volume). Although little is known about the physical state of the latent mCMV genome, analogy suggests that it also exists as an episome associated with cellular histones. Obviously, MIE locus desilencing by local opening of the viral chromatin is a prerequisite for MIE gene transcription, and completion of the productive cycle for virus recurrence requires desilencing of all essential genes, at least temporarily. In theory, an essential locus does not need to remain open all the time, as transcripts and proteins are often quite stable. For instance, IE1 and IE2 mRNAs were found to have half-lives of more than 24 h and of 6 (5-7) h, respectively (Simon et al. 2007), so that a single window of open chromatin conformation at the MIE locus could possibly provide MIE mRNA sufficient for the whole productive cycle of approximately 24 h. Hummel et al. (2007) have used the spleen explant model of mCMV reactivation to show that inhibition of DNA methylation and inhibition of the histone deacetylase, the latter of which favors hyperacetylated loose chromatin, act synergistically in inducing *ie1* gene expression as well as virus recurrence. These findings support the conclusion that gene silencing and chromatin structure, as in hCMV, are also involved in the regulation of mCMV latency. Notably, at least under the conditions of tissue explant cultures, TNF-α signaling did not appear to play a crucial role (Hummel et al. 2007).

Dynamic Control of Latency at Immunological Checkpoints: The Immune Sensing Hypothesis

Besides epigenetic control of latency, it is likely that the host immune system will take note of antigens presented on the cell surface of latently infected cells following reactivated gene expression. An immune response may terminate reactivation at various steps prior to virion morphogenesis and release. Thus, although the immune system may not recognize latently infected cells in a state of complete transcriptional latency, it contributes to the maintenance of replicative latency by stopping the reactivation. An involvement of the host immune system in latency control was indicated early on by the clinically relevant hCMV reactivation and recurrence in the immunocompromised patient. Proof of concept was provided in the mouse model by mCMV recurrence after depletion of lymphocyte subsets involved in innate and adaptive immunity, with a hierarchy pointing to a pivotal role for CD8 T cells (Polic et al. 1998). Likewise, global ablation of cellular immunity by total-body γ-irradiation was found to facilitate virus recurrence in vivo (Balthesen et al. 1993; Kurz and Reddehase 1999; Kurz et al. 1997; Reddehase et al. 1994; Steffens et al. 1998).

Astoundingly, although virus recurrence became detectable in literally all lungs of latently infected mice after total-body γ-irradiation, detailed and quantitative statistical analysis of the patterns of transcriptional reactivation and virus recurrence revealed that by far most of the viral genomes still remained in a state of MIE locus latency, indicating that epigenetic control dominates over immune control. Loss of immune control, however, led to an increased number of progressing transcriptional reactivations with prevalences of IE1 transcripts > IE1 plus IE3 transcripts > IE1 plus IE3 plus M55 (gB) transcripts, but only a few transcriptional reactivations culminated in the recurrence of infectious virus (Kurz and Reddehase 1999). Thus, even in the absence of cellular immune control, recurrence is a rare end point of transcriptional reactivation.

The observation of recurrence after lymphocyte subset depletion or after γ-irradiation is not a formal proof for an immune control of latency, since in vivo cell depletion is associated with a cytokine storm, and since γ-irradiation represents genotoxic stress inducing DNA repair, which both could trigger virus recurrence by pathways unrelated to the ablation of cellular immune control.

That immune sensing of early stages of virus reactivation, specifically the recognition of MHC class I-presented antigenic peptides by CD8 T cells, may indeed be involved in the control of latency was first suggested in the BALB/c mouse model of pulmonary latency by the finding that activated, CD62L^{low} effector-memory CD8 T cells (CD8-T$_{EM}$) specific for the immunodominant L^d-presented IE1 protein-derived peptide 168-YPHFMPTNL-176 expand in latently infected lungs and accumulate there over time (Holtappels et al. 2000). This immunological finding in combination with the molecular finding of MIE locus gene expression during pulmonary latency (Kurz et al. 1999; Grzimek et al. 2001) led to the idea that presentation of the IE1 peptide in cells with desilenced MIE locus iteratively restimulates IE1 epitope-specific CD8-T$_{EM}$, which then terminate the reactivation event by virtue of their effector functions. This is called the *immune sensing hypothesis of latency control* (Fig. 2) (Holtappels et al. 2000; Simon et al. 2006a, 2006b).

Evidence in support of this hypothesis was provided recently with recombinant virus mCMV-IE1-L176A, in which antigenicity and immunogenicity of the IE1 peptide are specifically wiped out by the amino acid point mutation Leu replaced with Ala at the C-terminal MHC class I anchor residue. When compared with the corresponding revertant virus mCMV-IE1-A176L, the mutant virus showed a pronounced transcriptional phenotype in latently infected lungs in that the prevalence of IE1 transcripts was fivefold increased in accordance with the lack of IE1 epitope-specific antiviral effector cells (Simon et al. 2006a). This clearly implies that the true incidence of MIE locus desilencing is underestimated in lungs latently infected with wild type or epitope-rescued mCMVs due to the termination of a large fraction of MIE locus reactivation events by the IE1 epitope-specific CD8-T$_{EM}$.

Fig. 2 (continued) which terminate the transcription and expand clonally. This defines the first immunological checkpoint. **b** Viral gene transcription and immune sensing during latency of mutant virus mCMV-IE1-L176A. Replacing Leu with Ala at the C-terminal MHC anchor position of the IE peptide eliminates IE1 antigenicity and thereby inactivates this checkpoint. Transcription/splicing proceeds to IE3 (which does not contain a CD8 T cell epitope for *H-2^d*) and to the E-phase transcript m164, which specifies the MHC class I D^d-restricted peptide 257-AGPPRYSRI-265 defining a proposed second immunological checkpoint. **c** Transcriptional reactivation and virus recurrence after global immune cell depletion. Absence of immune sensing by CD8 T cells allows desilencing of viral genomes to proceed to open viral chromatin structure at all essential loci, including the origin of replication, OriLyt, which leads to recurrence of infectious virions (*virion symbol*). *NC* nuclear compartment, *CYC* cytoplasmic compartment, *ECC* extracellular compartment. *Black arrowhead*, position of ORF *m01*. *Green, yellow, blue, and red wavy lines* symbolize IE1, IE3, m164, and downstream further transcripts, respectively. The mutation in the IE1 transcript is marked by a *red dot*. *Green, blue, and red triangles* symbolize the corresponding antigenic peptides presented by the respective MHC class I molecules to cognate T cell receptors

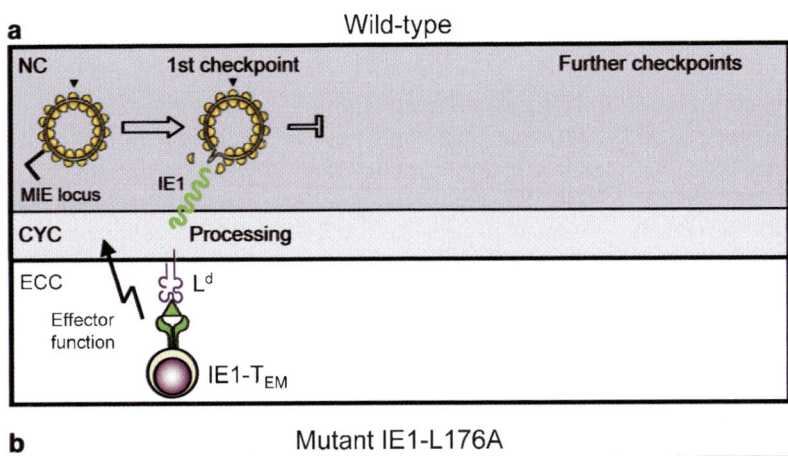
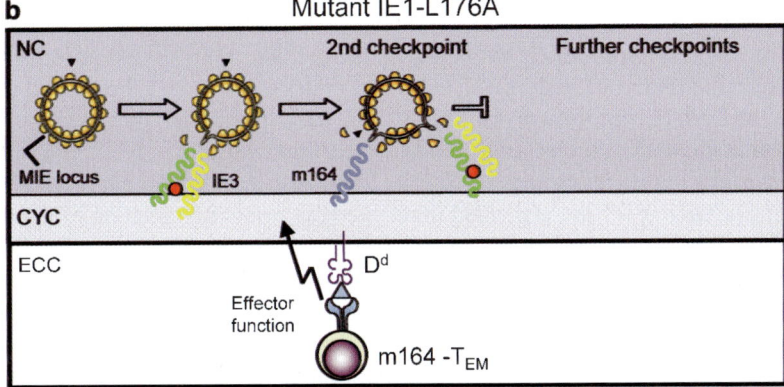
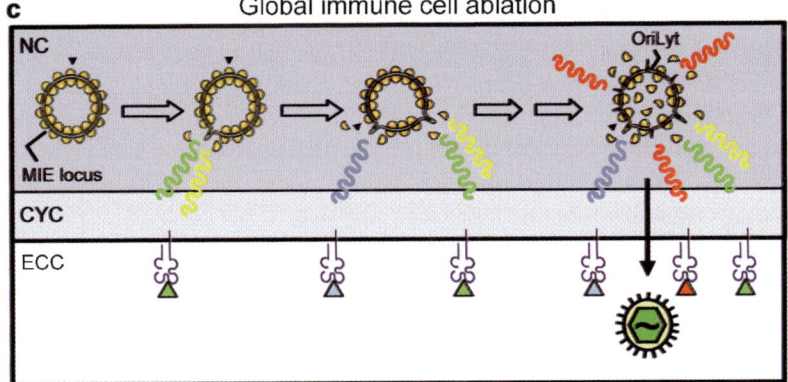

Fig. 2 Immune sensing hypothesis of cytomegalovirus latency control illustrated for pulmonary latency in the BALB/c (H-2^d haplotype) mouse model. **a** Viral gene transcription and immune sensing during latency of wild type mCMV. Most viral genomes, symbolized as episomes associated with cellular histones in a closed higher-order chromatin-like structure, are in a state of MIE locus silencing. Stochastic episodes of open viral chromatin structure at the MIE locus result in IE1 transcription, IE1 protein processing, and MHC class I L^d-restricted presentation of the IE1 peptide 168-YPHFMPTNL-176 to IE1 epitope-specific effector-memory CD8 T cells (IE1-T_{EM}),

Concluding Thoughts and Perspectives

Although transition to IE3 was observed with the mutant virus, reactivation did not proceed to M55 (gB) transcripts and virus recurrence. It is proposed that immunological checkpoints in series, defined by presentation of antigenic peptides, exist downstream of IE1. Expansion in latently infected lungs of CD8-T_{EM} specific for the D^d-presented peptide 257-AGPPRYSRI-265 (Holtappels et al. 2002; Simon et al. 2006a), which is derived from the E-phase protein m164/gp36.5, predicts a second immunological checkpoint, and further checkpoints may wait in line in case the first two checkpoints fail due to mutations in virus variants. Simultaneous expansion of IE1-specific and m164-specific CD8-T_{EM} during latency of wild type mCMV (Holtappels et al. 2002) predicted a leakiness of the IE1 checkpoint; nevertheless, we were so far unable to detect m164 mRNA during latency. This paradox might be explained by effective elimination of cells presenting the m164 epitope, which limits the chance for detection. A mutant virus with engineered deletion of both the IE1 and the m164 epitope, i.e., the recombinant virus mCMV-IE1-L176A+m164-I265A, should confirm the proposed second immunological checkpoint and might reveal the existence of a third one. Experiments in progress have shown that CD8 T cells specific for a newly defined epitope in ORF *m145* are expanded in the memory pool after ablation of epitopes IE1 and m164 (Holtappels et al., unpublished data), and viral transcription during pulmonary latency of this mutant is currently under investigation.

It is important to note that number and type of viral ORFs determining immunological checkpoints are not fixed but vary between individuals in reflection of MHC polymorphism in the population, depending upon the usually diploid set of peptide-presenting MHC class I molecules specified by the individual host's genome as well as on the proteome(s) of the latent virus variant(s) specifying the repertoire of potentially antigenic peptides. Thus, the model proposes a private pattern of immunological checkpoints for each latently infected individual.

A putative role for viral immunoevasins in canceling immunological checkpoints and facilitating virus recurrence is an intriguing aspect for future research in this field.

Acknowledgements The authors are supported by the Deutsche Forschungsgemeinschaft, Sonderforschungsbereich 490, individual project E2 "Immunological control of latent cytomegalovirus infection".

References

Adachi N, Lieber MR (2002) Bidirectional gene organization: a common architectural feature of the human genome. Cell 109:807–809

Angulo A, Ghazal P, Messerle M (2000) The major immediate-early gene *ie3* of mouse cytomegalovirus is essential for viral growth. J Virol 74:11129–11136

Bain M, Reeves M, Sinclair J (2006) Regulation of human cytomegalovirus gene expression by chromatin remodeling. In: Reddehase MJ (ed) Cytomegaloviruses: molecular biology and immunology. Caister Academic Press, Wymondham, Norfolk, pp 167–183

Balthesen M, Messerle M, Reddehase MJ (1993) Lungs are a major organ site of cytomegalovirus latency and recurrence. J Virol 67:5360–5366

Benedict CA, Angulo A, Patterson G, Ha S, Huang A, Messerle M, Ware CF, Ghazal P (2004) Neutrality of the canonical NF-kappaB-dependent pathway for human and murine cytomegalovirus transcription and replication in vitro. J Virol 78:741–750

Bevan IS, Sammons CC, Sweet C (1996) Investigation of murine cytomegalovirus latency and reactivation in mice using viral mutants and the polymerase chain reaction. J Med Virol 48:308–320

Brune W, Wagner M, Messerle M (2006) Manipulating cytomegalovirus genomes by BAC mutagenesis: strategies and applications. In: Reddehase MJ (ed) Cytomegaloviruses: molecular biology and immunology. Caister Academic Press, Wymondham, Norfolk, pp 63–89

Cardin RD, Abenes GB, Stoddard CA, Mocarski ES (1995) Murine cytomegalovirus IE2, an activator of gene expression, is dispensable for growth and latency in mice. Virology 209:236–241

Cavanaugh VJ, Deng Y, Birkenbach MP, Slater JS, Campbell AE (2003) Vigorous innate and virus-specific cytotoxic T-lymphocyte responses to murine cytomegalovirus in the submaxillary salivary gland. J Virol 77:1703–1717

Chatellard P, Pankiewicz R, Meier E, Durrer L, Sauvage C, Imhof MO (2007) The IE2 promoter/enhancer region from mouse CMV provides high levels of therapeutic protein expression in mammalian cells. Biotechnol Bioeng 96:106–117

Cook CH, Zhang Y, McGuinness BJ, Lahm MC, Sedmak DD, Ferguson RM (2002) Intra-abdominal bacterial infection reactivates latent pulmonary cytomegalovirus in immunocompetent mice. J Infect Dis 185:1395–1400

Cook CH, Trgovcich J, Zimmermann PD, Zhang Y, Sedmak DD (2006) Lipopolysaccharide, tumor necrosis factor alpha, or interleukin-1beta triggers reactivation of latent cytomegalovirus in immunocompetent mice. J Virol 80:9151–9158

Davis-Poynter NJ, Lynch DM, Vally H, Shellam GR, Rawlinson WD, Barrell BG, Farrell HE (1997) Identification and characterization of a G protein-coupled receptor homolog encoded by murine cytomegalovirus. J Virol 71:1521–1529

Dorsch-Häsler K, Keil GM, Weber F, Jasin M, Schaffner W, Koszinowski UH (1985) A long and complex enhancer activates transcription of the gene coding for the highly abundant immediate early mRNA in murine cytomegalovirus. Proc Natl Acad Sci U S A 82:8325–8329

Fiering S, Whitelaw E, Martin DIK (2000) To be or not to be active: the stochastic nature of enhancer action. BioEssays 22:381–387

Ghazal P, Messerle M, Osborn K, Angulo A (2003) An essential role of the enhancer for murine cytomegalovirus in vivo growth and pathogenesis. J Virol 77:3217–3228

Gribaudo G, Riera L, Lembo D, De Andrea M, Gariglio M, Rudge TL, Johnson LF, Landolfo S (2000) Murine cytomegalovirus stimulates cellular thymidylate synthase gene expression in quiescent cells and requires the enzyme for replication. J Virol 74:4979–4987

Grzimek NKA, Dreis D, Schmalz S, Reddehase MJ (2001) Random, asynchronous, and asymmetric transcriptional activity of enhancer-flanking major immediate-early genes *ie1/3* and *ie2* during murine cytomegalovirus latency in the lungs. J Virol 75:2692–2705

Holtappels R, Pahl-Seibert MF, Thomas D, Reddehase MJ (2000) Enrichment of immediate-early 1 (m123/pp89) peptide-specific CD8 T-cells in a pulmonary CD62L^{lo} memory-effector cell pool during latent murine cytomegalovirus infection of the lungs. J Virol 74:11495–11503

Holtappels R, Thomas D, Podlech J, Reddehase MJ (2002) Two antigenic peptides from genes *m123* and *m164* of murine cytomegalovirus quantitatively dominate CD8 T-cell memory in the *H-2^d* haplotype. J Virol 76:151–164

Hummel M, Abecassis MM (2002) A model for reactivation of CMV from latency. J Clin Virol 25:S123–S136

Hummel M, Zhang Z, Yan S, Deplaen I, Golia P, Varghese T, Thomas G, Abecassis MI (2001) Allogeneic transplantation induces expression of cytomegalovirus immediate-early genes in vivo: a model for reactivation from latency. J Virol 75:4814–4822

Hummel M, Yan S, Li Z, Varghese TK, Abecassis M (2007) Transcriptional reactivation of murine cytomegalovirus ie gene expression by 5-aza-2'-deoxycytidine and trichostatin A in latently infected cells despite lack of methylation of the major immediate-early promoter. J Gen Virol 88:1097–1102

Humphreys IR, de Trez C, Kinkade A, Benedict CA, Croft M, Ware CF (2007) Cytomegalovirus exploits IL-10-mediated immune regulation in the salivary glands. J Exp Med 204:1217–1225

Ishiwata M, Baba S, Kawashima M, Kosugi I, Kawasaki H, Kaneta M, Tsuchida T, Kozuma S, Tsutsui Y (2006) Differential expression of the immediate-early 2 and 3 proteins in developing mouse brains infected with murine cytomegalovirus. Arch Virol 151:2181–2196

Jonjic S, Mutter W, Weiland F, Reddehase MJ, Koszinowski UH (1989) Site-restricted persistent cytomegalovirus infection after selective long-term depletion of $CD4^+$ T-lymphocytes. J Exp Med 169:1199–1212

Jordan MC (1983) Latent infection and the elusive cytomegalovirus. Rev Infect Dis 5:205–215

Kulesza CA, Shenk T (2006) Murine cytomegalovirus encodes a stable intron that facilitates persistent replication in the mouse. Proc Natl Acad Sci U S A 103:18302–18307

Kurz SK, Reddehase MJ (1999) Patchwork pattern of transcriptional reactivation in the lungs indicates sequential checkpoints in the transition from murine cytomegalovirus latency to recurrence. 73:8612–8622

Kurz SK, Steffens HP, Mayer A, Harris JR, Reddehase MJ (1997) Latency versus persistence or intermittent recurrences: evidence for a latent state of murine cytomegalovirus in the lungs. J Virol 71:2980–2987

Kurz SK, Rapp M, Steffens HP, Grzimek NKA, Schmalz S, Reddehase MJ (1999) Focal transcriptional activity of murine cytomegalovirus during latency in the lungs. J Virol 73:482–494

Lagenaur LA, Manning WC, Vieira J, Martens CL, Mocarski ES (1994) Structure and function of the murine cytomegalovirus sgg1 gene: a determinant of viral growth in salivary gland acinar cells. J Virol 68:7717–7727

Lembo D, Gribaudo G, Hofer A, Riera L, Cornaglia M, Mondo A, Angeretti A, Gariglio M, Thelander L, Landolfo S (2000) Expression of an altered ribonucleotide reductase activity associated with the replication of murine cytomegalovirus in quiescent fibroblasts. J Virol 74:11557–11565

Li YY, Yu H, Guo ZM, Guo TQ, Tu K, Li YX (2006) Systematic analysis of head-to-head gene organization: evolutionary conservation and potential biological relevance. PLoS Comput Biol 2(7):e74 doi:10.1371/journal.pcbi.0020074

Meier JL, Stinski MF (2006) Major immediate-early enhancer and its gene products. In: Reddehase MJ (ed) Cytomegaloviruses: molecular biology and immunology. Caister Academic Press, Wymondham, Norfolk, pp 151–166

Messerle M, Keil GM, Koszinowski UH (1991) Structure and expression of murine cytomegalovirus immediate-early gene 2. J Virol 65:1638–1643

Polic B, Hengel H, Krmpotic A, Trgovcich J, Pavic I, Lucin P, Jonjic S, Koszinowski UH (1998) Hierarchical and redundant T-lymphocyte subset control precludes cytomegalovirus replication during latent infection. J Exp Med 188:1047–1054

Pollock JL, Virgin HW IV (1995) Latency, without persistence, of murine cytomegalovirus in the spleen and kidney. J Virol 69:1762–1768

Prösch S, Staak K, Stein J, Liebenthal C, Stamminger T, Volk HD, Krüger DH (1995) Stimulation of the human cytomegalovirus IE enhancer/promoter in HL-60 cells by TNFalpha is mediated via induction of NF-kappaB. Virology 208:197–206

Rawlinson WD, Farrell HE, Barrell BG (1996) Analysis of the complete DNA sequence of murine cytomegalovirus. J Virol 70:8833–8849

Reddehase MJ (2002) Antigens and immunoevasins: opponents in cytomegalovirus immune surveillance. Nat Immunol 2:831–844

Reddehase MJ, Balthesen M, Rapp M, Jonjic S, Pavic I, Koszinowski UH (1994) The conditions of primary infection define the load of latent viral genome in organs and the risk of recurrent cytomegalovirus disease. J Exp Med 179:185–193

Reddehase MJ, Podlech J, Grzimek NKA (2002) Mouse models of cytomegalovirus latency: overview. J Clin Virol 25:S23–S36

Simon CO, Seckert CK, Dreis D, Reddehase MJ, Grzimek NKA (2005) Role for tumor necrosis factor alpha in murine cytomegalovirus transcriptional reactivation in latently infected lungs. J Virol 79:326–340

Simon CO, Holtappels R, Tervo HM, Böhm V, Däubner T, Oehrlein-Karpi SA, Kühnapfel B, Renzaho A, Strand D, Podlech J, Reddehase MJ, Grzimek NKA (2006a) CD8 T cells control cytomegalovirus latency by epitope-specific sensing of transcriptional reactivation. J Virol 80:10436–10456

Simon CO, Seckert CK, Grzimek NKA, Reddehase MJ (2006b) Murine model of cytomegalovirus latency and reactivation: the silencing/desilencing and immune sensing hypothesis. In: Reddehase MJ (ed) Cytomegaloviruses: molecular biology and immunology. Caister Academic Press, Wymondham, Norfolk, pp 483–500

Simon CO, Kühnapfel B, Reddehase MJ, Grzimek NKA (2007) Murine cytomegalovirus major immediate early enhancer region operating as a genetic switch in bidirectional gene pair transcription. J Virol 81:7805–7810

Steffens HP, Kurz S, Holtappels R, Reddehase MJ (1998) Preemptive CD8 T-cell immunotherapy of acute cytomegalovirus infection prevents lethal disease, limits the burden of latent viral genomes, and reduces the risk of virus recurrence. J Virol 72:1797–1804

Tang Q, Maul GG (2006) Immediate-early interactions and epigenetic defense mechanisms. In: Reddehase MJ (ed) Cytomegaloviruses: molecular biology and immunology. Caister Academic Press, Wymondham, Norfolk, pp 131–149

Trinklein ND, Aldred SF, Hartman SJ, Schroeder DL, Otillar RP, Myers RM (2004) An abundance of bidirectional promoters in the human genome. Genome Res 14:62–66

Xiao J, Tong T, Zhan X, Haghjoo E, Liu F (2000) In vitro and in vivo characterization of a murine cytomegalovirus with a transposon insertional mutation at open reading frame M43. J Virol 74:9488–9497

Xu Y, Cei SA, Rodriguez Huete A, Colletti KS, Pari GS (2004) Human cytomegalovirus DNA replication requires transcriptional activation via an IE2-and UL84-responsive bidirectional promoter element within oriLyt. J Virol 78:11664–11677

Cytomegalovirus Immune Evasion

C. Powers, V. DeFilippis, D. Malouli, K. Früh(✉)

Contents

The Type I Interferon System and Its Effects on CMV Replication. 334
    Induction of IFN 334
    CMV Interference with the Induction of IFN and ISGs 335
    CMV Interference with JAK/STAT Signal Transduction 336
    CMV Interference with ISG Function 337
Interference with Antigen Presentation by Cytomegalovirus 337
    Inhibition of Antigen Presentation to CD8+ T Lymphocytes. 337
    MCMV as a Model System for CMV Interference with MHC-I Expression 340
    CMV Interference with T Cell Costimulation 341
Escaping the Natural Killer Cell Response 342
    MHC Class I Homologs and NK Cell Evasion Proteins 343
    The UL40 Protein 344
    NK Cell Modulation by UL14 and UL141 345
    Retention of NKG2D Ligands by UL16 345
    Downregulation of the NKG2D Ligand MICA by UL142 346
    Downregulation of Murine NKG2D Ligands by m138, m145, m152, and m155. 346
    Activation of Ly49H by m157 347
    The Rat Cytomegalovirus RCTL Protein. 348
Perspectives 348
References 349

Abstract Human cytomegalovirus (HCMV) has become a paradigm for viral immune evasion due to its unique multitude of immune-modulatory strategies. HCMV modulates the innate as well as adaptive immune response at every step of its life cycle. It dampens the induction of antiviral interferon-induced genes by several mechanisms. Further striking is the multitude of genes and strategies devoted to modulating and escaping the cellular immune response. Several genes are independently capable of inhibiting antigen presentation to cytolytic T cells by downregulating

---

K. Früh
Vaccine and Gene Therapy Institute, Oregon Health and Science University, Portland, OR 97201, USA
fruehk@ohsu.edu

MHC class I. Recent data revealed an astounding variety of methods in triggering or inhibiting activatory and inhibitory receptors found on NK cells, NKT cells, T cells as well as auxiliary cells of the immune system. The multitude and complexity of these mechanisms is fascinating and continues to reveal novel insights into the host–pathogen interaction and novel cell biological and immunological concepts.

The Type I Interferon System and Its Effects on CMV Replication

Induction of IFN

Interferons (IFNs) are powerful antimicrobial cytokines that induce the expression of numerous IFN-stimulated genes (ISGs). The actual antiviral effects are conferred by ISG products that target different stages of virus replication (Samuel 2001). CMV replication is susceptible to the effects of ISGs and IFN-induced cellular states in vitro (Gribaudo et al. 1993; Torigoe et al. 1993; Chin and Cresswell 2001; Sainz et al. 2005; DeFilippis et al. 2006) and in vivo for murine CMV (MCMV) (Yeow et al. 1998; Cull et al. 2002; Salazar-Mather et al. 2002).

IFN-dependent ISG transcription is induced by signal transduction triggered by IFNα or β binding to the type I IFN receptor, resulting in phosphorylation of tyrosine kinase 2 (Tyk2) and Janus kinase 1 (JAK1) that subsequently phosphorylate STAT (signal transducers and activators of transcription) 2 (Fig. 1). STAT2 phosphorylation leads to its heterodimerization with STAT1, association with IFN regulatory factor 9 (IRF9), and nuclear accumulation of the complex (termed IFN-stimulated gene factor 3; ISGF3). ISGF3 binds to IFN-stimulated response elements (ISREs) contained in the promoters of numerous ISGs, resulting in their transcriptional upregulation (Shuai and Liu 2003).

Induction of IFNβ itself (reviewed in Hiscott et al. 2003) occurs following stimulation of pattern recognition receptors (PRRs) and involves signaling separate from the JAK/STAT pathway. PRRs detect molecular components of pathogens such as LPS and double-stranded RNA (dsRNA). The best-studied PRRs, Toll-like receptors (TLRs), are expressed predominantly in immune surveillance cells such as DCs and macrophages (Martin and Wesche 2002; Kawai and Akira 2006). TLR activation triggers secretion of inflammatory cytokines and expression of co-stimulatory molecules that activate innate and adaptive immune responses (Takeuchi and Akira 2001). A separate class of cytoplasmic PRRs detects virus-specific molecules and is also capable of inducing IFNβ (Kato et al. 2005). These include retinoic acid inducible gene I (RIG-I) and melanoma differentiation-associated gene 5 (MDA-5) (Andrejeva et al. 2004; Yoneyama et al. 2004), which react to dsRNA (Gitlin et al. 2006; Kato et al. 2006). The recently identified receptor DAI also recognizes dsDNA (Takaoka et al. 2007). Two ubiquitous transcription factors, IFN regulatory factor 3 (IRF3) and nuclear factor kappa B (NFκB), are terminal in the PRR-triggered signaling cascades and are both required for transcription of the IFNβ gene. Both reside in the cytoplasm but accumulate in the nucleus upon activation, where they bind to the IFNβ promoter. IRF3 also induces a subset of ISGs (Wathelet et al. 1998; Hiscott et al. 2003).

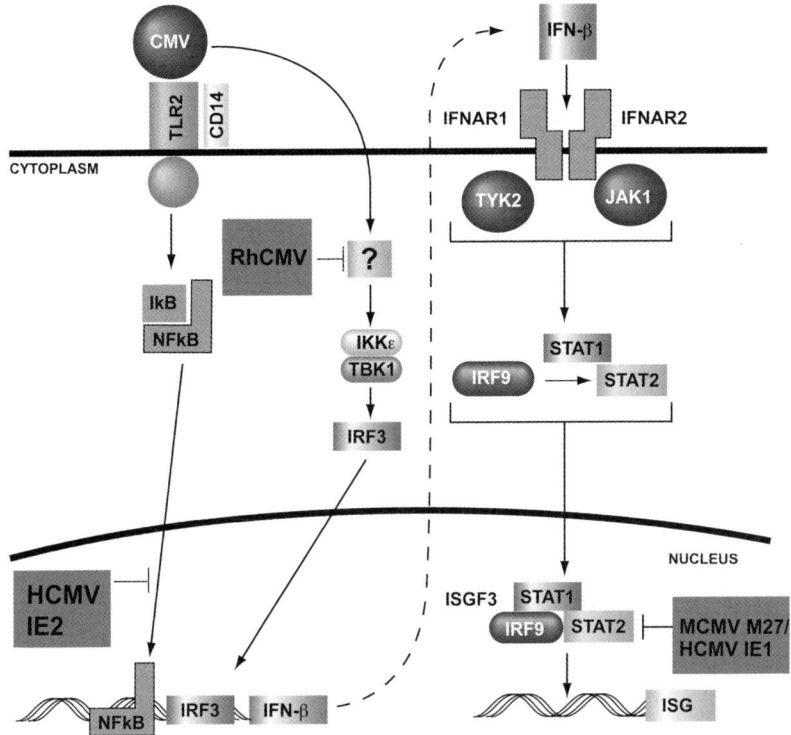

Fig. 1 Induction of IFNβ and ISG expression by CMV. Virus entry triggers the activation of NFκB and IRF3 by way of TLRs (NFκB) and an unknown pattern recognition receptor (IRF3). Following association with other proteins these transcription factors accumulate in the nucleus and bind to promoter elements upstream of IFNβ and specific ISGs, thereby stimulating their expression. HCMV IE2 is known to block binding of NFκB to DNA, while RhCMV has been shown to block IRF3 activation via and unknown mechanism. Expression and secretion of IFNβ further induces expression of ISGs by binding to the IFN receptor, which leads to dimerization of the receptor subunits 1 and 2, resulting in phosphorylation of tyrosine kinase 2 (Tyk2) and Janus kinase 1 (JAK). Tyk2 and JAK phosphorylate signal transducer and activator of transcription (STAT) 2. STAT2 phosphorylation leads to its heterodimerization with STAT1 followed by nuclear accumulation and association with IFN regulatory factor 9 (IRF9). This complex (termed IFN-stimulated gene factor 3; ISGF3) binds to IFN stimulated response elements (ISREs) upstream of numerous ISGs. Both the MCMV M27 protein and HCMV IE1 protein have been shown to impair DNA binding of this complex

CMV Interference with the Induction of IFN and ISGs

Despite their antiviral effects, HCMV infection strikingly activates IRF3 and NFκB, thus inducing ISGs and IFN (see the chapter by M.K. Isaacson et al., this volume). However, infections performed in the absence of viral gene expression induce these genes more strongly (Browne et al. 2001; Taylor and Bresnahan 2005; DeFilippis et al. 2006). Yet even UV-inactivated virus does not induce full activation of the IFN pathway (Browne et al. 2001; Simmen et al. 2001). These observations suggest that HCMV limits the induction of IFNβ and ISG transcription and perhaps even JAK/STAT signaling through the actions of viral products synthesized during infection or introduced with the virus particle. Unlike HCMV, in vitro infection with rhesus CMV (RhCMV) fails to induce innate antiviral gene transcription or activate IRF3 or NFκB (DeFilippis and Früh 2005). In addition, RhCMV infection prevents IRF3 activation by UV-inactivated HCMV particles (DeFilippis and Früh 2005). Interestingly, lack of ISG-induction is also observed when RhCMV particles were UV-inactivated or in the presence of cycloheximide, indicating that perhaps a component of the RhCMV virion is responsible for this interference.

Much attention has focused on components of the CMV virion tegument, a proteinaceous region composed of virus-encoded proteins located between the nucleocapsid and the envelope (Varnum et al. 2004). A major tegument constituent that has been shown to possess immunomodulatory capability is phosphoprotein pp65 (Odeberg et al. 2003; Arnon et al. 2005). Infection with a pp65-deleted HCMV (RVAd65) induces higher levels of IFNβ and ISGs than infection with wild type (WT) HCMV (Browne and Shenk 2003; Abate et al. 2004). However, the molecular basis for this increase remains controversial. Browne et al. showed increased nuclear translocation and DNA binding by NFκB and increased DNA binding by IRF1 following infection with RVAd65, whereas IRF3 activation was similar to WT. Abate et al. did not observe IRF3 activation by WT HCMV (contrary to numerous other reports), but infection with RVAd65 induced IRF3 nuclear accumulation, whereas NFκB was similarly activated by both viruses.

Recent work (Taylor and Bresnahan 2005, 2006a, 2006b) showed that the HCMV immediate early 86 kDa protein (IE2-86) inhibits transcription of IFNβ and RANTES by interfering with NFκB activation (Fig. 1) (Taylor and Bresnahan 2006a). The same group showed that deletion of the pp65 ORF (UL83) interferes with expression of the adjacent protein pp71 (ORF UL82), which is transcribed as a bicistronic mRNA with pp65 (UL83) (Ruger et al. 1987; Taylor and Bresnahan 2006b). Since pp71 is a positive regulator of IE promoters (Bresnahan and Shenk 2000; Cantrell and Bresnahan 2005), the increased ISG induction by RVAd65 is likely the result of reduced IE2-86 expression. In contrast, inserting stop codons into the pp65 ORF does not affect pp71 expression and the resulting virus does not differ from WT with respect to expression of IFNβ and RANTES (Taylor and Bresnahan 2006b).

CMV Interference with JAK/STAT Signal Transduction

HCMV has been shown to block IFNα-stimulated gene expression (Miller et al. 1999; Paulus et al. 2006) following viral gene expression and an initial ISG activation phase (Paulus et al. 2006). A recent study uncovered an association between HCMV IE 72 kDa protein and STAT2 that prevented DNA binding by

ISGF3 and subsequent ISG induction (Paulus et al. 2006). In MCMV, the M27 protein blocks both type I and type II IFN signaling by binding and degrading STAT2, a protein that, until this study, was not believed to be involved in IFNγ-dependent responses (Zimmermann et al. 2005). The importance of this phenotype is illustrated by the indispensability of M27 for MCMV growth in vivo (Abenes et al. 2001; Zimmermann et al. 2005). Interestingly, both HCMV and MCMV have been shown to target STAT2 albeit via different mechanisms.

CMV Interference with ISG Function

The antiviral effects of the IFN response are ultimately the result of IFN- or IRF3-induced host cell proteins. Protein synthesis is shut off during viral infection by protein kinase R (PKR) and 2'-5' oligoadenylate synthetase (OAS)/RNaseL (reviewed in Schneider and Mohr 2003). OAS and PKR are induced by type I and type II IFNs and the proteins are activated following exposure to viral dsRNA. Vaccinia virus (VV) prevents activation of PKR and OAS and VV lacking this function could be complemented by infection with live, but not UV-inactivated HCMV (Child et al. 2002). Subsequent work identified the HCMV proteins pTRS1 and pIRS1 to block OAS-mediated eIF2α phosphorylation and to reduce RNA degradation by RNase L (Child et al. 2004).

Interference with Antigen Presentation by Cytomegalovirus

Among the best studied immune modulators of CMVs are viral inhibitors of antigen presentation (VIPRs) (Johnson and Hegde 2002; Reddehase 2002; Yewdell and Hill 2002; Basta and Bennink 2003; Mocarski 2004; Reddehase et al. 2004; Pinto and Hill 2005). While nonessential in vitro, VIPRs might enable CMV to establish chronic infection despite a tremendous cellular immune response. The presence of multiple genes and mechanisms in all CMV species indicates that inhibiting antigen presentation is critical to the success of the virus. CMV affects both antigen presentation by major histocompatibility complex class I (MHC-I) and class II (MHC-II). Since modulation of MHC-II has been recently reviewed (Wiertz et al. 2007), we will focus on new findings on MHC-I downregulation.

Inhibition of Antigen Presentation to CD8⁺ T Lymphocytes

CD8⁺ T cells defend the host against viruses. They are able to kill infected cells upon recognizing virus-derived peptides displayed by MHC-I molecules at the cell surface.

MHC-I complexes are heterotrimers consisting of a 45-kDa type 1 transmembrane glycoprotein heavy chain (HC), the 12-kDa light chain β2-microglobulin (β2m), and a nine amino acid peptide. MHC-I-bound peptides are proteasomal breakdown products of both host and viral proteins. Proteolytic fragments are transported into the endoplasmic reticulum (ER) by the transporter associated with antigen presentation (TAP) and loaded onto MHC-I with help from the chaperones tapasin, calreticulin, Erp57, and protein disulfide isomerase (PDI).

HCMV encodes at least four VIPRs targeting MHC-I: US2, US3, US6, and US11 (see Fig. 2); single transmembrane, immunoglobulin (Ig) domain superfamily glycoproteins (Gewurz et al. 2001). US8 and US10 interact with MHC-I but they do not inhibit antigen presentation (Furman et al. 2002a; Huber et al. 2002; Tirabassi and Ploegh 2002). US2 and US11 retrotranslocate the HC from the ER to the cytosol for proteasomal degradation (reviewed in van der Wal et al. 2002). US3 binds MHC-I and causes ER retention (Ahn et al. 1996; Jones et al. 1996). US6 inhibits peptide transport and prevents ATP hydrolysis by TAP (Hewitt et al. 2001). Since the molecular function of these molecules has been reviewed extensively in

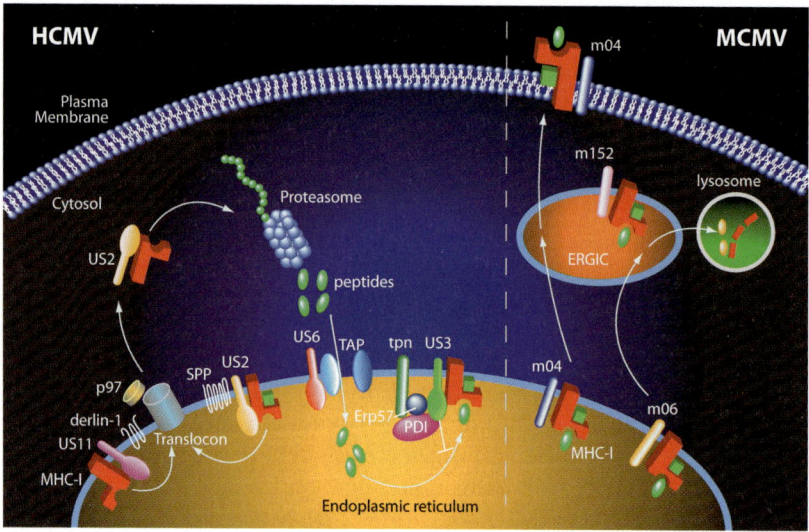

Fig. 2 CMVs encode multiple MHC-I modulators. An illustration of HCMV and MCMV mechanisms of MHC-I modulation. For details and references see the text. Both US2 and US11 cause the retrotranslocation of MHC-I heavy chains from the ER to the cytosol where they are degraded by the proteasome. Retrotranslocation is achieved by different mechanisms with US11 employing derlin-1, whereas US2 interacts with SPP. The transport of peptides via TAP is blocked by US6, inhibiting the ATP-hydrolysis step. US3 associates with the peptide loading complex and prevents optimal peptide loading in a tapasin (tpn)-dependent manner by causing the degradation of PDI. MCMV m04 forms a complex with MHC-I in the ER and at the cell surface, preventing T cell recognition. m06 redirects MHC-I to lysosomes for degradation. m152 retains MHC-I molecules in the ERGIC by an as yet unknown mechanism

the past, here we focus on recent findings regarding host molecules interacting with US2, US11, and US3, along with studies in MCMV characterizing the role of VIPRs in vivo.

US2 and US11 Cause Retrotranslocation of MHC-I Heavy Chains by Distinct Means

Both US2 and US11 interact with BiP (Hegde et al. 2006) and require a functional ubiquitin system (Hassink et al. 2006), but each has a distinct HLA allele specificity (van der Wal et al. 2002; Barel et al. 2006) and different requirements for function (Furman et al. 2002b). HC is ubiquitinated during US2, but not during US11-mediated degradation (Hassink et al. 2006), and each requires different cellular interactors for function (Lilley and Ploegh 2004; Hassink et al. 2006; Loureiro et al. 2006). HC dislocation by US11 is mediated by its transmembrane domain (Lilley et al. 2003), which contains a Gln residue essential for dislocation but not for the interaction with MHC-I (Lilley et al. 2003). Screening for cellular proteins interacting with US11 but not with the Gln-mutant identified Derlin-1, whose yeast homolog is required for the degradation of a subset of ER proteins (Lilley and Ploegh 2004). Independently, Derlin-1 was identified as a multiple transmembrane domain protein responsible for recruiting to the ER the cytosolic ATPase p97, a protein required for retrotranslocation (Ye et al. 2004). Both studies further proposed that Derlin-1 is a component of the retrotranslocation channel.

Interestingly, a dominant negative Derlin-1 failed to prevent dislocation by US2 (Lilley and Ploegh 2004). A screen for cellular proteins interacting with wild type but not dislocation-defective US2 implicated signal peptide peptidase (SPP) in HC dislocation by US2 but not US11 (Loureiro et al. 2006). While the cytosolic tail of US2 is required for SPP binding, it is not sufficient for dislocation since US2 containing the CD4 transmembrane domain was unable to cause dislocation. This indicates a necessary interaction between the US2 transmembrane domain and either SPP or some other protein (Loureiro et al. 2006). Thus, US2 and US11 might have evolved independently to achieve MHC-I destruction by different molecular means.

US3 Inhibits Optimal Peptide Loading

Recently two studies have further investigated the molecular mechanisms by which US3 retains MHC-I in the ER (Park et al. 2004, 2006). Both studies revealed that US3 prevented the optimization of peptide loading onto MHC-I heterodimers. Peptide loading is optimized by Tapasin, which forms a transient complex with empty MHC-I, and TAP and releases MHC-I peptide complexes (Schoenhals et al. 1999; Grandea and Van Kaer 2001; Purcell et al. 2001; Williams et al. 2002; Cresswell et al. 2005). The availability of MHC-I binding peptides regulates the duration of this transient complex resulting in fast (tapasin-independent) and slowly exiting (tapasin-dependent)

MHC-I alleles (Thammavongsa et al. 2006). US3 was shown to preferentially retain tapasin-dependent MHC-I alleles by inhibiting their acquisition of high-affinity peptides, whereas tapasin-independent alleles were not affected (Park et al. 2004). The same group recently identified a critical role of PDI in stabilizing the peptide-receptive site of MHC-I by regulating the oxidation of the α2 disulfide bond in the peptide-binding groove (Park et al. 2006). Interestingly, PDI protein levels were decreased in the presence of US3 and a complex between US3 and PDI is stabilized by proteasome inhibitors. By degrading PDI, US3 inhibits the binding of high-affinity peptides to tapasin-dependent alleles of MHC-I. Since PDI and tapasin are part of the peptide loading complex and can be co-immunoprecipitated, it is likely that previously observed interactions between US3 and MHC-I or tapasin are the result of US3 entering the peptide loading complex (Park et al. 2006).

MCMV as a Model System for CMV Interference with MHC-I Expression

MCMV does not encode homologs of the HCMV VIPRs, but encodes its own VIPRs m04, m06, and m152 (see Fig. 2; recently reviewed in Reddehase et al. 2004; Pinto and Hill 2005). The gp34 protein encoded by m04 does not reduce MHC-I surface levels but forms a tight association in the ER and accompanies MHC-I to the cell surface (Kleijnen et al. 1997; Kavanagh et al. 2001) where it is able to inhibit cytotoxic T cell lysis by an unknown mechanism (Pinto and Hill 2005). The m06-encoded gp48 associates with MHC-I and directs this complex to the lysosomes where both are destroyed (Reusch et al. 1999). Lastly, m152 encodes gp40, which retains MHC-I in the ERGIC (Ziegler et al. 1997). Interestingly, MHC-I retention occurs in the absence of a detectable biochemical interaction (Ziegler et al. 1997; Pinto and Hill 2005). Since HCMV cannot infect immunocompetent experimental animals, the MCMV VIPR system has been used to determine the role of MHC-I inhibitors in the context of CMV infection in vivo.

Initial studies in immunocompromised mice have suggested a role for m152/gp40 in controlling the CD8+ T cell response to the virus and being responsible for increased viral titers (Krmpotic et al. 1999). It was further demonstrated that m152/gp40 protected MCMV from adoptively transferred epitope-specific T cells (Holtappels et al. 2004), the first (and so far only) experiment showing a VIPR preventing viral peptide presentation in vivo. In contrast, the m152/gp40-deleted virus induced a CD8+ T cell response to the immunodominant M45 epitope that was similar to WT-MCMV in C57BL/6 mice (Gold et al. 2002). Moreover, the CD8+ T cell response to MCMV lacking all three VIPRs was very similar to that induced by WT (Gold et al. 2004; Munks et al. 2007). An important conclusion from these studies was that VIPRs do not seem to influence the induction of CD8+ T cell responses, suggesting that passive presentation of infected cells (cross-presentation) rather than direct presentation of virus-derived peptides by infected cells induces the T cell response.

Unexpectedly, the genome copy numbers of the triple-deleted virus were comparable to WT MCMV during the acute phase of infection, although this could be explained by NK cell control early in infection prior to the robust CD8+ T cell response (Gold et al. 2004). Even more surprising was the ability of the triple knockout to establish infection for at least 6 weeks, after which it was able to reactivate upon immunosuppression the mice (Gold et al. 2004). Perhaps the redundancy of immune evasion mechanisms, most notably the modulation of T cell and NK cell activating signals (see Sects. 2.3 and 3 below), enables this deletion virus to survive. It is also possible that VIPRs facilitate transmission since salivary gland titers of the triple-deleted virus were lower than WT virus (Lu et al. 2006). It is further conceivable that VIPRs are required for superinfection of CMV-immune individuals (Booth et al. 1993; Boppana et al. 2001; Rizvanov et al. 2003). Finally, the use of inbred laboratory mice might not accurately reflect the infection and spread of CMV in an outbred population. Thus, additional studies of animal CMVs in outbred populations might help to establish the role of VIPRs in vivo.

CMV Interference with T Cell Costimulation

HCMV UL144

UL144 encodes the only known tumor necrosis factor receptor superfamily member (TNFRSF) among herpesviruses (Locksley et al. 2001; Croft 2003; Ware 2003) (TNFR-homologs are widespread in poxviruses). UL144 shows strong sequence similarity to the herpes simplex virus entry mediator (HVEM) (Benedict et al. 1999). Recently, both HVEM and UL144 have been shown to interact with B and T lymphocyte attenuator (BTLA), a member of the Ig superfamily that negatively regulates T cell proliferation (Cheung et al. 2005). The interaction with BTLA suggests that UL144 mimics the inhibitory co-signaling function of HVEM. UL144 was further shown to induce NFκB-dependent transcription by sequestering TRAF6 (Poole et al. 2006). The UL144 gene shows significant strain-specific variability of up to 21% difference in the nucleotide as well as the amino acid sequence (Lurain et al. 1999; Arav-Boger et al. 2002). The polymorphism of UL144 was investigated in the context of congenital CMV disease. Although some groups reported a significant association between the UL144 subtype and the outcome of the CMV infection (Arav-Boger et al. 2002, 2006; He et al. 2004; Tanaka et al. 2005), there are other reports claiming the opposite (Bale et al. 2001; Murayama et al. 2005; Picone et al. 2005; Mao et al. 2007). Interestingly, a natural variant of RhCMV, strain 180.92, was recently reported to lack UL144, which is present in strain 68.1 (Hansen et al. 2003; Rivailler et al. 2006). Natural UL144 variants as well as UL144 deletion might enable in vivo studies to address the role of UL144 in immune evasion and CMV disease.

MCMV m147.5 and m138

MCMV was recently shown to downregulate two co-stimulatory molecules in T cell activation. B7.1 (CD80) is targeted to lysosomes by the Fc receptor m138 (Mintern et al. 2006) and B7.2 (CD86) cell surface expression is reduced by m147.5 (Loewendorf et al. 2004). m147.5 encodes a predicted 23-kDa type IIIb transmembrane protein named modB7.2, which is encoded in two exons in ORF m147 and complementary to ORF m149. m147.5 is highly conserved among different MCMV strains (Smith et al. 2006), which correlates with B7.2 conservation. m138 encodes a type 1 glycoprotein Fc receptor measured at 75–80 kDa (Mintern et al. 2006). Also implicated in modulating NKG2D ligand expression (see above in this section), m138 is able to redirect B7.1 to lysosomes independent of its cytoplasmic tail and transmembrane domain (Mintern et al. 2006). It is interesting to note that although deletion of the Fc receptor attenuates MCMV in vivo, this is not due to antibody control (Crnkovic-Mertens et al. 1998). Rather, these studies suggest that it may be due to a combination of other evasion strategies by m138. Downregulation of the B7 costimulatory family has so far not been reported for primate CMVs that do not contain m147.5 or m138 homologs, although HCMV does contain two distinct Fcγ receptors (Atalay et al. 2002).

Escaping the Natural Killer Cell Response

While the epitope-specific adaptive T cell response is central to the long-term immune control of HCMV, natural killer cells (NK cells) are important during primary infection prior to the onset of the adaptive immune response (reviewed by Tay et al. 1998). There is also some evidence that humans with defects in their NK cell response are extremely susceptible to infections by herpesviruses (Biron et al. 1989). NK cells are regulated by the integration of activatory and inhibitory signals generated by a set of cell surface receptors that is unique to each NK cell clone (reviewed by Lanier 1998). In primates, the killer cell immunoglobulin-like receptor (KIR) family binds to HLA-A,B,C, and G. KIRs are composed of activating and inhibitory molecules and the clonotypic array of receptors determines the specificity of each NK cell clone with regard to their MHC-I recognition. Mice do not have KIR receptors. Instead, KIR-analogous functions are performed by the Ly49 family of lectin-like receptors. A general rule is that NK cells cannot kill cells expressing a full complement of autologous MHC-I allotypes, but can kill cells lacking MHC-I (missing self). As a result of MHC-I downregulation, CMV-infected cells should be excellent targets for NK cells. Indeed, increased lysis of HCMV-infected cells, as well as cells transfected with viral MHC-evasion genes, can be observed for some human NK cell clones (Huard and Fruh 2000). However, HCMV has also devised clever strategies to prevent NK cell activation.

MHC Class I Homologs and NK Cell Evasion Proteins

One strategy used by CMVs to prevent missing self is to express decoy MHC-I like molecules that trigger inhibitory NK cell receptors.

The MHC-I homology of UL18 was originally discovered during the sequencing of the AD169 genome (Beck and Barrell 1988). UL18 was later shown to bind β2-microglobulin (Browne et al. 1990) and endogenous peptides (Fahnestock et al. 1995). This ORF is conserved in chimpanzee CMV (ChCMV) (Davison et al. 2003), but not in rhesus CMV (RhCMV) (Hansen et al. 2003). UL18 is a ligand for the leukocyte immunoglobulin-like receptor 1 (LIR-1) (Cosman et al. 1997), an inhibitory receptor that is widely expressed in macrophages, dendritic cells, as well as subsets of NK cells (Colonna 1998), and is the only LIR family member expressed by T cells (Borges and Cosman 2000). LIR-1 is highly expressed on HCMV-specific cytotoxic T lymphocytes (Antrobus et al. 2005; Northfield et al. 2005) and on lymphocytes in lung-transplanted patients weeks before development of CMV disease (Berg et al. 2003). The widespread expression of LIR-1 suggests multiple possible functions of UL18, but also complicates the interpretation of results obtained with UL18-deleted viruses or UL18-transfected cells. LIR-1 interacts with the α3 domain of MHC-I (Willcox et al. 2003) but binds to UL18 1,000-fold stronger than to MHC-I (Borges et al. 1997; Cosman et al. 1997; Chapman et al. 1999). However, the UL18 gene shows substantial strain variations that differ in their affinity to LIR-1 (Vales-Gomez et al. 2005). UL18 is expressed on the surface of HCMV-infected cells (Park et al. 2002; Griffin et al. 2005) and escapes the MHC-retaining or -degrading functions of the US6 family (Park et al. 2002). However, initial results that UL18 expression inhibited NK cell lysis through the lectin-like receptor CD94 (Reyburn et al. 1997) were later disputed since UL18 actually increased the susceptibility of infected or transfected fibroblasts to NK cell killing (Leong et al. 1998; Odeberg et al. 2002). Moreover, the response of CD94/NKG2C$^+$ NK cells was independent of UL16, UL18, and UL40, as well as 19 clinical strain-specific viral genes, but was impaired when cells were infected with a mutant lacking the US2–11 gene region (Guma et al. 2006). However, fibroblasts infected with an HCMV UL18 deletion mutant exhibited enhanced susceptibility to NKL killing (Robertson et al. 1996) relative to cells infected with the parental virus (Prod'homme et al. 2007). While this study supports a role of UL18 in NK evasion, others reported a T cell stimulatory function of UL18. Co-incubation of PMBC from CMV-seropositive donors with virus-infected lung fibroblasts leads to T cell-dependent secretion of IFN-γ, which was significantly reduced when the lung fibroblasts were infected with a UL18 deletion mutant (Wagner et al. 2007). Moreover, resting and activated CD8$^+$ T cells lysed UL18-expressing cells irrespective of their Ag-specificity, in a non-MHC-restricted fashion, whereas cells infected with CMV defective for UL18 were not killed (Saverino et al. 2004). Thus, the role of UL18 in viral immune evasion is still unclear.

Sequencing of the MCMV genome also revealed an MCMV-MHC I homolog, m144, (Rawlinson et al. 1996), which also binds to β2-microglobulin but does not seem to bind peptides (Chapman and Bjorkman 1998) due to a misfolded peptide binding domain (Natarajan et al. 2006). m144 is polymorphic (Smith et al. 2006) and partially inhibits NK cell responses in vitro (Kubota et al. 1999). A role of m144 in NK cell evasion is supported by the finding that replication of m144-deleted MCMV is restricted in vivo, but normal upon NK cell depletion (Farrell et al. 1997). However, the receptor for m144 is currently unknown.

The UL40 Protein

The inhibitory CD94/NKG2A receptor recognizes the nonpolymorphic human HLA-E or mouse Qa-1b molecule (Lee et al. 1998). HLA-E displays signal sequence peptides of classical MHC-I molecules at the cell surface, thus allowing NK cells to monitor MHC-I expression in parallel to interaction of MHC-I with the KIR/Ly49 family (Braud et al. 1997; Vance et al. 1998; Ulbrecht et al. 2000). Remarkably, the glycoprotein UL40 of HCMV encodes a signal sequence that is identical to that of classical HLA-C molecules and can thus be presented by HLA-E (Tomasec et al. 2000; Ulbrecht et al. 2000). Loading of the UL40 signal peptide is independent of TAP, whereas the loading of HLA-signal sequences is TAP-dependent. Thus, HCMV ensures surface expression of high levels of peptide-loaded HLA-E, even upon TAP inhibition by US6 and MHC-I destruction by US2-US11. In vitro experiments showed that fibroblasts infected with wild type, but not UL40-deleted HCMV were protected against lysis by CD94/NKG2A-positive NK cell lines as well primary NK cells (Wang et al. 2002). Thus it was concluded that HCMV counteracts NK cell activation by the US6 family, at least for NKG2A-expressing NK cell clones. However, this was disputed by another study that found increased recognition of US2–11-deleted HCMV regardless of the presence or absence of UL40 (Falk et al. 2002). Moreover, it was recently shown that NK-CTLs (CD3$^+$CD8$^+$TCRαβ$^+$) recognize peptides in an HLA-E restricted fashion (Pietra et al. 2001; Romagnani et al. 2002, 2004; Moretta et al. 2003). NK-CTLs efficiently killed CMV-infected cells by recognizing the HLA-E presented UL40-derived peptide VMAPRTLIL (identical to HLA-C) resulting in the induction of cytotoxicity, IFN-γ production, and cell proliferation (Pietra et al. 2003; Mazzarino et al. 2005). Strain variants of UL40 further encode other HLA-E-binding peptides (Tomasec et al. 2000; Cerboni et al. 2001). The VMAPRTLIL-sequence is identical in chimpanzee and gorilla MHC-I, but it is not known whether the slightly altered sequence (TMAPKTLLI) of the ChCMV UL40 homolog can serve as a decoy for HLA-E. There is currently no evidence that MCMV or RhCMV displays a similar evasion mechanism.

NK Cell Modulation by UL14 and UL141

UL14 and UL141 show significant homology and have been grouped into the UL14 family (Davison et al. 2003). UL141 resides in the ER and retains the poliovirus receptor (CD155) (Tomasec et al. 2005), a ligand for the activating NK receptors DNAM-1 (CD226) and TACTILE (CD96). UL141 showed specificity for CD155 in that it did not influence the expression of nectin-2 (CD112), another reported ligand for CD226 (Bottino et al. 2003).

UL14 also encodes an EndoH-sensitive, ER-resident glycoprotein that provides efficient protection to NK cell-mediated attack by NK cell lines and interferon-activated bulk cultures in allogeneic and autologous assays. Although exhibiting homology to UL141, the mechanism of UL14-mediated protection seems to be different from UL141. In contrast to several other NK cell evasion molecules, the UL14 family is highly conserved among different HCMV strains (Ma et al. 2006) (although UL141 is often lost in laboratory isolates). Moreover, homologous proteins are found in both human and nonhuman primate CMVs (Davison et al. 2003; Hansen et al. 2003). This conservation suggests that retaining the respective NK cell ligands is important for the pathogenesis of primate CMVs.

Retention of NKG2D Ligands by UL16

During a search for ligands of the nonessential HCMV-glycoprotein UL16, MIC-B, a nonpolymorphic, stress-induced MHC-I-like protein, but not MIC-A, was observed to bind to UL16 (Cosman et al. 2001). In addition, two members of a novel family of GPI-linked proteins, termed UL-binding proteins (ULBP1 and ULBP2) were observed to bind to UL16 (Kubin et al. 2001), whereas the other two members of that family (ULBP3 and ULBP4) did not (Rolle et al. 2003; Welte et al. 2003). The specificity of UL16 in binding only some of these closely related family members seems to be due to sequence differences in their $\alpha 2$ domain, which could be shown for MIC-A and MIC-B (Spreu et al. 2006). ULBPs show a low homology to MHC-I as well as to MIC. However, they lack the $\alpha 3$ domain and do not bind $\beta 2m$. Importantly, ULBP1 and ULBP2 bind to NKG2D and thus seem to be able to stimulate NK cell as well as T cell activity in a manner that is similar to MIC (Sutherland et al. 2002). This binding can be inhibited by UL16, which counteracts the cell surface expression of the NKG2D ligands by retaining them in the ER (Dunn et al. 2003; Rolle et al. 2003; Welte et al. 2003; Wu et al. 2003). ER localization of UL16 requires the transmembrane and cytoplasmic domains (Vales-Gomez and Reyburn 2006), while the ectodomain is associated with ULBP1, ULBP2, and MICB.

Downregulation of the NKG2D Ligand MICA by UL142

The HCMV MHC-class-I-like molecule UL142, which has a homolog in ChCMV but not in RhCMV (Davison et al. 2003; Wills et al. 2005), inhibits NK cell lysis by downregulating MIC-A (Wills et al. 2005; Chalupny et al. 2006). Interestingly, only the full-length alleles of MIC-A are targeted by the viral protein, whereas the MICA 008 allele, which has a truncated transmembrane region, is not affected by UL142 (Zou et al. 2005; Chalupny et al. 2006). MIC-A is the ligand of the activating NK cell receptor NKG2D, a lectin-like receptor that is not only expressed on NK cells, but also on $CD8^+$ T cells as well as $\gamma\delta^+$ T cells (Bauer et al. 1999). Since NKG2D exists only as an activating receptor, the missing self principle does not apply given that stimulation of NK cells is observed even if a full set of MHC I molecules is expressed on the target cells (Bauer et al. 1999). Thus, downregulation of MIC-A increases the activation threshold of NKG2D-expressing NK cells and T cells.

Downregulation of Murine NKG2D Ligands by m138, m145, m152, and m155

MIC proteins are not conserved in rodents. The NKG2D ligands identified in mice include retinoic acid early inducible gene-1 (RAE-1) (Cerwenka et al. 2000), which has five different isoforms (RAE-1α, β, γ, δ, and ϵ), the minor histocompatibility antigen H60 (Malarkannan et al. 1998; Diefenbach and Raulet 2003) and the murine UL16-binding protein-like transcript (MULT)-1 glycoprotein (Carayannopoulos et al. 2002; Diefenbach et al. 2003). All three of these ligands are targeted by MCMV to prevent NK cell activation through NKG2D. Three of the four MCMV genes identified thus far that act on NKG2D ligands are members of the m145 gene family (m145, m152, and m155). An initial report of the classical MHC-evasion gene m152 playing a role in NK cell evasion in vivo (Krmpotic et al. 2002) led to the discovery that m152 reduced surface expression of all five isoforms of RAE-1, but not H60 (Lodoen et al. 2003). H60 was later shown to be targeted for downregulation by the m155 gene (Lodoen et al. 2004; Hasan et al. 2005) and by the Fc receptor m138 (Lenac et al. 2006). Lastly, MULT-1 is also targeted by m138 (Lenac et al. 2006) as well as m145 (Krmpotic et al. 2005). Importantly, each study showed that viruses that independently lack m138, m145, m152, or m155 have an NK cell-dependent attenuation in vivo compared to wild type and revertant viruses. The fact that MCMV targets each of the NKG2D ligands and that viruses lacking any one of these modulators is attenuated in vivo underlines the critical importance of this innate response to viral infection. Interestingly, m145 and m155 display significant sequence variations among different MCMV strains, while the protein m152 is highly conserved, which could reflect polymorphism of their respective targets (Smith et al. 2006).

Activation of Ly49H by m157

C57BL/6 mice are relatively resistant to MCMV, whereas Balb/c mice are more susceptible due to the cmv-1 genetic locus (Scalzo et al. 1990) encoding the Ly49 NK-cell receptor complex (Scalzo et al. 1995). Further mapping revealed that the cmv-1 resistance trait corresponded to the Ly49H NK cell receptor gene (Brown et al. 2001; Lee et al. 2001; Scalzo 2002; Dimasi and Biassoni 2005) and depletion of NK cells expressing Ly49H-enhanced MCMV titers in the resistant C57BL/6 mouse strain (Daniels et al. 2001). Of note, the clearance of MCMV by Ly49H NK cells cannot be overcome by the effects of m152 (Krmpotic et al. 2002). Since Ly49H is an activating receptor, it seemed plausible that the ligand was expressed on MCMV-infected cells. It turned out that Ly49H directly recognizes a viral protein (Arase et al. 2002; Smith et al. 2002), the ORF m157 encoding a predicted GPI-linked protein with structural homology to nonclassical MHC molecules. m157 is a ligand for both the activating receptor Ly49H and inhibitory receptor Ly49I (Arase et al. 2002; Smith et al. 2002). Thus, the effects of m157 expression depend on Ly49 receptor expression, which varies in mouse strains. Ly49 homodimerization also seems to be crucial for the interaction with m157, which is not the case for the interaction with MHC class I (Kielczewska et al. 2007). Deletion of the m157 gene from the MCMV genome results in a virus that is less susceptible to the host immune system compared to WT in Ly49H+ mice (Bubic et al. 2004). m157 is highly variable in wild mice and several of these isolates are unable to bind to Ly49H (Voigt et al. 2003).

Using 3D homology searches, Smith et al. predict that as many as 12 open reading frames in the genome of MCMV show limited homology to nonclassical MHC molecules (Arase et al. 2002; Smith et al. 2002). Moreover, MCMV mutants lacking genes of the m02 gene family encoding additional putative type I membrane glycoproteins were attenuated but grew normally upon NK cell depletion (Oliveira et al. 2002). Thus, it seems that ligands for activating and inhibitory NK cell receptors might be a widespread feature of the MCMV genome and that many of the unknown open reading frames encoding glycoproteins might encode immune-regulatory proteins. In fact, stocks of wild-derived inbred mice are naturally resistant to MCMV due to NK cells, and multiple loci are responsible for this resistance. One NK cell-dependent viral resistance mechanism was mapped to Cmv4, which most likely encodes for a new NK activating receptor (Adam et al. 2006). Non-Ly49-depedendent MCMV resistance was also reported for NZ mice (Rodriguez et al. 2004). In addition, MCMV resistance of MA/My mice was mapped to H2k-linked nonclassical MHC-I genes (Dighe et al.; Xie et al. 2007). Thus, CMVs might encode multiple, polymorphic NK cell evasion molecules to counteract the natural resistance mediated by a host of activation receptor combinations.

The Rat Cytomegalovirus RCTL Protein

Within the numerous NK- and NKT-cell receptors, the c-type lectin-like NKP-P1 family recognizes members of the Clr-family (Iizuka et al. 2003). Clr-b, which is expressed on a wide range of cells, serves as a ligand for NKR-P1 (Carlyle et al. 2004). This interaction transmits an inhibitory NK cell signal. Recently Hao et al. were able to show that infection with RCMV leads to a nearly complete loss of Clr-b expression (Hao et al. 2006), which should normally lead to an enhanced susceptibility of the host cell to NK cell-mediated killing. Interestingly, this is not the case. In vivo studies with a RCTL knockout mutant of RCMV further showed that the virus displays a diminished virulence in a strain-dependent manner controlled by host NKR-P1 polymorphism (Voigt et al. 2007). RCTL exhibits homology to Clr-b (Voigt et al. 2001; Mesci et al. 2006), and it is most likely that it serves as a decoy for NKR-P1, sustaining the inhibitory NK cell signaling.

Perspectives

CMVs encode more than 100 genes that are nonessential for growth in vitro (Yu et al. 2003) and hence are likely to modulate the virus–host interaction in vivo. This list includes all of the immune modulatory genes discussed here, but also a number of genes with unknown function. Thus, one of the continuing goals of research in immune modulation of HCMV will be to determine the function of nonessential ORFs. However, defining the effects of unknown ORFs on the host cell or host organisms can be daunting, particularly since HCMV does not infect immunocompetent animals. These obstacles can be, at least partially, overcome by using high throughput technologies to identify host protein interacting with or destroyed by CMV ORFs. Examples are quantitative proteomics technologies (Bartee et al. 2006), yeast two-hybrid screening (Uetz et al. 2006), or protein profiling using TAP tags (Holowaty et al. 2003). Additional information can be obtained by studying the role of homologous ORFs in murine, rat, and rhesus CMV in acute and latent infection. Since many of the HCMV immune modulators have homologs in the RhCMV genome, this emerging model will be particularly useful to provide new information on unknown ORFs. In addition, the RhCMV model will be used to study the function of homologs of HCMV immune modulators with known functions in vitro, but not in vivo (Lockridge et al. 2000; Pande et al. 2005).

The demonstration that viral microRNAs (miRs) inhibit the expression of the NKG2D-ligand MicB (Stern-Ginossar et al. 2007) recently also implicated noncoding parts of the viral genome in immune evasion. Since HCMV encodes a panel of miRs (see the chapter by P.J.F Rider et al., this volume), it can be expected that additional host cell miR targets involved in immune stimulation will be discovered. Thus, the immune evasion toolbox of CMVs still holds many surprises.

References

Abate DA, Watanabe S, Mocarski ES (2004) Major human cytomegalovirus structural protein pp65 (ppUL83) prevents interferon response factor 3 activation in the interferon response. J Virol 78:10995–11006

Abenes G, Lee M, Haghjoo E, Tong T, Zhan X, Liu F (2001) Murine cytomegalovirus open reading frame M27 plays an important role in growth and virulence in mice. J Virol 75:1697–1707

Adam SG, Caraux A, Fodil-Cornu N, Loredo-Osti JC, Lesjean-Pottier S, Jaubert J, Bubic I, Jonjic S, Guenet JL, Vidal SM, Colucci F (2006) Cmv4, a new locus linked to the NK cell gene complex, controls innate resistance to cytomegalovirus in wild-derived mice. J Immunol 176:5478–5485

Ahn K, Angulo A, Ghazal P, Peterson PA, Yang Y, Fruh K (1996) Human cytomegalovirus inhibits antigen presentation by a sequential multistep process. Proc Natl Acad Sci U S A 93:10990–10995

Andrejeva J, Childs KS, Young DF, Carlos TS, Stock N, Goodbourn S, Randall RE (2004) The V proteins of paramyxoviruses bind the IFN-inducible RNA helicase, mda-5, and inhibit its activation of the IFN-beta promoter. Proc Natl Acad Sci U S A 101:17264–17269

Antrobus RD, Khan N, Hislop AD, Montamat-Sicotte D, Garner LI, Rickinson AB, Moss PA, Willcox BE (2005) Virus-specific cytotoxic T lymphocytes differentially express cell-surface leukocyte immunoglobulin-like receptor-1, an inhibitory receptor for class I major histocompatibility complex molecules. J Infect Dis 191:1842–1853

Arase H, Mocarski ES, Campbell AE, Hill AB, Lanier LL (2002) Direct recognition of cytomegalovirus by activating and inhibitory NK cell receptors. Science 296:1323–1326

Arav-Boger R, Willoughby RE, Pass RF, Zong JC, Jang WJ, Alcendor D, Hayward GS (2002) Polymorphisms of the cytomegalovirus (CMV)-encoded tumor necrosis factor-alpha and beta-chemokine receptors in congenital CMV disease. J Infect Dis 186:1057–1064

Arav-Boger R, Battaglia CA, Lazzarotto T, Gabrielli L, Zong JC, Hayward GS, Diener-West M, Landini MP (2006) Cytomegalovirus (CMV)-encoded UL144 (truncated tumor necrosis factor receptor) and outcome of congenital CMV infection. J Infect Dis 194:464–473

Arnon TI, Achdout H, Levi O, Markel G, Saleh N, Katz G, Gazit R, Gonen-Gross T, Hanna J, Nahari E, Porgador A, Honigman A, Plachter B, Mevorach D, Wolf DG, Mandelboim O (2005) Inhibition of the NKp30 activating receptor by pp65 of human cytomegalovirus. Nat Immunol 6:515–523

Atalay R, Zimmermann A, Wagner M, Borst E, Benz C, Messerle M, Hengel H (2002) Identification and expression of human cytomegalovirus transcription units coding for two distinct Fcgamma receptor homologs. J Virol 76:8596–8608

Bale JF Jr, Petheram SJ, Robertson M, Murph JR, Demmler G (2001) Human cytomegalovirus a sequence and UL144 variability in strains from infected children. J Med Virol 65:90–96

Barel MT, Pizzato N, Le Bouteiller P, Wiertz EJ, Lenfant F (2006) Subtle sequence variation among MHC class I locus products greatly influences sensitivity to HCMV US2- and US11-mediated degradation. Int Immunol 18:173–182

Bartee E, McCormack A, Fruh K (2006) Quantitative membrane proteomics reveals new cellular targets of viral immune modulators. PLoS Pathog 2:e107

Basta S, Bennink JR (2003) A survival game of hide and seek: cytomegaloviruses and MHC class I antigen presentation pathways. Viral Immunol 16:231–242

Bauer S, Groh V, Wu J, Steinle A, Phillips JH, Lanier LL, Spies T (1999) Activation of NK cells and T cells by NKG2D, a receptor for stress-inducible MICA. Science 285:727–729

Beck S, Barrell BG (1988) Human cytomegalovirus encodes a glycoprotein homologous to MHC class-I antigens. Nature 331:269–272

Benedict CA, Butrovich KD, Lurain NS, Corbeil J, Rooney I, Schneider P, Tschopp J, Ware CF (1999) Cutting edge: a novel viral TNF receptor superfamily member in virulent strains of human cytomegalovirus. J Immunol 162:6967–6970

Berg L, Riise GC, Cosman D, Bergstrom T, Olofsson S, Karre K, Carbone E (2003) LIR-1 expression on lymphocytes, and cytomegalovirus disease in lung-transplant recipients. Lancet 361:1099–1101

Biron CA, Byron KS, Sullivan JL (1989) Severe herpesvirus infections in an adolescent without natural killer cells. N Engl J Med 320:1731–1735

Booth TW, Scalzo AA, Carrello C, Lyons PA, Farrell HE, Singleton GR, Shellam GR (1993) Molecular and biological characterization of new strains of murine cytomegalovirus isolated from wild mice. Arch Virol 132:209–220

Boppana SB, Rivera LB, Fowler KB, Mach M, Britt WJ (2001) Intrauterine transmission of cytomegalovirus to infants of women with preconceptional immunity. N Engl J Med 344:1366–1371

Borges L, Hsu ML, Fanger N, Kubin M, Cosman D (1997) A family of human lymphoid and myeloid Ig-like receptors, some of which bind to MHC class I molecules. J Immunol 159:5192–5196

Borges L, Cosman D (2000) LIRs/ILTs/MIRs, inhibitory and stimulatory Ig-superfamily receptors expressed in myeloid and lymphoid cells. Cytokine Growth Factor Rev 11:209–217

Bottino C, Castriconi R, Pende D, Rivera P, Nanni M, Carnemolla B, Cantoni C, Grassi J, Marcenaro S, Reymond N, Vitale M, Moretta L, Lopez M, Moretta A (2003) Identification of PVR (CD155) and Nectin-2 (CD112) as cell surface ligands for the human DNAM-1 (CD226) activating molecule. J Exp Med 198: 557–567

Braud V, Jones EY, McMichael A (1997) The human major histocompatibility complex class Ib molecule HLA-E binds signal sequence-derived peptides with primary anchor residues at positions 2 and 9. Eur J Immunol 27:1164–1169

Bresnahan WA, Shenk TE (2000) UL82 virion protein activates expression of immediate early viral genes in human cytomegalovirus-infected cells. Proc Natl Acad Sci USA 97:14506–14511

Brown MG, Dokun AO, Heusel JW, Smith HR, Beckman DL, Blattenberger EA, Dubbelde CE, Stone LR, Scalzo AA, Yokoyama WM (2001) Vital involvement of a natural killer cell activation receptor in resistance to viral infection. Science 292:934–937

Browne EP, Shenk T (2003) Human cytomegalovirus UL83-coded pp65 virion protein inhibits antiviral gene expression in infected cells. Proc Natl Acad Sci USA 100:11439–11444

Browne EP, Wing B, Coleman D, Shenk T (2001) Altered cellular mRNA levels in human cytomegalovirus-infected fibroblasts: viral block to the accumulation of antiviral mRNAs. J Virol 75:12319–12330

Browne H, Smith G, Beck S, Minson T (1990) A complex between the MHC class I homologue encoded by human cytomegalovirus and beta 2 microglobulin. Nature 347:770–772

Bubic I, Wagner M, Krmpotic A, Saulig T, Kim S, Yokoyama WM, Jonjic S, Koszinowski UH (2004) Gain of virulence caused by loss of a gene in murine cytomegalovirus. J Virol 78:7536–7544

Cantrell SR, Bresnahan WA (2005) Interaction between the human cytomegalovirus UL82 gene product (pp71) and hDaxx regulates immediate-early gene expression and viral replication. J Virol 79:7792–7802

Carayannopoulos LN, Naidenko OV, Fremont DH, Yokoyama WM (2002) Cutting edge: murine UL16-binding protein-like transcript 1: a newly described transcript encoding a high-affinity ligand for murine NKG2D. J Immunol 169:4079–4083

Carlyle JR, Jamieson AM, Gasser S, Clingan CS, Arase H, Raulet DH (2004) Missing self-recognition of Ocil/Clr-b by inhibitory NKR-P1 natural killer cell receptors. Proc Natl Acad Sci USA 101:3527–3532

Cerboni C, Mousavi-Jazi M, Wakiguchi H, Carbone E, Karre K, Soderstrom K (2001) Synergistic effect of IFN-gamma and human cytomegalovirus protein UL40 in the HLA-E-dependent protection from NK cell-mediated cytotoxicity. Eur J Immunol 31:2926–2935

Cerwenka A, Bakker AB, McClanahan T, Wagner J, Wu J, Phillips JH, Lanier LL (2000) Retinoic acid early inducible genes define a ligand family for the activating NKG2D receptor in mice. Immunity 12:721–727

Chalupny NJ, Rein-Weston A, Dosch S, Cosman D (2006) Down-regulation of the NKG2D ligand MICA by the human cytomegalovirus glycoprotein UL142. Biochem Biophys Res Commun 346:175–181

Chapman TL, Bjorkman PJ (1998) Characterization of a murine cytomegalovirus class I major histocompatibility complex (MHC) homolog: comparison to MHC molecules and to the human cytomegalovirus MHC homolog. J Virol 72:460–466

Chapman TL, Heikeman AP, Bjorkman PJ (1999) The inhibitory receptor LIR-1 uses a common binding interaction to recognize class I MHC molecules and the viral homolog UL18. Immunity 11:603–613

Cheung TC, Humphreys IR, Potter KG, Norris PS, Shumway HM, Tran BR, Patterson G, Jean-Jacques R, Yoon M, Spear PG, Murphy KM, Lurain NS, Benedict CA, Ware CF (2005) Evolutionarily divergent herpesviruses modulate T cell activation by targeting the herpesvirus entry mediator cosignaling pathway. Proc Natl Acad Sci USA 102:13218–13223

Child SJ, Hakki M, De Niro KL, Geballe AP (2004) Evasion of cellular antiviral responses by human cytomegalovirus TRS1 and IRS1. J Virol 78:197–205

Child SJ, Jarrahian S, Harper VM, Geballe AP (2002) Complementation of vaccinia virus lacking the double-stranded RNA-binding protein gene E3L by human cytomegalovirus. J Virol 76:4912–4918

Chin KC, Cresswell P (2001) Viperin (cig5), an IFN-inducible antiviral protein directly induced by human cytomegalovirus. Proc Natl Acad Sci USA 98:15125–15130

Colonna M (1998) Immunology. Unmasking the killer's accomplice. Nature 391:642–643

Cosman D, Fanger N, Borges L, Kubin M, Chin W, Peterson L, Hsu ML (1997) A novel immunoglobulin superfamily receptor for cellular and viral MHC class I molecules. Immunity 7:273–282

Cosman D, Mullberg J, Sutherland CL, Chin W, Armitage R, Fanslow W, Kubin M, Chalupny NJ (2001) ULBPs, novel MHC class I-related molecules, bind to CMV glycoprotein UL16 and stimulate NK cytotoxicity through the NKG2D receptor. Immunity 14:123–133

Cresswell P, Ackerman AL, Giodini A, Peaper DR, Wearsch PA (2005) Mechanisms of MHC class I-restricted antigen processing and cross-presentation. Immunol Rev 207:145–157

Crnkovic-Mertens I, Messerle M, Milotic I, Szepan U, Kucic N, Krmpotic A, Jonjic S, Koszinowski UH (1998) Virus attenuation after deletion of the cytomegalovirus Fc receptor gene is not due to antibody control. J Virol 72:1377–1382

Croft M (2003) Co-stimulatory members of the TNFR family: keys to effective T-cell immunity? Nat Rev Immunol 3:609–620

Cull VS, Bartlett EJ, James CM (2002) Type I interferon gene therapy protects against cytomegalovirus-induced myocarditis. Immunology 106:428–437

Daniels KA, Devora G, Lai WC, O'Donnell CL, Bennett M, Welsh RM (2001) Murine cytomegalovirus is regulated by a discrete subset of natural killer cells reactive with monoclonal antibody to Ly49H. J Exp Med 194:29–44

Davison AJ, Dolan A, Akter P, Addison C, Dargan DJ, Alcendor DJ, McGeoch DJ, Hayward GS (2003) The human cytomegalovirus genome revisited: comparison with the chimpanzee cytomegalovirus genome. J Gen Virol 84:17–28

DeFilippis VR, Früh KJ (2005) Rhesus cytomegalovirus particles prevent activation of interferon regulatory factor 3. J Virol 79:6419–6431

DeFilippis VR, Robinson B, Keck TM, Hansen SG, Nelson JA, Früh K (2006) Interferon regulatory factor 3 is necessary for induction of antiviral genes during human cytomegalovirus infection. J Virol 80:1032–1037

Diefenbach A, Raulet DH (2003) Innate immune recognition by stimulatory immunoreceptors. Curr Opin Immunol 15:37–44

Diefenbach A, Hsia JK, Hsiung MY, Raulet DH (2003) A novel ligand for the NKG2D receptor activates NK cells and macrophages and induces tumor immunity. Eur J Immunol 33:381–391

Dighe A, Rodriguez M, Sabastian P, Xie X, McVoy M, Brown MG (2005) Requisite H2k role in NK cell-mediated resistance in acute murine cytomegalovirus-infected MA/My mice. J Immunol 175:6820–6828

Dimasi N, Biassoni R (2005) Structural and functional aspects of the Ly49 natural killer cell receptors. Immunol Cell Biol 83:1–8

Dunn C, Chalupny NJ, Sutherland CL, Dosch S, Sivakumar PV, Johnson DC, Cosman D (2003) Human cytomegalovirus glycoprotein UL16 causes intracellular sequestration of NKG2D ligands, protecting against natural killer cell cytotoxicity. J Exp Med 197:1427–1439

Fahnestock ML, Johnson JL, Feldman RM, Neveu JM, Lane WS, Bjorkman PJ (1995) The MHC class I homolog encoded by human cytomegalovirus binds endogenous peptides. Immunity 3:583–590

Falk CS, Mach M, Schendel DJ, Weiss EH, Hilgert I, Hahn G (2002) NK cell activity during human cytomegalovirus infection is dominated by US2–11-mediated HLA class I down-regulation. J Immunol 169:3257–3266

Farrell HE, Vally H, Lynch DM, Fleming P, Shellam GR, Scalzo AA, Davis-Poynter NJ (1997) Inhibition of natural killer cells by a cytomegalovirus MHC class I homologue in vivo. Nature 386:510–514

Furman MH, Dey N, Tortorella D, Ploegh HL (2002a) The human cytomegalovirus US10 gene product delays trafficking of major histocompatibility complex class I molecules. J Virol 76:11753–11756

Furman MH, Ploegh HL, Tortorella D (2002b) Membrane-specific, host-derived factors are required for US2- and US11-mediated degradation of major histocompatibility complex class I molecules. J Biol Chem 277:3258–3267

Gewurz BE, Gaudet R, Tortorella D, Wang EW, Ploegh HL, Wiley DC (2001) Antigen presentation subverted: structure of the human cytomegalovirus protein US2 bound to the class I molecule HLA-A2. Proc Natl Acad Sci USA 98:6794–6799

Gitlin L, Barchet W, Gilfillan S, Cella M, Beutler B, Flavell RA, Diamond MS, Colonna M (2006) Essential role of mda-5 in type I IFN responses to polyriboinosinic:polyribocytidylic acid and encephalomyocarditis picornavirus. Proc Natl Acad Sci USA 103:8459–8464

Gold MC, Munks MW, Wagner M, Koszinowski UH, Hill AB, Fling SP (2002) The murine cytomegalovirus immunomodulatory gene m152 prevents recognition of infected cells by M45-specific CTL but does not alter the immunodominance of the M45-specific CD8 T cell response in vivo. J Immunol 169:359–365

Gold MC, Munks MW, Wagner M, McMahon CW, Kelly A, Kavanagh DG, Slifka MK, Koszinowski UH, Raulet DH, Hill AB (2004) Murine cytomegalovirus interference with antigen presentation has little effect on the size or the effector memory phenotype of the CD8 T cell response. J Immunol 172:6944–6953

Grandea AG 3rd, Van Kaer L (2001) Tapasin: an ER chaperone that controls MHC class I assembly with peptide. Trends Immunol 22:194–199

Gribaudo G, Ravaglia S, Caliendo A, Cavallo R, Gariglio M, Martinotti MG, Landolfo S (1993) Interferons inhibit onset of murine cytomegalovirus immediate-early gene transcription. Virology 197:303–311

Griffin C, Wang EC, McSharry BP, Rickards C, Browne H, Wilkinson GW, Tomasec P (2005) Characterization of a highly glycosylated form of the human cytomegalovirus HLA class I homologue gpUL18. J Gen Virol 86:2999–3008

Guma M, Budt M, Saez A, Brckalo T, Hengel H, Angulo A, Lopez-Botet M (2006) Expansion of CD94/NKG2C+ NK cells in response to human cytomegalovirus-infected fibroblasts. Blood 107:3624–3631

Hansen SG, Strelow LI, Franchi DC, Anders DG, Wong SW (2003) Complete sequence and genomic analysis of rhesus cytomegalovirus. J Virol 77:6620–6636

Hao L, Klein J, Nei M (2006) Heterogeneous but conserved natural killer receptor gene complexes in four major orders of mammals. Proc Natl Acad Sci USA 103:3192–3197

Hasan M, Krmpotic A, Ruzsics Z, Bubic I, Lenac T, Halenius A, Loewendorf A, Messerle M, Hengel H, Jonjic S, Koszinowski UH (2005) Selective down-regulation of the NKG2D ligand H60 by mouse cytomegalovirus m155 glycoprotein. J Virol 79:2920–2930

Hassink GC, Barel MT, Van Voorden SB, Kikkert M, Wiertz EJ (2006) Ubiquitination of MHC class I heavy chains is essential for dislocation by human cytomegalovirus-encoded US2 but not US11. J Biol Chem 281:30063–30071

He R, Ruan Q, Xia C, Liu LQ, Lu SM, Lu Y, Qi Y, Ma YP, Liu Q, Ji YH (2004) Sequence variability of human cytomegalovirus UL144 open reading frame in low-passage clinical isolates. Chin Med Sci J 19:293–297

Hegde NR, Chevalier MS, Wisner TW, Denton MC, Shire K, Frappier L, Johnson DC (2006) The role of BiP in endoplasmic reticulum-associated degradation of major histocompatibility complex class I heavy chain induced by cytomegalovirus proteins. J Biol Chem 281:20910–20919

Hewitt EW, Gupta SS, Lehner PJ (2001) The human cytomegalovirus gene product US6 inhibits ATP binding by TAP. EMBO J 20:387–396

Hiscott J, Grandvaux N, Sharma S, Tenoever BR, Servant MJ, Lin R (2003) Convergence of the NF-kappaB and interferon signaling pathways in the regulation of antiviral defense and apoptosis. Ann NY Acad Sci 1010:237–248

Holowaty MN, Zeghouf M, Wu H, Tellam J, Athanasopoulos V, Greenblatt J, Frappier L (2003) Protein profiling with Epstein-Barr nuclear antigen-1 reveals an interaction with the herpesvirus-associated ubiquitin-specific protease HAUSP/USP7. J Biol Chem 278:29987–29994

Holtappels R, Podlech J, Pahl-Seibert MF, Julch M, Thomas D, Simon CO, Wagner M, Reddehase MJ (2004) Cytomegalovirus misleads its host by priming of CD8 T cells specific for an epitope not presented in infected tissues. J Exp Med 199:131–136

Huard B, Fruh K (2000) A role for MHC class I down-regulation in NK cell lysis of herpes virus-infected cells. Eur J Immunol 30:509–515

Huber MT, Tomazin R, Wisner T, Boname J, Johnson DC (2002) Human cytomegalovirus US7, US8, US9, and US10 are cytoplasmic glycoproteins, not found at cell surfaces, and US9 does not mediate cell-to-cell spread. J Virol 76:5748–5758

Iizuka K, Naidenko OV, Plougastel BF, Fremont DH, Yokoyama WM (2003) Genetically linked C-type lectin-related ligands for the NKRP1 family of natural killer cell receptors. Nat Immunol 4:801–807

Johnson DC, Hegde NR (2002) Inhibition of the MHC class II antigen presentation pathway by human cytomegalovirus. Curr Top Microbiol Immunol 269:101–115

Jones TR, Wiertz EJ, Sun L, Fish KN, Nelson JA, Ploegh HL (1996) Human cytomegalovirus US3 impairs transport and maturation of major histocompatibility complex class I heavy chains. Proc Natl Acad Sci U S A 93:11327–11333

Kato H, Sato S, Yoneyama M, Yamamoto M, Uematsu S, Matsui K, Tsujimura T, Takeda K, Fujita T, Takeuchi O, Akira S (2005) Cell type-specific involvement of RIG-I in antiviral response. Immunity 23:19–28

Kato H, Takeuchi O, Sato S, Yoneyama M, Yamamoto M, Matsui K, Uematsu S, Jung A, Kawai T, Ishii KJ, Yamaguchi O, Otsu K, Tsujimura T, Koh CS, Reis e Sousa C, Matsuura Y, Fujita T, Akira S (2006) Differential roles of MDA5 and RIG-I helicases in the recognition of RNA viruses. Nature 441:101–105

Kavanagh DG, Koszinowski UH, Hill AB (2001) The murine cytomegalovirus immune evasion protein m4/gp34 forms biochemically distinct complexes with class I MHC at the cell surface and in a pre-Golgi compartment. J Immunol 167:3894–3902

Kawai T, Akira S (2006) TLR signaling. Cell Death Differ 13:816–825

Kielczewska A, Kim HS, Lanier LL, Dimasi N, Vidal SM (2007) Critical residues at the Ly49 natural killer receptor's homodimer interface determine functional recognition of m157, a mouse cytomegalovirus MHC class I-like protein. J Immunol 178:369–377

Kleijnen MF, Huppa JB, Lucin P, Mukherjee S, Farrell H, Campbell AE, Koszinowski UH, Hill AB, Ploegh HL (1997) A mouse cytomegalovirus glycoprotein, gp34, forms a complex with folded class I MHC molecules in the ER which is not retained but is transported to the cell surface. EMBO J 16:685–694

Krmpotic A, Messerle M, Crnkovic-Mertens I, Polic B, Jonjic S, Koszinowski UH (1999) The immunoevasive function encoded by the mouse cytomegalovirus gene m152 protects the virus against T cell control in vivo. J Exp Med 190:1285–1296

Krmpotic A, Busch DH, Bubic I, Gebhardt F, Hengel H, Hasan M, Scalzo AA, Koszinowski UH, Jonjic S (2002) MCMV glycoprotein gp40 confers virus resistance to CD8+ T cells and NK cells in vivo. Nat Immunol 3:529–535

Krmpotic A, Hasan M, Loewendorf A, Saulig T, Halenius A, Lenac T, Polic B, Bubic I, Kriegeskorte A, Pernjak-Pugel E, Messerle M, Hengel H, Busch DH, Koszinowski UH, Jonjic S

(2005) NK cell activation through the NKG2D ligand MULT-1 is selectively prevented by the glycoprotein encoded by mouse cytomegalovirus gene m145. J Exp Med 201:211–220

Kubin M, Cassiano L, Chalupny J, Chin W, Cosman D, Fanslow W, Mullberg J, Rousseau AM, Ulrich D, Armitage R (2001) ULBP1, 2, 3: novel MHC class I-related molecules that bind to human cytomegalovirus glycoprotein UL16, activate NK cells. Eur J Immunol 31:1428–1437

Kubota A, Kubota S, Farrell HE, Davis-Poynter N, Takei F (1999) Inhibition of NK cells by murine CMV-encoded class I MHC homologue m144. Cell Immunol 191:145–151

Lanier LL (1998) NK cell receptors. Annu Rev Immunol 16:359–393

Lee N, Llano M, Carretero M, Ishitani A, Navarro F, Lopez-Botet M, Geraghty DE (1998) HLA-E is a major ligand for the natural killer inhibitory receptor CD94/NKG2A. Proc Natl Acad Sci U S A 95:5199–5204

Lee SH, Girard S, Macina D, Busa M, Zafer A, Belouchi A, Gros P, Vidal SM (2001) Susceptibility to mouse cytomegalovirus is associated with deletion of an activating natural killer cell receptor of the C-type lectin superfamily. Nat Genet 28:42–45

Lenac T, Budt M, Arapovic J, Hasan M, Zimmermann A, Simic H, Krmpotic A, Messerle M, Ruzsics Z, Koszinowski UH, Hengel H, Jonjic S (2006) The herpesviral Fc receptor fcr-1 down-regulates the NKG2D ligands MULT-1 and H60. J Exp Med 203:1843–1850

Leong CC, Chapman TL, Bjorkman PJ, Formankova D, Mocarski ES, Phillips JH, Lanier LL (1998) Modulation of natural killer cell cytotoxicity in human cytomegalovirus infection: the role of endogenous class I major histocompatibility complex and a viral class I homolog. J Exp Med 187:1681–1687

Lilley BN, Ploegh HL (2004) A membrane protein required for dislocation of misfolded proteins from the ER. Nature 429:834–840

Lilley BN, Tortorella D, Ploegh HL (2003) Dislocation of a type I membrane protein requires interactions between membrane-spanning segments within the lipid bilayer. Mol Biol Cell 14:3690–3698

Lockridge KM, Zhou SS, Kravitz RH, Johnson JL, Sawai ET, Blewett EL, Barry PA (2000) Primate cytomegaloviruses encode and express an IL-10-like protein. Virology 268:272–280

Locksley RM, Killeen N, Lenardo MJ (2001) The TNF and TNF receptor superfamilies: integrating mammalian biology. Cell 104:487–501

Lodoen M, Ogasawara K, Hamerman JA, Arase H, Houchins JP, Mocarski ES, Lanier LL (2003) NKG2D-mediated natural killer cell protection against cytomegalovirus is impaired by viral gp40 modulation of retinoic acid early inducible 1 gene molecules. J Exp Med 197:1245–1253

Lodoen MB, Abenes G, Umamoto S, Houchins JP, Liu F, Lanier LL (2004) The cytomegalovirus m155 gene product subverts natural killer cell antiviral protection by disruption of H60-NKG2D interactions. J Exp Med 200:1075–1081

Loewendorf A, Kruger C, Borst EM, Wagner M, Just U, Messerle M (2004) Identification of a mouse cytomegalovirus gene selectively targeting CD86 expression on antigen-presenting cells. J Virol 78:13062–13071

Loureiro J, Lilley BN, Spooner E, Noriega V, Tortorella D, Ploegh HL (2006) Signal peptide peptidase is required for dislocation from the endoplasmic reticulum. Nature 441:894–897

Lu X, Pinto AK, Kelly AM, Cho KS, Hill AB (2006) Murine cytomegalovirus interference with antigen presentation contributes to the inability of CD8 T cells to control virus in the salivary gland. J Virol 80:4200–4202

Lurain NS, Kapell KS, Huang DD, Short JA, Paintsil J, Winkfield E, Benedict CA, Ware CF, Bremer JW (1999) Human cytomegalovirus UL144 open reading frame: sequence hypervariability in low-passage clinical isolates. J Virol 73:10040–10050

Ma YP, Ruan Q, He R, Qi Y, Sun ZR, Ji YH, Huang YJ, Liu Q, Chen SR, Wang JD (2006) Sequence variability of the human cytomegalovirus UL141 open reading frame in clinical strains. Arch Virol 151:827–835

Malarkannan S, Shih PP, Eden PA, Horng T, Zuberi AR, Christianson G, Roopenian D, Shastri N (1998) The molecular and functional characterization of a dominant minor H antigen, H60. J Immunol 161:3501–3509

Mao ZQ, He R, Sun M, Qi Y, Huang YJ, Ruan Q (2007) The relationship between polymorphisms of HCMV UL144 ORF and clinical manifestations in 73 strains with congenital and/or perinatal HCMV infection. Arch Virol 152:115–124

Martin MU, Wesche H (2002) Summary and comparison of the signaling mechanisms of the Toll/interleukin-1 receptor family. Biochim Biophys Acta 1592:265–280

Mazzarino P, Pietra G, Vacca P, Falco M, Colau D, Coulie P, Moretta L, Mingari MC (2005) Identification of effector-memory CMV-specific T lymphocytes that kill CMV-infected target cells in an HLA-E-restricted fashion. Eur J Immunol 35:3240–3247

Mesci A, Ljutic B, Makrigiannis AP, Carlyle JR (2006) NKR-P1 biology: from prototype to missing self. Immunol Res 35:13–26

Miller DM, Zhang Y, Rahill BM, Waldman WJ, Sedmak DD (1999) Human cytomegalovirus inhibits IFN-alpha-stimulated antiviral and immunoregulatory responses by blocking multiple levels of IFN-alpha signal transduction. J Immunol 162:6107–6113

Mintern JD, Klemm EJ, Wagner M, Paquet ME, Napier MD, Kim YM, Koszinowski UH, Ploegh HL (2006) Viral interference with B7–1 costimulation: a new role for murine cytomegalovirus fc receptor-1. J Immunol 177:8422–8431

Mocarski ES Jr (2004) Immune escape and exploitation strategies of cytomegaloviruses: impact on and imitation of the major histocompatibility system. Cell Microbiol 6:707–717

Moretta L, Romagnani C, Pietra G, Moretta A, Mingari MC (2003) NK-CTLs, a novel HLA-E-restricted T-cell subset. Trends Immunol 24:136–143

Munks MW, Pinto AK, Doom CM, Hill AB (2007) Viral interference with antigen presentation does not alter acute or chronic CD8 T cell immunodominance in murine cytomegalovirus infection. J Immunol 178:7235–7241

Murayama T, Takegoshi M, Tanuma J, Eizuru Y (2005) Analysis of human cytomegalovirus UL144 variability in low-passage clinical isolates in Japan. Intervirology 48:201–206

Natarajan K, Hicks A, Mans J, Robinson H, Guan R, Mariuzza RA, Margulies DH (2006) Crystal structure of the murine cytomegalovirus MHC-I homolog m144. J Mol Biol 358:157–171

Northfield J, Lucas M, Jones H, Young NT, Klenerman P (2005) Does memory improve with age? CD85j (ILT-2/LIR-1) expression on CD8 T cells correlates with 'memory inflation' in human cytomegalovirus infection. Immunol Cell Biol 83:182–188

Odeberg J, Cerboni C, Browne H, Karre K, Moller E, Carbone E, Soderberg-Naucler C (2002) Human cytomegalovirus (HCMV)-infected endothelial cells and macrophages are less susceptible to natural killer lysis independent of the downregulation of classical HLA class I molecules or expression of the HCMV class I homologue, UL18. Scand J Immunol 55:149–161

Odeberg J, Plachter B, Branden L, Soderberg-Naucler C (2003) Human cytomegalovirus protein pp65 mediates accumulation of HLA-DR in lysosomes and destruction of the HLA-DR alpha-chain. Blood 101:4870–4877

Oliveira SA, Park SH, Lee P, Bendelac A, Shenk TE (2002) Murine cytomegalovirus m02 gene family protects against natural killer cell-mediated immune surveillance. J Virol 76:885–894

Pande NT, Powers C, Ahn K, Fruh K (2005) Rhesus cytomegalovirus contains functional homologues of US2, US3, US6, and US11. J Virol 79:5786–5798

Park B, Oh H, Lee S, Song Y, Shin J, Sung YC, Hwang SY, Ahn K (2002) The MHC class I homolog of human cytomegalovirus is resistant to down-regulation mediated by the unique short region protein (US)2, US3, US6, and US11 gene products. J Immunol 168:3464–3469

Park B, Kim Y, Shin J, Lee S, Cho K, Fruh K, Ahn K (2004) Human cytomegalovirus inhibits tapasin-dependent peptide loading and optimization of the MHC class I peptide cargo for immune evasion. Immunity 20:71–85

Park B, Lee S, Kim E, Cho K, Riddell SR, Cho S, Ahn K (2006) Redox regulation facilitates optimal peptide selection by MHC class I during antigen processing. Cell 127:369–382

Paulus C, Krauss S, Nevels M (2006) A human cytomegalovirus antagonist of type I IFN-dependent signal transducer and activator of transcription signaling. Proc Natl Acad Sci U S A 103:3840–3845

Picone O, Costa JM, Chaix ML, Ville Y, Rouzioux C, Leruez-Ville M (2005) Human cytomegalovirus UL144 gene polymorphisms in congenital infections. J Clin Microbiol 43:25–29

Pietra G, Romagnani C, Falco M, Vitale M, Castriconi R, Pende D, Millo E, Anfossi S, Biassoni R, Moretta L, Mingari MC (2001) The analysis of the natural killer-like activity of human cytolytic T lymphocytes revealed HLA-E as a novel target for TCR alpha/beta-mediated recognition. Eur J Immunol 31:3687–3693

Pietra G, Romagnani C, Mazzarino P, Falco M, Millo E, Moretta A, Moretta L, Mingari MC (2003) HLA-E-restricted recognition of cytomegalovirus-derived peptides by human CD8+ cytolytic T lymphocytes. Proc Natl Acad Sci U S A 100:10896–10901

Pinto AK, Hill AB (2005) Viral interference with antigen presentation to CD8+ T cells: lessons from cytomegalovirus. Viral Immunol 18:434–444

Poole E, King CA, Sinclair JH, Alcami A (2006) The UL144 gene product of human cytomegalovirus activates NFkappaB via a TRAF6-dependent mechanism. EMBO J 25:4390–4399

Prod'homme V, Griffin C, Aicheler RJ, Wang EC, McSharry BP, Rickards CR, Stanton RJ, Borysiewicz LK, Lopez-Botet M, Wilkinson GW, Tomasec P (2007) The human cytomegalovirus MHC class I homolog UL18 inhibits LIR-1+ but activates LIR-1-NK cells. J Immunol 178:4473–4481

Purcell AW, Gorman JJ, Garcia-Peydro M, Paradela A, Burrows SR, Talbo GH, Laham N, Peh CA, Reynolds EC, Lopez De Castro JA, McCluskey J (2001) Quantitative and qualitative influences of tapasin on the class I peptide repertoire. J Immunol 166:1016–1027

Rawlinson WD, Farrell HE, Barrell BG (1996) Analysis of the complete DNA sequence of murine cytomegalovirus. J Virol 70:8833–8849

Reddehase MJ (2002) Antigens and immunoevasins: opponents in cytomegalovirus immune surveillance. Nat Rev Immunol 2:831–844

Reddehase MJ, Simon CO, Podlech J, Holtappels R (2004) Stalemating a clever opportunist: lessons from murine cytomegalovirus. Hum Immunol 65:446–455

Reusch U, Muranyi W, Lucin P, Burgert HG, Hengel H, Koszinowski UH (1999) A cytomegalovirus glycoprotein re-routes MHC class I complexes to lysosomes for degradation. EMBO J 18:1081–1091

Reyburn HT, Mandelboim O, Vales-Gomez M, Davis DM, Pazmany L, Strominger JL (1997) The class I MHC homologue of human cytomegalovirus inhibits attack by natural killer cells. Nature 386:514–517

Rivailler P, Kaur A, Johnson RP, Wang F (2006) Genomic sequence of rhesus cytomegalovirus 180.92: insights into the coding potential of rhesus cytomegalovirus. J Virol 80:4179–4182

Rizvanov AA, van Geelen AG, Morzunov S, Otteson EW, Bohlman C, Pari GS, St Jeor SC (2003) Generation of a recombinant cytomegalovirus for expression of a hantavirus glycoprotein. J Virol 77:12203–12210

Robertson MJ, Cochran KJ, Cameron C, Le JM, Tantravahi R, Ritz J (1996) Characterization of a cell line, NKL, derived from an aggressive human natural killer cell leukemia. Exp Hematol 24:406–415

Rodriguez M, Sabastian P, Clark P, Brown MG (2004) Cmv1-independent antiviral role of NK cells revealed in murine cytomegalovirus-infected New Zealand White mice. J Immunol 173:6312–6318

Rolle A, Mousavi-Jazi M, Eriksson M, Odeberg J, Soderberg-Naucler C, Cosman D, Karre K, Cerboni C (2003) Effects of human cytomegalovirus infection on ligands for the activating NKG2D receptor of NK cells: up-regulation of UL16-binding protein (ULBP)1 and ULBP2 is counteracted by the viral UL16 protein. J Immunol 171:902–908

Romagnani C, Pietra G, Falco M, Millo E, Mazzarino P, Biassoni R, Moretta A, Moretta L, Mingari MC (2002) Identification of HLA-E-specific alloreactive T lymphocytes: a cell subset that undergoes preferential expansion in mixed lymphocyte culture and displays a broad cytolytic activity against allogeneic cells. Proc Natl Acad Sci U S A 99:11328–11333

Romagnani C, Pietra G, Falco M, Mazzarino P, Moretta L, Mingari MC (2004) HLA-E-restricted recognition of human cytomegalovirus by a subset of cytolytic T lymphocytes. Hum Immunol 65:437–445

Ruger B, Klages S, Walla B, Albrecht J, Fleckenstein B, Tomlinson P, Barrell B (1987) Primary structure and transcription of the genes coding for the two virion phosphoproteins pp65 and pp71 of human cytomegalovirus. J Virol 61:446–453

Sainz B Jr, Lamarca HL, Garry RF, Morris CA (2005) Synergistic inhibition of human cytomegalovirus replication by interferon-alpha/beta and interferon-gamma. Virol J 2:14

Salazar-Mather TP, Lewis CA, Biron CA (2002) Type I interferons regulate inflammatory cell trafficking and macrophage inflammatory protein 1alpha delivery to the liver. J Clin Invest 110:321–330

Samuel CE (2001) Antiviral actions of interferons. Clin Microbiol Rev 14:778–809

Saverino D, Ghiotto F, Merlo A, Bruno S, Battini L, Occhino M, Maffei M, Tenca C, Pileri S, Baldi L, Fabbi M, Bachi A, De Santanna A, Grossi CE, Ciccone E (2004) Specific recognition of the viral protein UL18 by CD85j/LIR-1/ILT2 on CD8+ T cells mediates the non-MHC-restricted lysis of human cytomegalovirus-infected cells. J Immunol 172:5629–5637

Scalzo AA (2002) Successful control of viruses by NK cells – a balance of opposing forces? Trends Microbiol 10:470–474

Scalzo AA, Fitzgerald NA, Simmons A, La Vista AB, Shellam GR (1990) Cmv-1, a genetic locus that controls murine cytomegalovirus replication in the spleen. J Exp Med 171:1469–1483

Scalzo AA, Lyons PA, Fitzgerald NA, Forbes CA, Yokoyama WM, Shellam GR (1995) Genetic mapping of Cmv1 in the region of mouse chromosome 6 encoding the NK gene complex-associated loci Ly49 and musNKR-P1. Genomics 27:435–441

Schneider RJ, Mohr I (2003) Translation initiation and viral tricks. Trends Biochem Sci 28: 130–136

Schoenhals GJ, Krishna RM, Grandea AG 3rd, Spies T, Peterson PA, Yang Y, Fruh K (1999) Retention of empty MHC class I molecules by tapasin is essential to reconstitute antigen presentation in invertebrate cells. EMBO J 18:743–753

Shuai K, Liu B (2003) Regulation of JAK-STAT signalling in the immune system. Nat Rev Immunol 3:900–911

Simmen KA, Singh J, Luukkonen BG, Lopper M, Bittner A, Miller NE, Jackson MR, Compton T, Fruh K (2001) Global modulation of cellular transcription by human cytomegalovirus is initiated by viral glycoprotein B. Proc Natl Acad Sci U S A 98:7140–7145

Smith HR, Heusel JW, Mehta IK, Kim S, Dorner BG, Naidenko OV, Iizuka K, Furukawa H, Beckman DL, Pingel JT, Scalzo AA, Fremont DH, Yokoyama WM (2002) Recognition of a virus-encoded ligand by a natural killer cell activation receptor. Proc Natl Acad Sci USA 99:8826–8831

Smith LM, Shellam GR, Redwood AJ (2006) Genes of murine cytomegalovirus exist as a number of distinct genotypes. Virology 352:450–465

Spreu J, Stehle T, Steinle A (2006) Human cytomegalovirus-encoded UL16 discriminates MIC molecules by their alpha2 domains. J Immunol 177:3143–3149

Stern-Ginossar N, Elefant N, Zimmermann A, Wolf DG, Saleh N, Biton M, Horwitz E, Prokocimer Z, Prichard M, Hahn G, Goldman-Wohl D, Greenfield C, Yagel S, Hengel H, Altuvia Y, Margalit H, Mandelboim O (2007) Host immune system gene targeting by a viral miRNA. Science 317:376–381

Sutherland CL, Chalupny NJ, Schooley K, VandenBos T, Kubin M, Cosman D (2002) UL16-binding proteins, novel MHC class I-related proteins, bind to NKG2D and activate multiple signaling pathways in primary NK cells. J Immunol 168:671–679

Takaoka A, Wang Z, Choi MK, Yanai H, Negishi H, Ban T, Lu Y, Miyagishi M, Kodama T, Honda K, Ohba Y, Taniguchi T (2007) DAI (DLM-1/ZBP1) is a cytosolic DNA sensor and an activator of innate immune response. Nature 448:501–505

Takeuchi O, Akira S (2001) Toll-like receptors; their physiological role and signal transduction system. Int Immunopharmacol 1:625–635

Tanaka K, Numazaki K, Tsutsumi H (2005) Human cytomegalovirus genetic variability in strains isolated from Japanese children during 1983–2003. J Med Virol 76:356–360

Tay CH, Szomolanyi-Tsuda E, Welsh RM (1998) Control of infections by NK cells. Curr Top Microbiol Immunol 230:193–220

Taylor RT, Bresnahan WA (2005) Human cytomegalovirus immediate-early 2 gene expression blocks virus-induced beta interferon production. J Virol 79:3873–3877

Taylor RT, Bresnahan WA (2006a) Human cytomegalovirus IE86 attenuates virus- and tumor necrosis factor alpha-induced NFkappaB-dependent gene expression. J Virol 80:10763–10771

Taylor RT, Bresnahan WA (2006b) Human cytomegalovirus immediate-early 2 protein IE86 blocks virus-induced chemokine expression. J Virol 80:920–928

Thammavongsa V, Raghuraman G, Filzen TM, Collins KL, Raghavan M (2006) HLA-B44 polymorphisms at position 116 of the heavy chain influence TAP complex binding via an effect on peptide occupancy. J Immunol 177:3150–3161

Tirabassi RS, Ploegh HL (2002) The human cytomegalovirus US8 glycoprotein binds to major histocompatibility complex class I products. J Virol 76:6832–6835

Tomasec P, Braud VM, Rickards C, Powell MB, McSharry BP, Gadola S, Cerundolo V, Borysiewicz LK, McMichael AJ, Wilkinson GW (2000) Surface expression of HLA-E, an inhibitor of natural killer cells, enhanced by human cytomegalovirus gpUL40. Science 287:1031

Tomasec P, Wang EC, Davison AJ, Vojtesek B, Armstrong M, Griffin C, McSharry BP, Morris RJ, Llewellyn-Lacey S, Rickards C, Nomoto A, Sinzger C, Wilkinson GW (2005) Downregulation of natural killer cell-activating ligand CD155 by human cytomegalovirus UL141. Nat Immunol 6:181–188

Torigoe S, Campbell DE, Torigoe F, Michelson S, Starr SE (1993) Cytofluorographic analysis of effects of interferons on expression of human cytomegalovirus proteins. J Virol Methods 45:219–228

Uetz P, Dong YA, Zeretzke C, Atzler C, Baiker A, Berger B, Rajagopala SV, Roupelieva M, Rose D, Fossum E, Haas J (2006) Herpesviral protein networks and their interaction with the human proteome. Science 311:239–242

Ulbrecht M, Martinozzi S, Grzeschik M, Hengel H, Ellwart JW, Pla M, Weiss EH (2000) Cutting edge: the human cytomegalovirus UL40 gene product contains a ligand for HLA-E and prevents NK cell-mediated lysis. J Immunol 164:5019–5022

Vales-Gomez M, Reyburn HT (2006) Intracellular trafficking of the HCMV immunoevasin UL16 depends on elements present in both its cytoplasmic and transmembrane domains. J Mol Biol 363:908–917

Vales-Gomez M, Shiroishi M, Maenaka K, Reyburn HT (2005) Genetic variability of the major histocompatibility complex class I homologue encoded by human cytomegalovirus leads to differential binding to the inhibitory receptor ILT2. J Virol 79:2251–2260

van der Wal FJ, Kikkert M, Wiertz E (2002) The HCMV gene products US2 and US11 target MHC class I molecules for degradation in the cytosol. Curr Top Microbiol Immunol 269:37–55

Vance RE, Kraft JR, Altman JD, Jensen PE, Raulet DH (1998) Mouse CD94/NKG2A is a natural killer cell receptor for the nonclassical major histocompatibility complex (MHC) class I molecule Qa-1(b). J Exp Med 188:1841–1848

Varnum SM, Streblow DN, Monroe ME, Smith P, Auberry KJ, Pasa-Tolic L, Wang D, Camp DG 2nd, Rodland K, Wiley S, Britt W, Shenk T, Smith RD, Nelson JA (2004) Identification of proteins in human cytomegalovirus (HCMV) particles: the HCMV proteome. J Virol 78:10960–10966

Voigt S, Sandford GR, Ding L, Burns WH (2001) Identification and characterization of a spliced C-type lectin-like gene encoded by rat cytomegalovirus. J Virol 75:603–611

Voigt V, Forbes CA, Tonkin JN, Degli-Esposti MA, Smith HR, Yokoyama WM, Scalzo AA (2003) Murine cytomegalovirus m157 mutation and variation leads to immune evasion of natural killer cells. Proc Natl Acad Sci U S A 100:13483–13488

Voigt S, Mesci A, Ettinger J, Fine JH, Chen P, Chou W, Carlyle JR (2007) Cytomegalovirus evasion of innate immunity by subversion of the NKR-P1B:Clr-b missing-self axis. Immunity 26:617–627

Wagner CS, Riise GC, Bergstrom T, Karre K, Carbone E, Berg L (2007) Increased expression of leukocyte Ig-like receptor-1 and activating role of UL18 in the response to cytomegalovirus infection. J Immunol 178:3536–3543

Wang EC, McSharry B, Retiere C, Tomasec P, Williams S, Borysiewicz LK, Braud VM, Wilkinson GW (2002) UL40-mediated NK evasion during productive infection with human cytomegalovirus. Proc Natl Acad Sci U S A 99:7570–7575

Ware CF (2003) The TNF superfamily. Cytokine Growth Factor Rev 14:181–184

Wathelet MG, Lin CH, Parekh BS, Ronco LV, Howley PM, Maniatis T (1998) Virus infection induces the assembly of coordinately activated transcription factors on the IFN-beta enhancer in vivo. Mol Cell 1:507–518

Welte SA, Sinzger C, Lutz SZ, Singh-Jasuja H, Sampaio KL, Eknigk U, Rammensee HG, Steinle A (2003) Selective intracellular retention of virally induced NKG2D ligands by the human cytomegalovirus UL16 glycoprotein. Eur J Immunol 33:194–203

Wiertz EJ, Devlin R, Collins HL, Ressing ME (2007) Herpesvirus interference with major histocompatibility complex class II-restricted T-cell activation. J Virol 81:4389–4396

Willcox BE, Thomas LM, Bjorkman PJ (2003) Crystal structure of HLA-A2 bound to LIR-1, a host and viral major histocompatibility complex receptor. Nat Immunol 4:913–919

Williams AP, Peh CA, Purcell AW, McCluskey J, Elliott T (2002) Optimization of the MHC class I peptide cargo is dependent on tapasin. Immunity 16:509–520

Wills MR, Ashiru O, Reeves MB, Okecha G, Trowsdale J, Tomasec P, Wilkinson GW, Sinclair J, Sissons JG (2005) Human cytomegalovirus encodes an MHC class I-like molecule (UL142) that functions to inhibit NK cell lysis. J Immunol 175:7457–7465

Wu J, Chalupny NJ, Manley TJ, Riddell SR, Cosman D, Spies T (2003) Intracellular retention of the MHC class I-related chain B ligand of NKG2D by the human cytomegalovirus UL16 glycoprotein. J Immunol 170:4196–4200

Xie X, Dighe A, Clark P, Sabastian P, Buss S, Brown MG (2007) Deficient major histocompatibility complex-linked innate murine cytomegalovirus immunity in MA/My.L-H2b mice and viral downregulation of H-2k class I proteins. J Virol 81:229–236

Ye Y, Shibata Y, Yun C, Ron D, Rapoport TA (2004) A membrane protein complex mediates retrotranslocation from the ER lumen into the cytosol. Nature 429:841–847

Yeow WS, Lawson CM, Beilharz MW (1998) Antiviral activities of individual murine IFN-alpha subtypes in vivo: intramuscular injection of IFN expression constructs reduces cytomegalovirus replication. J Immunol 160:2932–2939

Yewdell JW, Hill AB (2002) Viral interference with antigen presentation. Nat Immunol 3:1019–1025

Yoneyama M, Kikuchi M, Natsukawa T, Shinobu N, Imaizumi T, Miyagishi M, Taira K, Akira S, Fujita T (2004) The RNA helicase RIG-I has an essential function in double-stranded RNA-induced innate antiviral responses. Nat Immunol 5:730–737

Yu D, Silva MC, Shenk T (2003) Functional map of human cytomegalovirus AD169 defined by global mutational analysis. Proc Natl Acad Sci U S A 100:12396–12401

Ziegler H, Thale R, Lucin P, Muranyi W, Flohr T, Hengel H, Farrell H, Rawlinson W, Koszinowski UH (1997) A mouse cytomegalovirus glycoprotein retains MHC class I complexes in the ERGIC/cis-Golgi compartments. Immunity 6:57–66

Zimmermann A, Trilling M, Wagner M, Wilborn M, Bubic I, Jonjic S, Koszinowski U, Hengel H (2005) A cytomegaloviral protein reveals a dual role for STAT2 in IFN-{gamma} signaling and antiviral responses. J Exp Med 201:1543–1553

Zou Y, Bresnahan W, Taylor RT, Stastny P (2005) Effect of human cytomegalovirus on expression of MHC class I-related chains A. J Immunol 174:3098–3104

Cytomegalovirus Vaccine Development

M. R. Schleiss

Contents

Spectrum of HCMV Disease, Rationale for Vaccine, and Target Population............ 362
    Congenital HCMV Infection: A Major Public Health Problem 362
Healthcare Costs Associated with Congenital HCMV Infection:
    A Compelling Argument for Vaccine Development 363
    HCMV Vaccine: What Is the Ideal Target Population?........................... 363
Evidence That Immunity Protects Against HCMV Infection and Disease.............. 365
    Role of Preconception Maternal Immunity in Protection Against Congenital
        HCMV Transmission and HCMV Disease in the Newborn 365
    Lessons from Adoptive Transfer Studies....................................... 366
HCMV Vaccines in Clinical Trials 367
    Live, Attenuated HCMV Vaccines.. 368
    Subunit Vaccines 369
HCMV Vaccine Approaches in Preclinical Development........................... 372
    Alternative Expression Strategies for HCMV gB, pp65, and IE1 372
    Potential Role of Other Viral Proteins in HCMV Vaccine Design 373
    Dense Body Vaccines... 375
    Peptide-Based Vaccines... 375
    Novel Vaccine Approaches 376
Perspectives 377
References 378

Abstract Although infection with human cytomegalovirus (HCMV) is ubiquitous and usually asymptomatic, there are individuals at high risk for serious HCMV disease. These include solid organ and hematopoietic stem cell (HSC) transplant patients, individuals with HIV infection, and the fetus. Since immunity to HCMV ameliorates the severity of disease, there have been efforts made for over 30 years to develop vaccines for use in these high-risk settings. However, in spite of these efforts, no HCMV vaccine appears to be approaching imminent licensure. The

M.R. Schleiss
Division of Pediatric Infectious Diseases, Department of Pediatrics, Center for Infectious Diseases and Microbiology Translational Research, University of Minnesota Medical School, 2001 6th Street SE, Minneapolis, MN 55455, USA
schleiss@umn.edu

reasons for the failure to achieve the goal of a licensed HCMV vaccine are complex, but several key problems stand out. First, the host immune correlates of protective immunity are not yet clear. Secondly, the viral proteins that should be included in a HCMV vaccine are uncertain. Third, clinical trials have largely focused on immunocompromised patients, a population that may not be relevant to the problem of protection of the fetus against congenital infection. Fourth, the ultimate target population for HCMV vaccination remains unclear. Finally, and most importantly, there has been insufficient education about the problem of HCMV infection, particularly among women of child-bearing age and in the lay public. This review considers the strategies that have been explored to date in development of HCMV vaccines, and summarizes both active clinical trials as well as novel technologies that merit future consideration toward the goal of prevention of this significant public health problem.

Spectrum of HCMV Disease, Rationale for Vaccine, and Target Population

Congenital HCMV Infection: A Major Public Health Problem

The problem of congenital HCMV infection is unquestionably the major driving force behind efforts to develop a HCMV vaccine. In the developed world, HCMV is the most common congenital viral infection (Whitley 1994). Estimates of the prevalence of congenital HCMV infection suggest that between 0.5% and 2% of all newborns in the developed world are infected in utero (Demmler 1996). In the United States alone, this corresponds to approximately 40,000 infected newborn infants born annually with HCMV infection. The concern is particularly acute for HCMV-seronegative women of child-bearing age. Based on recent HCMV incidence estimates, approximately 27,000 new infections are believed to occur among seronegative pregnant women in the United States each year (Colugnati et al. 2007). Approximately 10% of congenitally infected infants have clinically evident disease in the newborn period, including visceral organomegaly, microcephaly with intracranial calcifications, chorioretinitis, and skin lesions including petechiae and purpura. Although the majority of congenitally infected infants appear normal at birth, these children are nonetheless at risk for neurodevelopmental sequelae, in particular sensorineural hearing loss (SNHL). Antiviral therapy in infected newborns with neurologic involvement is of value in ameliorating the severity and progression of SNHL (Kimberlin et al. 2003), but the toxicities of available antiviral agents are of concern, and the benefits of therapy are limited. Therefore, there are few medical interventions currently available to prevent or limit HCMV-induced neurological morbidity in infants, underscoring the urgent need for vaccine development.

Healthcare Costs Associated with Congenital HCMV Infection: A Compelling Argument for Vaccine Development

The economic burden on the healthcare system in caring for neurodevelopmental disability in early childhood caused by congenital HCMV infection is substantial. Congenital HCMV infection is the most common infectious cause of brain damage in children, and HCMV causes more hearing loss in children than did *Haemophilus influenzae* meningitis in the pre-Hib vaccine era (Pass 1996). The economic costs to society associated with congenital HCMV infection present a compelling argument for vaccine development. In the early 1990s, the expense to the US healthcare system associated with congenital HCMV infection was estimated at approximately $1.9 billion annually, with an average cost per child of over $300,000 (Arvin et al. 2004). Children with congenital HCMV infection often require long-term custodial care and extensive medical and surgical interventions. A recent economic analysis by the Institute of Medicine (IOM) examined the theoretical cost-effectiveness of a hypothetical HCMV vaccine based on quality adjusted life years (QALYs). QALYs quantify the acute and chronic problems caused by an illness. Employing this model, the more severe or permanent the sequelae, the larger the potential benefit conferred by an effective intervention will be. Not surprisingly, a hypothetical HCMV vaccine administered to 12-year-olds was in the level 1 group (the group for which a vaccine development strategy would save society money), and in fact was the single most cost-effective vaccine identified (Stratton et al. 1999). Thus, the economic benefit of HCMV vaccination holds the highest priority for any hypothetical new vaccine.

HCMV Vaccine: What Is the Ideal Target Population?

Perinatal and Early Childhood HCMV Infection

One strategy for vaccine-mediated prevention of HCMV would be to target acquisition of primary infection in infancy and early childhood. Perinatal acquisition of HCMV may occur by one of three different routes: exposure to HCMV in the birth canal during labor and delivery, transmission of HCMV by blood transfusion, or transmission by breast-feeding. In a prospective study in premature infants receiving breast milk containing HCMV, transmission was observed in 33 of 87 exposed infants, and approximately half of these babies developed disease, including hepatitis, neutropenia, thrombocytopenia, and sepsis-like state (Maschmann et al. 2001). It is uncertain if HCMV infection of low-birth-weight premature infants by this route carries any risk of long-term sequelae, and highly speculative as to whether maternal immunization programs would play a role in elimination of transmission by breast milk in this vulnerable population.

Beyond the immediate neonatal period, an extremely important population for primary HCMV infection - and a potential target for implementation of a vaccine program - is the early childhood population, in particular infants and toddlers attending group daycare. Although primary HCMV infection may occasionally cause mild disease in the toddler, a far greater concern is that the child may serve as a vehicle for subsequent infection of a parent. Should a pregnant mother become infected, the resulting newborn would then be at significant risk of HCMV disease and its attendant sequelae. Such child-to-parent transmission of HCMV has been well documented: daycare workers are in particular at increased risk for primary HCMV infection (Pass et al. 1990; Murph et al. 1991). Thus, interruption of HCMV transmission in the daycare environment could serve as an important efficacy endpoint for vaccine programs. Behavioral interventions and improved education about the risks of transmission can also likely play a role in decreasing the likelihood of this mode of transmission (Cannon and Davis 2005), but behavioral interventions alone are unlikely to completely eliminate the risk of transmission in this setting. Development and implementation of HCMV serologic screening programs for women of child-bearing age may be of benefit in identifying those women and families who might benefit most from behavioral and vaccination strategies aimed at interrupting this type of transmission.

Clearly, immunization of infants and toddlers prior to acquisition of primary HCMV infection is a strategy that should be considered for HCMV disease control. Immunization of the infant or toddler could result in secondary benefits for adult subjects, particularly mothers, who would be at decreased risk for acquiring infection from their child. Such an immunization approach has been utilized for rubella vaccine, which is routinely administered to young children; in this setting, the primary benefit of vaccination is not the prevention of rubella per se in the child, but the prevention of rubella transmission to young women, with the secondary benefit of prevention of congenital rubella syndrome in subsequent pregnancies. This occurs, in part, through herd immunity, which ultimately benefits all women of child-bearing age. Use of mathematical modeling suggests that such an approach for a HCMV vaccine would produce benefits similar to those realized by rubella vaccination (Griffiths et al. 2001). Thus, the strategy of universal immunization of young children against HCMV deserves further consideration.

Moreover, universal immunization against HCMV in early life may confer health benefits that extend ultimately to men as well as women of child-bearing age. Increasingly, HCMV infection has been tied to an increased lifetime risk of illnesses such as atherosclerosis, malignancies, inflammatory and autoimmune diseases, and the phenomenon of immune senescence in later life (Soderberg-Naucler 2006). Prevention of HCMV infection, and conceivably elimination of infection through herd immunity, could provide widespread benefits for human health.

Adolescent HCMV Infections

Adolescents acquire primary HCMV infections at a high frequency. In a prospective study of HCMV-seronegative adolescents, an annual HCMV infection rate of 13.1% was observed (Zhanghellini et al. 1999). Onset of sexual activity and exposure to young children, particularly in childcare settings, have been proposed as potential sources of primary HCMV infection in this population. Therefore, adolescence may also be an important target population for eventual implementation of HCMV vaccination programs. In recognition of the importance of the adolescent period in acquisition of primary HCMV infection, the Institute of Medicine (IOM) modeled its analysis of the potential benefits of a HCMV vaccine program upon hypothetical administration of vaccine to the adolescent patient (Stratton et al. 1999).

HCMV Infection and Disease in the Immunocompromised Patient

The potential value of HCMV vaccines is not limited to prevention of congenital infection. Bone marrow/stem cell transplant and solid organ transplant patients are at high risk for HCMV disease, pneumonitis, enteritis, retinitis, and viremia. Although the availability of effective prophylactic and preemptive antiviral therapy has made HCMV a rare cause of mortality in the HSC transplantation setting, HCMV-seropositive transplant recipients and seronegative recipients of a positive graft have a mortality disadvantage when compared with seronegative recipients with a seronegative donor (Boeckh et al. 2003). HCMV seropositivity is an important risk factor for impaired graft survival, increased risk of graft-versus-host disease, and other opportunistic infections such as invasive fungal infections. Therefore, prevention strategies that employ vaccines capable of stimulating both humoral and cell-mediated immune responses to HCMV may be of value in further decreasing the incidence and severity of HCMV disease, as well as these other complications of transplantation. Such vaccines could be administered to either the transplant recipient or to the HSC donor prior to transplantation. Whether the same vaccines that might prove successful in this patient population would protect against HCMV transmission in women of child-bearing age is uncertain.

Evidence That Immunity Protects Against HCMV Infection and Disease

Role of Preconception Maternal Immunity in Protection Against Congenital HCMV Transmission and HCMV Disease in the Newborn

Preconceptual maternal immunity to HCMV clearly provides some degree of protection against the most devastating forms of congenital infection. In a comparison of outcomes of HCMV-infected infants born to mothers who acquired primary infection during pregnancy with those of infected infants born to mothers with preconception immunity, only infants born in the primary-infection group had symptomatic disease at birth. These infants were at the highest risk for long-term sequelae (Fowler et al. 1992). However, a recent study suggests that preconceptual immunity does not completely eliminate the risk of symptomatic congenital transmission. In this study, some women who were seropositive for HCMV were nonetheless susceptible to reinfection with a new HCMV strain during pregnancy, and such reinfections did lead in some cases to symptomatic disease in the neonate (Boppana et al. 2001). In light of these data, a HCMV vaccine may not completely eliminate the potential for congenital HCMV transmission. Substantial evidence nonetheless strongly suggests that a HCMV vaccine program would protect many newborns. Other lines of evidence indicate that preconceptual immunity reduces both the incidence of congenital transmission and the severity of disease if transmission occurs. In a recent study that followed over 3,000 women from one pregnancy to the subsequent pregnancy and delivery, the rate of congenital HCMV infection was three times higher in offspring of women who initially were HCMV seronegative (Fowler et al. 2003). Preconception immunity was clearly protective in this study and resulted in a 69% reduction of congenital HCMV infection. Protection against congenital HCMV infection is enhanced by longer time intervals between pregnancies (Fowler et al. 2004). This effect is likely due to maturation of antibody avidity against HCMV (Revello and Gerna 2002). Therefore, emphasis should be placed on developing HCMV vaccines that are capable of mimicking the protective components of natural immunity, particularly antibody avidity, and such vaccines would likely have a significant impact on preventing symptomatic congenital HCMV infections.

Lessons from Adoptive Transfer Studies

Additional evidence supporting the protective role of immunity in preventing symptomatic congenital HCMV transmission comes from recently described passive immunization studies, using high-titer anti-HCMV immunoglobulin. In this study (Nigro et al. 2005), pregnant women with a primary HCMV infection were offered intravenous HCMV hyperimmune globulin, in two different dose regimens (therapy and prevention groups). In the therapy group, only 1 of 31 women gave birth to an infant with HCMV disease (defined as an infant who was symptomatic at birth and handicapped at 2 or more years of age), compared with 7 of 14 women in an untreated control group. In the prevention group, 6 of 37 women who received hyperimmune globulin during pregnancy had infants with congenital HCMV infection, compared with 19 of 47 women who did not receive the high-titer HCMV globulin. Although uncontrolled, these data support the protective effect of humoral immunity in prevention of fetal HCMV-associated disease. Additional randomized controlled trials of immune globulin are warranted in high-risk pregnancies, to further validate the protective effect of passive immunization.

HCMV Vaccines in Clinical Trials

A number of HCMV vaccines have been evaluated in clinical trials. These vaccine candidates are summarized in Table 1. A variety of strategies have been employed, but generally HCMV vaccines can be conceptually subdivided into the categories of live, attenuated vaccines, and subunit vaccines that target individual proteins (see the chapter by W. Gibson, this volume). Progress in study of these vaccines is considered the next section.

Table 1 HCMV vaccines that have undergone evaluation in clinical trials

Live, attenuated vaccines	
AD169 vaccine	Elicited HCMV-specific antibody responses in seronegative vaccine recipients
	Significant injection-site and systemic reactogenicity
	No ongoing studies active
Towne (±rhIL12)	Elicits humoral and cellular immune responses
	Favorable safety profile; no evidence for latency or viral shedding in recipients
	Lack of efficacy for HCMV infection; reduced HCMV disease in renal transplant recipients
	Augmentation of immunogenicity by inclusion of recombinant IL–12 in phase 1 studies
Towne/Toledo chimera vaccines	Favorable safety profile; no evidence for latency or viral shedding in recipients
	Attenuated compared to Toledo strain of HCMV
	No efficacy data available
Subunit vaccines	
Glycoprotein B/MF59 adjuvant (CHO cell expression)	Favorable safety profile
	High-titer neutralizing antibody and strong cell-mediated immune responses
	Efficacy studies ongoing in young women, adolescents, renal transplant patients
Glycoprotein B/canarypox vector	Favorable safety profile
	Suboptimal immunogenicity
	Prime–boost effect when administered in combination with Towne vaccine
pp65 (U83)/canarypox vector	Favorable safety profile
	Strong antibody and cell-mediated immune responses
	No efficacy data available
gB/pp65/IE1 trivalent DNA vaccine gB/pp65 bivalent DNA vaccine	DNA vaccine with poloxamer adjuvant
	Phase I studies completed
	Phase 2 study ongoing with bivalent gB/pp65 vaccine in HSC transplant recipients
gB/pp65/IE1 alphavirus replicon trivalent vaccine	Based on replication-deficient alphavirus technology
	Generation of virus-like replicon particles (VRPs)
	Phase I clinical trial recently initiated

Live, Attenuated HCMV Vaccines

HCMV has been the target of live, attenuated vaccine development efforts since the 1970s (reviewed in Schleiss and Heineman 2005). The first live, attenuated HCMV vaccine candidate tested in humans was based on the laboratory-adapted AD169 strain. Subsequent trials with another laboratory-adapted clinical isolate, the Towne strain, confirmed that live attenuated vaccines could elicit neutralizing antibodies, as well as $CD4^+$ and $CD8^+$ T lymphocyte responses. The efficacy of Towne vaccine was tested in a series of studies in renal transplant recipients. Although Towne failed to prevent HCMV infection after transplantation, vaccination did provide a protective impact on HCMV disease (Plotkin et al. 1994). Towne vaccine was also evaluated in a placebo-controlled study in seronegative mothers who had children attending group daycare. This study indicated that immunization with Towne failed to protect these women from acquiring HCMV infection from their children. The apparent failure of Towne vaccine was in contrast to the protection against reinfection observed in women with preexisting immunity, who were protected against acquiring a new strain of HCMV from their children (Adler et al. 1995). One interpretation of this study is that a HCMV vaccine that induced immune responses comparable to natural infection could provide protection of a high-risk patient population, but that the Towne vaccine may be overattenuated for this purpose. The molecular basis for the apparent overattenuation of the Towne vaccine remains unknown. Recent evidence suggests that the relative defect in Towne vaccine may be related to inadequate antigen-specific interferon gamma responses by $CD4^+$ and $CD8^+$ cells following vaccination (Jacobsen et al. 2006a). An approach to improve the immunogenicity of the Towne vaccine is currently being explored, in which recombinant interleukin-12 (rhIL-12) is co-administered with Towne vaccine. The adjuvant effect of rhIL-12 was associated with increases in antibody titer to glycoprotein B and improved $CD4^+$ T cell proliferation responses in this recently reported phase I study(Jacobsen et al. 2006b).

Another approach to improve the immunogenicity of the Towne vaccine has recently been reported, in which a series of genetic recombinant vaccines were generated containing regions from the genome of the unattenuated Toledo strain of HCMV, substituted for the corresponding regions of the Towne genome (see the chapter by E. Murphy and T. Shenk, this volume). These Towne/Toledo chimeras retain some, but not all, of the mutations that apparently contribute to Towne vaccine attenuation and were hypothesized to be less attenuated, and hence presumably more immunogenic, than the Towne vaccine. Four independent chimeric vaccines were produced and tested in a double-blinded, placebo-controlled study (Heineman et al. 2006). All of the vaccines were well tolerated, and none were shed by vaccinees, as assessed by viral culture and PCR analyses of blood and body fluids. Thus, these vaccines are sufficiently attenuated to warrant future studies in seronegative individuals. Concerns about the potential risk of establishing a latent HCMV infection have hindered the progress of live, attenuated vaccine studies, although to date there has been no evidence that any of these approaches have resulted in latent or persistent infections in any subject.

Subunit Vaccines

Subunit vaccine approaches emphasize specific immunogenic viral proteins, expressed by a variety of techniques, and administered either singly or in combination. The candidate subunit vaccines that are in clinical or preclinical development are described in the following sections, along with an overview of the expression techniques being employed.

Glycoprotein B (gpUL55) Vaccine

The humoral immune response to HCMV is dominated by responses to viral glycoproteins, present in the outer envelope of the virus particle (see the chapters by W. Gibson, this volume and M.K. Isaacson et al., this volume). Of these, the most fully characterized is the glycoprotein complex I (gcI) consisiting of gB (gB; UL55). All sera from HCMV-seropositive individuals contain antibodies to gB, and up to 70% of the neutralizing antibody response is gB-specific (Britt et al. 1990). Recombinant vaccines based on gB demonstrate efficacy against disease in murine and guinea pig models of cytomegalovirus infection (Rapp et al. 1993; Schleiss et al. 2004), providing further support for human efficacy testing. Accordingly, the gB protein is the leading candidate for subunit vaccine development and testing and the vaccine currently most actively studied in clinical trials.

One formulation of HCMV gB currently being explored in clinical trials is a recombinant protein expressed in Chinese hamster ovary (CHO) cells. In contrast to native gB, this formulation of gB is a truncated, secreted form of the protein, modified in two ways to facilitate its expression and purification. First, the proteolytic cleavage site, R-T-K-R, at which gB is normally cleaved into its amino and carboxyl moieties, was modified to prevent cleavage of the protein; secondly, a stop mutation was introduced prior to its hydrophobic transmembrane domain, resulting in a truncated, soluble form of gB (Spaete 1991). The resulting secreted protein is purified from CHO cell culture supernatants and used, with adjuvant, as a vaccine. Purified recombinant gB vaccine has undergoing safety, immunogenicity, and efficacy testing in several clinical trials. The first study of this vaccine was a phase I randomized, double-blind, placebo-controlled trial, in adults, in which recombinant gB was combined with one of two adjuvants, MF59 or alum (Pass et al. 1999). Levels of gB-specific antibodies and total virus-neutralizing activity after the third dose of vaccine exceeded those observed in HCMV-seropositive controls. Antigen dose was evaluated in a phase I study of 95 HCMV-seronegative adult volunteers (Frey et al. 1999), and the immunogenicity and safety of the vaccine has been studied in a limited number of toddlers (Mitchell et al. 2002). In all studies reported to date, the safety profile of the vaccine has been favorable, although injection-site discomfort has been observed. There is currently a double-blinded, placebo-controlled phase II study of gB/MF59 vaccine ongoing in young HCMV-seronegative women who are at high risk for acquisition of primary infection (Zhang et al. 2006). This study should provide insights into the potential protective efficacy of this vaccine in young women. A recombinant gB study is also currently in progress in renal transplant patients. In this study, subjects awaiting transplantation receive gB vaccine, to test whether the antibody responses engendered will contribute to reduction of HCMV viral load following transplantation (P.D. Griffiths, personal communication).

Another formulation of recombinant gB has also been evaluated in clinical trials using a vectored vaccine expression system based on a canarypox vector, ALVAC, an attenuated poxvirus that replicates abortively in mammalian cells. Clinical trials have focused on using ALVAC-gB in a prime-boost approach, in which ALVAC vaccine is administered to prime immune responses for subsequent boost with live, attenuated vaccine, or recombinant protein. In the first such prime-boost study, ALVAC-gB was evaluated alone or in combination with live, attenuated Towne vaccine. ALVAC-gB vaccine induced low neutralizing and ELISA antibodies in seronegative adults, but subjects primed with ALVAC-gB and then boosted with a single dose of Towne developed binding and neutralizing antibody titers comparable to naturally seropositive individuals (Adler et al. 1999). A subsequent study compared three immunization regimens: subunit gB vaccine, ALVAC-gB followed by gB/MF59, or both vaccines administered concomitantly (Bernstein et al. 2002). All three vaccine approaches induced high-titer antibody and lymphoproliferative responses, but no benefit for priming was detected. Thus, ALVAC-gB priming appears to result in augmented gB-specific responses following a boost with Towne vaccine, but not subunit gB/MF59.

Another approach used to express gB as a vaccine is the use of an alphavirus replicon system. This approach results in generation of virus-like replicon particles (VRPs) based on an attenuated Venezuelan equine encephalitis (VEE) expression system. Advantages of the VRP approach include the expression of high levels of heterologous proteins, the targeting of expression to dendritic cells, and the induction of both humoral and cellular immune responses to the vectored gene products of interest. HCMV gB has been expressed in the VRP system, and these VRPs have undergone protein expression analyses in cell culture as well as immunogenicity studies in mice. These studies demonstrated that protein expression levels are highest in VRPs expressing the extracellular domain of gB. BALB/c mice immunized with VRP expressing gB developed high titers of neutralizing antibody to HCMV (Reap et al. 2007). Based on these encouraging results, a phase I study of a trivalent vaccine including gB has recently been commenced in humans.

A final expression approach that has been applied to HCMV gB is DNA vaccination. In preclinical studies of HCMV gB DNA vaccines in mice, both the full-length gB, as well as a truncated, secreted form expressing amino acids 1-680 (of a total of 906 gB residues), were evaluated. Immunization with both constructs induced neutralizing antibodies, but titers were higher in mice immunized with the DNA encoding the truncated form of gB, which predominately elicited IgG1 antibody. In contrast, the full-length gB construct primarily elicited IgG2a antibodies (Endresz et al. 1999). The gB plasmid vaccine that has moved forward in human clinical trials is accordingly based on a construct encoding a truncated, secreted form of the protein. For clinical trials, HCMV

DNA vaccines are currently formulated using the poloxamer adjuvant, CRL1005 and benzalkonium chloride. In a phase I trial using a bivalent vaccine consisting of gB and pp65 (see the next section), 1-mg and 5-mg doses were studied in HCMV-seropositive and -seronegative subjects, and appeared to be safe and well tolerated (Evans et al. 2004; R. Moss, personal communication). There is currently an ongoing multicenter study in HSC patients of this adjuvanted bivalent HCMV DNA vaccine, toward the goal of reducing HCMV viremia and disease in this high-risk patient population.

pp65 (ppUL83) Vaccines

The cellular immune response to HCMV infection includes MHC class II restricted $CD4^+$ and MHC class I restricted, cytotoxic $CD8^+$ T lymphocyte responses to a number of viral antigens, many of which are found in the viral tegument, the region of the viral particle that lies between the envelope and nucleocapsid (see the chapter by R. Kalejta, this volume). For vaccination strategies aimed at eliciting T cell responses, most attention has focused on the pp65 protein (ppUL83). This is in part based on the apparent dominance of pp65 in the cellular response to HCMV: this protein elicits the majority of $CD8^+$ T lymphocyte responses following HCMV infection (McLaughlin-Taylor et al. 1994; Wills et al. 1996). The observation that adoptive transfer of pp65-specific CTL ameliorates HCMV disease in high-risk transplant patients provides further support for the study of pp65-based vaccines (Walter et al. 1995).

Many of the same expression strategies described for development of candidate gB vaccines in clinical trials have been employed for generation of pp65-based vaccines. An ALVAC vaccine expressing pp65 was administered to HCMV seronegative adult volunteers in a placebo-controlled trial (Berencsi et al. 2001). The ALVAC/pp65 recipients developed HCMV-specific $CD8^+$ CTL responses at frequencies comparable to those seen in naturally seropositive individuals. A pp65-based alphavirus/VRP vaccine has also been developed, using the approach described above for VRP-gB (Reap et al. 2007). Support for a VRP-pp65 vaccine approach was garnered in a recent guinea pig study, in which the guinea pig CMV (GPCMV) homolog of pp65, the GP83 gene product, was studied as a vaccine against congenital GPCMV infection. In this study, the VRP-GP83 vaccine improved pregnancy outcomes and reduced maternal viral load following early third-trimester viral challenge (Schleiss et al. 2007). The VRP-pp65 vaccine has entered phase 1 trials in humans, administered in a trivalent formulation with gB and IE1 (UL123) VRPs. As noted above, pp65 has also been expressed in a DNA vaccine, and is currently being evaluated in a phase II study in BMT recipients, co-administered in a bivalent formulation with gB DNA vaccine.

IE1 Vaccines

Based on the observation that the HCMV IE1 gene is an important target of the $CD8^+$ T cell response to HCMV infection, with IE1-specific responses being identified in up to 40% of HCMV-seropositive subjects (Slezak et al. 2007), this gene product is also being evaluated in a number of clinical trials. These studies to date have not involved administration of IE1 vaccine alone, but have consisted of trivalent vaccines that also contain pp65 and gB. As noted, one expression approach that has undergone phase 1 evaluation is that of DNA vaccination. A trivalent DNA vaccine targeting gB, pp65, and IE1 (Vilalta et al. 2005) was evaluated in a phase 1 trial involving a total of 40 healthy adult subjects (24 HCMV-seronegative, 16 HCMV-seropositive). Subjects received a 1-mg or 5-mg dose of trivalent vaccine in several multidose regimens; safety and immunogenicity studies are ongoing (R. Moss, personal communication). IE1 has also been expressed using the alphavirus/VRP approach. A trivalent VRP vaccine, consisting of the IE1 gene product along with gB and pp65, is currently being evaluated in a phase I study (Reap et al. 2007).

HCMV Vaccine Approaches in Preclinical Development

Alternative Expression Strategies for HCMV gB, pp65, and IE1

In addition to the expression strategies outlined above that have made their way into human clinical trials, there are other modes of expression of HCMV subunit vaccine candidates that appear useful in preclinical study. These approaches are summarized in Table 2. One particularly promising approach is based on a recombinant attenuated poxvirus, modified vaccinia virus Ankara. A recombinant Ankara vaccine has been constructed that expresses a soluble, secreted form of

Table 2 Alternative subunit vaccine expression strategies proposed for HCMV gB, pp65, and IE1

Modified vaccinia virus Ankara (MVA)	High-level protein expression
	Excellent immunogenicity (humoral and cellular responses) in mice
	Ability to express multiple immunogens (bivalent or trivalent vaccines) in single construct
	Preexisting immunity to poxvirus does not limit immune response (utility for vaccinees who have received smallpox vaccine)
Recombinant adenovirus	Potential for induction of mucosal immune responses
	Replication-deficient adenoviruses available to ensure vector safety
	Efficacy in murine model using MCMV gB homolog
Transgenic plants	Recombinant HCMV gB successfully expressed in transgenic rice
	Offers potential for oral vaccination
	Potential for induction of mucosal immune responses
	No animal immunogenicity data yet reported

HCMV gB, based on the AD169 strain sequence (Wang et al. 2004). In preclinical studies, high levels of gB-specific neutralizing antibodies, equivalent to those induced by natural HCMV infection, were induced in immunized mice. Recombinant MVA have similarly been generated expressing pp65 and IE1, and are capable of inducing robust cell-mediated immune responses in preclinical studies in mice. A trivalent MVA expressing gB, pp65, and IE1 has been developed and proposed for clinical studies (Wang et al. 2006).

Another vectored vaccine approach that has been pursued in preclinical studies is that of utilization of a recombinant adenovirus vector for expression of HCMV subunit vaccine candidates. The observation that a recombinant adenovirus vaccine expressing the related murine CMV (MCMV) gB demonstrated protection in a mouse model of MCMV disease provides support for further studies of this strategy in human clinical trials (Shanley and Wu 2003).

In addition, HCMV gB has been successfully expressed in transgenic plants (Tackaberry et al. 1999), offering the possibility of a novel vaccination approach through oral/mucosal immunization.

Potential Role of Other Viral Proteins in HCMV Vaccine Design

As noted, most efforts in clinical trials of candidate subunit HCMV vaccine development and testing have focused on the envelope glycoprotein gB and the T cell targets, pp65 and IE1. However, a plethora of other HCMV-encoded proteins play key roles in the host immune response and these warrant consideration in future clinical trials. To date, only animal model data are available to validate the potential role of these proteins as vaccines. This information is summarized in Table 3.

HCMV Glycoproteins

In addition to gB, other envelope glycoproteins have been considered for vaccine development, although to date no candidates have been tested in human trials. Among the other HCMV glycoproteins, the gcII complex, consisting of gN (UL73) and gM (UL100), is of particular interest. Proteomic analyses of the HCMV virion have demonstrated that gcII is the most abundantly expressed glycoprotein in virus particles, emphasizing its potential importance in protective immunity (Varnum et al. 2004). HCMV infection elicits a gcII-specific antibody response in a majority of seropositive individuals (Shimamura et al. 2006), and DNA vaccines consisting of gcII antigens gM and gN are able to elicit neutralizing antibody responses in rabbits and mice (Shen et al. 2007).

Constituents of the gcIII complex, consisting of glycoproteins gH (UL75), gO (UL74), and gL (UL115), are also targets of neutralizing antibody responses, and these proteins may also merit consideration in future vaccine studies. Recently the Shenk and Koszinowski laboratories have shown that there are two gH/gL

Table 3 Potential novel HCMV vaccine strategies that have been explored in preclinical/animal model studies

gM/gN (gcII complex)	Major glycoprotein constituent of virion
	Majority of human sera contain anti-gcII antibodies
	gM/gN DNA vaccine immunogenic in mice
gH/gL/gO (gcIII complex)	Target of neutralizing antibody response in setting of HCMV infection
	gH Vaccine based on MCMV homolog protective in murine model when expressed using recombinant adenovirus technique
Essential/nonstructural gene products as novel CTL targets	DNA polymerase (UL54) and helicase (UL105) as novel T cell targets
	Protective in MCMV model when expressed as DNA vaccines
Prime–boost strategy	Prime with cocktail of plasmid DNA vaccines
	Boost with formalin-inactivated viral particles
	Induces sterilizing immunity in MCMV model
Bacterial artificial chromosome (BAC) vaccines	Protective in MCMV model following delivery in bacteria with reconstitution of virus in vivo
	Protective in GPCMV model when administered as replication-disabled DNA vaccine
	Offers theoretical potential for specifically engineered vaccines with deletions in putative pathogenesis or immune evasion genes: improved immunogenicity
Peptide vaccines	Effective in MCMV model following mucosal immunization with cholera toxin
	Allows simultaneous immunization against broad range of CTL epitopes: polyepitope vaccine
	Requires knowledge of HLA status of vaccine recipient; best suited to HCMV vaccination in transplantation setting?

complexes in virions from clinical isolates: gH/gL and gO is one; the other is gH/gL, pUL128, pUL130 (Wang and Shenk 2005; Adler et al. 2006). Antibodies to pUL128, pUL130, or pUL131 can neutralize infection of epithelial cells but have no effect on infection of fibroblasts. Whether these observations will be translated into an expanded list of candidate subunit targets remains to be determined. Since acquisition of new antibody specificities to gH in the setting of reinfection with a new strain of HCMV is associated with symptomatic congenital transmission in pregnant patients (Boppana et al. 2001) and with adverse outcomes following renal transplantation (Ishibashi et al. 2007), these observations might be important for potential gcIII-based vaccine design. Proof-of-concept has been studied with vaccines based on the MCMV gH homolog using recombinant vaccinia (Rapp et al. 1993) and adenovirus (Shanley and Wu 2005) vectors; of these two approaches, the adenovirus-expressed gH vaccine was more effective as a vaccine against MCMV.

Regulatory/Structural HCMV Proteins Involved in T Cell Response

Most attention on HCMV vaccine candidates that elicit potentially protective T cell responses has been focused on the pp65 and IE1 gene products. However, recent evaluation of the T cell responses in HCMV-seropositive individuals identified a plethora of additional, previously unrecognized CD4+ and CD8+ T lymphocyte targets encoded by the HCMV genome. Other potential CD8+ T cell targets, in addition to pp65 and IE1, are just beginning to be investigated. A recently described bioinformatics and ex vivo functional T cell assay approach revealed that CD8+ T cell responses to HCMV often contained multiple antigen-specific reactivities, which were not just constrained to pp65 or IE1 antigens. These studies identified structural, early/late antigens, and HCMV-encoded immunomodulators (pp28, pp50, gH, gB, US2, US3, US6, and UL18) as potential targets for HCMV-specific CD8+ T cell immunity (Elkington et al. 2003). An elegant and comprehensive analysis of T cell responses to HCMV infection was recently conducted using cytokine flow cytometry in conjunction with overlapping peptides comprising 213 HCMV open reading frames: this study demonstrated that 151 HCMV ORFs were immunogenic for CD4+ and/or CD8+ T cells (Sylwester et al. 2005). Recently, an approach to vaccination has been examined in the MCMV model in which essential, nonstructural proteins that are highly conserved among the CMVs were explored as a novel class of T cell targets. These studies found that DNA immunization of mice with the murine CMV (MCMV) homologs of HCMV DNA polymerase (M54) or helicase (M105) was protective against virus replication following systemic challenge, and that gamma interferon staining of CD8+ T cells from mice immunized with either the M54 or M105 DNAs showed strong primary responses that recalled rapidly after viral challenge (Morello et al. 2007). These conserved, essential proteins thus may represent a novel class of CD8+ T cell targets that could contribute to successful HCMV vaccine design.

Dense Body Vaccines

A novel candidate for vaccination against HCMV currently in preclinical development is the dense body vaccine. Dense bodies (DBs) are enveloped, replication-defective particles formed during replication of CMVs in cell culture. These structures are a potentially promising vaccine because they contain both envelope glycoproteins and large quantities of pp65 protein, two key targets of the protective immune response to infection. DBs are noninfectious and therefore, in principle, would be immunogenic, but incapable of establishing latent HCMV infection in the vaccine recipient, providing a useful safety feature. DBs have been shown to be capable of inducing virus neutralizing antibodies and T cell responses after immunization of mice, including human HLA-A2.K(b) transgenic mice, in the absence of viral gene expression. Based on these studies, the utilization of DBs may represent a promising, novel approach to the development of a subunit vaccine against HCMV infection (Pepperl et al. 2000).

Peptide-Based Vaccines

Another potential approach to HCMV vaccination is the use of peptide vaccination employing synthetic peptides comprising immunodominant cytotoxic T cell epitopes. This approach may ultimately prove to be most useful in the vaccine-mediated prevention of HCMV disease in the transplant setting, where specific peptide vaccine regimens could be tailored for donor-recipient pairs based on HLA genetics. Nasal peptide vaccination with the immunodominant MCMV IE1 epitope, YPHFMPTNL, in combination with cholera toxin adjuvant, protected mice against virulent MCMV challenge (Gopal et al. 2005). In a preclinical study relevant to HCMV vaccines, a pp65 human leukocyte antigen (HLA)-A2.1-restricted CTL epitope corresponding to an immunodominant region spanning amino acid residues 495-503, fused to the carboxyl terminus of a pan-DR T-help epitope, was capable of eliciting CTL responses from mice transgenic for the same human HLA molecule (BenMohamed et al. 2000). Since this epitope is highly conserved in clinical isolates, this vaccine could conceivably be broadly protective against multiple HCMV strains. Other peptide-based vaccine studies are envisioned in future clinical trials. A subset of epitopes from a group of important CTL targets has been nominated for inclusion in a polyepitope HCMV vaccine on the basis of human immune responsiveness and population coverage (Khanna and Diamond 2006).

Novel Vaccine Approaches

Several other vaccination approaches have been proposed for HCMV and have been validated in varying degrees in animal models. One approach is based on exploitation of viral genomes cloned in *Escherichia coli* as bacterial artificial chromosomes (BACs). Vaccination of mice with bacteria containing the MCMV genome cloned as a BAC conferred protective immunity against subsequent challenge (Cicin-Sain et al. 2003). In guinea pigs, a noninfectious BAC generated by transposon mutagenesis induced immune responses that protected against congenital GPCMV infection and disease (Schleiss et al. 2006). Given the ease of manipulation of BACs using mutagenesis techniques available for *E. coli*, future BAC studies provide the opportunity to generate recombinant, designer vaccines with specific genomic deletions or insertions that could modify the immune response or improve the safety profile of the candidate vaccine. Such modified BACs are being employed as immunocontraceptive vaccines for population control in mice, with insertion of heterologous proteins that can serve as targets for immune responses that result in decreased fertility in vaccinated animals (Hardy 2007).

Another novel strategy that has been validated in the MCMV model and proposed for clinical trial evaluation is a prime-boost approach, in which priming with DNA vaccination is followed by boosting with formalin-inactivated viral particles. Mice were immunized with a cocktail of 13 MCMV-containing plasmids followed by boosting with formalin-inactivated, alum-adjuvanted MCMV. This approach elicited high levels of neutralizing antibodies as well as CD8$^+$ T cells specific for the virion-associated antigen (Morello et al. 2002). Subsequent studies examined whether similar protection levels could be achieved by priming with a pool of plasmids encoding MCMV proteins IE1, M84, and gB. This approach was found to elicit CD8+ T lymphocyte responses and, following boost with inactivated MCMV, high levels of virus-neutralizing antibody. Following MCMV challenge, titers of virus were either at or below the detection limits for the salivary glands, liver, and spleen of most immunized animals (Morello et al. 2005). These results support further study of prime-boost approaches for vaccination in other animal models, and, potentially, for future clinical trials of HCMV vaccines.

Perspectives

Although it is appropriate to consider multiple potential options for HCMV vaccination, the principle barrier to licensure of a vaccine is not a paucity of acceptable candidate vaccines from which to currently choose. Rather, the major barrier to progress is the lack of knowledge about the public health significance of congenital HCMV infection and the disabilities it produces in children. Increased public awareness about the risks of HCMV is urgently needed: this in turn will drive the social, political, and economic forces necessary to increase the pace of progress of clinical trials. Vaccine manufacturers need to increase emphasis on research and development of novel strategies at the same time that clinical trials of existing candidates move forward.

Although it has been asserted that a better understanding of the correlates of protective immunity must be achieved before the goal of a HCMV vaccine can be realized, in fact licensure and implementation of any of the vaccines tested to date would likely impact significantly on disease. A recent analysis by the Cannon group at the CDC examined the force of infection of HCMV, defined as the instantaneous per capita rate of acquisition of infection that approximates the incidence of infection in the seronegative population. Among individuals of reproductive age in the United States, the force of infection was 1.6 infections per 100 susceptible persons per year. Comparison of this to the basic reproductive rate of 1.7 indicated that, on average, an infected person transmits CMV to nearly two susceptible people - a relatively low force of infection when compared to other vaccine-preventable infectious diseases (Colugnati et al. 2007). This analysis suggested that even a modestly effective vaccine and rate of vaccination compliance could significantly reduce CMV transmission. Sterilizing immunity conferred by a vaccine is therefore not likely to be necessary for prevention of disease and disability in newborns due to congenital HCMV infection. Indeed, insights from the GPCMV model of congenital infection suggest that reduction of viral load below a critical threshold, and not sterilizing immunity, is sufficient to ensure improved pregnancy outcomes following vaccination in the setting of maternal viremia. Therefore, the pace of research should be substantially accelerated with existing vaccine candidates, even as research continues to more fully explore the correlates of protective immunity. Although expensive to perform and logistically challenging to conduct, it is imperative that trials be powered to examine symptomatic congenital HCMV infection as an efficacy endpoint.

Industry-sponsored clinical trials are currently largely focusing on evaluation of HCMV vaccines in HSC and solid organ transplant patients at high risk for HCMV viremia and disease. While such studies advance the field, negative data from these studies should be interpreted cautiously, and such data cannot automatically be extrapolated toward the problem of prevention of congenital HCMV infection and disease. Industry sponsors, funding agencies, and regulatory bodies must work together to dramatically accelerate the pace of clinical trials for this urgent public health priority.

References

Adler SP, Starr SE, Plotkin SA, Hempfling SH, Buis J, Manning ML, Best AM (1995) Immunity induced by primary human cytomegalovirus infection protects against secondary infection among women of childbearing age. J Infect Dis 171:26-32

Adler SP, Plotkin SA, Gonczol E, Cadoz M, Meric C, Wang JB, Dellamonica P, Best AM, Zahradnik J, Pincus S, Berencsi K, Cox WI, Gyulai Z (1999) A canarypox vector expressing cytomegalovirus (CMV) glycoprotein B primes for antibody responses to a live attenuated CMV vaccine (Towne). J Infect Dis 180:843-846

Adler B, Scrivano L, Ruzcics Z, Rupp B, Sinzger C, Koszinowski U (2006) Role of human cytomegalovirus UL131A in cell type-specific virus entry and release. J Gen Virol 87:2451-2460

Arvin AM, Fast P, Myers M, Plotkin S, Rabinovich R; National Vaccine Advisory Committee (2004) Vaccine development to prevent cytomegalovirus disease: report from the National Vaccine Advisory Committee. Clin Infect Dis 39:233-239

BenMohamed L, Krishnan R, Longmate J, Auge C, Low L, Primus J, Diamond DJ (2000) Induction of CTL response by a minimal epitope vaccine in HLA A*0201/DR1 transgenic mice: dependence on HLA class II restricted T(H) response. Hum Immunol 61:764-779

Berencsi K, Gyulai Z, Gonczol E, Pincus S, Cox WI, Michelson S, Kari L, Meric C, Cadoz M, Zahradnik J, Starr S, Plotkin S (2001) A canarypox vector-expressing cytomegalovirus (CMV) phosphoprotein 65 induces long-lasting cytotoxic T cell responses in human CMV-seronegative subjects. J Infect Dis 183:1171-1179

Bernstein DI, Schleiss MR, Berencsi K, Gonczol E, Dickey M, Khoury P, Cadoz M, Meric C, Zahradnik J, Duliege AM, Plotkin S (2002) Effect of previous or simultaneous immunization with canarypox expressing cytomegalovirus (CMV) glycoprotein B (gB) on response to subunit gB vaccine plus MF59 in healthy CMV-seronegative adults. J Infect Dis 185:686-690

Boeckh M, Nichols WG, Papanicolaou G, Rubin R, Wingard JR, Zaia J (2003) Cytomegalovirus in hematopoietic stem cell transplant recipients: current status, known challenges, and future strategies. Biol Blood Marrow Transplant 9:543-558

Boppana SB, Rivera LB, Fowler KB, Mach M, Britt WJ (2001) Intrauterine transmission of cytomegalovirus to infants of women with preconceptional immunity. N Engl J Med 344:1366-1371

Britt WJ, Vugler L, Butfiloski EJ, Stephens EB (1990) Cell surface expression of human cytomegalovirus (HCMV) gp55-116 (gB): use of HCMV-recombinant vaccinia virus-infected cells in analysis of the human neutralizing antibody response. J Virol 64:1079-1085

Cannon MJ, Davis KF (2005) Washing our hands of the congenital cytomegalovirus disease epidemic. BMC Public Health 5:70

Cicin-Sain L, Brune W, Bubic I, Jonjic S, Koszinowski UH (2003) Vaccination of mice with bacteria carrying a cloned herpesvirus genome reconstituted in vivo. J Virol 77:8249-8255

Colugnati FA, Staras SA, Dollard SC, Cannon MJ (2007) Incidence of cytomegalovirus infection among the general population and pregnant women in the United States. BMC Infect Dis 7:71

Demmler GJ (1996) Congenital cytomegalovirus infection and disease. Adv Pediatr Infect Dis 11:135-162

Elkington R, Walker S, Crough T, Menzies M, Tellam J, Bharadwaj M, Khanna R (2003) Ex vivo profiling of CD8+-T-cell responses to human cytomegalovirus reveals broad and multispecific reactivities in healthy virus carriers. J Virol 77:5226-5240

Endresz V, Kari L, Berencsi K, Kari C, Gyulai Z, Jeney C, Pincus S, Rodeck U, Meric C, Plotkin SA, Gonczol E (1999) Induction of human cytomegalovirus (HCMV)-glycoprotein B (gB)-specific neutralizing antibody and phosphoprotein 65 (pp65)-specific cytotoxic T lymphocyte responses by naked DNA immunization. Vaccine 17:50-58

Evans TG, Wloch M, Hermanson G, Selinsky C, Geall A, Kaslow D (2004) Phase 1 trial of a bivalent, formulated plasmid DNA CMV vaccine for use in the transplant population. Abstracts of the 44th ICAAC; G-543

Fowler KB, Stagno S, Pass RF, Britt WJ, Boll TJ, Alford CA (1992) The outcome of congenital cytomegalovirus infection in relation to maternal antibody status. N Engl J Med 326:663-667

Fowler KB, Stagno S, Pass RF (2003) Maternal immunity and prevention of congenital cytomegalovirus infection. JAMA 289:1008-1011

Fowler KB, Stagno S, Pass RF (2004) Interval between births and risk of congenital cytomegalovirus infection. Clin Infect Dis 38:1035-1037

Frey SE, Harrison C, Pass RF, Yang E, Boken D, Sekulovich RE, Percell S, Izu AE, Hirabayashi S, Burke RL, Duliege AM (1999) Effects of antigen dose and immunization regimens on antibody responses to a cytomegalovirus glycoprotein B subunit vaccine. J Infect Dis 180:1700-1703

Gopal IN, Quinn A, Henry SC, Hamilton JD, Staats HF, Frothingham R (2005) Nasal peptide vaccination elicits CD8 responses and reduces viral burden after challenge with virulent murine cytomegalovirus. Microbiol Immunol 49:113-119

Griffiths PD, McLean A, Emery VC (2001) Encouraging prospects for immunisation against primary cytomegalovirus infection. Vaccine 19:1356-1362

Hardy CM (2007) Current status of virally vectored immunocontraception for biological control of mice. Soc Reprod Fertil Suppl 63:495-506

Heineman TC, Schleiss M, Bernstein DI, Spaete RR, Yan L, Duke G, Prichard M, Wang Z, Yan Q, Sharp MA, Klein N, Arvin AM, Kemble G (2006) A phase 1 study of 4 live, recombinant human cytomegalovirus Towne/Toledo chimeric vaccines. J Infect Dis 193:1350-1360

Ishibashi K, Tokumoto T, Tanabe K, Shirakawa H, Hashimoto K, Kushida N, Yanagida T, Inoue N, Yamaguchi O, Toma H, Suzutani T (2007) Association of the outcome of renal transplantation with antibody response to cytomegalovirus strain-specific glycoprotein H epitopes. Clin Infect Dis 45:60-67

Jacobson MA, Sinclair E, Bredt B, Agrillo L, Black D, Epling CL, Carvidi A, Ho T, Bains R, Adler SP (2006a) Antigen-specific T cell responses induced by Towne cytomegalovirus (CMV) vaccine in CMV-seronegative vaccine recipients. J Clin Virol 35:332-337

Jacobson MA, Sinclair E, Bredt B, Agrillo L, Black D, Epling CL, Carvidi A, Ho T, Bains R, Girling V, Adler SP (2006b) Safety and immunogenicity of Towne cytomegalovirus vaccine with or without adjuvant recombinant interleukin-12. Vaccine 24:5311-5319

Khanna R, Diamond DJ (2006) Human cytomegalovirus vaccine: time to look for alternative options. Trends Mol Med 12:26-33

Kimberlin DW, Lin CY, Sanchez PJ, Demmler GJ, Dankner W, Shelton M, Jacobs RF, Vaudry W, Pass RF, Kiell JM, Soong SJ, Whitley RJ; National Institute of Allergy and Infectious Diseases Collaborative Antiviral Study Group (2003) Effect of ganciclovir therapy on hearing in symptomatic congenital cytomegalovirus disease involving the central nervous system: a randomized, controlled trial. J Pediatr 143:16-25

Maschmann J, Hamprecht K, Dietz K, Jahn G, Speer CP (2001) Cytomegalovirus infection of extremely low-birth weight infants via breast milk. Clin Infect Dis 33:1998-2003

McLaughlin-Taylor E, Pande H, Forman SJ, Tanamachi B, Li CR, Zaia JA, Greenberg PD, Riddell SR (1994) Identification of the major late human cytomegalovirus matrix protein pp65 as a target antigen for CD8+ virus-specific cytotoxic T lymphocytes. J Med Virol 43:103-110

Mitchell DK, Holmes SJ, Burke RL, Duliege AM, Adler SP (2002) Immunogenicity of a recombinant human cytomegalovirus gB vaccine in seronegative toddlers. Pediatr Infect Dis J 21:133-138

Morello CS, Ye M, Spector DH (2002) Development of a vaccine against murine cytomegalovirus (MCMV), consisting of plasmid DNA and formalin-inactivated MCMV, that provides long-term, complete protection against viral replication. J Virol 76:4822-4835

Morello CS, Ye M, Hung S, Kelley LA, Spector DH (2005) Systemic priming-boosting immunization with a trivalent plasmid DNA and inactivated murine cytomegalovirus (MCMV) vaccine provides long-term protection against viral replication following systemic or mucosal MCMV challenge. J Virol 79:159-175

Morello CS, Kelley LA, Munks MW, Hill AB, Spector DH (2007) DNA immunization using highly conserved murine cytomegalovirus genes encoding homologs of human cytomegalovirus UL54 (DNA polymerase) and UL105 (helicase) elicits strong CD8 T-cell responses and is protective against systemic challenge. J Virol 81:7766-7775

Murph JR, Baron JC, Brown CK, Ebelhack CL, Bale JF Jr (1991) The occupational risk of cytomegalovirus infection among day-care providers. JAMA 265:603-608

Nigro G, Adler SP, La Torre R, Best AM; Congenital Cytomegalovirus Collaborating Group (2005) Passive immunization during pregnancy for congenital cytomegalovirus infection. N Engl J Med 353:1350-1362

Pass RF (1996) Immunization strategy for prevention of congenital cytomegalovirus infection. Infect Agents Dis 5:240-244

Pass RF, Hutto C, Lyon MD, Cloud G (1990) Increased rate of cytomegalovirus infection among day care center workers. Pediatr Infect Dis J 9:465-467

Pass RF, Duliege AM, Boppana S, Sekulovich R, Percell S, Britt W, Burke RL (1999) A subunit cytomegalovirus vaccine based on recombinant envelope glycoprotein B and a new adjuvant. J Infect Dis 180:970-975

Pepperl S, Munster J, Mach M, Harris JR, Plachter B (2000) Dense bodies of human cytomegalovirus induce both humoral and cellular immune responses in the absence of viral gene expression. J Virol 74:6132-6146

Plotkin SA, Higgins R, Kurtz JB, Morris PJ, Campbell DA Jr, Shope TC, Spector SA, Dankner WM (1994) Multicenter trial of Towne strain attenuated virus vaccine in seronegative renal transplant recipients. Transplantation 58:1176-1178

Rapp M, Messerle M, Lucin P, Koszinowski UH (1993) In vivo protection studies with mCMV glycoproteins gB and gH expressed by vaccinia virus. In: Michelson S, Plotkin SA (eds) Multidisciplinary approach to understanding cytomegalovirus disease. Elsevier, Amsterdam, pp 327-332

Reap EA, Dryga SA, Morris J, Rivers B, Norberg PK, Olmsted RA, Chulay JD (2007) Cellular and humoral immune responses to alphavirus replicon vaccines expressing cytomegalovirus pp65, IE1, and gB proteins. Clin Vaccine Immunol 14:748-755

Revello MG, Gerna G (2002) Diagnosis and management of human cytomegalovirus infection in the mother, fetus, and newborn infant. Clin Microbiol Rev 15:680-715

Schleiss MR, Heineman TC (2005) Progress toward an elusive goal: current status of cytomegalovirus vaccines. Expert Rev Vaccines 4:381-406

Schleiss MR, Bourne N, Stroup G, Bravo FJ, Jensen NJ, Bernstein DI (2004) Protection against congenital cytomegalovirus (CMV) infection and disease in guinea pigs conferred by a purified recombinant glycoprotein B (gB) vaccine. J Infect Dis 189:1374-1381

Schleiss MR, Stroup G, Pogorzelski K, McGregor A (2006) Protection against congenital cytomegalovirus (CMV) disease, conferred by a replication-disabled, bacterial artificial chromosome (BAC)-based DNA vaccine. Vaccine 24:6175-6186

Schleiss MR, Lacayo JC, Belkaid Y, McGregor A, Stroup G, Rayner J, Alterson K, Chulay JD, Smith JF (2007) Preconceptual administration of an alphavirus replicon UL83 (pp65 homolog) vaccine induces humoral and cellular immunity and improves pregnancy outcome in the guinea pig model of congenital cytomegalovirus infection. J Infect Dis 195:789-798

Shanley JD, Wu CA (2003) Mucosal immunization with a replication-deficient adenovirus vector expressing murine cytomegalovirus glycoprotein B induces mucosal and systemic immunity. Vaccine 21:2632-2642

Shanley JD, Wu CA (2005) Intranasal immunization with a replication-deficient adenovirus vector expressing glycoprotein H of murine cytomegalovirus induces mucosal and systemic immunity. Vaccine 23:996-1003

Shen S, Wang S, Britt WJ, Lu S (2007) DNA vaccines expressing glycoprotein complex II antigens gM and gN elicited neutralizing antibodies against multiple human cytomegalovirus (HCMV) isolates. Vaccine 25:3319-3327

Shimamura M, Mach M, Britt WJ (2006) Human cytomegalovirus infection elicits a glycoprotein M (gM)/gN-specific virus-neutralizing antibody response. J Virol 80:4591-4600

Slezak SL, Bettinotti M, Selleri S, Adams S, Marincola FM, Stroncek DF (2007) CMV pp65 and IE-1 T cell epitopes recognized by healthy subjects. J Transl Med 5:17

Soderberg-Naucler C (2006) Does cytomegalovirus play a causative role in the development of various inflammatory diseases and cancer? J Intern Med 259:219-246

Spaete RR (1991) A recombinant subunit vaccine approach to HCMV vaccine development. Transplant Proc 23 [Suppl 3]:90-96

Stratton KR, Durch JS, Lawrence RS (1999) Vaccines for the 21st century: a tool for decisionmaking, Committee to Study Priorities for Vaccine Development, Division of Health Promotion and Disease Prevention, Institute of Medicine, Washington, DC

Sylwester AW, Mitchell BL, Edgar JB, Taormina C, Pelte C, Ruchti F, Sleath PR, Grabstein KH, Hosken NA, Kern F, Nelson JA, Picker LJ (2005) Broadly targeted human cytomegalovirus-specific CD4+ and CD8+ T cells dominate the memory compartments of exposed subjects. J Exp Med 202:673-685

Tackaberry ES, Dudani AK, Prior F, Tocchi M, Sardana R, Altosaar I, Ganz PR (1999) Development of biopharmaceuticals in plant expression systems: cloning, expression and immunological reactivity of human cytomegalovirus glycoprotein B (UL55) in seeds of transgenic tobacco. Vaccine 17:3020-3029

Varnum SM, Streblow DN, Monroe ME, Smith P, Auberry KJ, Pasa-Tolic L, Wang D, Camp DG 2nd, Rodland K, Wiley S, Britt W, Shenk T, Smith RD, Nelson JA (2004) Identification of proteins in human cytomegalovirus (HCMV) particles: the HCMV proteome. J Virol 78:10960-10966

Vilalta A, Mahajan RK, Hartikka J, Rusalov D, Martin T, Bozoukova V, Leamy V, Hall K, Lalor P, Rolland A, Kaslow DC (2005) I. Poloxamer-formulated plasmid DNA-based human cytomegalovirus vaccine: evaluation of plasmid DNA biodistribution/persistence and integration. Hum Gene Ther 16:1143-1150

Walter EA, Greenberg PD, Gilbert MJ, Finch RJ, Watanabe KS, Thomas ED, Riddell SR (1995) Reconstitution of cellular immunity against cytomegalovirus in recipients of allogeneic bone marrow by transfer of T-cell clones from the donor. N Engl J Med 333:1038-1044

Wang D, Shenk T (2005) Human cytomegalovirus virion protein complex required for epithelial and endothelial cell tropism. Proc Natl Acad Sci USA 102:18153-18158

Wang Z, La Rosa C, Maas R, Ly H, Brewer J, Mekhoubad S, Daftarian P, Longmate J, Britt WJ, Diamond DJ (2004) Recombinant modified vaccinia virus Ankara expressing a soluble form of glycoprotein B causes durable immunity and neutralizing antibodies against multiple strains of human cytomegalovirus. J Virol 78:3965-3976

Wang Z, La Rosa C, Lacey SF, Maas R, Mekhoubad S, Britt WJ, Diamond DJ (2006) Attenuated poxvirus expressing three immunodominant CMV antigens as a vaccine strategy for CMV infection. J Clin Virol 35:324-331

Whitley RJ (2004) Congenital cytomegalovirus infection: epidemiology and treatment. Adv Exp Med Biol 549:155-160

Wills MR, Carmichael AJ, Mynard K, Jin X, Weekes MP, Plachter B, Sissons JG (1996) The human cytotoxic T-lymphocyte (CTL) response to cytomegalovirus is dominated by structural protein pp65: frequency, specificity, and T-cell receptor usage of pp65-specific CTL. J Virol 70:7569-7579

Zanghellini F, Boppana SB, Emery VC, Griffiths PD, Pass RF (1999) Asymptomatic primary cytomegalovirus infection: virologic and immunologic features. J Infect Dis 180:702-707

Zhang C, Buchanan H, Andrews W, Evans A, Pass RF (2006) Detection of cytomegalovirus infection during a vaccine clinical trial in healthy young women: seroconversion and viral shedding. J Clin Virol 35:338-342

Cytomegalovirus Infection in the Human Placenta: Maternal Immunity and Developmentally Regulated Receptors on Trophoblasts Converge

L. Pereira(✉), E. Maidji

Contents

Introduction 384
Spatially Distinct Infection in the Developing Placenta 384
Villous Cytotrophoblasts Express EGFR and Upregulate Integrin αV 386
Cell Column Cytotrophoblasts Induce Integrin α1β1 Expression. 388
Replication in Differentiating/Invading Cytotrophoblasts. 389
Infection Impairs Cell Functions Through Diverse Membrane Proximal Events. 391
Implications for Congenital CMV Infection 392
References 393

Abstract During human pregnancy, CMV infects the uterine-placental interface with varied outcomes from fetal intrauterine growth restriction to permanent birth defects, depending on the level of maternal immunity and gestational age. Virus spreads from infected uterine blood vessels, amplifies by replicating in decidual cells, and disseminates to the placenta in immune complexes. Cytotrophoblasts – epithelial cells of the placenta – differentiate along two distinct pathways. In the first, cells fuse into syncytiotrophoblasts covering the surface of chorionic villi that transport substances from the maternal to fetal bloodstream. In the second, cells invade the uterine interstitium and blood vessels, remodel the vasculature and form anchoring villi. CMV initiates replication in cytotrophoblast progenitor cells of floating villi, whereas syncytiotrophoblasts are spared. This extraordinary pattern of focal infection in underlying cells hinges on virion receptors being upregulated as villous cytotrophoblasts begin to differentiate. Expression of developmentally regulated receptors could explain viral replication in spatially distinct maternal and fetal compartments. Reduced invasiveness of infected cells could impair remodeling of the uterine vasculature, restrict maternal blood flow and access of the fetus to nutrients causing intrauterine growth restriction.

L. Pereira
Department of Cell and Tissue Biology, School of Dentistry, University of California San Francisco, 513 Parnassus, C-734, Box-0640, San Francisco, 94143, USA
lenore.pereira@ucsf.edu

T. E. Shenk and M.F. Stinski (eds.), *Human Cytomegalovirus*.
Current Topics in Microbiology and Immunology 325.
© Springer-Verlag Berlin Heidelberg 2008

Introduction

Primary maternal CMV infection during gestation poses a 40%–50% risk of intrauterine transmission (Britt 1999), whereas reactivated infection in seropositive women rarely causes symptomatic disease, highlighting the central role for maternal humoral immunity in fetal protection (Fowler et al. 1992, 2003). It has long been recognized that congenital disease and placental damage are more severe when primary maternal infection occurs in the first trimester (Stagno et al. 1986; Garcia et al. 1989; Muhlemann et al. 1992; Benirschke and Kaufmann 2000). Several groups have reported that cytotrophoblasts, the specialized cells that compose the developing placenta, support CMV replication in utero (Sinzger et al. 1993; Fisher et al. 2000; Kumazaki et al. 2002; Pereira et al. 2003; McDonagh et al. 2004; Trincado et al. 2005). Cells isolated from early gestation (Fisher et al. 2000) and term placentas (Halwachs-Baumann et al. 1998; Hemmings et al. 1998; Tabata et al. 2007) are susceptible to CMV infection. Studies in early gestation revealed that virus replicates in the pregnant uterus and spreads from decidua to the adjacent placenta (Pereira et al. 2003; McDonagh et al. 2004). Islands of infection are found in cytotrophoblasts, whereas intercalating macrophages and dendritic cells internalize virions without replication (Fisher et al. 2000; Pereira et al. 2003; Maidji et al. 2006). In the placenta, infection occurs in small foci of cytotrophoblasts underlying syncytiotrophoblasts when maternal antibody has low neutralizing titer, but not high titer. Here we consolidate recent studies designed to clarify the synergy between the neonatal Fc receptor for IgG in syncytiotrophoblasts and virion receptors induced by villous cytotrophoblasts and invasive cells in the uterine interstitium and vasculature. Infection in early gestation could undermine placental development and lead to complications including fetal intrauterine growth restriction, a hallmark of congenital infection.

Spatially Distinct Infection in the Developing Placenta

Placentation is a stepwise process whereby specialized cytotrophoblast progenitor cells leave the basement membrane, differentiating along two pathways depending on their location (Fig. 1). In floating villi, cells fuse to form a multinucleate syncytial covering, syncytiotrophoblasts, attached at one end to the tree-like fetal portion of the placenta. These villi float in a stream of maternal blood, the source of nutrients and IgG transported by the neonatal Fc receptor for passive immunity of the fetus (Story et al. 1994; Simister et al. 1996). In anchoring villi, cytotrophoblasts switch from an epithelial to an endothelial phenotype controlled through the coordinated actions of numerous interrelated factors (Zhou et al. 1997; Janatpour et al. 2000; Genbacev et al. 2001). Cytotrophoblasts remodel uterine arteries, replacing endothelial cells and creating a hybrid vasculature that increases maternal blood flow to supply the developing placenta (Damsky et al. 1994; Damsky and Fisher 1998). The cells express adhesion molecules – integrins, Ig superfamily members

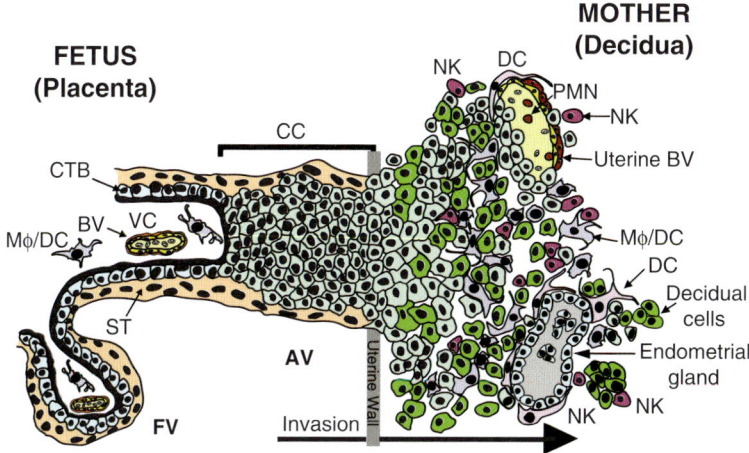

Fig. 1 Diagram of the placental (fetal)–decidual (maternal) interface near the end of the first trimester of human pregnancy (10 weeks of gestational age). A longitudinal section includes floating and anchoring chorionic villi. The floating villus (*FV*) is bathed by maternal blood. The anchoring villus (*AV*) functions as a bridge between the fetal and maternal compartments. Cytotrophoblasts in AV form cell columns that attach to the uterine wall. They invade the decidua and uterine vasculature, anchoring the placenta and gaining access to the maternal circulation. Colors illustrate different cell types: syncytiotrophoblasts (*ST*) (*beige*), cytotrophoblast (*CTB*) progenitors and invasive cells (*light green*), decidual cells (*green*), endothelial cells (*yellow*), smooth muscle cells (*brown*), glandular epithelial cells (*gray*), macrophage/dendritic cells (*Mϕ/DC*) (*light purple/pink*), polymorphonuclear neutrophils (*PMN*) (*red*), and natural killer cells (*NK*) (*fuchsia*)

and proteinases that enable invasiveness – and immune-modulating factors for maternal tolerance of the hemiallogeneic fetus (Cross et al. 1994; Damsky and Fisher 1998; Norwitz et al. 2001). Villous cytotrophoblasts express certain integrin subunits (Zhou et al. 1997), and interstitial invasive cells upregulate integrin $\alpha 1\beta 1$ expression (Damsky et al. 1994). Endovascular cytotrophoblasts express $\alpha V\beta 3$ and vasculogenic factors and receptors that mimic the surface of vascular cells (Zhou et al. 1997; Damsky and Fisher 1998). Invasive cytotrophoblasts also upregulate matrix metalloproteinase 9 (MMP-9), which degrades the extracellular matrix of the uterine stroma (Librach et al. 1991), and interleukin 10 (IL-10) for immune tolerance and modulation of metalloproteinases and invasiveness (Roth et al. 1996; Roth and Fisher 1999).

CMV is especially adapted to replicate in the immune tolerant microenvironment at the maternal–fetal interface (Pereira et al. 2005). Examination of intrauterine CMV infection in early gestation revealed patterns of virus replication in the decidua, mirrored in the placenta, and dependent in part on maternal immune responses (Fisher et al. 2000; Pereira et al. 2003; Maidji et al. 2006). With low neutralizing antibody titers, areas of infection are found in the decidua and adjacent placenta. With moderate neutralizing titers, cytotrophoblast infection is reduced in the decidua, and occasional focal infection is found in floating villi. With high neutralizing titers, few cytotrophoblasts contain viral replication proteins in the decidua, and none stain in the placenta. In syncytiotrophoblasts, the neonatal Fc receptor transcytoses complexes of IgG and virions that potentially infect underlying cytotrophoblasts (Maidji et al. 2006). Consistent patterns of replication in cytotrophoblast progenitors in villi and invasive cells in the decidua suggest that virion receptors are regulated during development.

Human fibroblasts that support CMV replication express the epidermal growth factor receptor (EGFR) (Wang et al. 2003) and co-receptors such as integrins αVβ3, α2β1 and α6β1 (Feire et al. 2004; Wang et al. 2005). In human umbilical vein endothelial cells (HUVEC), outcomes of infection differ depending on EGFR expression levels (E. Huang, personal communication). Coordinated signals from high-level EGFR expression and integrin αVβ3 result in rapid delivery of virion capsids to the nucleus. Internalization and nuclear transport involve caveolin-mediated endosomal pathways that are pH independent. Productive infection entails the inhibition of cofilin phosphorylation and increased tubulin acetylation, which are hallmarks of actin disassembly and microtubule assembly. Detailed analysis of naturally infected placentas support strong expression of EGFR and induction of specific integrins as developmentally induced CMV receptors in cytotrophoblasts that enable intrauterine replication in sites optimal for virus amplification and fetal transmission.

Villous Cytotrophoblasts Express EGFR and Upregulate Integrin αV

In naturally infected placentas, CMV replicates in underlying cytotrophoblast progenitors but not in syncytiotrophoblasts covering the villus surface (Fisher et al. 2000; Pereira et al. 2003; Maidji et al. 2007). CMV gB was repeatedly detected in a punctate staining pattern in syncytiotrophoblasts of immune donors, but viral replication proteins were absent. Notably, syncytiotrophoblasts expressed EGFR in apical microvilli, but neither integrin α1β1 nor αV was detected. Occasionally, gB appeared in a punctate pattern in cytotrophoblast progenitors reactive with antibodies to EGFR and integrin αV, but α1β1 was not detected. Frequently, virion capsids were found clustered near the microvillous surface of syncytiotrophoblasts, and CMV DNA was detected by in situ hybridization. These findings indicated that viral capsids, DNA and gB were present without replication in placentas that contained high-titer neutralizing antibodies, sometimes in the presence of other pathogens (McDonagh et al. 2004).

Examination of CMV receptors expressed in cytotrophoblasts of villus explants infected with a recombinant Toledo[rec] virus encoding green fluorescent protein in the villus explant model (Fig. 2a) confirmed that the receptors were expressed in infected cells. Virions attached in a punctate (green) pattern to apical microvilli of syncytiotrophoblasts, and some internalized but did not replicate (Fig. 2b). Although syncytiotrophoblasts express EGFR, functional coreceptors are missing. Even so, virion-containing immune complexes internalize and colocalize with caveolin-1-containing vesicular compartments that were not detected

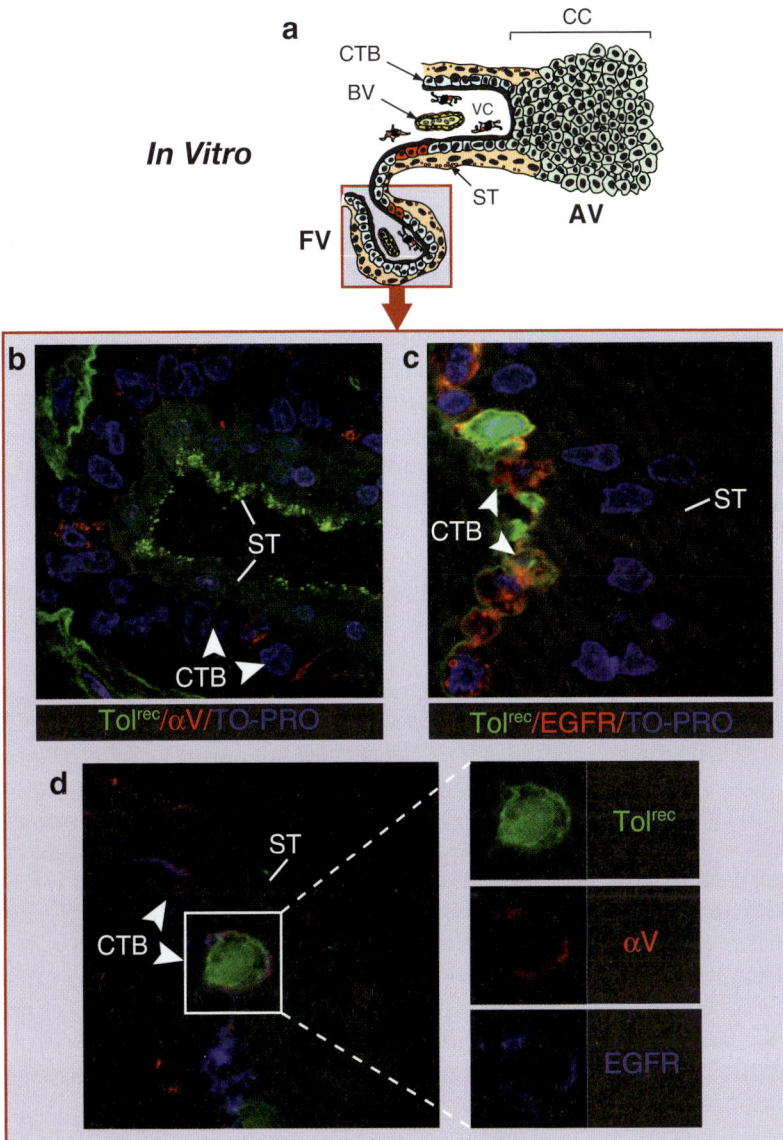

Fig. 2 CMV replicates in cytotrophoblasts expressing EGFR and αV integrin in villous explants infected in vitro. **a** Diagram shows focal infection in a floating villus. The *blue field* represents the localization of CMV structural and replication proteins in panels **b** and **c**. Explants were fixed at 24 h postinfection. **b** Toledo[rec] virions bound to apical microvilli of syncytiotrophoblasts (*ST*). Selected cytotrophoblasts (*CTB*), but not syncytiotrophoblasts, reacted with antibody to αV integrin (*red*). **c** Underlying cytotrophoblasts stained intensely with EGFR-specific antibody (*red*); replicating Toledo[rec] broadly expressed green fluorescent protein in nuclei of infected cells. Nuclei were counterstained with TO-PRO-3 iodide (*blue*). **d** Villous cytotrophoblasts infected with Toledo[rec] simultaneously expressed αV integrin (*red*) and EGFR (*blue*)

with IgG alone, suggesting that virions drive caveolae formation (Maidji et al. 2006). CMV gB accumulates in caveosomes, but nucleocapsids and viral DNA remain cytoplasmic without replication, suggesting endocytosis of immune complexes in syncytiotrophoblasts.

Clusters of adjacent cytotrophoblasts underlying syncytiotrophoblasts expressed integrin αV (Fig. 2b). When infection was detected in villous cytotrophoblasts, as indicated by the nuclear (green) staining pattern, the cells stained intensely for EGFR (Fig. 2c). Notably, integrin α1β1 was absent (not shown), which confirms earlier reports (Damsky et al. 1992; Zhou et al. 1997). Double immunostaining showed that Toledo[rec] replicated in cytotrophoblasts that expressed both integrin αV (red) and EGFR (blue) (Fig. 2d). Together these results indicated that CMV virions internalize in syncytiotrophoblasts with EGFR alone. Inasmuch as more than half of the population was immune to CMV, and neutralizing antibodies are secreted into conditioned medium from villus explants (Maidji et al. 2006), viral replication occurred infrequently (seven of 42 placentas). When placentas contained IgG–CMV virion complexes with low-avidity antibodies, focal replication occurred in susceptible cytotrophoblast progenitor cells that expressed both EGFR and integrin αV, providing an explanation for earlier observations (Fisher et al. 2000; Pereira et al. 2003). Likewise, flow cytometric analysis of cytotrophoblasts isolated from placentas at term confirmed that EGFR and selected integrins were expressed (Tabata et al. 2007).

Villous cytotrophoblasts express integrin αV but not the β3 subunit, suggesting that other partners could function as coreceptors, particularly subunits β5 and β6 (Zhou et al. 1997; Tabata et al. 2007). CMV infection in epithelial-like progenitor cells could resemble foot-and-mouth disease virus, which binds integrins αVβ3 (Berinstein et al. 1995) and αVβ6 in epithelia (Jackson et al. 2000). We recently observed integrin αVβ6 expression in cytotrophoblasts of floating villi contiguous with large deposits of extracellular matrix (Tabata et al., unpublished observations) and integrin β6 was expressed in freshly isolated villous cytotrophoblasts of placentas at term (Tabata et al. 2007). Particular expression of FcRn, endocytosis of immune complexes, coupled with temporal and spatial regulation of virion receptors as cytotrophoblasts differentiate, could limit virus access to the villous core. These findings help explain the discrepancies between frequent detection of viral DNA in the decidua (80%), reduced levels in the placenta (60%), and a low rate of congenital infection in seropositive women (1%–3%) (Fowler et al. 2003; McDonagh et al. 2004; 2006).

Cell Column Cytotrophoblasts Induce Integrin α1β1 Expression

When early gestation biopsy specimens were stained for potential CMV receptors, proximal cell columns did not express EGFR, but punctate staining for gB was sometimes detected at the cell surface, suggesting that virions were clustered in membranes (Maidji et al. 2007). CMV receptors expressed on cytotrophoblasts in villus explants were studied in model cell columns of anchoring villi infected with Toledo[rec] ex vivo (Fig. 3a). Like cell columns in intact placentas, extravillous

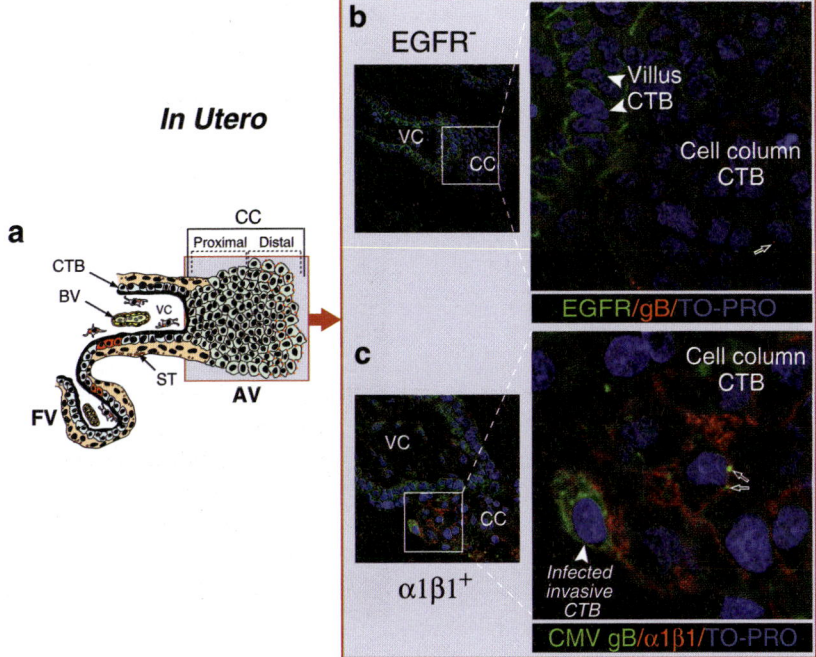

Fig. 3 CMV virions bind to cytotrophoblast outgrowths in a model of anchoring villus development in vitro. **a** Diagram of an anchoring villus cell column. The *blue field* represents the localization of virions in panels **b** and **c**. The insets in panels **b** and **c** correspond to the outlined areas in each panel. Explants were fixed at 24 h postinfection. **b** Cell column cytotrophoblasts did not react with antibodies to EGFR (*red*). **c** Toledo^{rec} virions (*green*) bind cytotrophoblasts expressing α1β1 integrin (*red*) on the edge of a distal cell column. Nuclei were counterstained with TO-PRO-3 iodide (*blue*)

cytotrophoblasts in these outgrowths did not express EGFR (red) (Fig. 3b). Nonetheless, virion attachment was observed as punctate (green) staining in membranes of distal columns where potential integrin receptors αV and α1β1 were expressed. Cytotrophoblasts on the edge of distal portions of columns upregulated α1β1 (red), and Toledo^{rec} virions were bound to surface membranes (Fig. 3c). Integrin αV was also present (data not shown). The results suggest that cytotrophoblast expression of integrin coreceptors without EGFR enables virion binding but is insufficient for entry and replication.

Replication in Differentiating/Invading Cytotrophoblasts

Analysis of first- and second-trimester placentas from women with moderate CMV neutralizing titers showed that interstitial cytotrophoblasts in decidua are infected in utero (Pereira et al. 2003). As cytotrophoblasts differentiate and progress along

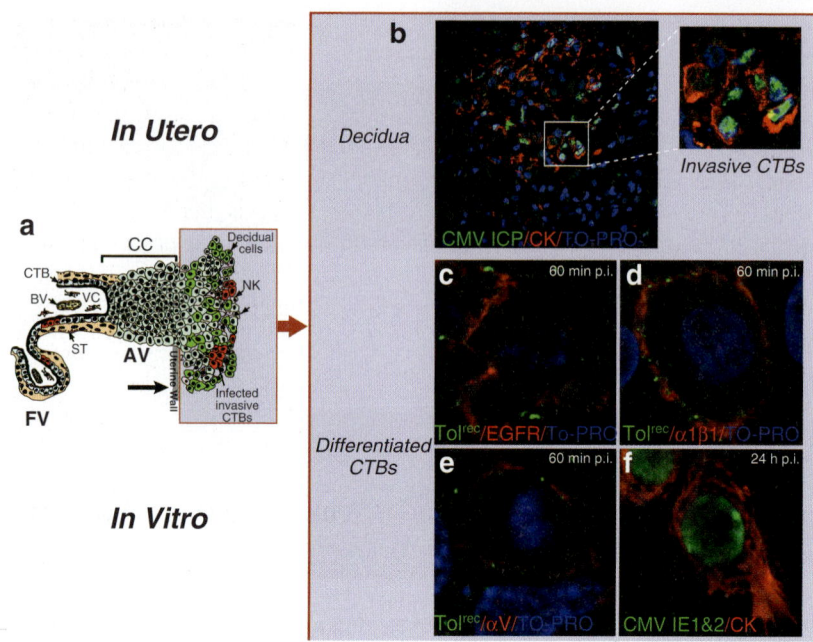

Fig. 4 CMV replicates in invasive cytotrophoblasts that express virion receptors. **a** Diagram of differentiating/invading cytotrophoblasts. The *blue field* represents the localization of viral proteins in **b, c, d, e,** and **f**. **b** Invasive cells in decidua infected in utero. The inset corresponds to the outlined area in the panel. Toledo[rec] virions (*green*) bind to cytotrophoblast membranes with EGFR (**c**), integrin α1β1 (**d**), and integrin αV (**e**) (*red*) at 1 h after infection. **f** IE1 and IE2 proteins (*green*) in cytokeratin (*CK*)-stained cells (*red*) at 24 h after infection. Nuclei were counterstained with TO-PRO-3 iodide (*blue*)

the invasive pathway, integrins α1β1 and αVβ3 are upregulated (Damsky et al. 1994; Zhou et al. 1997). Location of CMV-infected invasive cytotrophoblasts is illustrated in Fig. 4a. Analysis of early gestation decidua from a placenta with low neutralizing titers showed that infected cytotrophoblasts express viral replication proteins (Fig. 4b). Cultures of differentiated cytotrophoblasts on filters coated with Matrigel were used to monitor these cells for expression of potential receptors, binding of Toledo[rec] virions and productive infection. Virions (punctate green) adsorbed to membranes of cells expressing EGFR (Fig. 4c) and integrins α1β1 (Fig. 4d) and αV (Fig. 4e) at 1 h after infection. In addition, CMV IE1- and IE2-infected cell proteins were detected in nuclei of Toledo-infected cytotrophoblasts at 24 h, suggesting early-stage viral replication (Fig. 4f). These results showed that virion attachment and infection of invasive cytotrophoblasts occur in vitro in cells that express CMV receptors, EGFR and integrins αV and α1β1.

Using function-perturbing antibodies and soluble integrins, we evaluated the role of molecules that function as potential receptors in differentiating/invasive cytotrophoblasts (Maidji et al. 2007). Intense expression of EGFR and integrins α1β1 and αVβ3 was found as cytotrophoblasts differentiated and invaded. Moreover, caveolin-1 was expressed, as reported for syncytiotrophoblasts and villous cytotrophoblasts where CMV virion gB was localized (Maidji et al. 2006). Function-blocking antibody to EGFR blocked 63% of infectivity, whereas antibody to integrin β1 blocked 80% and antibody to integrin αV blocked 43%. Function-blocking antibody to integrin α5 and isotype controls failed to reduce viral replication. When CMV virions were pretreated with soluble forms of integrins before infection, α1β1 blocked 100% of virion infectivity. Pretreatment with soluble integrin αVβ3 reduced infectivity by approximately 37%, and pretreatment with soluble integrin α3β1 reduced infectivity by 13%. Together the results of function-perturbing experiments confirmed that EGFR and integrins α1β1 and αVβ3 – developmentally regulated molecules – function as CMV receptors as cytotrophoblasts progress along the differentiation pathway.

Infection Impairs Cell Functions Through Diverse Membrane Proximal Events

CMV replication in early gestation differentiating/invading cytotrophoblasts radically impacts cell function and development. Infected cytotrophoblasts dysregulate expression of stage-specific adhesion molecules, HLA-G and MMP-9 activity (Fisher et al. 2000; Yamamoto-Tabata et al. 2004; Tabata et al. 2007). Importantly, cytotrophoblasts' central function, invasiveness, is significantly impaired through membrane-proximal defects. This includes downregulation of integrins α1β1 and α9 and VE-cadherin (Fisher et al. 2000; Tabata et al. 2007), stage-specific proteins that mediate cell–matrix and cell–cell adhesion and invasion. In contrast, integrin α5, which counterbalances invasion, was not affected. Activated MMP-9 is required for cytotrophoblast invasion, whereas pro-MMP-9, the uncleaved precursor, is associated with noninvasive cells and preeclampsia (Librach et al. 1991). Using zymography, we examined MMP-9 protein and gelatinase activity of cytotrophoblasts infected with the pathogenic strain VR1814 and observed reduced MMP-9 protein and activity (Yamamoto-Tabata et al. 2004).

Functional assays of the cells' invasiveness showed significant impairment. Human IL-10, a pleiotropic cytokine, downregulates MMP-9 activity and impairs cytotrophoblast invasiveness (Roth et al. 1996; Roth and Fisher 1999). cmvIL-10 (Kotenko et al. 2000) has comparable immunosuppressive activity (Spencer et al. 2002) and impairs immune cell functions in vitro (Chang et al. 2004). Examination of MMP-9 production and activity in differentiating cytotrophoblasts treated with recombinant proteins showed that cmvIL-10-treated cytotrophoblasts contain less MMP-9 protein and activity and upregulate human IL-10 (Yamamoto-Tabata et al. 2004). The impact of viral infection on cytotrophoblast invasiveness was evaluated using a functional assay that tests the ability of differentiating cells plated on the upper surfaces of Matrigel-coated filters to penetrate the surface, pass through pores in the underlying filter and emerge on the lower surface of the membrane

(Fisher et al. 2000; Yamamoto-Tabata et al. 2004). Invasiveness of infected cytotrophoblasts and cell aggregates was dramatically impaired. In addition, CMV exploits the similarity between viral and human IL-10 to impair the cells' invasion functions as a consequence of reduced proteinase activity. Cytotrophoblast invasiveness was impaired by treatment with human IL-10 or cmvIL-10 in accord with decreased MMP-9 production and activity. These results showed that cmvIL-10, like human IL-10, impairs cytotrophoblast invasiveness indirectly by impairing MMP-9 activity. The effect was greater than what could be accounted for by the number of infected cytotrophoblasts, suggesting that secreted viral and/or cellular factors could affect uninfected cells in the aggregates and influence the behavior of the population as a whole.

Implications for Congenital CMV Infection

Recent gene expression profiling of conjoined areas of the human placenta and decidua that compose the maternal–fetal interface revealed dramatic changes between midgestation and term (Winn et al. 2007). Many of the developmentally expressed genes are known in other contexts to be involved in differentiation, motility, transcription, immunity, angiogenesis and extracellular matrix modulation. We recently examined placentas from pregnancies complicated by symptomatic congenital infection and fetal intrauterine growth restriction (Nigro et al. 1999, 2005). Although sites of viral replication were rare, fibrinoid deposition and necrosis resulting from viral damage in the placenta and markedly reduced cytotrophoblast invasion in decidua were evident as compared with uninfected healthy controls (E. Maidji, G. Nigro, S. Muci and L. Pereira, unpublished observations). Moreover, expression of genes associated with inflammatory responses, oxidative stress, chemotaxis, extracellular matrix deposition and angiogenesis was broadly dysregulated. These results suggest that early gestation intrauterine infection alters the local environment directly and through paracrine mechanisms, with long-term effects that could result in placental insufficiency as gestation progresses. Low birth weight is associated with increased rates of cardiovascular disease and non-insulin-dependent diabetes in adult life (Lau and Rogers 2004; Reyes and Manalich 2005). The fetal origins hypothesis proposes that these diseases originate through fetal adaptation – cardiovascular, metabolic or endocrine in nature – made when the fetus is undernourished (Barker 1999). A better understanding of the molecular mechanisms underlying placental dysfunction could allow formulation of novel treatment for congenital infection, of benefit to women with pregnancy disorders and fetal intrauterine growth restriction (Cetin et al. 2004).

Acknowledgements These studies were supported by grants from the National Institutes of Health (AI46657, AI53782) and the Thrasher Research Fund 02821-7 and from the Academic Senate of the University of California, San Francisco.

References

Barker DJ (1999) Fetal origins of cardiovascular disease. Ann Med 31 Suppl 1:3–6

Benirschke K, Kaufmann P (2000) Pathology of the human placenta. Springer, Berlin, New York, Heidelberg

Berinstein A, Roivainen M, Hovi T, Mason PW, Baxt B (1995) Antibodies to the vitronectin receptor (integrin alpha V beta 3) inhibit binding and infection of foot-and-mouth disease virus to cultured cells. J Virol 69:2664–2666

Britt WJ (1999) Congenital cytomegalovirus infection. In: Hitchcock PJ, MacKay HT, Wasserheit JN (eds) Sexually transmitted diseases and adverse outcomes of pregnancy. ASM Press, Washington, DC, pp 269–281

Cetin I, Foidart JM, Miozzo M, Raun T, Jansson T, Tsatsaris V, Reik W, Cross J, Hauguel-De-Mouzon S, Illsley N, Kingdom J, Huppertz B (2004) Fetal growth restriction: a workshop report. Placenta 25:753–757

Chang WL, Baumgarth N, Yu D, Barry PA (2004) Human cytomegalovirus-encoded interleukin-10 homolog inhibits maturation of dendritic cells and alters their functionality. J Virol 78:8720–8731

Cross JC, Werb Z, Fisher SJ (1994) Implantation and the placenta: key pieces of the development puzzle. Science 266:1508–1518

Damsky CH, Fisher SJ (1998) Trophoblast pseudo-vasculogenesis: faking it with endothelial adhesion receptors. Curr Opin Cell Biol 10:660–666

Damsky CH, Fitzgerald ML, Fisher SJ (1992) Distribution patterns of extracellular matrix components and adhesion receptors are intricately modulated during first trimester cytotrophoblast differentiation along the invasive pathway, in vivo. J Clin Invest 89:210–222

Damsky CH, Librach C, Lim KH, Fitzgerald ML, McMaster MT, Janatpour M, Zhou Y, Logan SK, Fisher SJ (1994) Integrin switching regulates normal trophoblast invasion. Development 120:3657–3666

Feire AL, Koss H, Compton T (2004) Cellular integrins function as entry receptors for human cytomegalovirus via a highly conserved disintegrin-like domain. Proc Natl Acad Sci U S A 101:15470–15475

Fisher S, Genbacev O, Maidji E, Pereira L (2000) Human cytomegalovirus infection of placental cytotrophoblasts in vitro and in utero: implications for transmission and pathogenesis. J Virol 74:6808–6820

Fowler KB, Stagno S, Pass RF, Britt WJ, Boll TJ, Alford CA (1992) The outcome of congenital cytomegalovirus infection in relation to maternal antibody status. N Engl J Med 326:663–667

Fowler KB, Stagno S, Pass RF (2003) Maternal immunity and prevention of congenital cytomegalovirus infection. JAMA 289:1008–1011

Garcia AG, Fonseca EF, Marques RL, Lobato YY (1989) Placental morphology in cytomegalovirus infection. Placenta 10:1–18

Genbacev O, Krtolica A, Kaelin W, Fisher SJ (2001) Human cytotrophoblast expression of the von Hippel-Lindau protein is downregulated during uterine invasion in situ and upregulated by hypoxia in vitro. Dev Biol 233:526–536

Halwachs-Baumann G, Wilders-Truschnig M, Desoye G, Hahn T, Kiesel L, Klingel K, Rieger P, Jahn G, Sinzger C (1998) Human trophoblast cells are permissive to the complete replicative cycle of human cytomegalovirus. J Virol 72:7598–7602

Hemmings DG, Kilani R, Nykiforuk C, Preiksaitis J, Guilbert LJ (1998) Permissive cytomegalovirus infection of primary villous term and first trimester trophoblasts. J Virol 72:4970–4979

Jackson T, Sheppard D, Denyer M, Blakemore W, King AM (2000) The epithelial integrin alphavbeta6 is a receptor for foot-and-mouth disease virus. J Virol 74:4949–4956

Janatpour MJ, McMaster MT, Genbacev O, Zhou Y, Dong J, Cross JC, Israel MA, Fisher SJ (2000) Id-2 regulates critical aspects of human cytotrophoblast differentiation, invasion and migration. Development 127:549–558

Kotenko SV, Saccani S, Izotova LS, Mirochnitchenko OV, Pestka S (2000) Human cytomegalovirus harbors its own unique IL-10 homolog (cmvIL-10). Proc Natl Acad Sci USA 97:1695–1700

Kumazaki K, Ozono K, Yahara T, Wada Y, Suehara N, Takeuchi M, Nakayama M (2002) Detection of cytomegalovirus DNA in human placenta. J Med Virol 68:363–369

Lau C, Rogers JM (2004) Embryonic and fetal programming of physiological disorders in adulthood. Birth Defects Res C Embryo Today 72:300–312

Librach CL, Werb Z, Fitzgerald ML, Chiu K, Corwin NM, Esteves RA, Grobelny D, Galardy R, Damsky CH, Fisher SJ (1991) 92-kD type IV collagenase mediates invasion of human cytotrophoblasts. J Cell Biol 113:437–449

Maidji E, McDonagh S, Genbacev O, Tabata T, Pereira L (2006) Maternal antibodies enhance or prevent cytomegalovirus infection in the placenta by neonatal fc receptor-mediated transcytosis. Am J Pathol 168:1210–1226

Maidji E, Genbacev O, Chang HT, Pereira L (2007) Developmental regulation of human cytomegalovirus receptors in cytotrophoblasts correlates with distinct replication sites in the placenta. J Virol 81:4701–4712

McDonagh S, Maidji E, Ma W, Chang HT, Fisher S, Pereira L (2004) Viral and bacterial pathogens at the maternal-fetal interface. J Infect Dis 190:826–834

McDonagh S, Maidji E, Chang H-T, Pereira L (2006) Patterns of human cytomegalovirus infection in term placentas: a preliminary analysis. J Clin Virol 35:210–215

Muhlemann K, Miller RK, Metlay L, Menegus MA (1992) Cytomegalovirus infection of the human placenta: an immunocytochemical study. Human Pathol 23:1234–1237

Nigro G, La Torre R, Anceschi MM, Mazzocco M, Cosmi EV (1999) Hyperimmunoglobulin therapy for a twin fetus with cytomegalovirus infection and growth restriction. Am J Obstet Gynecol 180:1222–1226

Nigro G, Adler SP, La Torre R, Best AM (2005) Passive immunization during pregnancy for congenital cytomegalovirus infection. N Engl J Med 353:1350–1362

Norwitz ER, Schust DJ, Fisher SJ (2001) Implantation and the survival of early pregnancy. N Engl J Med 345:1400–1408

Pereira L, Maidji E, McDonagh S, Genbacev O, Fisher S (2003) Human cytomegalovirus transmission from the uterus to the placenta correlates with the presence of pathogenic bacteria and maternal immunity. J Virol 77:13301–13314

Pereira L, Maidji E, McDonagh S, Tabata T (2005) Insights into viral transmission at the uterine-placental interface. Trends Microbiol 13:164–174

Reyes L, Manalich R (2005) Long-term consequences of low birth weight. Kidney Int Suppl: S107–S111

Roth I, Fisher SJ (1999) IL-10 is an autocrine inhibitor of human placental cytotrophoblast MMP-9 production and invasion. Dev Biol 205:194–204

Roth I, Corry DB, Locksley RM, Abrams JS, Litton MJ, Fisher SJ (1996) Human placental cytotrophoblasts produce the immunosuppressive cytokine interleukin 10. J Exp Med 184:539–548

Simister NE, Story CM, Chen HL, Hunt JS (1996) An IgG-transporting Fc receptor expressed in the syncytiotrophoblast of human placenta. Eur J Immunol 26:1527–1531

Sinzger C, Müntefering H, Löning T, Stöss H, Plachter B, Jahn G (1993) Cell types infected in human cytomegalovirus placentitis identified by immunohistochemical double staining. Virchows Archiv A Pathol Anat Histopathol 423:249–256

Spencer JV, Lockridge KM, Barry PA, Lin G, Tsang M, Penfold ME, Schall TJ (2002) Potent immunosuppressive activities of cytomegalovirus-encoded interleukin-10. J Virol 76:1285–1292

Stagno S, Pass RF, Cloud G, Britt WJ, Henderson RE, Walton PD, Veren DA, Page F, Alford CA (1986) Primary cytomegalovirus infection in pregnancy. Incidence, transmission to fetus, and clinical outcome. JAMA 256:1904–1908

Story CM, Mikulska JE, Simister NE (1994) A major histocompatibility complex class I-like Fc receptor cloned from human placenta: possible role in transfer of immunoglobulin G from mother to fetus. J Exp Med 180:2377–2381

Tabata T, McDonagh S, Kawakatsu H, Pereira L (2007) Cytotrophoblasts infected with a pathogenic human cytomegalovirus strain dysregulate cell-matrix and cell-cell adhesion molecules: a quantitative analysis. Placenta 28:527–537

Trincado DE, Munro SC, Camaris C, Rawlinson WD (2005) Highly sensitive detection and localization of maternally acquired human cytomegalovirus in placental tissue by in situ polymerase chain reaction. J Infect Dis 192:650–657

Wang X, Huong SM, Chiu ML, Raab-Traub N, Huang ES (2003) Epidermal growth factor receptor is a cellular receptor for human cytomegalovirus. Nature 424:456–461

Wang X, Huang DY, Huong SM, Huang ES (2005) Integrin alphavbeta3 is a coreceptor for human cytomegalovirus. Nat Med 11:515–521

Winn VD, Haimov-Kochman R, Paquet AC, Yang YJ, Madhusudhan MS, Gormley M, Feng KT, Bernlohr DA, McDonagh S, Pereira L, Sali A, Fisher SJ (2007) Gene expression profiling of the human maternal–fetal interface reveals dramatic changes between midgestation and term. Endocrinology 148:1059–1079

Yamamoto-Tabata T, McDonagh S, Chang H-T, Fisher S, Pereira L (2004) Human cytomegalovirus interleukin-10 downregulates matrix metalloproteinase activity and impairs endothelial cell migration and placental cytotrophoblast invasiveness in vitro. J Virol 78:2831–2840

Zhou Y, Fisher SJ, Janatpour M, Genbacev O, Dejana E, Wheelock M, Damsky CH (1997) Human cytotrophoblasts adopt a vascular phenotype as they differentiate. A strategy for successful endovascular invasion? J Clin Invest 99:2139–2151

Mechanisms of Cytomegalovirus-Accelerated Vascular Disease: Induction of Paracrine Factors That Promote Angiogenesis and Wound Healing

D. N. Streblow(✉), J. Dumortier, A. V. Moses, S. L. Orloff, J. A. Nelson

Contents

Introduction 398
Tissue Repair and Angiogenic Factors Mediate TVS 398
Animal Models of CMV-Accelerated Graft Rejection 400
    RCMV Accelerates TVS and CR in a Rat Heart Transplantation Model 400
    RCMV Induces Allograft AG and WH Genes During the Acceleration of TVS 401
In Vitro Models of HCMV-Mediated Wound Healing and Angiogenesis 402
    What Factors Constitute the HCMV Secretome? 405
    HCMV Secretome Induces Angiogenesis in Endothelial Cells..................... 407
    HCMV Secretome Induces Wound Healing in Endothelial Cells.................... 408
Conclusions 411
References 412

Abstract Human cytomegalovirus (HCMV) is associated with the acceleration of a number of vascular diseases such as atherosclerosis, restenosis, and transplant vascular sclerosis (TVS). All of these diseases are the result of either mechanical or immune-mediated injury followed by inflammation and subsequent smooth muscle cell (SMC) migration from the vessel media to the intima and proliferation that culminates in vessel narrowing. A number of epidemiological and animal studies have demonstrated that CMV significantly accelerates TVS and chronic rejection (CR) in solid organ allografts. In addition, treatment of human recipients and animals alike with the antiviral drug ganciclovir results in prolonged survival of the allograft, indicating that CMV replication is a requirement for acceleration of disease. However, although virus persists in the allograft throughout the course of disease, the number of directly infected cells does not account for the global effects that the virus has on the acceleration of TVS and CR. Recent investigations of up- and downregulated cellular genes in infected allografts in comparison to native heart has demonstrated that rat CMV (RCMV) upregulates genes involved in wound healing (WH) and angiogenesis (AG). Consistent with this result,

---

D.N. Streblow
Vaccine and Gene Therapy Institute and Department of Molecular Microbiology
and Immunology, Oregon Health and Science University, Portland, OR 97201, USA
streblow@ohsu.edu

we have found that supernatants from HCMV-infected cells (HCMV secretome) induce WH and AG using in vitro models. Taken together, these findings suggest that one mechanism for HCMV acceleration of TVS is mediated through induction of secreted cytokines and growth factors from virus-infected cells that promote WH and AG in the allograft, resulting in the acceleration of TVS. We review here the ability of CMV infection to alter the local environment by producing cellular factors that act in a paracrine fashion to enhance WH and AG processes associated with the development of vascular disease, which accelerates chronic allograft rejection.

Introduction

The importance and interest of HCMV as a pathogen has increased over the last few decades, with the escalation in the number of immunosuppressed patients, either undergoing immunosuppressive therapy following solid organ or bone marrow transplantation, or with AIDS patients. Primary HCMV infection is followed by a lifelong persistence of the virus in a latent state, and reactivation of latent virus is considered to be the major source of virus in immunocompromised individuals (see the chapter by M. Reeves and J. Sinclair, this volume). HCMV is linked to the development of arterial restenosis following angioplasty, atherosclerosis, and solid organ TVS (Melnick et al. 1983; Speir et al. 1994; Melnick et al. 1998). HCMV infection nearly doubles the 5-year rate of cardiac graft failure due to accelerated TVS (Grattan et al. 1989), and prior to the advent of ganciclovir therapy, it doubled the rate of liver graft loss at 3 years (Deotero et al. 1998; Rubin 1999). In recipients of heart transplants, treatment with ganciclovir, a potent inhibitor of viral replication and CMV disease, delayed the time to allograft rejection (Merigan et al. 1992). A subsequent post-hoc analysis of these data confirmed that prophylactic ganciclovir treatment delayed graft rejection compared to controls (Valantine et al. 1999). Moreover, the early control of subclinical HCMV replication after cardiac transplantation by T cell immunity reduces allograft TVS and CR (Tu et al. 2006). Additional proof to HCMV's affect on graft TVS comes from the finding of a higher incidence of viral DNA detected in the explant vascular intima of those patients with cardiac allograft TVS than in those explants without vasculopathy (Wu et al. 1992). In fact, the mere presence of HCMV infection in kidney transplant patients, whether displaying asymptomatic or overt symptoms, was shown to negatively impact allograft survival (Fitzgerald et al. 2004). While a number of studies have provided strong evidence for a role of HCMV in the development of TVS and accelerated CR, the precise mechanisms involved in this process are still unknown.

Tissue Repair and Angiogenic Factors Mediate TVS

Despite recent medical advances, the long-term survival of solid organ allografts has not improved, largely due to CR. The high prevalence of CR is of particular concern given that to date the only effective therapy is retransplantation. The primary component of CR is an accelerated form of arteriosclerosis known as transplant vascular sclerosis (TVS). TVS is characterized histologically by diffuse concentric intimal proliferation that ultimately occludes the vessel (Billingham 1992). A growing body of evidence supports a role for angiogenesis and tissue repair processes in the development of vascular disease (Shibata et al. 2001; Khurana and Simons 2003). A sparse number of macrophages, T cells, NK cells, and B cells are seen in early lesions, while late lesions are associated with a thickened intima containing SMC interspersed with macrophages (Clinton and Libby 1992). Activated inflammatory cells and SMC within and near vascular lesions are important local sources of pro-angiogenic factors (Bouis et al. 2006). TVS development involves chronic perivascular inflammation, endothelial cell (EC) dysfunction, and SMC migration from the media to the intima and proliferation that result in deposition of extracellular matrix (ECM) and neointimal thickening of the allograft arterial wall (Libby et al. 1989; Billingham 1992; Hosenpud et al. 1992). These events result in vessel narrowing, occlusion, and graft failure. The pathological processes involved in TVS and other vascular diseases are akin to many of the cellular events that mediate normal tissue repair/angiogenesis. In all cases, a complex interaction between cells and surrounding regulatory factors, enzymes, and ECM components leads to cellular migration, proliferation, and tissue remodeling.

The formation of new blood vessels or angiogenesis is broadly divided into the following three phases: vessel destabilization, proliferation/migration, and vessel maturation (Carmeliet 2003). In normal adults, angiogenesis is normally restricted to formation of placental and endometrial tissue, hair follicle vascularization, and tissue repair. During these processes, the endothelium remains inactive due to a balance of positive and negative regulatory factors. When vessel growth is required, the regulatory balance tips toward pro-angiogenic factors. Restoration of steady state is achieved by increasing angiogenesis inhibitors and vessel stabilization factors. Breakdown of the tightly regulated angiogenic balance leads to abnormal angiogenesis and contributes to a variety of pathological disorders, including cancer, autoimmune conditions, and cardiovascular disease (Carmeliet 2003). Leukocytes play a role in normal angiogenesis through contribution of pro- and anti-angiogenic factors, but are particularly important in pathological (immune- or tumor-driven) angiogenesis (Kent and Sheridan 2003).

Wound healing (WH) is a complex process involving the sequence of inflammation, tissue formation, and tissue remodeling, which results in the re-establishment of an anatomical or physiological barrier (Kent and Sheridan 2003). Wound healing involves a local milieu created by the coordinated action of growth factors, cytokines, enzymes, and extracellular matrix (ECM) components, interacting with the injured tissue. In addition, an influx of inflammatory cells is involved in this process. Angiogenesis itself is an important component of WH. Given the role of inflammation, new vessel growth and tissue remodeling in WH, it is not surprising that many of the same stimulatory and inhibitory factors promote both angiogenesis and WH (Werner and Grose 2003).

Animal Models of CMV-Accelerated Graft Rejection

Determining the mechanisms involved in the development of HCMV-associated TVS has been difficult because the etiology of this disease is multifactorial. In addition, HCMV is ubiquitous throughout the human population, and hence negative controls are rare. Furthermore, HCMV infections are lifelong, during which time the virus infects all of the cell types involved in TVS formation including SMC, EC, macrophages, and fibroblasts. Along these lines, HCMV evades the immune system by remaining latent in monocytes, and HCMV reactivation in immunocompetent hosts is difficult to detect when clinically silent (Lemstrom et al. 1993, 1995; Bruning et al. 1994; Orloff 1999; Orloff et al. 2000). These undeniable factors make it difficult to determine a temporal relationship between the virus infection and TVS, and for obvious ethical reasons, human studies are impossible. Therefore, animal models provide an ideal tool to study the association between CMV and TVS. In fact, the most compelling evidence that herpesvirus infections play a role in the vascular disease process is exemplified through the use of animal models. For example, in rat solid organ transplantation, acute infection with rat (R)CMV infection accelerates TVS, which leads to untimely graft failure (Orloff et al. 2002; Streblow et al. 2003; Soule et al. 2006). Similar to the human transplantation setting, antiviral therapy reduces the acceleration of rejection in rat transplant models, demonstrating that active CMV replication is required for these disease processes (Tikkanen et al. 2001b; Zeng et al. 2005). Importantly, the effects of RCMV on the acceleration of TVS are not organ-type-specific but can occur in a broad range of solid organ transplants, including heart, kidney, lung, and small bowel (Tikkanen et al. 2001a; Orloff et al. 2002; Streblow et al. 2003; Soule et al. 2006).

RCMV Accelerates TVS and CR in a Rat Heart Transplantation Model

In order to study the role of CMV in the development of TVS, we have taken advantage of the F344 into Lewis rat heterotopic solid organ transplant model (Ely et al. 1983; Klempnauer and Marquarding 1989; Lubaroff et al. 1989), which we have used to study TVS/CR in transplanted heart, kidney, and small bowel grafts (Orloff et al. 2002; Streblow et al. 2003, 2005, 2007; Soule et al. 2006). Since this strain combination exhibits reduced allogenicity, acute rejection is prevented by a short regimen of cyclosporine A, resulting in long-term surviving allografts developing histological evidence of CR (Orloff et al. 2002). Heart allograft recipients not treated with CsA acutely rejected, and syngeneic transplants failed to develop TVS/CR. In the CsA-treated heart allograft recipients, the mean time to develop CR was 90 days, as determined by palpation of the abdomen for induration of the graft and diminished pulsation of the heartbeat. The majority of the vessels in the rejecting cardiac allografts showed the presence of TVS.

Using this rat heart transplant model, we have determined that RCMV accelerated the time to develop TVS and graft failure from 90 to 45 days and increased the degree of TVS from a mean neointimal index (NI)=42.9 to 82.9 (Streblow et al. 2003). RCMV infection failed to induce TVS in syngeneic cardiac grafts by the study endpoint of 120 days (mean, NI=4.2 vs uninfected controls, mean, NI=8.8). Assessment of grafts at earlier times before CR has revealed an even more pronounced effect of RCMV infection on TVS progress. RCMV infection significantly increased the severity of TVS in heart allografts at days 28, 35 and 45 (mean, NI=48, 70, and 82) compared to uninfected recipients (mean, NI=31, 30 and 43). Graft hearts from infected but not from uninfected recipients showed an early (days 7 and 14) presence of endothelialitis (Streblow et al. 2003). In these animals, PCR identified RCMV DNA in all submandibular glands (SMG) at days 7, 14, 21, 28, 35, and 45. In native and graft hearts, RCMV DNA was highest at postoperation days 7 and 14, corresponding to the time when endothelialitis is present within the graft vessels. After this time, virus production is low but detectable throughout the development of TVS (Streblow et al. 2003). However, RCMV was only detected in the blood until 14 days. The presence of virus in the heart allografts at the later time points postinfection was confirmed by immunostaining for RCMV-IE proteins. The number of RCMV-IE-positive cells present in the allografts at days 21 and 28 was low; however, positive cells are observed as single, infected cells or in small 10-30 infected cell foci. To determine whether virus replication was necessary for RCMV-accelerated TVS, we treated allograft recipients with ganciclovir (20 mg/kg/day for 45 days), which increased the mean time to allograft CR from 45 to 75 days ($p<0.001$) and decreased TVS formation from a mean neointimal index of 80 to 65 ($p<0.001$).

In the rat heart transplant model, using a bone marrow chimera model of tolerance induction, the recipient alloreactive immune response was shown to be required for RCMV acceleration of graft loss (Orloff et al. 2002). These data suggest that RCMV infection requires an allogeneic immune environment for the acceleration of TVS. In this rat cardiac transplant CR model, chemokine expression was increased in virus-infected recipient grafts compared to uninfected controls (Streblow et al. 2003). RCMV-infected graft tissues also contained increased numbers of immune cell infiltrates including macrophages, CD4 and CD8 T cells, and the presence of these cells paralleled the upregulation of chemokines. Similar findings were observed in a rat CR small bowel and renal transplant models (Orloff et al. 2002; Soule et al. 2006). Our results indicate that while not all allograft cells are infected, viral replication is important to drive the acceleration of TVS. Taken together, our findings suggest that the virus-mediated acceleration of TVS occurs through altered regulation of inflammation and wound healing processes.

RCMV Induces Allograft AG and WH Genes During the Acceleration of TVS

To determine the mechanisms by which RCMV accelerates the development of allograft TVS, we analyzed the cellular RNA expression in allograft hearts with and without CMV infection using DNA microarrays. For these experiments, Lewis recipients of F344 allograft hearts were infected with RCMV, and graft and native hearts were harvested at 21 and 28 days after transplantation (the critical time of RCMV-accelerated TVS). The allograft hearts from three recipients (±RCMV) were analyzed using the Affymetrix Rat Genome 230 2.0 arrays with probe sets for 30,000 individual transcripts. The gene expression was analyzed and fold changes were calculated by comparing the native hearts to the infected or uninfected allografts. As anticipated, we identified a number of cellular genes expressed in allografts (infected and uninfected) that were significantly up- or downregulated (more than twofold) compared to native hearts. While comparing the infected vs uninfected allografts, we found 385 cellular genes significantly deregulated (more than twofold) at day 21 and 143 such genes at day 28. Interestingly, many of the upregulated genes are involved in WH, including genes associated with tumor invasion, cytokines/chemokines, cytoskeleton, signaling, adhesion, and ECM. Approximately 134 of the upregulated genes are known mediators of AG/WH, including angiopoietin, cathepsins, chemokines, the CNN family, endothelin, ECM/BM components (laminin, fibronectin, osteopontin, tenascin); the EGF family; hematopoietic growth factors, the insulin growth factor binding protein (IGFBP) family, matrix metalloproteinases (MMPs); platelet derived growth factor (PDGF); TGF-β; the tumor necrosis factor receptor (TNFR) superfamily; urokinase-type plasminogen activator and receptor (uPA/uPAR), and vascular endothelial growth factor (VEGF). Shown in Table 1 are the average fold changes of the AG/WH genes from the infected or uninfected allografts compared to the average intensities of the native hearts. The most highly induced genes are part of a signaling complex involved in ECM modification during WH. Of these genes, MMP12 and osteopontin were upregulated at day 28 in the infected allografts compared to uninfected allografts (367- and 395-fold vs 6- and 12-fold, respectively, compared to native hearts). Another set of key players of WH is urokinase-type plasminogen activator (uPA) and receptor (uPAR), which were also highly upregulated in infected allografts compared to uninfected (65- and 12-fold vs nine- and threefold, respectively). In addition, MCP-1 and IL-1, which are potent inducers of MMP12, osteopontin, and the uPA system, are highly represented in the infected allografts. Overall, our findings support a hypothesis that RCMV infection tips the balance of activator and inhibitory effectors that drive WH, the result of which is acceleration of vascular disease.

In Vitro Models of HCMV-Mediated Wound Healing and Angiogenesis

While *in vivo* animal models have provided solid evidence for the link between CMV and the acceleration of vascular disease processes, *in vitro* models allow one to explore underlying molecular and cellular mechanisms associated with this link. Vascular tissue repair during allograft rejection involves the cellular processes of migration, activation, proliferation, and differentiation, and these events occur in multiple cell types including macrophages, EC, SMC, and fibroblasts. *In vivo* and

Table 1 Cellular genes induced in infected vs. uninfected allografts at POD 21 ($*p<0.05$ infected vs uninfected)

Description	POD 21		POD 28	
Pro-angiogenic	Mock	RCMV	Mock	RCMV
Angiotensinogen	1.86	1.95	1.98	3.04*
Angiopoietin-like 2	3.29	1.86	1.88	3.53*
Angiotensin II receptor, type 1 (AT1A)	3.55	3.09	2.94	4.53*
Cathepsin B	1.76	2.38	1.70	3.45*
Cathepsin C	1.60	3.63	2.06	7.50*
Cathepsin E	6.62	27.26*	14.68	31.62*
Cathepsin S	5.54	12.86	7.40	19.31*
Cathepsin W	3.10	34.34*	11.11	30.74*
Cathepsin Y	13.04	58.28	26.13	76.60*
Chemokine (C-C motif) ligand 2	12.98	50.08*	40.79	69.13
Chemokine (C-C motif) ligand 4	11.90	66.04*	30.69	137.02
Chemokine (C-X-C motif) ligand 1	4.52	4.03	8.25	19.34
Chemokine (C-X-C motif) ligand 2	2.67	2.46	2.80	4.79
Chemokine (C-X-C motif) ligand 12	4.13	5.62	3.64	10.85*
Cysteine rich protein 61	27.47	37.53	22.21	19.56
Nephroblastoma overexpressed gene	11.39	9.65	4.02	8.11
WNT1 inducible signaling pathway protein 2	34.70	5.24	5.68	15.03
Laminin, alpha 3	2.71	2.48	2.49	4.20*
Laminin receptor-like protein LAMRL5	17.31	81.01*	52.92	106.35
Fibronectin 1	3.82	3.01	2.38	12.11*
Secreted phosphoprotein 1	32.17	6.44	12.14	395.56*
Epidermal growth factor receptor	4.49	5.95*	5.15	10.81*
EGF-like-domain, multiple 3	6.17	5.29	4.75	7.06*
Fibroblast growth factor 17	5.13	5.83	5.46	4.45
Fibroblast growth factor receptor-like 1	2.64	2.90	2.60	3.77*
FGF receptor activating protein 1	2.97	3.62	3.27	5.74*
Hepatocyte growth factor (scatter factor)	1.94	2.16	2.05	2.44
Hepatocyte growth factor activator	2.03	2.64*	2.56	3.29*
Hypoxia inducible factor 1, alpha subunit	2.75	5.24	4.00	8.37
Insulin 2	8.65	16.68	13.67	12.30
Insulin-like 6	2.29	2.39	2.35	2.51*
Insulin receptor-related receptor	4.82	4.41	4.38	4.50
Insulin receptor substrate 3	4.15	3.16	3.27	4.59*
Insulin-like growth factor 2	2.57	2.46	2.09	2.87*
IGF 2, binding protein 1	2.92	2.93	3.43	3.48
IGF binding protein 5	3.34	3.16	2.84	3.05
Integrin alpha 1	12.64	8.06	10.15	4.86
Integrin alpha 8	4.34	2.89	2.89	3.78
Integrin alpha 9	2.69	3.39	2.78	2.93
Integrin alpha M	4.70	15.03*	7.80	34.54*
Integrin alpha v	2.23	2.99	2.65	3.43
Integrin alpha X	3.13	4.92	3.70	6.19*
Integrin beta 1	1.27	2.28*	1.59	1.57
Integrin beta 7	7.16	17.27	10.24	19.29*
Interleukin 1 alpha	2.34	4.38*	4.21	4.96
Interleukin 1 beta	16.00	50.91	31.78	82.71*
Interleukin 6	1.42	2.13*	2.76	4.59
Midkine	3.94	3.27	2.62	2.33
Colony stimulating factor 1(macrophage)	12.42	10.84	11.33	10.81

(continued)

Table 1 (continued)

Description	POD 21		POD 28	
Pro-angiogenic	Mock	RCMV	Mock	RCMV
M-CSF I receptor	3.07	6.38	3.86	11.31*
CSF 2 receptor, beta 1 (gran-mac)	3.27	5.21	4.14	6.59
Matrix metalloproteinase 3	2.13	2.39	2.30	4.86
Matrix metalloproteinase 12	17.23	4.17	8.46	367.09*
Matrix metalloproteinase 14	8.63	4.20	3.37	11.00*
Matrix metalloproteinase 23	4.51	3.29	3.32	6.19
Plasminogen activator inhibitor 2 type A	1.55	2.08	1.80	2.93
Ser (or cys) proteinase inhibitor, member 1	8.88	10.34	15.03	20.39
Plasminogen activator, urokinase	6.62	12.91*	9.36	65.34*
Plasminogen activator, urokinase receptor	3.46	5.98	3.85	12.38*
Platelet-derived growth factor, C	2.36	2.48	2.47	3.89*
Transforming growth factor, beta 1	2.96	4.69	3.93	9.78*
Transforming growth factor, beta 2	6.57	10.85	13.96	4.69
Transforming growth factor, beta receptor 1	3.61	5.78*	4.73	6.36*
Transforming growth factor, beta receptor 2	3.95	5.90	4.38	6.45*
Bone morphogenetic protein 2	4.24	5.78	5.00	6.82
Bone morphogenetic protein 7	3.57	3.89	3.89	3.73
Growth differentiation factor 15	3.03	2.83	4.62	5.17
Lymphotoxin A	3.71	5.17*	6.28	5.10*
TNF superfamily member 6	3.96	11.71*	6.77	17.88*
TNF superfamily member 11	2.50	3.66	2.97	3.92*
TNF superfamily member 13	4.71	12.04*	7.05	17.88*
TNF receptor superfamily member 1b	3.71	11.08*	6.53	13.83*
TNF receptor superfamily member 4	3.47	8.75	6.13	10.48
TNF receptor superfamily member 11b	3.96	18.64*	5.10	40.22*
TNF receptor superfamily member 12a	5.72	7.46	8.32	4.03
Vascular endothelial growth factor D	3.33	3.33	3.26	3.18
fms-related tyrosine kinase 4	3.99	3.56	3.75	3.92
Anti-Angiogenic				
Interferon gamma	1.64	4.79*	4.33	10.78*
Interleukin 4 receptor	3.11	6.32*	3.69	6.02*
Interleukin 10	1.38	3.07*	2.11	3.86*
Interleukin 12b	3.65	5.35	4.54	3.58
Chemokine (C-X-C motif) ligand 10	28.49	222.51*	143.15	178.07*
Thrombospondin 2	5.95*	2.93	3.05	5.24
Tissue inhibitor of metalloproteinase 1	17.75	33.59	33.75	41.64
Tissue inhibitor of metalloproteinase 2	2.33	1.53	1.67	3.41*

in vitro, HCMV infects all of these cell types, and aside from immunologic clearance, viral infection modifies many of the host cellular functions that promote tissue repair. For example, CMV infection resulting in acceleration of CR is the increase in the host immune response to the allograft and the virus resulting in recruitment of inflammatory cells, and inflammatory effectors such as chemokines and cytokines including interferon-gamma (IFN-γ), TNF-α, interleukin-4 (IL-4), IL-18, RANTES, MCP-1, MIP-1α, IL-8, and IP-10 (Almeida et al. 1994; Almeida-Porada et al. 1997; Vieira et al. 1998; Humar et al. 1999; Streblow et al. 1999, 2003;

Vliegen et al. 2004). CMV also encodes CC and CXC chemokine homologs (see the chapter by P.S. Beisser et al., this volume) that recruit a multitude of host cellular infiltrates (Sparer et al. 2004; Noda et al. 2006). In addition, CMV modifies a number of other cellular factors involved in angiogenesis and wound repair processes, including adhesion molecules (ICAM-1, VCAM-1, VAP-1, and E-selectin) and growth factors and receptors (TGF-β, PDGF-AA, VEGF, and PDGFR) (Shahgasempour et al. 1997; Burns et al. 1999; Zhou et al. 1999; Inkinen et al. 2003, 2005; Helantera et al. 2005, 2006; Reinhardt et al. 2005a, 2005b). In addition, increased matrix metalloproteinase (MMP)-2 activity is observed in HCMV-infected cells in conjunction with a reduction in matrix gene expression, resulting in a malleability to SMC migration, an alteration in vessel remodeling that promotes a vasculopathy (Schaarschmidt et al. 1999; Reinhardt et al. 2006). The upregulation of agents that initiate endothelial adhesion and those that promote wound healing provides a means for CMV to enhance the adherence and infiltration of inflammatory cells that drive vascular disease.

What Factors Constitute the HCMV Secretome?

We and others have demonstrated that HCMV infection alters the types and quantities of bioactive proteins released from infected cells, which we designate as the HCMV secretome. Many of these factors have important roles in vascular disease, and we hypothesize that a major role of CMV infection in the acceleration of TVS occurs through the increased production of WH and AG factors in the allograft. However, neither the complete proteome of the HCMV secretome nor its effects on WH and AG are known. Therefore, in order to determine the effects of HCMV infection on the extracellular milieu, we generated secretomes from HCMV-infected and mock-infected fibroblasts and determined their protein contents (proteomes) by gel-free LC-MS/MS at the Pacific Northwest National Laboratory (PNNL, Richland, WA) and by specific protein arrays. In our LC-MS/MS analysis of the HCMV-infected and mock-infected secretomes, we identified more than 1,200 proteins with 800 having two or more peptide hits. Of the proteins identified by MS/MS, more than 1,000 were specific or highly enriched in the HCMV secretome, more than 260 proteins were common to both the HCMV- and mock-infected secretomes, and more than 225 were specific for the mock-infected secretome. We detected ten viral proteins in the HCMV secretome of which only four were identified with more than peptides, including UL32 (pp150), UL44, UL122, and UL123. Overlaying pathway information and literature mining results, it was noted that a cluster of proteins were involved in integrin signaling and that this cluster was enriched for laminins. Laminins are widely distributed ECM proteins involved in cell adhesion signaling and have recently been implicated in playing a fundamental role in angiogenesis by directly affecting gene and protein expression profiles (Folkman 2003). TGF-β signaling and angiogenesis pathways were also identified in this initial screen.

We also assayed for changes in 174 common cytokines/growth factors present in the HCMV secretome using RayBio® Human Cytokine Array G Series 2000 antibody arrays. We analyzed secretomes from HCMV- and mock-infected fibroblasts. We detected 144 of the 174 factors in the HCMV secretome when our cut-off value

Table 2 Detection of factors present in the HCMV and mock secretomes by Raybiotech Protein Array Analysis

Induced or repressed factors				Most abundant factors in HCMV secretome	
Protein	Mock average	HCMV-infected average	Fold change	Protein	Average intensity
GDNF	25±6	5666±814*	225.3	IL–5	24633
GM-CSF	28±5	566±813*	199.4	IL–6	24618
MIP–1 alpha	336±117	22519±2926*	66.9	MIP–1 alpha	22519
MIP–3 alpha	17±12	896±92*	51.5	PECAM–1	20638
MIF	311±129	15732±2729*	50.6	PIGF	18107
MIP–1delta	26±13	915±104*	35.5	GRO	17656
GRO	614±143	17656±4136*	28.8	GITR-Ligand	17469
GITR-Ligand	655±209	17469±4198*	26.7	MIF	15732
IL–5	1037±443	24633±2638*	23.8	VE-Cadherin	15200
IL–6	1037±447	24617±2641*	23.7	TIMP–4	15136
MIP–1 beta	435±89	7058±238*	16.2	PDGF-AA	11622
RANTES	39±6	596±90*	15.2	PDGF-AB	11595
Angiogenin	87±53	676±47*	7.7	TIMP–1	8920
IL–8	234±162	1787±400*	7.6	PDGF-BB	8880
ITAC	251±189	1115±227*	4.4	TIMP–2	8839
Advin A	308±119	1290±229*	4.2	MIP–1 beta	7058
ALCAM	282±130	1181±264*	4.2	GDNF	5666
MCP–3	26±9	102±22*	4.0	GM-CSF	5661
MMP–9	35±14	113±39	3.3	TNF-RI	3574
MCP–4	35±10	111±24*	3.2	TNF RII	3451
IL–1alpha	123±38	372±48*	3.0	HCC–4	3193
IL–6 R	266±92	770±297	2.9	MCP–1	3186
IL–1beta	35±2	102±18*	2.9	LIGHT	3185
IGF-II	85±22	241±95	2.8	HGF	3174
TNF RII	1316±959	3451±867*	2.6	Oncostatin M	2724
bFGF	195±104	503±70*	2.6	Osteoprotegerin	2710
FGF–7	196±33	485±12*	2.5	IL–8	1787
Flt–3 Ligand	191±23	468±13*	2.4	EGF-R	1773
IL–16	130±35	307±36*	2.4	egp 130	1732
MCP–1	1388±926	3186±414	2.3	Advin A	1290
AgRP	181±88	416±68*	2.3	TRAIL R3	1266
LIGHT	1394±934	3185±411	2.3	Thrombopoietin	1221
ICAM–1	325±185	725±267	2.2	IL–1R II	1210
IL–9	271±62	559±122*	2.1	ALCAM	1181
TGF-alpha	220±137	433±42	2.0	ICAM–2	1123
IP–10	531±77	975±105*	1.8	ITAC	1115
				uPAR	1109
TECK	313±126	121±4	0.4	IL–2 R alpha	1029

(continued)

Table 2 (continued)

Protein	Induced or repressed factors			Most abundant factors in HCMV secretome	
	Mock average	HCMV-infected average	Fold change	Protein	Average intensity
ICAM–3	331±183	123±8	0.4	IP–10	975
IL–1R II	3387±570	1210±374*	0.4		
IL–2 alpha	341±159	112±19	0.3		
VEGF-D	374±188	119±21	0.3		
IL–2 R alpha	3287±635	1029±392*	0.3		
HCC–4	14014±11353	3193±1191	0.2		
HGF	13954±11376	3174±1194	0.2		

for detection was set at an average intensity of 500. Of these 144 proteins detected, 41 factors were significantly induced over the mock-infected secretome (Table 2). The 35 most predominant proteins detected in the HCMV secretome are listed in Table 2. The most highly abundant cellular factors present in the HCMV secretome that contribute to WH/AG include the cytokines/chemokines (IL-6, osteoprotegerin, GRO, CCL3, CCL5, CCL7, CCL20, CXCL-5, and CXCL-16), receptors (TNF-RI and II and ICAM-1), growth factors (TGF-β1 and HGF), ECM modifiers (MMP-1, TIMP-1, TIMP-2, TIMP-4), and the angiogenic RNase angiogenin. Interestingly, MCP-3 was induced over 85-fold in the HCMV secretome compared to the mock secretome. Most of the highly induced proteins detected in our assay (HCMV vs mock) included chemokines such as CCL1, CCL3, CCL4, CCL5, CCL20, CXCL10, CXCL11). Interestingly, many of the genes that were identified by microarray analysis in the rat allograft hearts were also found in the HCMV secretome, suggesting that the factors involved in HCMV-induced angiogenesis and wound healing are similar to those expressed in the RCMV-infected allografts and importantly, that these in vitro and in vivo processes are parallel to each other.

HCMV Secretome Induces Angiogenesis in Endothelial Cells

Angiogenesis leading to vessel formation in vivo consists of a growth phase followed by a stabilization phase (Auerbach et al. 2003; Guidolin et al. 2004). Growth phase events include proteolytic digestion of the basement membrane (BM) and extracellular matrix (ECM) of the existing vessel, migration and proliferation of ECs, lumen formation within the EC sprout, and anastomosis of sprouts to form neovessels. Stabilization involves arrest of EC proliferation, EC differentiation, intercellular adhesion and remodeling of the BM/ECM network to create an immature capillary. Key steps in both phases of angiogenesis can be modeled using an in vitro assay: the matrigel assay for capillary-like tubule formation (Wegener et al. 2000; Xiao et al. 2002). The extent of tubule formation and stabilization depends on factors produced by the EC themselves in coordination with exogenous angiogenic agonists or antagonists in the culture medium. We utilized the matrigel assay to test the angiogenic activity of the HCMV secretome. Low-passage primary human umbilical vein endothelial cells (HUVECs) were nutrient-starved in serum and growth factor-free endothelial medium (SFM) prior to harvest and resuspension in this same medium. Cells were introduced into 24-well trays containing polymerized plugs of growth-factor-reduced matrigel in the presence of 300 µl of control and test supernatants. Control supernatants included SF-SFM and complete SFM containing 10% human serum and endothelial cell growth supplement. Each supernatant was tested in quadruplicate. Cell phenotype was digitally recorded at 24 h to evaluate the degree of EC migration and differentiation into tubule structures, and again after 2 weeks to evaluate vessel survival and stability. Figure 1a graphically depicts two quantitative measures of angiogenesis: the number of enclosed polygonal spaces delimited by complete tubules (lumens) and the number of nodes where branching tubules meet (branch points). Figure 1b shows a representative example of each culture condition as a low-power image. Figure 1c shows high-power images to emphasize the differences in vessel integrity between conditions. Results with control supernatants confirmed that exogenous angiogenic factors are required to support the formation of a robust capillary network when ECs are plated on GFR-matrigel. Specifically, ECs cultured in complete medium for 24 h aligned to form a meshwork of anastomosing tubules with multinodal branch points and enclosed lumens. In contrast, ECs cultured in SFM for 24 h were unable to form a consistent network of interconnecting tubules, with many cells generating incomplete tubes or aggregating in clumps. In contrast, ECs cultured in the presence of the HCMV secretome supported the formation of an extensive polygonal capillary network. Vessels formed in the presence of the HCMV secretome appeared to be more defined than those formed in the presence of complete medium, presumably reflecting increased stability of intercellular junctions. To test the degree of stabilization afforded by the HCMV supernatants, trays were kept in culture for 2 weeks (Fig. 1d). By this time, networks induced by complete medium had degenerated, but those induced by the HCMV secretome remained intact. Collectively, we show that the HCMV secretome contains factors that promote angiogenesis and allow stabilization of neovessels and that generation of this active secretome requires HCMV replication (data not shown).

HCMV Secretome Induces Wound Healing in Endothelial Cells

For the WH assays we used an electric cell-substrate impedance sensing (ECIS) system available from Applied Biophysics Inc., to monitor cell behavior. In ECIS, cells are grown on eight-well chamber slides with 250-mm-diameter gold electrodes microfabricated onto each well bottom. A larger counter-electrode completes the circuit, using standard tissue culture medium as an electrolyte. When a weak AC signal is applied to the system, the presence of a confluent

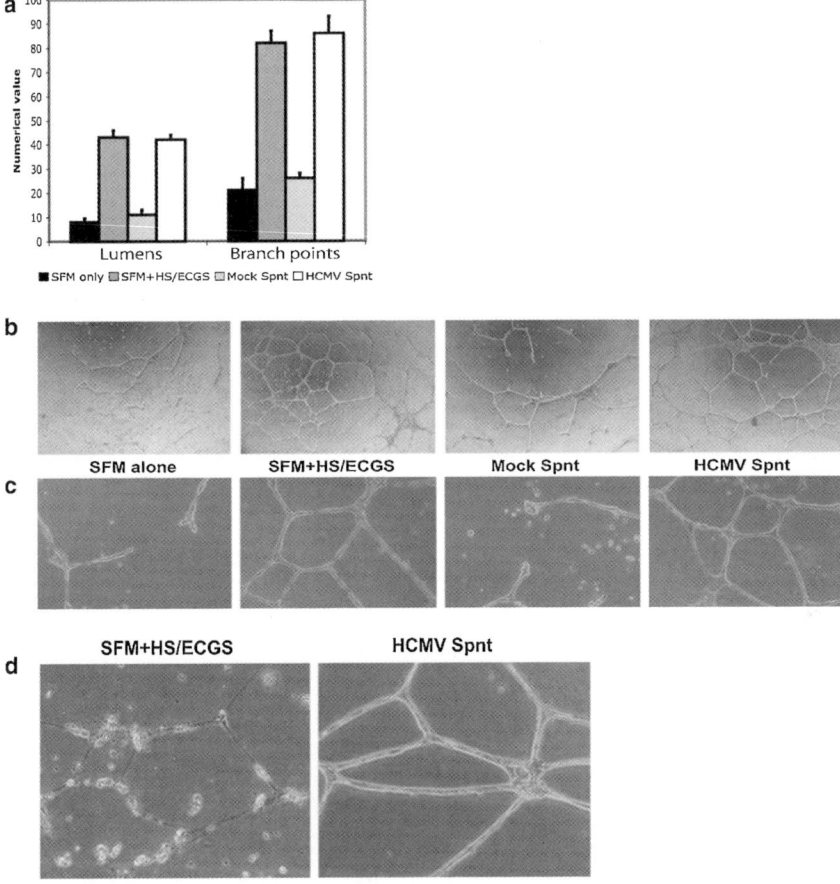

Fig. 1 HCMV secretome mediates EC tubule formation. **a** Quantitation of two parameters of angiogenesis, lumen formation, and number of branch points at 24 h after plating on matrigel in the presence of control supernatants (SFM alone or SFM + HS/ECGS) and test supernatants conditioned by factors secreted by Mock- and HCMV-infected cells. **b** Low-power images of EC differentiation on matrigel. **c** High-power images, conditions as for (**b**), illustrating the integrity of individual tubules. **d** Tubule survival after 2 weeks on matrigel in the presence of SFM plus HS/ECGS or supernatant conditioned by HCMV-infected cells

monolayer is reflected by a marked increase in impedance, since cells restrict the effective area available for current flow. Fluctuations in measured impedance occur in response to micromotions of cells and can be used as indications of cell viability or morphology change (Keese et al. 2004). ECIS technology has been adapted for wound healing assays (Charrier et al. 2005). Cells are grown to confluence on the arrays to achieve high impedance values and a transient voltage spike is applied to kill only the cells on the electrode. Normal ECIS measurements are then used to monitor of the rate of repopulation of the wounded area from cells surrounding the electrode, which is superior to traditional WH assays since it is entirely automated and allows continuous monitoring of the cellular response to wounding. We have used the ECIS system to measure the influence of the HCMV secretome on WH. For these assays, primary HUVECs were plated in two arrays in complete medium and incubated overnight to allow establishment of a confluent monolayer. The next day, cells were serum-starved prior to the addition of secretome samples from mock- and AD169- (plus or minus UV or foscarnet treatments) infected or HSV-1-infected HFs to duplicate wells. SFM and complete medium were used as negative and positive controls, respectively. Immediately after application of supernatants, arrays were placed in the electrode holders and impedance was measured for a few minutes. This confirmed the existence of confluent cell monolayers over each electrode and verified that the cells were healthy. All but one of the chambers of cells were wounded for 30 s. As expected, the electrical wounding resulted in an immediate drop in impedance to the level of an open electrode. The subsequent rise in impedance due to migration and repopulation of the wound was monitored. As shown in Fig. 2 under the presence of complete medium, the wound was repopulated in about 6 h. Similarly,

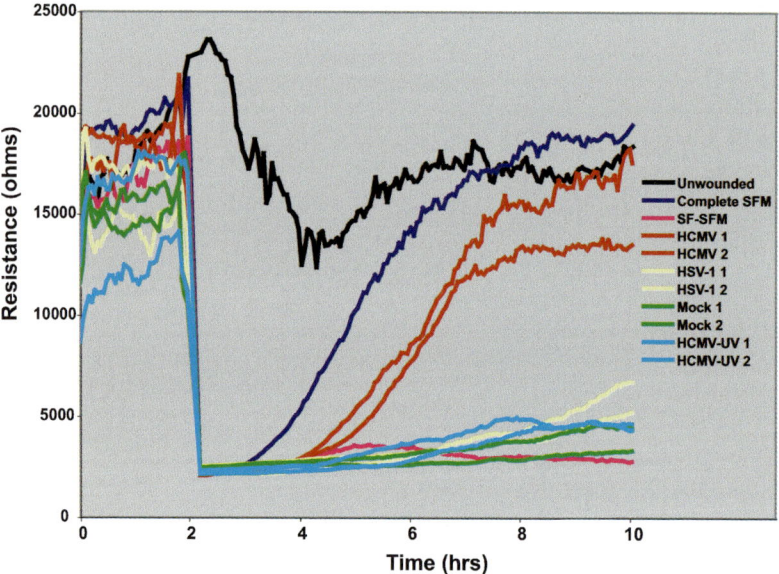

Fig. 2 HCMV secretome mediates EC wound healing. Wound healing activity of the HCMV secretome. ECs were grown to confluence on ECIS arrays and exposed to test supernatants prior to electrical wounding. Wound healing, as indicated by increasing resistance, is plotted as a function of time. Healing traces for duplicate wells are shown for the HSV-1 Secretome (*yellow*), mock (*green*), HCMV (*red*), or UV-inactivated HCMV (*light blue*). Control traces include a negative control (SSFM; *pink*), a positive control (Complete SFM; *dark blue*) and an unwounded control (*black*). Cells exposed to the HCMV secretome show wound repair within 6-10 h, whereas cells exposed to the HSV-1 or mock secretomes repopulate the wound inefficiently, indicating that the production of wound healing factors is specific for HCMV infections

cells cultured in the presence of the HCMV secretome were able to repopulate the wound with the same kinetics as with complete medium, although steady state resistance levels were slightly lower. In contrast, cells cultured in SFM lacking growth factors, the mock secretome, or the secretomes from HCMV UV- or foscarnet-treated cells all were unable to repopulate the electrode, even over the total 20 h measured. Similarly, cells incubated with a secretome derived from HSV-1 infected cells also failed to mediate WH, suggesting that the WH effects are specific for CMV. Importantly, these data clearly show that the HCMV secretome contains factors that promote cell migration into a mechanical wound. The ability of the HCMV secretome to mediate WH is due to active viral replication since both UV-treated virions and foscarnet-treated cells did not promote WH. Since foscarnet treatment of infected cells resulted in an inactive secretome, this observation suggests that a late kinetic class of HCMV gene(s) is involved in the generation of secretome WH factors. These observations correlate with studies in human heart transplant patients as well as our own observations in heart transplants in rats in which ganciclovir treatment prolongs graft survival.

Conclusions

There has been a steady progression to our understanding of role that HCMV plays during the acceleration of vascular diseases such as atherosclerosis, restenosis, and transplant vasculopathy associated with the development of chronic solid organ allograft rejection. This progress has been made through epidemiologic studies, the use of animal models of vascular disease and transplantation, as well as through the use of *in vitro* models that mimic the clinical scenario. While the precise mechanism(s) of action have yet to be fully elucidated, what has become clear over the last couple of decades of study is that HCMV is capable of modifying the extracellular host environment through the production and release of biologically active cellular factors, including growth factors, cytokines, and ECM-modifying enzymes. As we demonstrate here, the overall effect of this host manipulation is that the HCMV secretome is capable of mediating angiogenesis and wound healing, which are important processes that drive vascular disease formation. Indeed, many of the factors that we identified in the HCMV secretome were identified by microarray analysis in the rat allograft hearts, suggesting that the factors involved in these two processes are similar. Future research meant to identify the specific upstream gene targets that would make possible prevention or abrogation of the HCMV-associated vasculopathy and chronic allograft rejection will broaden the therapeutic profile used to combat this very important clinical problem.

Acknowledgements S.L. Orloff was supported by a grant from the Department of Veterans Affairs and from the National Institutes of Health (HL 66238-01); NIH grants were also awarded to D.N. Streblow (HL083194) and J.A. Nelson (AI21640, HL65754, and HL71695). D.N. Streblow is also supported by an AHA Scientist Development Grant.

References

Almeida GD, Porada CD, St Jeor S, Ascensao JL (1994) Human cytomegalovirus alters interleukin-6 production by endothelial cells. Blood 83:370-376

Almeida-Porada G, Porada CD, Shanley JD, Ascensao JL (1997) Altered production of GM-CSF and IL-8 in cytomegalovirus-infected, IL-1- primed umbilical cord endothelial cells. Exp Hematol 25:1278-1285

Auerbach R, Lewis R, Shinners B, Kubai L, Akhtar N (2003) Angiogenesis assays: a critical overview. Clin Chem 49:32-40

Billingham ME (1992) Histopathology of graft coronary disease. J Heart Lung Transplant 11:S38-S44

Bouis D, Kusumanto Y, Meijer C, Mulder NH, Hospers GA (2006) A review on pro- and anti-angiogenic factors as targets of clinical intervention. Pharmacol Res 53:89-103

Bruning JH, Persoons MCJ, Lemstrom KB, Stals FS, De Clereq E, Bruggeman CA (1994) Enhancement of transplantation associated atherosclerosis by CMV, which can be prevented by antiviral therapy in the form of HPMPC. Transplant Int 7:365-370

Burns LJ, Pooley JC, Walsh DJ, Vercellotti GM, Weber ML, Kovacs A (1999) Intercellular adhesion molecule-1 expression in endothelial cells is activated by cytomegalovirus immediate early proteins. Transplantation 67:137-144

Carmeliet P (2003) Angiogenesis in health and disease. Nat Med 9:653-660

Charrier L, Yan Y, Driss A, Laboisse CL, Sitaraman SV, Merlin D (2005) ADAM-15 inhibits wound healing in human intestinal epithelial cell monolayers. Am J Physiol Gastrointest Liver Physiol 288:G346-G353

Clinton SK, Libby P (1992) Cytokines and growth factors in atherogenesis. Arch Pathol Lab Med 116:1292

Deotero J, Gavalda J, Murio E, et al (1998) Cytomegalovirus disease as a risk factor for graft loss and death after orthotopic liver transplantation. Clin Invest Dis 26:865-870

Ely JM, Greiner DL, Lubaroff DM, Fitch FW (1983) Characterization of monoclonal antibodies that define rat T cell alloantigens. J Immunol 130:2798

Fitzgerald JT, Gallay B, Taranto SE, McVicar JP, Troppmann C, Chen X, McIntosh MJ, Perez RV (2004) Pretransplant recipient cytomegalovirus seropositivity and hemodialysis are associated with decreased renal allograft and patient survival. Transplantation 77:1405-1411

Folkman J (2003) Fundamental concepts of the angiogenic process. Curr Mol Med 3:643-651

Grattan MT, Moreno-Cabral CE, Starnes VA, Oyer PE, Stinson EB, Shumway NE (1989) Cytomegalovirus infection is associated with cardiac allograft rejection and atherosclerosis. JAMA 261:3561-3566

Guidolin D, Vacca A, Nussdorfer GG, Ribatti D (2004) A new image analysis method based on topological and fractal parameters to evaluate the angiostatic activity of docetaxel by using the matrigel assay in vitro. Microvasc Res 67:117-124

Helantera I, Loginov R, Koskinen P, Tornroth T, Gronhagen-Riska C, Lautenschlager I (2005) Persistent cytomegalovirus infection is associated with increased expression of TGF-beta1, PDGF-AA and ICAM-1 and arterial intimal thickening in kidney allografts. Nephrol Dial Transplant 20:790-796

Helantera I, Teppo AM, Koskinen P, Tornroth T, Gronhagen-Riska C, Lautenschlager I (2006) Increased urinary excretion of transforming growth factor-beta(1) in renal transplant recipients during cytomegalovirus infection. Transpl Immunol 15:217-221

Hosenpud JD, Shipley GD, Wagner CR (1992) Cardiac allograft vasculopathy: current concepts, recent developments and future directions. J Heart Lung Transplant 11:9-23

Humar A, St Louis P, Mazzulli T, McGeer A, Lipton J, Messner H, MacDonald KS (1999) Elevated serum cytokines are associated with cytomegalovirus infection and disease in bone marrow transplant recipients. J Infect Dis 179:484-488

Inkinen K, Holma K, Soots A, Krogerus L, Loginov R, Bruggeman C, Ahonen J, Lautenschlager I (2003) Expression of TGF-beta and PDGF-AA antigens and corresponding mRNAs in cytomegalovirus-infected rat kidney allografts. Transplant Proc 35:804-805

Inkinen K, Soots A, Krogerus L, Loginov R, Bruggeman C, Lautenschlager I (2005) Cytomegalovirus enhances expression of growth factors during the development of chronic allograft nephropathy in rats. Transpl Int 18:743-749

Keese CR, Wegener J, Walker SR, Giaever I (2004) Electrical wound-healing assay for cells in vitro. Proc Natl Acad Sci U S A 101:1554-1559

Kent D, Sheridan C (2003) Choroidal neovascularization: a wound healing perspective. Mol Vis 9:747-755

Khurana R, Simons M (2003) Insights from angiogenesis trials using fibroblast growth factor for advanced arteriosclerotic disease. Trends Cardiovasc Med 13:116-122

Klempnauer J, Marquarding E (1989) RT1.C and rat bone marrow transplantation. Transplant Proc 21:3292

Lemstrom KB, Bruning JH, Bruggeman CA, Lautenschlager IT, Hayry PJ (1993) Cytomegalovirus infection enhances smooth muscle cell proliferation and intimal thickening of rat aortic allografts. J Clin Invest 92:549-558

Lemstrom K, Koskinen P, Krogerus L, Daemen M, Bruggeman CA, Hayry PJ (1995) Cytomegalovirus antigen expression, endothelial cell proliferation, and intimal thickening in rat cardiac allografts after cytomegalovirus infection. Circulation 92:2594-2604

Libby P, Salomon RN, Payne DD, Schoen FJ, Pober JS (1989) Functions of vascular wall cells related to development of transplantation-associated coronary arteriosclerosis. Transplant Proc 21:3677-3684

Lubaroff DM, Rasmussen GT, Hunt HD (1989) The RT6 T cell antigen: its role in the identification of functional subsets and in T cell activation. Transplant Proc 21:3251

Melnick JL, Petrie BL, Dreesman GR, Burek J, McCollum CH, DeBakey ME (1983) Cytomegalovirus antigen within human arterial smooth muscle cells. Lancet 2:644-647

Melnick JL, Adam E, DeBakey ME (1998) The link between CMV and atherosclerosis. Infect Med 15:479-486

Merigan TC, Renlund DG, Keay S et al (1992) A controlled trial of ganciclovir to prevent cytomegalovirus disease after heart transplantation. N Engl J Med 326:1182-1186

Noda S, Aguirre SA, Bitmansour A, Brown JM, Sparer TE, Huang J, Mocarski ES (2006) Cytomegalovirus MCK-2 controls mobilization and recruitment of myeloid progenitor cells to facilitate dissemination. Blood 107:30-38

Orloff SL (1999) Elimination of donor-specific alloreactivity by bone marrow chimerism prevents cytomegalovirus accelerated transplant vascular sclerosis in rat small bowel transplants. J Clin Virol 12:142

Orloff SL, Yin Q, Corless CL, Orloff MS, Rabkin JM, Wagner CR (2000) Tolerance induced by bone marrow chimerism prevents transplant vascular sclerosis in a rat model of small bowel transplant chronic rejection. Transplantation 69:1295-1303

Orloff SL, Streblow DN, Soderberg-Naucler C, Yin Q, Kreklywich C, Corless CL, Smith PA, Loomis CB, Mills LK, Cook JW, Bruggeman CA, Nelson JA, Wagner CR (2002) Elimination of donor-specific alloreactivity prevents cytomegalovirus-accelerated chronic rejection in rat small bowel and heart transplants. Transplantation 73:679-688

Reinhardt B, Mertens T, Mayr-Beyrle U, Frank H, Luske A, Schierling K, Waltenberger J (2005a) HCMV infection of human vascular smooth muscle cells leads to enhanced expression of functionally intact PDGF beta-receptor. Cardiovasc Res 67:151-160

Reinhardt B, Schaarschmidt P, Bossert A, Luske A, Finkenzeller G, Mertens T, Michel D (2005b) Upregulation of functionally active vascular endothelial growth factor by human cytomegalovirus. J Gen Virol 86:23-30

Reinhardt B, Winkler M, Schaarschmidt P, Pretsch R, Zhou S, Vaida B, Schmid-Kotsas A, Michel D, Walther P, Bachem M, Mertens T (2006) Human cytomegalovirus-induced reduction of extracellular matrix proteins in vascular smooth muscle cell cultures: a pathomechanism in vasculopathies? J Gen Virol 87:2849-2858

Rubin RH (1999) Importance of CMV in the transplant population. Transplant Infect Dis 1:3-7

Schaarschmidt P, Reinhardt B, Michel D, Vaida B, Mayr K, Luske A, Baur R, Gschwend J, Kleinschmidt K, Kountidis M, Wenderoth U, Voisard R, Mertens T (1999) Altered expression of extracellular matrix in human-cytomegalovirus-infected cells and a human artery organ culture model to study its biological relevance. Intervirology 42:357-364

Shahgasempour S, Woodroffe SB, Garnett HM (1997) Alterations in the expression of ELAM-1, ICAM-1 and VCAM-1 after in vitro infection of endothelial cells with a clinical isolate of human cytomegalovirus. Microbiol Immunol 41:121-129

Shibata M, Suzuki H, Nakatani M, Koba S, Geshi E, Katagiri T, Takeyama Y (2001) The involvement of vascular endothelial growth factor and flt-1 in the process of neointimal proliferation in pig coronary arteries following stent implantation. Histochem Cell Biol 116:471-481

Soule JL, Streblow DN, Andoh TF, Kreklywich CN, Orloff SL (2006) Cytomegalovirus accelerates chronic allograft nephropathy in a rat renal transplant model with associated provocative chemokine profiles. Transplant Proc 38:3214-3220

Sparer TE, Gosling J, Schall TJ, Mocarski ES (2004) Expression of human CXCR2 in murine neutrophils as a model for assessing cytomegalovirus chemokine vCXCL-1 function in vivo. J Interferon Cytokine Res 24:611-620

Speir E, Modali R, Huang ES, Leon MB, Shawl F, Finkel T, Epstein SE (1994) Potential role of human cytomegalovirus and p53 interaction in coronary restenosis. Science 265:391-394

Streblow DN, Söderberg-Nauclér C, Vieira J, Smith P, Wakabayashi E, Rutchi F, Mattison K, Altschuler Y, Nelson JA (1999) The human cytomegalovirus chemokine receptor US28 mediates vascular smooth muscle cell migration. Cell 99:511-520

Streblow DN, Kreklywich C, Yin Q, De La Melena VT, Corless CL, Smith PA, Brakebill C, Cook JW, Vink C, Bruggeman CA, Nelson JA, Orloff SL (2003) Cytomegalovirus-mediated upregulation of chemokine expression correlates with the acceleration of chronic rejection in rat heart transplants. J Virol 77:2182-2194

Streblow DN, Kreklywich CN, Smith P, Soule JL, Meyer C, Yin M, Beisser P, Vink C, Nelson JA, Orloff SL (2005) Rat cytomegalovirus-accelerated transplant vascular sclerosis is reduced with mutation of the chemokine-receptor R33. Am J Transplant 5:436-442

Streblow DN, van Cleef KW, Kreklywich CN, Meyer C, Smith P, Defilippis V, Grey F, Fruh K, Searles R, Bruggeman C, Vink C, Nelson JA, Orloff SL (2007) Rat cytomegalovirus gene expression in cardiac allograft recipients is tissue specific and does not parallel the profiles detected in vitro. J Virol 81:3816-3826

Tikkanen JM, Kallio EA, Bruggeman CA, Koskinen PK, Lemstrom KB (2001a) Prevention of cytomegalovirus infection-enhanced experimental obliterative bronchiolitis by antiviral prophylaxis or immunosuppression in rat tracheal allografts. Am J Respir Crit Care Med 164:672-679

Tikkanen J, Kallio E, Pulkkinen V, Bruggeman C, Koskinen P, Lemstrom K (2001b) Cytomegalovirus infection-enhanced chronic rejection in the rat is prevented by antiviral prophylaxis. Transplant Proc 33:1801

Tu W, Potena L, Stepick-Biek P, Liu L, Dionis KY, Luikart H, Fearon WF, Holmes TH, Chin C, Cooke JP, Valantine HA, Mocarski ES, Lewis DB (2006) T-cell immunity to subclinical cytomegalovirus infection reduces cardiac allograft disease. Circulation 114:1608-1615

Valantine HA, Gao SZ, Santosh G et al (1999) Impact of prophylactic immediate post-transplant ganciclovir on development of transplant atherosclerosis. A post-hoc analysis of a randomized, placebo-controlled study. Circulation 100:61-66

Vieira J, Schall TJ, Corey L, Geballe AP (1998) Functional analysis of the human cytomegalovirus US28 gene by insertion mutagenesis with the green fluorescent protein gene. 72:8158-8165

Vliegen I, Duijvestijn A, Stassen F, Bruggeman C (2004) Murine cytomegalovirus infection directs macrophage differentiation into a pro-inflammatory immune phenotype: implications for atherogenesis. Microbes Infect 6:1056-1062

Wegener J, Keese CR, Giaever I (2000) Electric cell-substrate impedance sensing (ECIS) as a noninvasive means to monitor the kinetics of cell spreading to artificial surfaces. Exp Cell Res 259:158-166

Werner S, Grose R (2003) Regulation of wound healing by growth factors and cytokines. Physiol Rev 83:835-870

Wu TC, Hruban RH, Ambinder RF, Pizzorno M, Cameron DE, Baumgartner WA, Reitz BA, Hayward GS, Hutchins GM (1992) Demonstration of cytomegalovirus nucleic acids in the coronary arteries of transplanted hearts. Am J Pathol 140:739-747

Xiao C, Lachance B, Sunahara G, Luong JH (2002) An in-depth analysis of electric cell-substrate impedance sensing to study the attachment and spreading of mammalian cells. Anal Chem 74:1333-1339

Zeng H, Waldman WJ, Yin DP, Knight DA, Shen J, Ma L, Meister GT, Chong AS, Williams JW (2005) Mechanistic study of malononitrileamide FK778 in cardiac transplantation and CMV infection in rats. Transplantation 79:17-22

Zhou YF, Yu ZX, Wanishsawad C, Shou M, Epstein SE (1999) The immediate early gene products of human cytomegalovirus increase vascular smooth muscle cell migration, proliferation, and expression of PDGF beta-receptor. Biochem Biophys Res Commun 256:608-613

Manifestations of Human Cytomegalovirus Infection: Proposed Mechanisms of Acute and Chronic Disease

W. Britt

Contents

Introduction 418
Natural History of Acute CMV Infections in the Normal Host 421
Natural History of Acute CMV Infection in the Immunocompromised Host 426
   Congenital Infection. 427
   Allograft Recipients. 432
   Patients with AIDS 439
Diseases Associated with Chronic Infection. 442
   Chronic Vascular Disease in the Normal Host. 443
   Chronic Disease in the Transplanted Allograft 446
References 449

Abstract Infections with human cytomegalovirus (HCMV) are a major cause of morbidity and mortality in humans with acquired or developmental deficits in innate and adaptive immunity. In the normal immunocompetent host, symptoms rarely accompany acute infections, although prolonged virus shedding is frequent. Virus persistence is established in all infected individuals and appears to be maintained by both a chronic productive infections as well as latency with restricted viral gene expression. The contributions of the each of these mechanisms to the persistence of this virus in the individual is unknown but frequent virus shedding into the saliva and genitourinary tract likely accounts for the near universal incidence of infection in most populations in the world. The pathogenesis of disease associated with acute HCMV infection is most readily attributable to lytic virus replication and end organ damage either secondary to virus replication and cell death or from host immunological responses that target virus-infected cells. Antiviral agents limit the severity of disease associated with acute HCMV infections, suggesting a requirement for virus replication in clinical

W. Britt
Departments of Pediatrics, Microbiology, and Neurobiology, University of Alabama School of Medicine, Childrens Hospital, Harbor Bldg. 104, 1600 7th Ave. SouthBirmingham, AL 35233, USA
wbritt@peds.uab.edu

syndromes associated with acute infection. End organ disease secondary to unchecked virus replication can be observed in infants infected in utero, allograft recipients receiving potent immunosuppressive agents, and patients with HIV infections that exhibit a loss of adaptive immune function. In contrast, diseases associated with chronic or persistent infections appear in normal individuals and in the allografts of the transplant recipient. The manifestations of these infections appear related to chronic inflammation, but it is unclear if poorly controlled virus replication is necessary for the different phenotypic expressions of disease that are reported in these patients. Although the relationship between HCMV infection and chronic allograft rejection is well known, the mechanisms that account for the role of this virus in graft loss are not well understood. However, the capacity of this virus to persist in the midst of intense inflammation suggests that its persistence could serve as a trigger for the induction of host-vs-graft responses or alternatively host responses to HCMV could contribute to the inflammatory milieu characteristic of chronic allograft rejection.

Introduction

Cytomegaloviruses are readily transmissible beta-herpesviruses that establish persistent infections for the life of the host. CMVs have been isolated from almost all mammalian species that have been studied, including commonly used experimental laboratory animals. Several lines of evidence argue that CMVs co-evolved with their natural hosts (Britt et al. 2005). These include the finding of functional host cell genes incorporated into the viral genome, the near universal lack of clinically significant disease following primary infection of the natural host, and a highly restricted tropism for the homologous host that appears to be linked to blocks of genes specific for each cytomegalovirus (Britt et al. 2005). This later biological property of CMVs has limited the use of experimental animals to precisely model human disease syndromes. As a result, considerable efforts have been made to define the pathogenesis of human CMV (HCMV) infections by observational studies. Recapitulation of all aspects of HCMV infection has been impossible to achieve in experimental animal models, but important insights into general mechanisms of virus persistence and immune control of acute and chronic viral infections have been gained through the study of rodent and more recently, rhesus macaque CMV infections.

Well-recognized disease syndromes follow acute infection with CMV and after reactivation of latent infections in both immunocompromised human and animal hosts; however, chronic persistent infections with CMV and possibly chronic abortive infections with restricted viral gene expression have also been proposed to explain less recognizable disease syndromes associated with these viruses. Several hurdles continue to plague experimental study of chronic diseases associated with CMV, including the complex organization and regulation of viral gene expression and more recently the realization that commonly used laboratory virus strains of HCMV and recent clinical isolates differ not only in their overall genetic composition but also their host cell tropism and often induce variable cellular responses following infection (Plachter et al. 1996; Fish et al. 1998; Sinzger et al. 1999b; Prichard et al. 2001; Murphy et al. 2003). Experimental systems that permit efficient genetic manipulation of CMV together with the promise of genomewide monitoring of gene expression of CMVs will undoubtedly renew interest in the pathogenesis of infection with these viruses and potentially identify previously unrecognized contributions of this large DNA virus to human diseases (Zhu et al. 1998; Chambers et al. 1999; Hobom et al. 2000; Messerle et al. 2000; Simmen et al. 2001; Dunn et al. 2003; Yu et al. 2003; Rupp et al. 2005; Streblow et al. 2007). This brief discussion will focus on the pathogenesis of diseases associated with HCMV and attempt to integrate observations in relevant animal models to current understanding of human disease.

The predictable occurrence of CMV end-organ disease in significantly immunocompromised hosts clearly demonstrates the importance of the immune system in limiting virus replication and disease (Table 1). Although the risk of acute disease following HCMV infection in allograft recipients is well recognized, our understanding of the key elements of immune control of this infection has been based almost entirely on observations made in patients with significant and sometimes global deficiencies in immune reactivity. Animal models, most notably murine and

Table 1 Clinical manifestations of HCMV infection

Population	Diseases associated with acute infection	Diseases associated with chronic infection
Immunocompetent	Asymptomatic infection, mononucleosis-like illness	Atherosclerotic vascular disease; inflammatory bowel disease, periodontal disease, rheumatologic disorders
Immunocompromised fetal and newborn infants (congenital infection)	Asymptomatic (90%) infections; 10% can have hepatitis, retinitis, thrombocytopenia, neurologic disease	Hearing loss (5%–15%); neuron-developmental abnormalities
Allograft recipients	Fever, decreased bone marrow function, hepatitis, pneumonitis, bacterial superinfections	Vascular disease (transplant vasculopathy), interstitial tubulosclerosis (renal allografts); loss of bile ducts (liver allografts); chronic graft rejection and graft loss
Immunodeficiency syndromes (acquired and inherited)	Gastrointestinal disease (esophagitis, colitis), retinitis, encephalitis	Colitis reported in inherited immunodeficiencies
Aged	Unknown	Immune senescence with potential decrease in immune responsiveness

the rhesus macaque models, have provided data consistent with proposed mechanisms of immune-mediated protection from HCMV in the human host (Baroncelli et al. 1997; Kuhn et al. 1999; Podlech et al. 2000; Kaur et al. 2002, 2003; Reddehase et al. 2002; Barry et al. 2006; Holtappels et al. 2006). The critical importance of virus-specific CD4$^+$ and CD8$^+$ lymphocyte responses, and in some cases antiviral antibodies, to protective immunity is mirrored by observational studies in human allograft recipients and HIV-infected individuals (Reusser et al. 1991, 1997; Walter et al. 1995; Schoppel et al. 1998b; Sester et al. 2001). Natural killer cell responses have been demonstrated to play a major role in host resistance in mice infected with MCMV, and studies in this system have provided insights into the biology of NK cell receptors as well as the genetic control of NK responses in mice (Scalzo et al. 1992; Polic et al. 1998; Biassoni et al. 2001; Moretta et al. 2001; Krmpotic et al. 2002; Yokoyama and Scalzo 2002; French and Yokoyama 2003; Lodoen et al. 2003; Voigt et al. 2003; Bubic et al. 2004; French et al. 2004; Hasan et al. 2005; Adam et al. 2006). Although presumed to be of similar importance in humans, the evidence implicating NK responses in protective responses of humans infected with HCMV is more limited (Bowden et al. 1987; Biron et al. 1989; Lopez-Botet et al. 2001; Iversen et al. 2005; Wills et al. 2005; Cook et al. 2006; Prod'homme et al. 2007; Wagner et al. 2007). The importance of HCMV as an opportunistic pathogen in HIV-infected patients prior to the availability of active antiretroviral therapy (ART) provided new insights into the relationship between HCMV and the immune system. Several important concepts were derived from the myriad of studies of HCMV infection in patients with AIDS, including the observation that seemingly minimal adaptive cellular immunity was sufficient for the control of virus replication and the maintenance of the homeostasis between host and virus (Autran et al. 1997; Komanduri et al. 2001a, 2001b). Secondly, technologies developed to study cellular immune responses to HCMV in these patients indicated that even in normal individuals, the host immune system has allocated an unexpectedly large proportion of circulating T lymphocytes to the recognition of HCMV-infected cells (Waldrop et al. 1997; Sylwester et al. 2005). With the apparent limited pathogenicity of HCMV based on studies in both normal individuals and those with treated HIV infection, it is perplexing why the host has dedicated so much of its T lymphocyte response to this single virus. It has been difficult to reconcile that in some normal immunocompetent individuals up to 10% of their circulating T lymphocytes are HCMV-specific in light of the limited pathogenicity of this virus in all but the most immunocompromised hosts (Jin et al. 2000; Sester et al. 2002a, 2002b; Sylwester et al. 2005). Furthermore, understanding mechanisms of HCMV persistence in the normal host even in the face of the exuberant host immune response remains a major question in viral immunology. A frequently invoked explanation for this apparent paradox is that the complex network of viral genes that encode immune evasion functions and facilitates persistent infection with this virus have been countered by an exaggerated host T lymphocyte response. Whether this so-called memory expansion of T lymphocyte responses is provoked by continual low-level replication or frequent reactivations from latency is unknown. Studies in experimental animal models have provided data suggestive that resistance to CMV-induced disease requires a quantitative response and that immune evasion functions of this virus can dilute a normal protective response and foster persistence of this virus (Krmpotic et al. 2001; Holtappels 2002). Because of the close evolutionary relationship between the HCMV virus and its human host and the persistence of replicating virus in normal individuals, it appears that viral functions contribute to the détente reached by this virus and its host.

Natural History of Acute CMV Infections in the Normal Host

Human CMV infection is ubiquitous in the populations throughout the world (Weller 1971a, 1971b; Stagno and Britt 2006). Serological evidence of past infection ranges from nearly 100% in children and adults from undeveloped countries in Africa, Asia and South America to less than 30% in adults in some areas of North America and northern Europe. Acquisition of HCMV is age-related in both developed and undeveloped countries. In well-developed areas of the world such as the US and northern Europe, serologic reactivity to HCMV increases at an approximate rate of 2% per year after adolescence, whereas in the developing world near universal serologic reactivity can be documented by adolescence (Stagno et al. 1982a, 1983; Stagno and Britt 2006). Consistent with the spread of HCMV in populations, animal CMVs spread efficiently in laboratory colonies and CMV-free colonies of experimental animals such as guinea pigs and rhesus macaques must be actively managed by separation of infected from noninfected animals. Only limited information is available on the epidemiology of CMVs in wild populations of nonhuman primates, but a study has documented near universal seropositivity in at least one free-ranging population of these animals (Swack and Hsiung 1982; Kessler et al. 1989; Vogel et al. 1994). Thus, the high incidence of RhCMV infection in rhesus macaque colonies is more likely related to the natural spread of this virus than to factors associated with captivity. Similarly, MCMV infection is commonplace in wild mice captured in Australia and viruses isolated from wild mice quickly infect laboratory mice when the two populations are housed together, regardless of the preexisting infections in the laboratory animals (Gorman et al. 2006; Smith et al. 2006). Together, these data indicate that CMVs are readily transmitted within populations through close contact and that increasing seroprevalence in human populations indicates continued exposure to these viruses secondary to the endemic and persistent nature of CMV infection within its animal host.

Children have been proposed to be a major source of CMV infections in human populations. Documented routes of infection in children include perinatal exposure to infectious genital secretions, breast milk ingestion, and exposure to other children infected with HCMV (Stagno et al. 1980; Pass et al. 1987; Adler 1989, 1991; Minamishima et al. 1994; Bale et al. 1996, 1999; de Mello et al. 1996; Vochem et al. 1998; Hamprecht et al. 2001; Lawrence and Lawrence 2004; Willeitner 2004; Doctor et al. 2005; Miron et al. 2005). Because HCMV infection at an early age is followed by prolonged excretion of infectious virus, often persisting for several years, transmission is commonplace during childhood in human societies in which crowding and group care of children take place. In the developed world, this has been observed in children from lower socioeconomic groups and children attending group care centers (Adler 1989; de Mello et al. 1996; Bale et al. 1999). In that latter setting, infection may be acquired by over 50% of attendees and infections with multiple viral strains have been observed, presumably through exposure to virus containing saliva and urine (Hutto et al. 1986; Pass et al. 1987; Adler 1991; Bale et al. 1996). In older children and sexually active adolescents and adults, infection is presumed to follow sexual exposure as cervical secretions and semen are rich sources of virus (Lang and Kummer 1975; Pass et al. 1982; Drew et al. 1984; Rinaldo et al. 1992; Collier et al. 1995). Sexual transmission of HCMV is consistent with the extremely high rates of serological reactivity in sexually active populations such as gay men and individuals attending sexually transmitted disease clinics (Drew et al. 1981; Chandler et al. 1985, 1987; Handsfield et al. 1985; Berry et al. 1988; Collier et al. 1989; Fowler and Pass 1991; Sohn et al. 1991; Shen et al. 1994; Ross et al. 2005). Contact with young children excreting infectious virus represents another mode of acquisition of HCMV by adults. Reinfections with new strains of virus appears commonplace in normal hosts and in immunocompromised hosts (Drew and Mintz 1984; Chou 1986, 1987; Bale et al. 1996; Boppana et al. 2001). In the normal host, reinfections are more common in populations with increased risks of exposure such as children attending group care centers or sexually active populations. Whether the full pathogenic potential of HCMV can follow a reinfection in a normal host remains an important question; however, reinfections can lead to serologic reactivity specific for the new viral strain and in pregnant women, transmission to the fetus and fetal disease, indicating that the disease-inducing potential of reinfection is similar to that associated with primary infection in normal hosts and can be associated with significant morbidity in immunocompromised patients such as transplant recipients (Boppana et al. 2001). Reinfection has also been documented in previously infected and seroreactive laboratory rodents and nonhuman primates (J. Nelson, personal communication) (Gorman et al. 2006). Lastly, non-community-acquired HCMV infections are associated with exposure to blood and blood products from infected donors, transplantation of allografts from infected donors, and in the past by exposure of newborn infants to banked human breast milk (Chou 1986; Bowden 1995; Vochem et al. 1998; Lawrence and Lawrence 2004).

Disease associated with acute infections in the normal host is infrequent but when present is limited to nonspecific viral syndromes that are characterized by low-grade fevers, fatigue, evidence of mild hepatocellular damage, and occasionally transient bone marrow suppression (Cohen and Corey 1985; Pannuti et al. 1985; Horwitz et al. 1986; Porath et al. 1987). These clinical manifestations of end-organ disease are consistent with a disseminated infection and are mirrored by experimental infection of the mouse, rat, guinea pig and rhesus macaque by their respective CMVs (Griffith et al. 1981; Bia et al. 1983; Stals et al. 1990; Kern 1999; Lockridge et al. 1999; Stals 1999; Barry et al. 2006). More severe infections with significant end-organ dysfunction have been related to the inoculum size in experimental animals and also in humans following inoculation of vaccine trial volunteers with a clinical virus isolate (Plotkin et al. 1989; Bernstein 1999; Kern 1999). Following human infection acquired by exposure of mucosal sites, it is presumed that local replication occurs and the primary viremia leads to spread from these sites to other sites of virus amplification such as the liver and/or spleen. Studies in experimental animals have utilized parenteral inoculations in most cases; however, descriptions of infections following oral inoculations have been reported in both mice and nonhuman primates (Lockridge et al. 1999) (S. Jonjic, personal communication). Because the manifestations of infection and dissemination after oral infection are similar to those of parenteral infection, it is likely that virus first replicates locally regardless of site or route of inoculation and then spreads to visceral sites such as the liver, lung, and spleen where it further amplifies. It is likely that visceral organs serve as reservoirs for infectious virus during acute infections and presumably dissemination to other organs takes place, depending on the immune status of the host. Virus is cleared over a period of weeks to months as measured by the presence of virus or viral DNA in peripheral blood. In the mouse and guinea pig (and presumably in humans), persistent viral infection is established in the salivary gland within 2 weeks after infection and virus can be recovered from this organ long after replicating virus has been cleared from the blood as well as the liver and spleen (Bernstein 1999; Kern 1999).

The relative contribution of cell-free vs cell-associated virus to either the primary or secondary viremia following HCMV infection has not been fully defined. Several possible mechanisms for virus dissemination in the vasculature have been proposed and explored experimentally. Viral genes responsible for the extended cellular tropism of clinical viral isolates as compared to laboratory passaged viruses have been identified (Percivalle et al. 1993; Gerna et al. 2000; Hassan-Walker et al. 2001; Hahn et al. 2004; Wang and Shenk 20054a, 2005b). Specific cellular tropism of HCMV appears to be an important determinant in the spread of this virus within the infected host (MA et al. 2006). Cell-free virus transmission during dissemination is thought to be unlikely because HCMV replication is highly cell-associated and infectious virus is recovered only rarely from cell-free serum or plasma and then usually only in severely immunocompromised patients with extremely large amounts of virus in the peripheral blood (Lathey et al. 1994). HCMV DNA can often be detected in plasma by PCR in this latter group of patients (Boivin et al. 1998; Caliendo et al. 2001). Interestingly, cell-free virus is commonly found in body fluids such as urine, saliva and breast milk and often at high titers, indicating that cell-free virus is readily released depending on the site of infection. Early studies suggested that infected endothelial cells that detach from infected vessels and/or polymorphonuclear (PMN) leukocytes could carry HCMV to distal sites (Percivalle et al. 1993; Gerna et al. 1998; Sinzger et al. 1999a; Gerna et al. 2000). Undoubtedly, infected endothelial cells carry infectious virus but because of their size and probable limited half-life in the circulation, these cells would be rapidly cleared and less likely to efficiently disseminate virus widely in the host (Pooley et al. 1999). The presence of these cells may be more reflective of the severity of an acute infection, a finding consistent with the observation that such cells are detectable almost exclusively in only the most severely immunocompromised patients (Gerna et al. 1998; Kas-Deelen et al. 2000). Similarly, many investigators have suggested that PMN leukocytes may efficiently disseminate HCMV to distal sites, a hypothesis based on clinical observations of the transmissibility of HCMV from blood cells found in the buffy coat of peripheral blood (Gerna et al. 2000; Saez-Lopez et al. 2005). This hypothesis is also consistent with recent studies that have demonstrated that HCMV infection not only upregulates IL-8 expression but encodes an IL-8 like molecule, the UL146/vCXC-1, (Grundy et al. 1998; Penfold et al. 1999; Redman et al. 2002). Because IL-8 is a chemoattractant for neutrophils, infected cells could recruit PMN into the foci of infected cells and disseminate infectious virus passively acquired by the PMN. Results from studies in mice infected with MCMV are consistent with such a mechanism (Saederup et al. 1999; Noda et al. 2006). Lastly, cells of the monocyte/macrophage lineage can be infected with HCMV and thus could serve to disseminate HCMV (Taylor-Wiedeman et al. 1991, 1994; Kondo et al. 1996; Soderberg-Naucler et al. 1997; Bolovan-Fritts et al. 1999). Interestingly, studies in immunocompromised patients suggest that circulating monocytes contain quantitatively a similar number of viral genomes as PMN, suggesting that cells of this lineage may disseminate the HCMV as efficiently as PMN (Hassan-Walker et al. 2001).

End-organ disease following acute infection has been most closely correlated with virus replication and virus-induced cytopathology, suggesting that organ dysfunction and cellular damage are likely related to the lytic replicative cycle of the virus in most cell types found in every organ system, with the possible exception of the CNS. Observations from animal models are consistent with this proposed mechanism of disease, and disease appears to correlate with levels of replicating virus (Persoons et al. 1998; Kern 1999; Podlech et al. 2000). The mechanism of cell death following lytic infection is not well understood; however, lytically infected human fibroblasts develop significant morphologic changes, including a marked increase in the size of the nucleus, nuclear and cytoplasmic inclusions, blebbing and focal loss of nuclear membrane integrity, displacement of normal cellular secretory pathway organelles with virus assembly sites, and eventually disruption of the plasma membrane. In some cell types, productive infection appears to be nonlytic (Fish et al. 1998). Apoptosis is not a prominent component of the early cellular response to CMV infection and at least in the case of HCMV, anti-apoptotic functions encoded by several viral genes inhibit this cellular response (Goldmacher 2002; McCormick et al. 2003; Andoniou and Degli-Esposti 2006; Sharon-Friling et al. 2006). In addition, recent findings indicate that HCMV encodes viral functions that inhibit innate cellular responses to virus infection such as nuclear responses to viral DNA shortly after infection, induction of PKR and phosphorylation of eIF2-α and activation of RNAase L (Child 2002; Cassady 2005; Hakki and Geballe 2005; Cantrell and Bresnahan 2006; Child et al. 2006; Hakki et al. 2006; Saffert and Kalejta 2006). Murine CMV has been shown to encode similar functions (m142 and m143) that blunt PKR responses to dsRNA, and deletion of these viral genes cripples replication virus in vivo (Valchanova et al. 2006). Thus, HCMV encodes a number of viral functions to prevent both intrinsic and innate cellular responses to infection, presumably to facilitate virus replication and generation of progeny virions. Finally, it should be noted that the presence of infiltrating lymphocytes and other mononuclear cells and elevated levels of inflammatory cytokines in HCMV-infected tissue also raise the possibility that host-derived immune effector functions contribute to disease manifestations associated with acute infection. In support of this possibility is the observation that children infected as fetuses, newborns or during infancy excrete virus for prolonged periods of time, in many cases several years, and yet do not exhibit clinically observable organ dysfunction (Stagno and Britt 2006). Thus, viral replication apparently does not represent the sole determinant of disease in the normal host.

Resolution of acute infection in normal individuals is associated with persistent immunological reactivity for HCMV that is characterized by a high frequency of HCMV-specific CD4$^+$ and CD8$^+$ T lymphocytes and stable levels of antiviral antibodies. Even in the face of persistent immunological reactivity specific for HCMV, normal individuals, as noted previously, can be reinfected with heterologous strains of virus. Whether the persistently elevated levels of anti-HCMV antibodies and HCMV-specific T lymphocytes detected in some individuals result from frequent reinfection is unknown. Furthermore, immune individuals periodically excrete infectious virus and the infected host remains the main source of community exposure for the uninfected individual.

Virus has been demonstrated in a variety of cell types in normal individuals with acute infections, including epithelium, hepatocytes, smooth muscle cells, endothelial cells, circulating mononuclear cells, macrophages, astrocytes and dermal fibroblasts (Tumilowicz et al. 1985; Sinzger et al. 1995, 1996, 1999a; Ricotta et al. 2001). Similar findings have been described in experimental animals. Virus persists in a number of tissues following resolution of an acute infection and may persist either as a chronic productive infection or as a latent infection. In experimental rodents, virus can be consistently recovered from the salivary gland. Virus has also been recovered from latently infected blood monocytes following culture in supernatant fluids from cultures of lymphocytes undergoing an allogenic reaction (Soderberg-Naucler et al. 1997). This finding was an extension of early studies in mice infected with murine CMV and demonstrated that a latent infection could contribute to CMV persistence (Jordan and Mar 1982). In addition, the finding that latent HCMV present in mononuclear cells could be reactivated by an ongoing allogenic reaction provided an explanation for the role of the transplanted organ as a source of CMV reactivation in patients undergoing allotransplantation (Gnann et al. 1988; Soderberg-Naucler et al. 1997). It remains unclear if persistence of CMV infection in humans can be best described as a chronic productive infection in sequestered sites or as a true latent infection with periodic reactivations, or more likely by both mechanisms. Most investigators would favor the description of CMV persistence as a chronic persistent replication. In the postpartum period, it appears that a previously quiescent infection can be reactivated in both genital tract and in the secretory epithelium of the breast (Pass et al. 1982; Hamprecht et al. 2001; Stagno and Britt 2006). In fact, reactivation of CMV in the breast epithelium and excretion into breast milk is almost universal in previously infected women during the postpartum period (Stagno et al. 1980; Stagno and Britt 2006). Finally, Reddehase and colleagues carefully demonstrated in the murine model of CMV infection that true latency existed in the lungs of previously infected animals and that following loss of immune control, productive infection could be demonstrated (Kurz et al. 1999). Whether active immune surveillance contributes to the establishment or maintenance of CMV latency in humans is unknown.

Natural History of Acute CMV Infection in the Immunocompromised Host

Exaggerated end-organ disease in the immunocompromised host has offered a glimpse of the pathogenic potential of CMVs, yet extensive disease in the setting of an absent or depressed immune system obviously does not reflect the phenotype of this virus in the immunologically normal host. In fact, one could argue that CMVs have evolved to be nonpathogenic in their hosts to ensure persistence and spread in the population. Prior to the HIV pandemic, widespread multiorgan disease had been described consistently in only one naturally acquired infection: infants with congenital HCMV infection. Thus, in the context of the normal non-immunocompromised host, the pathogenic potential of CMVs may be restricted to those of chronic infection and not to the more recognizable disease manifestations that are seen in patients with deficits in immune responsiveness. Nevertheless, the manifestations of acute HCMV infection remain of considerable medical importance and unraveling its pathogenesis remains an important goal of current research. HCMV has been associated with disease in three groups of immunocompromised hosts: (a) fetuses presumably secondary to immunological immaturity, (b) allograft recipients secondary to cytotoxic antirejection agents and in some cases graft-vs-host disease, and (c) HIV infection with loss of CD4$^+$ lymphocytes and the resulting loss of adaptive immune responses (Table 1). Less commonly, CMV disease has been reported in patients undergoing cytotoxic chemotherapy or prolonged therapy with corticosteroids (Stagno and Britt 2006). An interesting association of invasive CMV disease has been recently described in patients undergoing anti-TNF antibody therapy of rheumatologic diseases (Haerter et al. 2004; Mizuta and Schuster 2005; Kohara and Blum 2006). Diseases in these patients ranged from hepatitis, retinitis, and enteritis. Descriptions of intestinal disease associated with CMV infection in older patients with inherited immunodeficiencies have also been reported (Raeiszadeh et al. 2007). Finally, the role of HCMV in immune senescence in the aged population secondary to the phenomena of memory inflation of the T lymphocyte response to this virus remains an area of active investigation.

Congenital Infection

Congenital infection with HCMV is an important cause of cognitive and perceptual disorders in children in North America and Western Europe (Yow et al. 1988; Ahlfors et al. 1999, 2001; Stagno and Britt 2006). In the US it is estimated that between 0.1% and 1% of all live births have an intrauterine infection with HCMV; however, this number may reflect the results of epidemiological studies in geographically and in some cases racially restricted populations (Stagno and Britt 2006). Thus, the estimated incidence of this intrauterine infection may not reflect that of the entire US population. Interestingly, in a limited number of studies from the developing world, the incidence of congenitally infected newborns varied between 0.5% and 2.0%, a finding suggesting that intrauterine transmission of this virus is a frequent event in all populations (Alford and Pass 1981; Stagno et al. 1982b; Sohn et al. 1992; Yamamoto et al. 2001; Stagno and Britt 2006). In fact, the rate of congenital HCMV infection increases as the prevalence of serological reactivity to HCMV increases in the maternal population, suggesting that exposure to this virus represents the most consistent risk factor for delivery of an infected infant (Fig. 1). This is in contrast to descriptions of the relationship between rates of maternal seroimmunity and the occurrence of other well-studied intrauterine infections such as that described for congenital rubella syndrome, an infection that

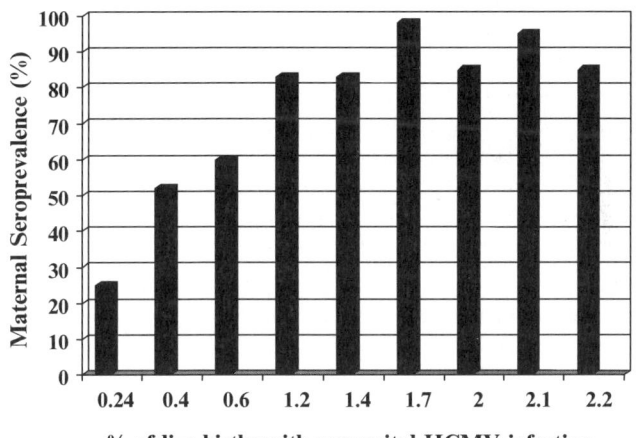

Fig. 1 Incidence of congenital CMV infection increases with maternal seroprevalence. The incidence of congenital HCMV infection and maternal seroprevalence rates from published studies carried out in North America, Europe, South America, and Africa were plotted as shown. Note that as rate of maternal seroprevalence increases so does incidence of congenital CMV infection, suggesting that a threshold level of immune individuals that will eliminate transmission within these population may not be reached

follows primary maternal rubella virus infection during pregnancy and transmission to the fetus. In the case of rubella infections, once the prevalence of maternal seroimmunity to rubella virus exceeds approximately 85%, the incidence of congenital rubella infection falls dramatically (Preblud and Alford 1990; MMWR 1997). Thus, the parameters of protective maternal immune responses to viruses such as rubella that do not establish persistent infections in the host or remain endemic in the population appear to differ substantially from those that are necessary to limit congenital CMV infections. It is of interest that aspects of the epidemiology of CMV mirror those seen in congenital syphilis infections, a sexually acquired infection that does not appear to induce protective immunity and whose incidence increases as the prevalence of the infection increases in the population (Anonymous 1999).

Transmission of HCMV to the developing fetus following maternal primary infection occurs in between 20% and 50% of cases (Stagno et al. 1982b; Griffiths and Baboonian 1984; Yow et al. 1988; Stagno and Britt 2006). Because fetal infection occurs in 0.1%-2.0% of women with preconceptional immunity to HCMV, it is clear that maternal immunity plays a major role in protecting the fetus from virus infection (Table 2). However, at what level this host response modulates intrauterine transmission of HCMV and fetal disease is not understood. Recent studies of parameters of HCMV infection and host immunity during primary HCMV infections in pregnancy have described increased levels of viremia and delayed development of CD4+ and CD8+ responses to HCMV that can be correlated with increased rates of fetal infection (Gibson et al. 2007). Although consistent with previous observations that the development of specific adaptive immune responses appeared delayed in primary HCMV infections, the interval between virus acquisition and the development of adaptive immunity in these women cannot be precisely determined (Gibson et al. 2007).

Table 2 Nonprimary maternal HCMV infection and outcome of congenital HCMV infection

	Type of maternal infection	
	Primary	Non-Primary
Incidence of congenital infection	13%[a]	87%
Transmission rate to fetus	20%–40%	0.1%–2.0%[b]
Incidence of congenitally infected infants with sequelae	24%[c]	5%–8%
Infants with sequelae following congenital HCMV infection[d]	31.2	43.5%–69.5

[a] Stagno et al. 1982
[b] Transmission rates following nonprimary infection varying depending on age and socioeconomic status of population
[c] Fowler et al. 1992
[d] Rate calculated per 1,000 infants with congenital HCMV infection following primary (130) and nonprimary infection (870)

Finally, a longstanding concept is that maternal immunity to HCMV prior to conception also provides protection to the developing fetus from damaging intrauterine infection with this virus; however, this protection is far from complete (Stagno et al. 1982b; Stagno and Britt 2006). More recent information suggests that the disease burden secondary to damaging congenital HCMV following nonprimary maternal infection (reactivation of existing infection or reinfection) is significant in delivery populations with an increased seroprevalence (Boppana et al. 1999). Although the frequency of infants with long-lasting sequelae following primary maternal infection is approximately two to three times higher than in infants infected as a result of nonprimary infection, overall in most populations, the incidence of congenitally infected infants born following nonprimary maternal infection is four to five times higher than that following primary maternal infection (Table 2). Thus, the absolute number of infants with long-lasting damage could be nearly the same for both groups (Table 2). These observations have suggested that current vaccine strategies developed for the prevention of neurodevelopmental sequelae associated with congenital HCMV infections that are targeted only at seronegative women may require re-evaluation (Ahlfors et al. 2001; Boppana et al. 2001).

A conventional view is that intrauterine transmission of virus results from viremic spread to the uterine-placental junction, infection of the uterine smooth muscle and endothelial cells, and then placental trophoblasts followed by entry into the fetal blood system (Muhlemann et al. 1992; Sinzger et al. 1993; Ozono et al. 1997; Halwachs-Baumann et al. 1998; Hemmings et al. 1998; Fisher et al. 2000; Pereira et al. 2005) (see the chapter by L. Pereira and E. Maidji, this volume). Histologic examination of placental sections often reveals focal evidence of villitis with evidence of HCMV infection, which would be consistent with this proposed mode of transmission; however, similar observations have been made in placental sections obtained following delivery of normal, uninfected babies. Thus, the host-derived responses that limit intrauterine transmission likely operate at several levels, including systemic responses to HCMV infection and possibly at local sites such as focal infections of the uterus and placenta. Virus-neutralizing antibodies, HCMV-specific CD8$^+$ T lymphocytes, NK cells and resident macrophages could act to prevent virus transmission to the fetus within the infected placenta, although studies supporting such a mechanism are based solely on in vitro activities of these immune effector functions. Interestingly, placental trophoblasts do not express class I HLA-A or HLA- B MHC molecules but do express HLA-G and -E antigens (Kovats et al. 1990; Lanier 1999; Le Bouteiller 2000). The HLA-G molecules have been shown to serve as weak restriction elements for CD8$^+$ T lymphocyte recognition of HCMV-encoded antigenic peptides, but it is unclear if they function similarly in vivo (Lenfant et al. 2003). However, it is also interesting to note that these particular MHC molecules are resistant to degradation induced by the HCMV US2 and US11 gene products and HLA-E can serve to present peptides derived from MHC molecules and presumably viral leader sequences to NK cells (King et al. 1997; Schust et al. 1998, 1999; Lanier 1999; Onno et al. 2000).

More recent findings have demonstrated that membrane-bound forms but not secreted HLA-G can be degraded by US2 (Barel et al. 2003). Decidual NK cells with cytotoxic activity have been described as well as inhibition of their activity by soluble HLA-G (Poehlmann et al. 2006; Tabiasco et al. 2006). These findings point to a complex and dynamic immunological relationship between host, trophoblast and virus, and suggest that understanding the role of the placental effector functions could lead to further understanding of how HCMV is transmitted to the developing fetus. Potential routes of transmission to the fetus has been studied in vivo in the guinea pig model of congenital CMV infection (Griffith et al. 1985, 1986). The finding that maternal viremia can be correlated with intrauterine transmission and that passively administered virus neutralizing antibodies could reduce placental infection and intrauterine transmission is consistent with conventional routes of fetal infection (Bourne et al. 2001; Chatterjee et al. 2001). In this model, placental infection has been correlated with both maternal disease and fetal wastage associated with placental infiltration with mononuclear cells (Harrison and Myers 1990; Harrison and Caruso 2000). Thus local inflammatory response secondary to virus infection can occur at the uterine-placental interface, arguing that either viral immune evasion functions or an ineffective host immune response could tip the balance toward fetal infection. Alternatively, the severity of maternal disease seen in the guinea pig model of congenital CMV infection also has raised the very distinct possibility that commonly observed disease manifestations in the guinea pig pup, such as runting, could be related to placental inflammation and insufficiency rather than a direct effect of virus infection on the developing fetal guinea pig. Similar studies have not been accomplished in the mouse model, presumably because of the multilayered structure of the murine placenta limits transplacental transfer of MCMV. Currently, rhesus CMV-free rhesus macaque colonies are being generated and these animals should provide an ideal experimental animal model for investigation of this human infection. Several key questions critical to the pathogenesis of congenital CMV infections remain unanswered, including:

1. What is the relationship between maternal viremia and seeding of the uterus?
2. Is seeding of the placenta associated with cell-free or cell-associated virus?
3. Can resident immune effector functions limit virus infection of the placenta or are circulating mononuclear cells required?
4. Can ascending infections from the uterine cervix infect the fetus?
5. What is the importance of placental inflammation and fetal outcome?
6. How does preexisting maternal immunity limit transmission?

This last question is important for the design of vaccines to limit damaging congenital HCMV infections.

Once the virus has entered the fetal circulation, it can in some instances replicate to high levels and damage a variety of organ systems presumably by lytic replication, although this has not been experimentally verified. Target organs most commonly damaged by severe intrauterine infection include the hepatobiliary system, the central nervous system, the lungs, and hematopoietic system (Becroft 1981;

Boppana et al. 1992, 1997; Perlman and Argyle 1992; Anderson et al. 1996). In almost all cases, infected infants resolve the infection but can be left with sequelae in organ systems with low regenerative capacity such as the brain and auditory system (Boppana et al. 1992, 1997; Dahle 2000). Brain damage and hearing loss can occur in up to 5%-20% of infants with congenital CMV (Fowler et al. 1992; Boppana et al. 1997, 1999; Dahle 2000; Noyola et al. 2001). Hearing loss is the most common long-term sequelae of infants with congenital HCMV infection and in the US and northern Europe may rank second only to familial or genetic causes of hearing loss (Harris et al. 1984; Hicks et al. 1993). Similarly, severe brain damage can result from intrauterine HCMV infection with loss of normal cortical architecture, intracranial calcium deposits following loss of the integrity of the endothelium, and loss of cognitive function (Becroft 1981; Perlman and Argyle 1992; Barkovich and Lindan 1994; Boppana et al. 1997).

Two histopathologic types have been described: a focal infection characterized by microglial nodules and more widespread involvement described as ventriculoencephalitis (Becroft 1981). In the former and more common presentation, virus is assumed to infect the parenchyma of the brain following viremic spread, whereas in infants with more severe disease characterized with ventriculoencephalitis, virus is thought to infect the ventricular epithelium and spread through the periventricular epithelium, possibly through the cerebrospinal fluid (Becroft 1981; Arribas et al. 1996).

The pathogenesis of brain damage following congenital HCMV infection is unknown, but several lines of evidence have suggested that it follows fetal infection early in gestation, and that the symmetry of involvement suggests that it is related to the infection and disruption of the microvasculature of the developing brain and/or disruption of the neuronal migration from the periventricular gray area (Becroft 1981; Perlman and Argyle 1992; Barkovich and Lindan 1994). Mechanisms of cell loss such as virus-induced apoptosis of neuronal stem cells have been suggested based on animal model systems, but only very limited information is available to support this mechanism. At this time, it is unknown whether cell death and/or cell dysfunction is a direct effect of virus infection or secondary to damage to supporting cells and structures from the associated inflammation. Animals models have provided only limited information, and to date, the rhesus macaque fetal model appears to most closely model human disease, although this model requires direct inoculation of the fetus with rhesus CMV (Tarantal et al. 1998). Findings from this model system indicate that gestational age of the fetus at the time of infection appears to determine the extent and severity of disease, a result consistent with the correlation between early gestational maternal seroconversion and central nervous system disease in congenital HCMV infections (Perlman and Argyle 1992; Barkovich and Lindan 1994; Stagno and Britt 2006). Once the fetus is infected, the CMV immune status of the mother appears to have only a limited role in the outcome of the fetal infection. Congenitally infected infants with evidence of end-organ disease and long-term sequelae have higher levels of replicating virus as well as a higher virus burden measured in peripheral blood (Fig. 2) (Stagno et al. 1975; Boppana et al. 2005; Stagno and Britt 2006). Interestingly, the strongest correlation between high viral burden in peripheral blood is the presence of hepatitis and in some infants with severe CNS involvement, the

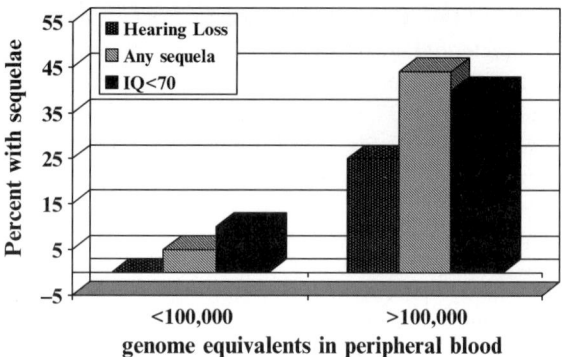

Fig. 2 Viral load and outcome of congenitally infected infants. DNA was extracted from 200 μl of peripheral blood and analyzed for HCMV viral genomes (reported as genome equivalents/ml) as described. Infants with congenital HCMV infection from a large natural history study of congenital HCMV infection were classified as having long-term neurologic sequelae, and more specifically hearing loss or diminished cognitive function (IQ<70). This study has been reported in more detail (Boppana et al. 2005)

viral burden is less than those presenting with only hepatitis (Boppana et al. 2005). This observation can be most readily explained by the duration of the congenital infection such that a predominance of CNS disease can reflect an infection of longer duration with resolution of the hepatic involvement.

Congenitally infected infants can excrete large amounts of virus, often reaching 4-5 logs of infectious virus per milliliter of urine, and can persistently excrete large amounts of virus for years. These same infants can resolve clinical evidence of end-organ disease within the first few months of life even with what is believed to be a limited T lymphocyte response to HCMV (Gehrz et al. 1977; Starr et al. 1979; Pass et al. 1983; Marchant et al. 2003; Gibson et al. 2004). It is of interest, however, that contrary to this previous dogma, newborn infants and fetuses can mount CMV-specific T cell responses, but whether these responses influence outcome of CMV infection in these infants is unknown. Infected infants act as viral reservoirs in their families and communities and serve as an important vector for spreading CMV in populations.

Allograft Recipients

The association of HCMV with disease in allograft recipients was described nearly 40 years ago (Rifkind 1965; Rubin et al. 1979). Current aggressive utilization of antiviral agents both as treatment and prophylaxis, screening of blood products, and whenever possible donor and recipient matching for CMV serological reactivity for CMV have decreased the incidence of severe HCMV infections in allograft recipients. Early reports in bone marrow allograft recipients described mortality rates of between 70% and 95% in patients with HCMV pneumonia (Meyers et al. 1982,

1986; Schmidt et al. 1991). Although HCMV remains the most common posttransplantation infection, mortality rates have fallen dramatically even in the most severely immunocompromised patients.

Of more recent concern has been the role of HCMV in infections in late times after transplantation and perhaps more importantly, chronic graft dysfunction. Infection can follow one of several routes, including:

1. Infection from the transplanted organ
2. Reactivation or dissemination of existing host infection
3. Through blood products required in the posttransplantation period
4. Hospital or community exposure to infectious virus

Replicating virus amplifies and disseminates in the absence of effective host immune responses and multiple organ systems become infected. Clinical symptoms are dependent on end-organ dysfunction such as hepatocellular damage, colitis, or interstitial pneumonitis.

A well-described febrile syndrome associated with laboratory abnormalities in peripheral hematological values and liver functions is a well-recognized clinical presentation of CMV infection (CMV syndrome) in the posttransplantation period is perhaps the most common clinical presentation of CMV infection in these populations. In contrast to patients with HIV infections and AIDS (see Sect. 3.3), HCMV infection of structures within the eye are uncommon in allograft recipients and occur at a very low rate of less than 1% (Chung et al. 2007).

In addition to the clinical manifestations of disseminated HCMV infection, HCMV infection in the donor or recipient has been associated with acute rejection episodes in up to 35% of renal allograft recipients, a finding that is consistent with the association between acute rejection events and the chronic graft rejection (Sola et al. 2003; Chen et al. 2005; Reischig et al. 2006).

Early studies documented the onset of virus replication in the posttransplantation period utilizing techniques of virus isolation. These studies indicated that patients at risk from infection from the transplanted organ or from reactivation/recurrence of persistent infection usually began excreting virus between 4 and 8 weeks after transplantation (Meyers et al. 1982; Rubin 1986). Studies using PCR have demonstrated virus replication as detected by viral DNA in blood within the first few weeks after transplantation, suggesting that infection in these patients likely results in virus replication, amplification and dissemination, a course similar to that seen in primary infection in normal hosts. It is important to note that these studies were done before the widespread use of antiviral agents as prophylaxis in the posttransplantation period to limit viral replication during the period of intensive immunosuppression. In many centers in North America, solid organ allograft recipients receive antiviral prophylaxis for up 100 days after transplantation. Thus, the kinetics of virus reactivation and replication has been modified such that HCMV infections more frequently occur late in the course of transplantation (late infections). Currently, it is estimated that late infections occur in approximately 20-30% of patients at risk for CMV infections, and up to 50% of these infections can be severe and invasive (Rubin 1986). In contrast to these rates of late infections, a recent study reported that 62% of renal allograft recipients who did not receive routine antiviral prophylaxis were infected with HCMV in the first 100 days after transplantation (Sagedal et al. 2004). This incidence of infection is significantly different than rates of less than 10% in patients receiving routine antiviral prophylaxis in the immediate posttransplantation period and has been used as evidence supporting the routine use of antiviral prophylaxis to limit HCMV disease in the early posttransplantation period and possibly decreasing chronic graft rejection that is associated with HCMV infection (Sagedal et al. 2004; Potena et al. 2006; Stoica et al. 2006; Fishman et al. 2007). The incidence of late infections with CMV in patients receiving antiviral prophylaxis has remained relatively steady and is a significant cause of morbidity and mortality in the transplant recipient, suggesting that antiviral prophylaxis does not eliminate the clinical impact of HCMV in allograft recipients (Singh 2005). Furthermore, quantitation of viral load in patients with late HCMV disease has been reported to be less predictive of disease susceptibility, suggesting that differences in the biology of the virus infection could exist between early and late infections (La Rosa et al. 2007). In hematopoietic allograft recipients, the most severe disease is associated with pulmonary infection, a site of disease that is observed less frequently in solid organ allograft recipients, with the exception of heart-lung transplant recipients (Wreghitt et al. 1988; Smyth et al. 1991; Sharples et al. 1996; Wreghitt et al. 1999). The mechanism for the severe lung disease seen in recipients of an allogenic bone marrow transplant is unclear but may involve concomitant graft-vs-host disease (host-vs-graft in heart-lung recipients), preexisting lung disease, or damage associated with pretransplant conditioning (Horak et al. 1992; Barry et al. 2000). In the mouse model of MCMV infection in the immunocompromised host, reactivation of MCMV following sublethal irradiation also involves the lung, and in this experimental model it has been argued that lung disease develops secondary to the lung being an important site of viral latency (Kurz et al. 1997; Reddehase et al. 2002).

Regardless of the exact mechanism of disease, HCMV infection is a necessary prerequisite for disease and effective therapy for this virus has reduced the mortality and morbidity of this posttransplantation complication. Finally, although lung disease observed in hematopoietic marrow allograft recipients is widely disseminated based on clinical parameters such as radiographic studies and clinical findings, early studies demonstrated only focal areas of virus infection (Myerson et al. 1984). This finding raised the possibility that HCMV may cause disease in this group of patients by a mechanism other than direct lytic infection, possibly by altering regulation of the inflammatory response in the infected host (Grundy et al. 1987).

Infection in solid organ allograft recipients is nearly universal if either the donor or the host has had a previous infection with HCMV. Epidemiological studies have consistently shown that transplantation of an organ from a previously infected donor (D^+) into an noninfected recipient (R^-), a D^+/R^- mismatch, can result in a primary and often severe infection in both the early and late posttransplantation period in the immunocompromised host. These individuals are at greatest risk for severe disease secondary to uncontrolled virus replication and have about a two- to threefold higher incidence of late disease following antiviral prophylaxis as compared to D^-/R^+, or D^+/R^- transplant recipients (Bonatti et al. 2004; Murray and Subramaniam 2004; Carstens et al. 2006; Lautenschlager et al. 2006; Potena et al. 2006). Infection by virus present in the transplanted organ also develops in patients with a past HCMV infection (superinfection or reinfection), further demonstrating the importance of the transplanted organ as a source of virus (Chou 1986; Gnann et al. 1998; Grundy et al. 1988). The incidence of clinically significant CMV infections in D^+/R^+ allografts has been reported to be increased as compared to D^-/R^+, suggesting that acquisition of a new strain of virus from the transplanted organ results in disease more frequently than reactivation of the recipient endogenous virus (Chou 1987; Grundy et al. 1988).

Several explanations have been offered for this interesting observation, yet none has been definitively investigated. One potential explanation is that the graft-acquired virus could be antigenically different than the endogenous virus, thus requiring a somewhat more primary-like immune response for virus control and clearance, similar to a response following primary infection in a D^+/R^- transplant situation.

Alternatively, the newly acquired virus could encode additional virulence characteristics not present in the endogenous virus of the recipient. Virus has been demonstrated in mononuclear cells present in transplanted organs, and studies have demonstrated that replicating virus can be recovered from mononuclear cells obtained from patients with past infection when these cells are exposed to a mixture of cytokines derived from allogenically stimulated lymphocytes (Soderberg-Naucler et al. 1997). Studies in transgenic mice with a reporter gene under control of the major immediate promoter of HCMV have demonstrated that this promoter-enhancer can be activated by inflammatory cytokines such as TNF that are produced in animals with transplanted allografts (Koffron et al. 1999; Hummel et al. 2001). In addition, more recent studies in an animal model system have shown that renal allograft ischemia and reperfusion injury activates the major immediate promoter of HCMV independently of TNF pathways, suggesting that ischemia and reperfusion alone are sufficient to reactivate HCMV from a transplanted allograft (Kim et al. 2005).

Infectious virus has been proposed to disseminate to distant sites in polymorphonuclear leukocytes and in infected endothelial cells detached from small vessels, as discussed in the previous section. End-organ disease following HCMV infection in both hematopoietic and solid organ allograft recipients has been most closely correlated with the presence of replicating virus in the host and by increased virus load as measured by PCR.

The quantity of virus measured in the blood or other body fluids has been reported to be a predictor of acute disease regardless of pretransplantation risk factors for HCMV infections in solid organ allograft recipients, such as an organ from a infected donor transplanted into a noninfected recipient (D^+/R^-) (Cope et al. 1997; Gor et al. 1998; Emery et al. 2000; Limaye et al. 2001; Martin-Davila et al. 2005; Gentile et al. 2006). Thus, monitoring patients with quantitative PCR techniques have provided a noninvasive method for identification of patients at risk for disseminated infection, and patients identified prior to the development of overt disease are more likely to benefit from antiviral therapy.

As was noted previously, many of these earlier studies were carried out in an era when widespread antiviral prophylaxis was limited and preemptive approaches for prevention of severe CMV infections represented standard of care. Antiviral prophylaxis, particularly in solid organ transplantation, has become widely employed and CMV infection in the early posttransplantation period is less frequent. As a result, late disease represents a common presentation of CMV infection in allograft recipients, and recent studies have suggested that previous approaches for monitoring CMV in peripheral blood as an indicator of patients at risk for invasive infection may not provide the same predictive value of invasive infection in these patients (Singh 2005; La Rosa et al. 2007). Studies in experimental models such as the immunocompromised mouse model of MCMV infection have also argued for a mechanism of disease which can be correlated with virus replication, although studies in mice have also shown that virus or viral DNA detected in the blood is only an indirect measure of the virus load in target organs (Brune et al. 2001).

Finally, it remains to be determined if the severe disease and organ failure seen in the immunocompromised human is due entirely to lytic virus replication, or if components of the antiviral immune response to this virus contribute to the observed organ damage. Studies in murine models of acute infection have demonstrated that effective immune control of virus replication often comes at the price of tissue damage. A clear distinction between virus replication and hepatocellular damage secondary to the inflammatory response in mice has been demonstrated utilizing transgenic mice with deletions in chemokine receptors (Salazar-Mather et al. 2002). Thus, the loss of normal immunoregulatory mechanisms in the CMV-infected host treated with immunosuppressive agents can lead not only to increased levels of virus replication but to unregulated inflammatory responses and tissue damage.

Immune reactivity to virus-encoded protein antigens appears to be key to the resistance to disease associated with HCMV as well as other CMV infection in their respective animal hosts. Elegant studies have been conducted in the murine model of disease in the immunocompromised host (Podlech et al. 2000; Krmpotic et al. 2003). These studies have documented that $CD8^+$ antiviral cytoxic T lymphocytes (CTL) can mediate resistance and effect recovery from MCMV infection in lethally irradiated mice (Podlech et al. 2000; Krmpotic et al. 2003). Studies have also shown that CTL directed against a single protein, pp89, encoded by the IE-1 gene of MCMV, can protect mice from lethal infection (Podlech et al. 2000). Additional targets of protective CTL have been identified suggesting that a polyclonal response to MCMV may be beneficial to an outbred population (Holtappels et al. 2001; Ye et al. 2002). Similarly, the loss of CTL responses to HCMV in human allograft recipients is associated with infection and disease (Reusser et al. 1991; Rubin 2002; La Rosa et al. 2007). In perhaps the most direct test of this hypothesis, Riddell and colleagues transferred in vitro derived HCMV-specific $CD8^+$ CTL into bone marrow allograft recipients and dramatically reduced invasive HCMV disease (Walter et al. 1995). This study demonstrated the critical importance of CTL in resistance to HCMV disease, but also indicated that maintenance of this immune reactivity required reconstitution of a CMV-specific $CD4^+$ lymphocyte response (Walter et al. 1995). The lack of a $CD4^+$ response appeared to correlate with the development of late disease (>100 days) in hematopoietic allograft recipients (Walter et al. 1995).

Subsequent studies from this institution and other groups have also documented the importance of T lymphocyte responses and late disease in hematopoietic cell transplant recipients and in solid organ recipients (Lacey et al. 2002; Boeckh et al. 2003; Bunde et al. 2005). A recent study argued that generation of CMV-specific CD8⁺ lymphocytes responses in hepatic allograft recipients failed to correlate with protection from late infection with CMV (La Rosa et al. 2007). Furthermore, it has been reported that development of detectable HCMV-specific CD4⁺ lymphocyte responses in the early posttransplantation period (1ˢᵗ month) lessened the risk for acute rejection, HCMV disease, and long-term loss of lumen cross-sectional area, suggesting that early control of virus replication was an important risk factor for late disease associated with HCMV infection (Tu et al. 2006). In contrast to this report, a very recent study suggested that increased CD4⁺ lymphocyte responses to HCMV are associated with chemokine-mediated endothelial damage, a process that is thought to be essential in the development of vascular disease in the transplanted allograft (Bolovan-Fritts et al. 2007).

The results of these studies have raised important questions about the nature of protective immunity against HCMV, the role of antiviral responses in inflammation associated with allograft damage, and the role of antiviral prophylaxis in increasing the risk for late disease associated with CMV infections. In addition, these studies have suggested that the presence of CD8⁺ responses may not be a surrogate of protection from late disease, a finding first observed by Riddell and colleagues nearly 15 years ago (Walter et al. 1995; La Rosa et al. 2007). Understanding what constitutes durable protective immunity in this population with exaggerated risks for invasive CMV infection secondary to immunosuppression could help elucidate the parameters of protective immunity in normal hosts, a prerequisite for design and testing vaccines to limit disease from congenital infections.

In agreement with what has been observed in the mouse, the predominant CTL response following HCMV infection was initially proposed to be directed against a limited set of virus-encoded antigens. The three dominant targets of HCMV-specific CD8⁺ CTL were pp65 (UL83), pp150 (UL32) and IE72 (UL123) based on studies reported from a number of different laboratories (McLaughlin-Taylor et al. 1994; Boppana and Britt 1996; Gillespie et al. 2000). A more definitive study defined the virus-specific CD8⁺ and CD4⁺ responses to all possible open reading frames of HCMV in a group of seropositive individuals (Sylwester et al. 2005). The results of this study provided evidence that the immune system has directed its dominant responses at the most abundant virus-encoded proteins, which in many cases are virion structural proteins (Sylwester et al. 2005). The breadth and magnitude of the T lymphocyte response to CMV was surprisingly large. In fact, in the normal immunocompetent host over 10% of circulating T lymphocytes are human CMV-specific (Sylwester et al. 2005). This finding suggested that in these donors, either the virus-specific response was completely ineffectual in the control of virus replication or conversely, it represented a host response that was required to overcome the vast number of immune modulating functions of the virus. Interestingly, two of the HCMV CTL targets are antigens encoded by early-late genes, and, thus, would presumably be expressed at the same time as the known immune evasion genes of HCMV, which inhibit CTL recognition of virus-infected cells (Tomazin et al. 1999; Alcami and Koszinowski 2000; Gewurz et al. 2001; Scalzo et al. 2007).

Using the mouse model, Holtappels and co-workers have provided evidence that immune evasion genes encoded by MCMV m04, m06 and m152 cannot overcome the CTL response to a peptide encoded by m164, even though the immune evasion genes could blunt the response to the immunodominant pp89 target peptide (Holtappels 2002). They have suggested that the m164 target peptide overcomes the immune evasion system simply by being present in saturating quantities in infected cells and thus remaining as a target peptide that can be presented to CTL (Holtappels 2002). The observation that the viral immune evasion functions do not significantly alter CD8 T cell priming in vivo has been further supported in studies utilizing a large number of MCMV antigens to examine CD8 T cell immunodominance over time in infected mice (Munks et al. 2007). Together with findings that CTL specific for the products of the m83 and m84 can provide protection from lethal MCMV infection in the mouse, even though these viral proteins do not induce dominant CTL responses, it was argued that a polyclonal CTL response to MCMV is generated and that this response can be protective even in the presence of virus-encoded immune evasion functions, an interpretation consistent with the limited pathogenicity of CMVs in the normal, immunocompetent host. Similar studies in humans have shown that even in the presence of known immune evasion functions encoded by US2,3,6, and 11 of HCMV (see the chapter by C. Powers et al., this volume), a broad $CD8^+$ cytotoxic T lymphocyte response was generated in immunocompetent individuals (Manley et al. 2004; Khan et al. 2005). The role of the immune modulatory functions of HCMV, immune evasion, and the focusing of the CTL responses continue to merit additional investigation. What is clear is that under normal circumstances, the host CTL response is sufficient to protect from uncontrolled virus replication in the presence of virus-encoded immune evasion functions and that exogenous immune suppression of host responses of significant magnitude can tip the balance in favor the virus. Finally, it should be emphasized that antiviral antibodies, particularly virus-neutralizing antibodies, almost certainly play a role in the protective immune response to CMV. Passive administration of virus-neutralizing antibodies can protect immunocompromised mice from disseminated MCMV infection and disease (Jonjic et al. 1994). Likewise studies in bone marrow and solid organ allograft recipients have demonstrated that passive administration of antiviral antibodies can protect from disease and that the quantity of virus neutralizing activity correlates with protection (Snydman et al. 1987; Schoppel et al. 1997, 1998a). Previous studies have argued that antibodies induced by gB (UL55), the gH/gL/gO complex (UL75,115,74) and the gM/gN (UL100,73) represent the bulk of virus-neutralizing antibodies in humans, whereas in the mouse only anti-gB antibodies appear protective (Rapp et al. 1992; Britt and Mach 1996). Studies in guinea pigs and rhesus macaques have also identified gB as target of virus-neutralizing antibodies (Britt and Harrison 1994; Kropff and Mach 1997).

Patients with AIDS

Human cytomegalovirus was one of the first opportunistic pathogens identified in patients with the acquired immunodeficiency syndrome (AIDS). In fact, the initial descriptions of patients with AIDS suggested that HCMV was an important cause of pneumonia in these patients, a claim that was not substantiated in later studies (Jacobson et al. 1991). Several unique features of the end-organ disease associated with HCMV infection in these patients extended the spectrum of disease phenotypes associated with HCMV infections. In contrast to infections associated with other opportunistic agents, patients with HIV infections frequently maintained sufficient immune responsiveness to HCMV until the very late stages of their immune deficiency (Komanduri et al. 1998). In fact, the manifestations of HCMV infections were routinely seen only after several more virulent pathogens such as *Pneumocystis carnii* and *Mycobacterium* were controlled by effective chemotherapy.

Another feature of HCMV infections in AIDS patient was that high levels of viral replication as detected by the presence of viral DNA in peripheral blood could exist for a considerable period of time prior to the onset of disease (Bowen et al. 1995; Spector et al. 1998; Emery et al. 1999). Disease manifestations of HCMV infection unique to AIDS patients included end-organ disease in two organ systems, the gastrointestinal tract and the eye, both of which were uncommon sites of end-organ disease in even the most immunocompromised allograft recipient. The pathogenesis of disease in the gastrointestinal tract is still unclear but included the presence of focal areas of virus replication, focal colitis, and chronic dysfunction of the absorptive functions of the intestinal tract (Francis et al. 1989; Wu et al. 1989; Dieterich and Rahmin 1991; Wilcox et al. 1998). Gastrointestinal disease secondary to HCMV is infrequently seen in transplant patients and is not well described in infants with congenital CMV infections, raising the possibility that other pathogens present in these patients contributed to CMV disease. Conversely, in vitro studies suggest that CMV infection of resident mononuclear cells in the intestinal tract could lead to enhanced production of inflammatory mediators and chronic inflammation in the intestinal tract (Smith et al. 1992). A similar disease is seen in rhesus macaques infected with rhesus CMV and simian immunodeficiency virus. However, the possibility that other pathogens are responsible for the gastrointestinal disease in these animals has been difficult to exclude, and a polymicrobial etiology was often invoked to explain the pathogenesis of disease in these animals (Kuhn et al. 1999).

A second unique manifestation of HCMV infection in AIDS patients is retinitis, a disease that rarely occurs in transplant patients but is a well recognized manifestation of congenital HCMV infection. However, the disease in AIDS patients differed both clinically and pathologically from the disease seen in infants with congenital HCMV infection. Retinitis was frequent in patients in the late stages of AIDS and in some studies was reported in as many as 25% of patients with AIDS (Gallant et al. 1992; Pecorella et al. 2000). This manifestation of AIDS is uncommonly reported in children with AIDS for reasons that have not been fully determined. Infection of the retina developed in patients with high HIV loads, nearly absent CD4+ lymphocytes, and importantly, periods of prolonged HCMV replication and presumably viremia. The disease was best characterized by an exuberant inflammation of the retina associated with infection of the vessels entering the retina from its anterior surface (Pepose et al. 1987; Glasgow and Weisberger 1994; Rao et al. 1998). Infection could be seen in the perivascular glial cells, neuronal cells, and pigmented retinal epithelium and associated with the loss of vascular integrity (Pepose et al. 1987; Glasgow and Weisberger 1994; Rao et al. 1998). Intense inflammation was associated with loss of retinal structure and in some cases, edema, detachment of the retina and loss of vision. Antiviral therapy, both local and systemic, that effectively inhibited virus replication also halted disease progression and led to resolution of the acute symptoms of the disease. Yet virus remained in the retina and disease recurrence and reinfection of other areas of the retina or the other eye were considered as part of the natural history of this infection. Prior to HAART, anti-HCMV drugs were continued for the life of the patient. Once HAART protocols that limited HIV replication and reversed the immunodeficiency associated with the late stages of AIDS were in widespread use, the incidence of HCMV retinitis fell dramatically, and today it is seen in patients who have failed HAART protocols either because of: (a) HIV drug resistance, (b) noncompliance, or (c) a first diagnosis in late stage AIDS (Whitcup et al. 1999; Torriani et al. 2000). In resource-limited regions of the world, CMV retinitis remains an important opportunistic infection in AIDS patients.

An interesting syndrome was reported by investigators who had successfully treated AIDS patients with HAART. In these patients, anti-HCMV antiviral therapy was discontinued once there was evidence of HIV suppression and immune reconstitution. These investigators noted that several patients with retinitis with improving immune function developed a recurrence of their retinitis that was associated with prominent findings of uveitis and less convincing findings of progressive HCMV retinitis; they coined the term immune recovery uveitis (IUR) for this syndrome (Karavellas et al. 1998; Holland 1999). It was suggested that following reconstitution of HCMV CD8+ responses in patients treated with HAART, clearance of the HCMV-infected cells in the retina by the reconstituted cytotoxic response could lead to an exacerbation of local disease by immunopathogenic mechanisms. The immunopathogenesis of this syndrome remains unclear at this time; however, several risk factors have been identified for development of IRU. These include response to HAART with drop in viral load, greater than 25%-30% surface area of retinal involvement at time of active retinitis, and treatment with the potent antiviral cidofovir (Karavellas et al. 2001; Kempen et al. 2006). More recent studies of one cohort of patients revealed that aqueous humor from the involved eye contained high levels of IL-12, moderate levels of IL-6, interferon gamma, but no viral DNA (Schrier et al. 2006). These latter findings together with early findings demonstrating the presence of CD8+ lymphocytes in the involved eyes and clinical response to intraocular corticosteroids suggested that IUR was an immunopathogenic response to HCMV infection.

Pathological descriptions of diseased tissue from patients with invasive HCMV infection and AIDS are most consistently interpreted as providing evidence of lytic infection leading to cell death and organ damage, except in the case of IUR. Studies of eye tissue from patients from retinitis also provided evidence of apoptosis in retinal cells but the cellular loss secondary to apoptosis failed to correlate with vision loss, suggesting other mechanisms accounted for loss of organ function (Buggage et al. 2000). A variety of cell types were demonstrated to be infected with HCMV in these patients including endothelial cells, epithelial cells in the gastrointestinal tract, neuronal cells in the retina and the brain and cells from the macrophage/ monocytic lineage (Pepose et al. 1987; Wiley and Nelson 1988; Francis et al. 1989; Wu et al. 1989; Glasgow and Weisberger 1994; Rao et al. 1998). Whether viral gene expression that was not associated with lytic infection contributed to disease has not been well studied due to the lack of convenient animal model. Numerous studies suggested that HCMV gene products could transactivate HIV LTRs, but the importance of this in vitro-defined phenomenon to the pathogenesis of HCMV disease in patients with AIDS has not been clarified (Moreno et al. 1997; Ranga et al. 1997; McCarthy et al. 1998).

Similarly, the possibility that active HCMV replication could contribute to the overall immunodeficiency in patients with AIDS has been suggested by the finding that persistent HCMV replication in these patients was associated with a more rapid decline in $CD4^+$ lymphocyte counts and decreased duration of survival (Webster et al. 1989; Spector et al. 1999). Animal models have provided some additional information, but it appears that only the rhesus macaque model of AIDS is sufficiently similar to human AIDS to permit investigators to address these questions. Concomitant infection with SIV and rhesus CMV led to a more rapid onset of disease in experimental animals as compared to animals given SIV alone (Sequar et al. 2002). Inoculation of animals with SIV prior to CMV infection lead to an uncontrolled CMV replication and the absence of a primary immune response to rhesus CMV (Sequar et al. 2002). In this model of acute infection, disease was correlated with high levels of rhesus CMV replication.

Finally, several studies have clearly demonstrated that HCMV infection of macrophage/monocytes can lead to the production of inflammatory mediators, which could recruit inflammatory cells as well as amplifying the inflammatory response of noninfected resident cells. The induction of IL-8 by infected lamina propria macrophages in AIDS patients could recruit and activate neutrophils to sites of virus infection and thus lead to focal inflammation in the colon, as observed in biopsy from patients with gastrointestinal disease thought to be caused by HCMV (Redman et al. 2002). Other mediators induced by HCMV infection including MIP-1α could recruit mononuclear cells, thus amplifying the inflammatory response in tissue (Redman et al. 2002). Without an effective $CD8^+$ response to clear virus-infected cells, the inflammatory response could be exaggerated in the absence of extensive virus replication and or local virus spread. This pathogenic mechanism may represent a bridge between the syndromes associated with an acute CMV infection that are best characterized by active virus replication and cell death secondary to lytic infection and those syndromes associated with chronic infection in which virus gene expression and virus replication are closely related to host inflammatory responses and likely promote host inflammatory responses leading to tissue damage.

Diseases Associated with Chronic Infection

Although the concept that chronic HCMV infection could be associated with identifiable disease syndromes seems likely because of the persistent nature of HCMV infection, definitive evidence linking HCMV with chronic disease is limited. Several characteristics of this virus-host relationship has made definitive studies difficult. First, the virus is ubiquitous in the population, with the incidence of infection exceeding 80% in some populations, often higher than the incidence of common chronic diseases. The high prevalence of infection has made epidemiological studies of disease association nearly impossible because of the need to study large numbers of patients to obtain sufficient statistical power. In addition, many studies have flaws in their design such as differing disease classifications, confounders such as race and other independent risk factors of disease, and differences in classifications of HCMV infection. Secondly, pathogenic mechanisms other than lytic virus replication are not well understood and animal model systems that may permit more informative studies of potential mechanisms have only recently been described. Moreover, the presence of viral nucleic acid or virus-encoded proteins in a tissue does not necessarily reflect disease association and may merely represent virus present as a passenger in a circulating mononuclear cells or tissue macrophages. Recent evidence has also indicated that viral strains from different patients may differ genetically and in their cellular tropism, thus offering another layer of complexity in the lytic and nonlytic effects of HCMV infections that could contribute to chronic disease. Finally, studies in animal models have demonstrated that the pattern of viral gene expression varies within the host depending on the site of infection (Streblow et al. 2007). This latter finding provides an especially strong argument for the bidirectional relationship between CMV and the host and suggests that merely detecting viral nucleic acids or viral protein expression will provide only a limited understanding of the contribution of this virus to organ-specific chronic diseases. Even with these limitations in mind, a myriad of studies from allograft transplant recipients and normal hosts have presented a strong case for HCMV as a co-factor in chronic inflammatory processes, particularly those resulting in vascular disease. Studies in experimental animals have also revealed several plausible mechanisms for a potential contribution of HCMV to chronic vascular disease. Lastly, HCMV has been suggested to be a co-factor in some human malignancies based on epidemiological studies in populations with a high incidence of HCMV infection (Huang et al. 1984; Shen et al. 1993; Han et al. 1997). More recent studies suggest that HCMV could also have a potential role in specific human malignancies, perhaps serving as a promoter for tumor invasion or proliferation. Several mechanisms, including its reported capacity to inhibit the function of tumor suppressor genes, induce chromosome breaks and possibly as a promoter of neoangiogenesis could contribute indirectly to the malignant phenotype of transformed cells (Cinatl et al. 1996, 1999; Moreno et al. 1997; Shen et al. 1997; Fortunato and Spector 1998; Zhu et al. 1998). HCMV infections have been previously implicated as a cause or co-factor of other diseases, including rheumatologic disorders such as rheumatoid arthritis and some rare autoimmune diseases; however, we will limit this brief overview to possible pathogenic mechanisms of vascular disease associated with HCMV infection and chronic allograft rejection because these groups of diseases have provided the most compelling evidence linking HCMV infection to disease.

Chronic Vascular Disease in the Normal Host

The association of a herpesvirus infection and vascular disease was made nearly 30 years ago when Marek's disease virus, an avian herpesvirus was shown to induce atherosclerotic disease in chickens (Fabricant et al. 1978; Fabricant and Fabricant 1999). Herpesviruses were detected in arterial endothelial cells and smooth muscle cells from patients with atherosclerotic heart disease and HCMV has been shown to replicate in arterial smooth muscle cells (Gyorkey et al. 1984; Tumilowicz et al. 1985; Hendrix et al. 1989, 1991; Wu et al. 1992; Melnick et al. 1993; Shi and Tokunaga 2002). More recently, findings in patients undergoing coronary artery angioplasty for atherosclerotic heart disease and studies using in vitro models of HCMV infection of endothelial and smooth muscle cells have provided additional evidence for the role of HCMV infection in vascular disease (Muhlestein et al. 2000; Mueller et al. 2003; Nerheim et al. 2004; Westphal et al. 2006). Epidemiological studies, including some with prospective study design, have suggested that individuals with serological reactivity to HCMV are at increased relative risk for coronary atherosclerotic disease as well as increased risk for arteriosclerotic disease (Nieto et al. 1996; Zhou et al. 1996, 2001; Drover et al. 1998; Sorlie et al. 2000; Grahame-Clarke et al. 2003; Horne et al. 2003; Mueller et al. 2003). In several of these studies, the risk for atherosclerotic disease attributable to HCMV infection was less than well-accepted risk factors such as serum lipid concentration, yet infection was consistently associated with atherosclerotic vascular disease. Infections with other herpesviruses such as HSV were not associated with a definable risk. Finally, animal models of atherosclerosis have demonstrated accelerated disease development following infection with CMV (Hsich et al. 2001; Vliegen et al. 2004). Overall there appears to be a body of provocative evidence implicating CMV infection as a contributor to atherosclerotic vascular disease in the normal host. This disease association is even more plausible when viewed together with the proposed role of this virus in the accelerated vascular disease observed in allograft recipients.

The presence of herpes virus-like particles in inflammatory cells from atheromatous plaques from patients with coronary artery disease initially suggested that the virus was an active participant in the disease process (Gyorkey et al. 1984). A study of atheromatous tissue from patients with restenosis following angioplasty for coronary artery atherosclerotic disease also demonstrated HCMV nucleic acids in a significant number of specimens, and more recent studies have shown that HCMV infection is more efficient in atherosclerotic blood vessels (Speir et al. 1994; Nerheim et al. 2004; Westphal et al. 2006). In one study, smooth muscle cells isolated from atherosclerotic plaques often contained cytoplasmic p53 and the protein product of the IE-2 gene of HCMV, findings consistent with the interpretation that HCMV IE-2 inhibited the normal functions of p53 in smooth muscle cells (Speir et al. 1994). This led to the hypothesis that HCMV could promote atherosclerotic vascular disease by inducing smooth muscle cell proliferation through its interaction with p53, particularly if mitogenic signals provided by growth factors and cytokines were also present as a result of ongoing inflammation and the accumulation of circulating mononuclear cells (Libby et al. 1988a, 1988b; Speir et al. 1994; Zhou et al. 1999). Subsequent studies from other laboratories have demonstrated inactivation of p53 function following HCMV infection and an anti-apoptotic effect provided by IE-2 expression (Zhu et al. 1995; Kovacs et al. 1996). It should be noted that a later study of diseased coronary arteries also detected HCMV nucleic acids in plaques in six of 13 specimens but detected cytoplasmic p53 in only two of 19 of specimens from the same study (Baas et al. 1996). Other mechanisms that have been proposed for the contribution of HCMV infection to arteriosclerosis include induction of expression of adhesion molecules such as ICAM-1 and growth factors such a PDGF and TGF by HCMV infection, elevation of IL-6 levels, induction of chemokine expression by endothelial cells, and endothelial cell dysfunction (Blankenberg et al. 2001; Grahame-Clarke et al. 2003; Petrakopoulou et al. 2004; Reinhardt et al. 2005; Westphal et al. 2006).

Subintimal infiltration by smooth muscle cells has been reported to be a critical feature of the arterial narrowing observed in atherosclerotic disease and in transplant vascular sclerosis, and when viewed together with the concept that atherosclerotic vascular disease is an inflammatory disease suggested several potential roles for HCMV in this disease (Lemstrom et al. 1993; Kloppenburg et al. 2005). Streblow and colleagues have reported that smooth muscle cells expressing the HCMV-encoded chemokine receptor, US28, will migrate in response to a CC chemokine gradient and in a rat model of transplant vascular sclerosis; similarly, the rat CMV encoded chemokine receptor, R33, has been shown to play a role in disease development presumably by acting as a chemoattractant for smooth muscle cell migration (Streblow et al. 1999; Melnychuk et al. 2005). Chemokines that have been shown to induce responses from US28 expressed in HCMV-infected cells include RANTES and MCP-1, both of which can be produced by resident macrophages and infiltrating mononuclear cells present within an inflammatory focus associated with an atheromatous plaque (Streblow et al. 1999). These findings together with previous observations that described increased expression of cell adhesion molecules such as ICAM-1 in endothelial cells infected with HCMV suggested that adherence of circulating mononuclear cells could initiate an inflammatory cascade leading to directional migration of HCMV-infected smooth muscle cells, subintimal thickening and arterial narrowing (Span et al. 1991; Sedmak et al. 1994; Yilmaz et al. 1996;

Knight et al. 1999; Dengler et al. 2000). Similarly, a nonspecific injury to the vascular endothelial wall could lead to expression of ICAM-1 and other cellular adhesion molecules that could allow circulating HCMV-infected mononuclear or polymorphonuclear leukocytes to attach and infect the vessel wall. Studies in a rat model of virus-induced vascular disease have demonstrated that nonspecific endothelial injury followed by immediate infection with rat CMV leads to subintimal thickening and narrowing of the artery to much greater degree than seen in control, noninfected animals (Persoons et al. 1994; Zhou et al. 1999; Kloppenburg et al. 2005). Thus, it appears that in several small animal models, CMV infection contributes to the development of vascular diseases. At the minimum, these models should help further our understanding of infection with this virus and the development and progression of vascular disease in the normal host.

Although several mechanisms that have been proposed for the pathogenesis of subintimal thickening and vessel narrowing do not require lytic virus replication, in vitro evidence suggests that viral gene expression is necessary for induction of host cytokines and cell adhesion molecules as well as expression virus-encoded chemokine receptors (Burns et al. 1999; Streblow et al. 1999). Other investigators have noted distal effects on surrounding uninfected cells, presumably from secreted cytokines and chemokines from infected endothelium, suggesting that viral genes need not be widely expressed in affected tissue to explain the contribution of HCMV infection to disease (Van Dam-Mieras et al. 1987; Stassen et al. 2006). In contrast to lytic infection in human fibroblasts, HCMV infection of aortic endothelial cells and smooth muscle cells is more prolonged and cell lysis either is not a characteristic of this infection or occurs at a reduced frequency (Tumilowicz et al. 1985; Tumilowicz 1990; Fish et al. 1998; Kahl et al. 2000). Moreover, in at least one experimental animal model system, subintimal thickening and vessel narrowing occurred after acute CMV infection, yet at the termination of the experiment, viral DNA could not be detected in vascular sites of disease (Zhou et al. 1999; Martelius et al. 2001). Although this result could be explained by lack of sufficient sensitivity in the detection system, it was argued that ongoing viral replication and gene expression were not required for disease. Potential explanations for this claim included early endothelial cell infection and initiation of the inflammatory cascade followed by virus clearance or alternatively, that distal noncardiac sites of virus infection were driving systemic inflammation and inflammation in the coronary arteries (Zhou et al. 1999; Blankenberg et al. 2001; Stassen et al. 2006). In other animal model systems, continued viral gene expression appears necessary for the development of disease, and the inhibition of pathogenic responses in experimental animals treated with ganciclovir argue that early and late viral genes are required for complete expression of CMV-associated vascular disease in these animal models and likely in humans (De La Melena et al. 2001; Valantine 2004; Mehra 2006; Potena et al. 2006). In summary, there is obviously a complex interaction between host and virus that leads to disease and to date only observational studies and fragments of a potential mechanism(s) have been defined. The available animal models will hopefully more completely define the parameters of virus-induced disease and point to relevant studies in humans.

Chronic Disease in the Transplanted Allograft

Chronic rejection is a leading cause of allograft loss and in many transplant centers is the most common reason that allograft recipients receive a second transplant (Evans et al. 1999; Hosenpud 1999). In the case of renal allograft recipients, the loss of the transplanted organs secondary to acute rejection has fallen dramatically with the availability of newer immunosuppressive agents, yet the incidence of chronic graft rejection and graft loss has remained relatively unchanged over the last decade. This complication of allograft transplantation is of utmost importance to recipients of cardiac and liver allografts because of the limited number of organs available for transplantation and the need for retransplatation if chronic rejection results in graft dysfunction and loss. A important observation in 1989 described the increased incidence of cardiac allograft rejection in patients with HCMV infection (Grattan et al. 1989). Subsequently, several other transplant centers reported this clinical association and distinctive histological changes in coronary arteries to patients with cardiac allograft rejection and HCMV infection (McDonald et al. 1989; Loebe et al. 1990; Koskinen et al. 1993; Paavonen et al. 1993). Ultimately, it has been reported that the most common cause of cardiac allograft loss is the development of cardiac allograft vasculopathy (CAV), a disease process most investigators believe to be a form of allograft rejection (Hosenpud et al. 1991; Hosenpud 1999; Valantine 2004; Mehra 2006). This disease is characterized by the progressive, diffuse concentric narrowing of allograft vasculature with loss of distal small vessels and by a clinical course that is relentlessly progressive. The histopathological findings in this disease are in contrast to focal narrowing of coronary vessels in normal individuals with atherosclerotic artery disease. Similarly, up to 10% of hepatic allografts are lost to chronic rejection in which hepatic endothelium is damaged and mononuclear cells infiltrate the intima. In addition, the bile duct epithelium is damaged and intrahepatic bile ducts are lost, a histopathologic finding that has been termed the vanishing bile duct syndrome (Hubscher et al. 1991). In both diseases, HCMV has been implicated in the acceleration of these processes. Similar vascular diseases are reported in renal allografts undergoing chronic rejection, although the association between HCMV infection and glomerular disease remains controversial (Richardson et al. 1981; Castro et al. 1983; Herrera et al. 1986). HCMV infection has been associated with chronic renal allograft rejection and graft dysfunction from interstitial fibrosis and loss of renal tubules (Vazquez-Martul et al. 2004). Histologically, chronic renal rejection is defined by the degree of tubulointerstitial fibrosis, indicating that, similar to observations in CAV in cardiac allografts, fibrosis is the end-stage outcome of chronic renal allograft rejection (Racusen et al. 1999; Vazquez-Martul et al. 2004; Hartmann et al. 2006). Rodent models of chronic renal allograft rejection have been developed and infection with rat CMV has been associated with acceleration of graft rejection in the rat model (Lautenschlager et al. 1997; Soots et al. 1998).

As a result of the clinical association of CMV infections with chronic allograft dysfunction, clinical trials with antiviral agents have been carried out in an attempt to limit this posttransplant complication. Although the results are far from definitive, evidence has been presented to suggest that prophylactic use of the antiviral ganciclovir in the immediate posttransplantation period can retard the development of CAV and prolong graft function (Valantine et al. 1999; Valantine 2004; Fishman et al. 2007). Similarly, extending the duration of treatment with prophylactic antiviral agents has been suggested as an intervention to prevent late-onset HCMV disease and the associated graft rejection in renal and liver allograft recipients (Razonable et al. 2001). Models of CAV have been developed in the rat and clearly implicate CMV as an important cofactor in disease. In agreement with the results of clinical trials in humans, treatment with antivirals at the time of virus infection in these animal models modified disease (Lemstrom et al. 1993, 1997; Koskinen et al. 1999; De La Melena et al. 2001).

The pathogenesis of CMV-associated chronic allograft rejection in humans is incompletely defined perhaps because of the complexity of the disease, the patient populations and the potent immunosuppressive agents that are required for their clinical management. Approximately 5%-20% of solid organ allografts are lost as a result of chronic rejection but other causes such as noncompliance with antirejection medications or donor-recipient MHC matching likely contribute to graft loss in these patients (Morris et al. 1993). Therefore, it has been difficult to perform well-controlled studies in these patients. However, the overwhelming majority of epidemiological studies have identified HCMV as an important risk factor for chronic rejection and graft loss and persistent virus excretion as a key risk factor for chronic rejection episodes associated with HCMV infection (Everett et al. 1992; Evans et al. 2000). Several risk associations of chronic graft rejection have been identified in allograft recipients, including:

1. HCMV infection in donor
2. HCMV infection and disease in the recipient in the 1st year after transplantation
3. Transplantation of an allograft from a noninfected donor into a HCMV-infected recipient
4. Prolonged viral replication
5. Episodes of acute rejection (Falagas et al. 1998; Evans et al. 1999, 2000; Tong et al. 2002; Fateh-Moghadam et al. 2003; Sola et al. 2003; Sagedal et al. 2004; Chen et al. 2005; Helantera et al. 2006; Potena et al. 2006; Reischig et al. 2006; Stoica et al. 2006; Hussain et al. 2007)

Persistence of virus replication in these patients likely is reflective of the failure of host immunological response to control HCMV infection rather than an arbitrary absolute level of virus replication as estimated by viral load in the blood. Clinical observations have suggested that both virus replication and graft rejection are necessary for accelerated vascular disease. Thus, if CMV promotes inflammatory vascular disease and atherosclerosis in the normal host, these mechanisms could be greatly amplified in allograft recipients secondary to an ongoing allogenic response, and the course of vascular disease in these patients could be expected to be accelerated. This appears to be the case in experimental animal models as well as in allograft recipients, findings that have been used to argue for a link between CMV replication, viral gene expression and inflammation.

The observation that persistent virus replication is correlated with CAV is consistent with stimulation of viral gene expression and replication by a milieu of mediators generated by allorecognition of the graft. Viral gene expression and/or replication could enhance the host inflammatory response and in effect amplify an ongoing inflammatory process. In support of this mechanism, animal models of transplant vascular sclerosis have been used to demonstrate that interruption of the virus-host stimulatory loop can also interrupt the development of vascular disease (Lemstrom et al. 1993). Either an increased level of immunosuppression or administration of an antiviral agent can prevent the development transplant vascular sclerosis in animals undergoing an allograft rejection. In cases in which virus replication is uncontrolled by host responses, virus infection of endothelial cells could induce a variety of inflammatory responses, as noted in Sect. 3.1 and the development of endotheliitis may potentially represent the initial insult leading to vascular sclerosis. Virus could directly infect and damage the endothelium, or more likely the endothelium could be damaged either by alloreactive T lymphocytes or as a result of ischemia/ reperfusion at the time of transplantation. Once the inflammatory response is initiated, a variety of cell adhesion molecules could be upregulated in the endothelium and recruit mononuclear and polymorphonuclear leukocytes, some of which may be carrying virus (Craigen et al. 1997). Infection of the endothelium followed by infection of smooth muscle cells could then lead to increased expression of MHC molecules, cytokines and chemokines such as MCP-1, MIP-1α and RANTES by the endothelium and IL-6 by smooth muscle cells (Taylor et al. 1992; Koskinen et al. 1993; Lemstrom et al. 1993; Arkonac et al. 1997; Grundy et al. 1998; Srivastava et al. 1999; Billstrom Schroeder and Worthen 2001; Froberg et al. 2001). In addition, CMVs have recently been shown to induce expression of host genes that promote host inflammatory responses such as Cox-2 or in the case of rhesus CMV actually encode this enzymatic activity (Zhu 2002; Rue et al. 2004). Interestingly, in a rat model of rat CMV accelerated graft loss, rat CMV has been shown to induce Cox-2 expression in the allograft (Martelius et al. 2002). Together, these pathways could create unregulated, autocrine and paracrine cascades in which viral gene expression could lead to an increased host inflammatory response. In turn, such a host response could promote virus replication and viral gene expression, thus creating an autostimulatory loop. It should be noted that other investigators have argued that the presence of virus-encoded chemokine receptors on infected cells could also function as immune evasion molecules and act to scavenge extracellular chemokines such as RANTES from areas of HCMV infection (Bodaghi et al. 1998; Billstrom et al. 1999). Although this mechanism may be operative during infection in vitro and possibly in normal hosts, it is likely that the outpouring of cytokines and chemokines during an allograft rejection would quickly saturate such functions.

Acknowledgements The author would like to thank Drs. Michael Mach and Suresh Boppana for helpful discussions about congenital HCMV infections and Drs. Michael Jarvis, Jay Nelson, and Dan Streblow for their contributions to sections on chronic diseases associated with HCMV infections.

References

Adam SG, Caraux A, Fodil-Cornu N, Loredo-Osti JC, Lesjean-Pottier S, Jaubert J, Bubic I, Jonjic S, Guenet JL, Vidal SM, Colucci F (2006) Cmv4, a new locus linked to the NK cell gene complex, controls innate resistance to cytomegalovirus in wild-derived mice. J Immunol 176:5478-5485

Adler SP (1989) Cytomegalovirus and child day care. Evidence for an increased infection rate among day-care workers. N Engl J Med 321:1290-1296

Adler SP (1991) Cytomegalovirus and child day care: risk factors for maternal infection. Pediatr Infect Dis J 10:590

Ahlfors K, Ivarsson SA, Harris S (1999) Report on a long-term study of maternal and congenital cytomegalovirus infection in Sweden. Review of prospective studies available in the literature. Scand J Infect Dis 31:443-457

Ahlfors K, Ivarsson SA, Harris S (2001) Secondary maternal cytomegalovirus infection--A significant cause of congenital disease. Pediatrics 107:1227-1228

Alcami A, Koszinowski UH (2000) Viral mechanisms of immune evasion. Immunol Today 21:447-455

Alford CA, Pass RF (1981) Epidemiology of chronic congenital and perinatal infections of man. Clin Perinatoly 8:397-414

Anderson KS, Amos CS, Boppana S, Pass R (1996) Ocular abnormalities in congenital cytomegalovirus infection. J Am Optometr Assoc 67:273-278

Andoniou CE, Degli-Esposti MA (2006) Insights into the mechanisms of CMV-mediated interference with cellular apoptosis. Immunol Cell Biol 84:99-106

Anonymous (1999) Congenital syphilis - United States, 1998. MMWR Morb Mortal Wkly Rep 48:757-761

Arkonac B, Mauck KA, Chou S, Hosenpud JD (1997) Low multiplicity cytomegalovirus infection of human aortic smooth muscle cells increases levels of major histocompatibility complex class I antigens and induces a proinflammatory cytokine milieu in the absence of cytopathology. J Heart Lung Transplant 16:1035-1045

Arribas JR, Storch GA, Clifford DB, Tselis AC (1996) Cytomegalovirus encephalitis. Ann Intern Med 125:577-587

Autran B, Carcelain G, Li TS, Blanc C, Mathez D, Tubiana R, Katlama C, Debre P, Leibowitch J (1997) Positive effects of combined antiretroviral therapy on CD4+ T cell homeostasis and function in advanced HIV disease. Science 277:112-116

Baas IO, Offerhaus JA, El-Deiry WS, Wu TC, Hutchins GM, Kasper EK, Baughman KL, Baumgartner WA, Chiou CJ, Hayward GS, Hruban RH (1996) The WAF1-mediated p53 growth-suppressor pathway is intact in the coronary arteries of heart transplant recipients. Human Pathol 27:324-329

Bale JF Jr, Petheram SJ, Souza IE, Murph JR (1996) Cytomegalovirus reinfection in young children. J Pediatr 128:347-352

Bale JF Jr, Zimmerman B, Dawson JD, Souza IE, Petheram SJ, Murph JR (1999) Cytomegalovirus transmission in child care homes. Arch Pediatr Adoles Med 153:75-79

Barel MT, Ressing M, Pizzato N, van Leeuwen D, Le Bouteiller P, Lenfant F, Wiertz EJ (2003) Human cytomegalovirus-encoded US2 differentially affects surface expression of MHC class I locus products and targets membrane-bound, but not soluble HLA-G1 for degradation. J Immunol 171:6757-6765

Barkovich AJ, Lindan CE (1994) Congenital cytomegalovirus infection of the brain: imaging analysis and embryologic considerations. Am J Neuroradiol 15:703-715

Baroncelli S, Barry PA, Capitanio JP, Lerche NW, Otsyula M, Mendoza SP (1997) Cytomegalovirus and simian immunodeficiency virus coinfection: longitudinal study of antibody responses and disease progression. J Acquir Immune Defic Syndr Human Retrovirol 15:5-15

Barry PA, Lockridge KM, Salamat S, Tinling SP, Yue Y, Zhou SS, Gospe SM Jr, Britt WJ, Tarantal AF (2006) Nonhuman primate models of intrauterine cytomegalovirus infection. ILAR J 47:49-64

Barry SM, Johnson MA, Janossy G (2000) Cytopathology or immunopathology? The puzzle of cytomegalovirus pneumonitis revisited. Bone Marrow Transplant 26:591-597

Becroft DM (1981) Prenatal cytomegalovirus infection: epidemiology, pathology and pathogenesis. Perspect Pediatr Pathol 6:203-241

Bernstein DI, Bourne N (1999) Animal models for cytomegalovirus infection: Guinea-pig CMV. In: Zak MS (ed) Handbook of animal models of infection. Academic Press, London, pp 935-941

Berry NJ, Burns DM, Wannamethee G, Grundy JE, Lui SF, Prentice HG, Griffiths PD (1988) Seroepidemiologic studies on the acquisition of antibodies to cytomegalovirus, herpes simplex virus, and human immunodeficiency virus among general hospital patients and those attending a clinic for sexually transmitted diseases. J Med Virol 24:385-393

Bia FJ, Griffith BP, Fong CK, Hsiung GD (1983) Cytomegaloviral infections in the guinea pig: experimental models for human disease. Rev Infect Dis 5:177-195

Biassoni R, Cantoni C, Pende D, Sivori S, Parolini S, Vitale M, Bottino C, Moretta A (2001) Human natural killer cell receptors and co-receptors. Immunol Rev 181:203-214

Billstrom MA, Lehman LA, Scott Worthen G (1999) Depletion of extracellular RANTES during human cytomegalovirus infection of endothelial cells. Am J Respiry Cell Mol Biol 21:163-167

Billstrom Schroeder M, Worthen GS (2001) Viral regulation of RANTES expression during human cytomegalovirus infection of endothelial cells. J Virol 75:3383-3390

Biron CA, Byron KS, Sullivan JL (1989) Severe herpesvirus infections in an adolescent without natural killer cells. N Engl J Med 320:1731-1735

Blankenberg S, Rupprecht HJ, Bickel C, Espinola-Klein C, Rippin G, Hafner G, Ossendorf M, Steinhagen K, Meyer J (2001) Cytomegalovirus infection with interleukin-6 response predicts cardiac mortality in patients with coronary artery disease. Circulation 103:2915-2921

Bodaghi B, Jones TR, Zipeto D, Vita C, Sun L, Laurent L, Arenzana-Seisdedos F, Virelizier JL, Michelson S (1998) Chemokine sequestration by viral chemoreceptors as a novel viral escape strategy: withdrawal of chemokines from the environment of cytomegalovirus-infected cells. J Exp Med 188:855-866

Boeckh M, Leisenring W, Riddell SR, Bowden RA, Huang ML, Myerson D, Stevens-Ayers T, Flowers ME, Cunningham T, Corey L (2003) Late cytomegalovirus disease and mortality in recipients of allogeneic hematopoietic stem cell transplants: importance of viral load and T-cell immunity. Blood 101:407-414

Boivin G, Handfield J, Toma E, Murray G, Lalonde R, Bergeron MG (1998) Comparative evaluation of the cytomegalovirus DNA load in polymorphonuclear leukocytes and plasma of human immunodeficiency virus-infected subjects. J Infect Dis 177:355-360

Bolovan-Fritts CA, Mocarski ES, Wiedeman JA (1999) Peripheral blood CD14(+) cells from healthy subjects carry a circular conformation of latent cytomegalovirus genome. Blood 93:394-398

Bolovan-Fritts CA, Trout RN, Spector SA (2007) High T-cell response to human cytomegalovirus induces chemokine-mediated endothelial cell damage. Blood 110:1857-1863

Bonatti H, Tabarelli W, Ruttmann E, Kafka R, Larcher C, Hofer D, Klaus A, Laufer G, Geltner C, Margreiter R, Muller L, Antretter H (2004) Impact of cytomegalovirus match on survival after cardiac and lung transplantation. Am Surg 70:710-714

Boppana SB, Britt WJ (1996) Recognition of human cytomegalovirus gene products by HCMV-specific cytotoxic T cells. Virology 222:293-296

Boppana SB, Fowler KB, Pass RF, Britt WJ, Stagno S, Alford CA (1992) Newborn findings and outcome in children with symptomatic congenital CMV infection. Pediatr Res 31:158A

Boppana SB, Fowler KB, Vaid Y, Hedlund G, Stagno S, Britt WJ, Pass RF (1997) Neuroradiographic findings in the newborn period and long-term outcome in children with symptomatic congenital cytomegalovirus infection. Pediatrics 99:409-414

Boppana SB, Fowler KB, Britt WJ, Stagno S, Pass RF (1999) Symptomatic congenital cytomegalovirus infection in infants born to mothers with preexisting immunity to cytomegalovirus. Pediatrics 104:55-60

Boppana SB, Rivera LB, Fowler KB, Mach M, Britt WJ (2001) Intrauterine transmission of cytomegalovirus to infants of women with preconceptional immunity. N Engl J Med 344:1366-1371

Boppana SB, Fowler KB, Pass RF, Rivera LB, Bradford RD, Lakeman FD, Britt WJ (2005) Congenital cytomegalovirus infection: association between virus burden in infancy and hearing loss. J Pediatr 146:817-823

Bourne N, Schleiss M, Bravo F, Bernstein D (2001) Preconception immunization with a cytomegalovirus (CMV) glycoprotein vaccine improves pregnancy outcome in a guinea pig model of congenital CMV infection. J Infect Dis 183:59-64

Bowden RA (1995) Transfusion-transmitted cytomegalovirus infection. Hematol Oncol Clin N Am 9:155-166

Bowden RA, Day LM, Amos DE, Meyers JD (1987) Natural cytotoxic activity against cytomegalovirus-infected target cells following marrow transplantation. Transplantation 44:504-508

Bowen EF, Wilson P, Atkins M, Madge S, Griffiths PD, Johnson MA, Emery VC (1995) Natural history of untreated cytomegalovirus retinitis. Lancet 346:1671-1673

Britt WJ, Harrison C (1994) Identification of an abundant disulfide-linked complex of glycoproteins in the envelope of guinea pig cytomegalovirus. Virology 201:294-302

Britt WJ, Mach M (1996) Human cytomegalovirus glycoproteins. Intervirology 39:401-412

Britt WJ, Jarvis MA, Drummond DD, Mach M (2005) Antigenic domain 1 is required for oligomerization of human cytomegalovirus glycoprotein B. J Virol 79:4066-4079

Brune W, Hasan M, Krych M, Bubic I, Jonjic S, Koszinowski UH (2001) Secreted virus-encoded proteins reflect murine cytomegalovirus productivity in organs. J Infect Dis 184:1320-1324

Bubic I, Wagner M, Krmpoti A, Saulig T, Kim S, Yokoyama WM, Jonji S, Koszinowski UH (2004) Gain of virulence caused by loss of a gene in murine cytomegalovirus. J Virol 78:7536-7544

Buggage RR, Chan CC, Matteson DM, Reed GF, Whitcup SM (2000) Apoptosis in cytomegalovirus retinitis associated with AIDS. Curr Eye Res 21:721-729

Bunde T, Kirchner A, Hoffmeister B, Habedank D, Hetzer R, Cherepnev G, Proesch S, Reinke P, Volk HD, Lehmkuhl H, Kern F (2005) Protection from cytomegalovirus after transplantation is correlated with immediate early 1-specific CD8 T cells. J Exp Med 201:1031-1036

Burns LJ, Pooley JC, Walsh DJ, Vercellotti GM, Weber ML, Kovacs A (1999) Intercellular adhesion molecule-1 expression in endothelial cells is activated by cytomegalovirus immediate early proteins. Transplantation 67:137-144

Caliendo AM, Schuurman R, Yen-Lieberman B, Spector SA, Andersen J, Manjiry R, Crumpacker C, Lurain NS, Erice A, The CMV Working Group of the Complications of HIV Disease Rac ACTG (2001) Comparison of quantitative and qualitative PCR assays for cytomegalovirus DNA in plasma. J Clin Microbiol 39:1334-1338

Cantrell SR, Bresnahan WA (2006) Human cytomegalovirus (HCMV) UL82 gene product (pp71) relieves hDaxx-mediated repression of HCMV replication. J Virol 80:6188-6191

Carstens J, Andersen HK, Spencer E, Madsen M (2006) Cytomegalovirus infection in renal transplant recipients. Transpl Infect Dis 8:203-212

Cassady KA (2005) Human cytomegalovirus TRS1 and IRS1 gene products block the double-stranded-RNA-activated host protein shutoff response induced by herpes simplex virus type 1 infection. J Virol 79:8707-8715

Castro LA, Gokel JM, Thoenes G, Frosner G, Land W, Hillebrand G, Gurland HJ (1983) Renal changes in cytomegalovirus infection. Proc Eur Dial Transplant Assoc 19:500-504

Chambers J, Angulo A, Amaratunga D, Guo H, Jiang Y, Wan JS, Bittner A, Frueh K, Jackson MR, Peterson PA, Erlander MG, Ghazal P (1999) DNA microarrays of the complex human cytomegalovirus genome: profiling kinetic class with drug sensitivity of viral gene expression. J Virol 73:5757-5766

Chandler SH, Holmes KK, Wentworth BB, Gutman LT, Wiesner PJ, Alexander ER, Handsfield HH (1985) The epidemiology of cytomegaloviral infection in women attending a sexually transmitted disease clinic. J Infect Dis 152:597-605

Chandler SH, Handsfield HH, McDougall JK (1987) Isolation of multiple strains of cytomegalovirus from women attending a clinic for sexually transmitted diseases. J Infect Dis 155:655-660

Chatterjee A, Harrison CJ, Britt WJ, Bewtra C (2001) Modification of maternal and congenital cytomegalovirus infection by anti-glycoprotein b antibody transfer in guinea pigs. J Infect Dis 183:1547-1553

Chen JH, Mao YY, He Q, Wu JY, Lv R (2005) The impact of pretransplant cytomegalovirus infection on acute renal allograft rejection. Transplant Proc 37:4203-4207

Child S, Jarrahian S, Harper V, Geballe A (2002) Complementation of vaccinia virus lacking the double stranded RNA-binding protein gene E3L by human cytomegalovirus. J Virol 76:4912-4918

Child SJ, Hanson LK, Brown CE, Janzen DM, Geballe AP (2006) Double-stranded RNA binding by a heterodimeric complex of murine cytomegalovirus m142 and m143 proteins. J Virol 80:10173-10180

Chou S (1986) Acquisition of donor strains of cytomegalovirus by renal-transplant recipients. N Engl J Med 314:1418-1423

Chou SW (1987) Cytomegalovirus infection and reinfection transmitted by heart transplantation. J Infect Dis 155:1054-1056

Chung H, Kim KH, Kim JG, Lee SY, Yoon YH (2007) Retinal complications in patients with solid organ or bone marrow transplantations. Transplantation 83:694-699

Cinatl J Jr, Cinatl J, Vogel JU, Rabenau H, Kornhuber B, Doerr HW (1996) Modulatory effects of human cytomegalovirus infection on malignant properties of cancer cells. Intervirology 39:259-269

Cinatl J Jr, Kotchetkov R, Scholz M, Cinatl J, Vogel JU, Driever PH, Doerr HW (1999) Human cytomegalovirus infection decreases expression of thrombospondin-1 independent of the tumor suppressor protein p53. Am J Pathol 155:285-292

Cohen JI, Corey GR (1985) Cytomegalovirus infection in the normal host. Medicine 64:100-114

Collier AC, Chandler SH, Handsfield HH, Corey L, McDougall JK (1989) Identification of multiple strains of cytomegalovirus in homosexual men. J Infect Dis 159:123-126

Collier AC, Handsfield HH, Ashley R, Roberts PL, DeRouen T, Meyers JD, Corey L (1995) Cervical but not urinary excretion of cytomegalovirus is related to sexual activity and contraceptive practices in sexually active women. J Infect Dis 171:33-38

Cook M, Briggs D, Craddock C, Mahendra P, Milligan D, Fegan C, Darbyshire P, Lawson S, Boxall E, Moss P (2006) Donor KIR genotype has a major influence on the rate of cytomegalovirus reactivation following T-cell replete stem cell transplantation. Blood 107:1230-1232

Cope AV, Sabin C, Burroughs A, Rolles K, Griffiths PD, Emery VC (1997) Interrelationships among quantity of human cytomegalovirus (HCMV) DNA in blood, donor-recipient serostatus, and administration of methylprednisolone as risk factors for HCMV disease following liver transplantation. J Infect Dis 176:1484-1490

Craigen JL, Yong KL, Jordan NJ, MacCormac LP, Westwick J, Akbar AN, Grundy JE (1997) Human cytomegalovirus infection up-regulates interleukin-8 gene expression and stimulates neutrophil transendothelial migration. Immunology 92:138-145

Dahle AF, Fowler KB, Wright JD, Boppana SB, Britt WJ, Pass RF (2000) Longitudinal investigation of hearing disorders in children with congenital cytomegalovirus. J Am Acad Audiol 11:283-290

De La Melena VT, Kreklywich CN, Streblow DN, Yin Q, Cook JW, Soderberg-Naucler C, Bruggeman CA, Nelson JA, Orloff SL (2001) Kinetics and development of CMV-accelerated transplant vascular sclerosis in rat cardiac allografts is linked to early increase in chemokine expression and presence of virus. Transplant Proc 33:1822-1823

de Mello AL, Ferreira EC, Vilas Boas LS, Pannuti CS (1996) Cytomegalovirus infection in a day-care center in the municipality of Sao Paulo. Rev Inst Med Trop Sao Paulo 38:165-169

Dengler TJ, Raftery MJ, Werle M, Zimmermann R, Schonrich G (2000) Cytomegalovirus infection of vascular cells induces expression of pro-inflammatory adhesion molecules by paracrine action of secreted interleukin-1beta. Transplantation 69:1160-1168

Dieterich DT, Rahmin M (1991) Cytomegalovirus colitis in AIDS: presentation in 44 patients and a review of the literature. J AIDS 4: S29-S35

Doctor S, Friedman S, Dunn MS, Asztalos EV, Wylie L, Mazzulli T, Vearncombe M, O'Brien K (2005) Cytomegalovirus transmission to extremely low-birthweight infants through breast milk. Acta Paediatr 94:53-58

Drew WL, Mintz L (1984) Cytomegalovirus infection in healthy and immune-deficient homosexual men. In: Ma P, Armstrong D (eds) The acquired immune deficiency syndrome and infections of homosexual men. Yorke Medical Books, New York, pp 117-123

Drew WL, Mintz L, Miner RC, Sands M, Ketterer B (1981) Prevalence of cytomegalovirus infection in homosexual men. J Infect Dis 143:188-192

Drew WL, Sweet ES, Miner RC, Mocarski ES (1984) Multiple infections by cytomegalovirus in patients with acquired immune deficiency syndrome: documentation by Southern blot hybridization. J Infect Dis 150:952-953

Drover S, Kovats S, Masewicz S, Blum JS, Nepom GT (1998) Modulation of peptide-dependent allospecific epitopes on HLA-DR4 molecules by HLA-DM. Human Immunol 59:77-86

Dunn W, Chou C, Li H, Hai R, Patterson D, Stolc V, Zhu H, Liu F (2003) Functional profiling of the human cytomegalovirus genome. Proc Natl Acad Sci USA 100:14223-14228

Emery VC, Sabin C, Feinberg JE, Grywacz M, Knight S, Griffiths PD (1999) Quantitative effects of valacyclovir on the replication of cytomegalovirus (CMV) in persons with advanced human immunodeficiency virus disease: baseline CMV load dictates time to disease survival. J Infect Dis 180:695-701

Emery VC, Sabin CA, Cope AV, Gor D, Hassan-Walker AF, Griffiths PD (2000) Application of viral-load kinetics to identify patients who develop cytomegalovirus disease after transplantation.[comment]. Lancet. 355:2032-2036

Evans PC, Coleman N, Wreghitt TG, Wight DG, Alexander GJ (1999) Cytomegalovirus infection of bile duct epithelial cells, hepatic artery and portal venous endothelium in relation to chronic rejection of liver grafts. J Hepatol 31:913-920

Evans PC, Soin A, Wreghitt TG, Taylor CJ, Wight DG, Alexander GJ (2000) An association between cytomegalovirus infection and chronic rejection after liver transplantation. Transplantation 69:30-35

Everett JP, Hershberger RE, Norman DJ, Chou S, Ratkovec RM, Cobanoglu A, Ott GY, Hosenpud JD (1992) Prolonged cytomegalovirus infection with viremia is associated with development of cardiac allograft vasculopathy. J Heart Lung Transplant 11:S133-S137

Fabricant CG, Fabricant J (1999) Atherosclerosis induced by infection with Marek's disease herpesvirus in chickens. Am Heart J 138: S465-S468

Fabricant CG, Fabricant J, Litrenta MM, Minick CR (1978) Virus-induced atherosclerosis. J Exp Med 148:335-340

Falagas ME, Paya C, Ruthazer R, Badley A, Patel R, Wiesner R, Griffith J, Freeman R, Rohrer R, Werner BG, Snydman DR (1998) Significance of cytomegalovirus for long-term survival after orthotopic liver transplantation: a prospective derivation and validation cohort analysis. Transplantation 66:1020-1028

Fateh-Moghadam S, Bocksch W, Wessely R, Jager G, Hetzer R, Gawaz M (2003) Cytomegalovirus infection status predicts progression of heart-transplant vasculopathy. Transplantation 76:1470-1474

Fish KN, Soderberg-Naucler C, Mills LK, Stenglein S, Nelson JA (1998) Human cytomegalovirus persistently infects aortic endothelial cells. J Virol 72:5661-5668

Fisher S, Genbacev O, Maidji E, Pereira L (2000) Human cytomegalovirus infection of placental cytotrophoblasts in vitro and in utero: implications for transmission and pathogenesis. J Virol 74:6808-6820

Fishman JA, Emery V, Freeman R, Pascual M, Rostaing L, Schlitt HJ, Sgarabotto D, Torre-Cisneros J, Uknis ME (2007) Cytomegalovirus in transplantation - challenging the status quo. Clin Transplant 21:149-158

Fortunato EA, Spector DH (1998) p53 and RPA are sequestered in viral replication centers in the nuclei of cells infected with human cytomegalovirus. J Virol 72:2033-2039

Fowler KB, Pass RF (1991) Sexually transmitted diseases in mothers of neonates with congenital cytomegalovirus infection. J Infect Dis 164:259-264

Fowler KB, Stagno S, Pass RF, Britt WJ, Boll TJ, Alford CA (1992) The outcome of congenital cytomegalovirus infection in relation to maternal antibody status. N Engl J Med 326:663-667

Francis ND, Boylston AW, Roberts AH, Parkin JM, Pinching AJ (1989) Cytomegalovirus infection in gastrointestinal tracts of patients infected with HIV-1 or AIDS. J Clin Pathol 42:1055-1064

French AR, Yokoyama WM (2003) Natural killer cells and viral infections. Curr Opin Immunol 15:45-51

French AR, Pingel JT, Wagner M, Bubic I, Yang L, Kim S, Koszinowski U, Jonjic S, Yokoyama WM (2004) Escape of mutant double-stranded DNA virus from innate immune control. Immunity 20:747-756

Froberg MK, Adams A, Seacotte N, Parker-Thornburg J, Kolattukudy P (2001) Cytomegalovirus infection accelerates inflammation in vascular tissue overexpressing monocyte chemoattractant protein-1. Circ Res 89:1224-1230

Gallant JE, Moore RD, Richman DD, Keruly J, Chaisson RE (1992) Incidence and natural history of cytomegalovirus disease in patients with advanced human immunodeficiency virus disease treated with zidovudine. The Zidovudine Epidemiology Study Group. J Infect Dis 166:1223-1227

Gehrz RC, Marker SC, Knorr SO, Kalis JM, Balfour HH Jr (1977) Specific cell-mediated immune defect in active cytomegalovirus infection of young children and their mothers. Lancet 2:844-847

Gentile G, Picardi A, Capobianchi A, Spagnoli A, Cudillo L, Dentamaro T, Tendas A, Cupelli L, Ciotti M, Volpi A, Amadori S, Martino P, de Fabritiis P (2006) A prospective study comparing quantitative cytomegalovirus (CMV) polymerase chain reaction in plasma and pp65 antigenemia assay in monitoring patients after allogeneic stem cell transplantation. BMC Infect Dis 6:167

Gerna G, Zavattoni M, Baldanti F, Furione M, Chezzi L, Revello MG, Percivalle E (1998) Circulating cytomegalic endothelial cells are associated with high human cytomegalovirus (HCMV) load in AIDS patients with late-stage disseminated HCMV disease. J Med Virol 55:64-74

Gerna G, Percivalle E, Baldanti F, Sozzani S, Lanzarini P, Genini E, Lilleri D, Revello MG (2000) Human cytomegalovirus replicates abortively in polymorphonuclear leukocytes after transfer from infected endothelial cells via transient microfusion events. J Virol 74:5629-5638

Gewurz BE, Gaudet R, Tortorella D, Wang EW, Ploegh HL (2001) Virus subversion of immunity: a structural perspective. Curr Opin Immunol 13:442-450

Gibson L, Piccinini G, Lilleri D, Revello MG, Wang Z, Markel S, Diamond DJ, Luzuriaga K (2004) Human cytomegalovirus proteins pp65 and immediate early protein 1 are common targets for CD8+ T cell responses in children with congenital or postnatal human cytomegalovirus infection. J Immunol 172:2256-2264

Gibson L, Dooley S, Trzmielina S, Somasundaran M, Fisher D, Revello MG, Luzuriaga K (2007) Cytomegalovirus (CMV) IE1- and pp65-specific CD8+ T cell responses broaden over time after primary CMV infection in infants. J Infect Dis 195:1789-1798

Gillespie GM, Wills MR, Appay V, O'Callaghan C, Murphy M, Smith N, Sissons P, Rowland-Jones S, Bell JI, Moss PA (2000) Functional heterogeneity and high frequencies of cytomegalovirus-specific CD8(+) T lymphocytes in healthy seropositive donors. J Virol 74:8140-8150

Glasgow BJ, Weisberger AK (1994) A quantitative and cartographic study of retinal microvasculopathy in acquired immunodeficiency syndrome. Am J Ophthalmol 118:46-56

Gnann JW, Ahlmen J, Svalander C et al (1988) Inflammatory cells in transplanted kidneys are infected by human cytomegalovirus. Am J Pathol 132:239-248

Goldmacher VS (2002) vMIA, a viral inhibitor of apoptosis targeting mitochondria. Biochimie 84:177-185

Gor D, Sabin C, Prentice HG, Vyas N, Man S, Griffiths PD, Emery VC (1998) Longitudinal fluctuations in cytomegalovirus load in bone marrow transplant patients: relationship between peak virus load, donor/recipient serostatus, acute GVHD and CMV disease. Bone Marrow Transplant 21:597-605

Gorman S, Harvey NL, Moro D, Lloyd ML, Voigt V, Smith LM, Lawson MA, Shellam GR (2006) Mixed infection with multiple strains of murine cytomegalovirus occurs following simultaneous or sequential infection of immunocompetent mice. J Gen Virol 87:1123-1132

Grahame-Clarke C, Chan NN, Andrew D, Ridgway GL, Betteridge DJ, Emery V, Colhoun HM, Vallance P (2003) Human cytomegalovirus seropositivity is associated with impaired vascular function. Circulation 108:678-683

Grattan MT, Moreno-Cabral CE, Starnes VA, Oyer PE, Stinson EB, Shumway NE (1989) Cytomegalovirus infection is associated with cardiac allograft rejection and atherosclerosis. JAMA 261:3561-3566

Griffith BP, Lucia HL, Bia FJ, Hsiung GD (1981) Cytomegalovirus-induced mononucleosis in guinea pigs. Infect Immun 32:857-863

Griffith BP, McCormick SR, Fong CK, Lavallee JT, Lucia HL, Goff E (1985) The placenta as a site of cytomegalovirus infection in guinea pigs. J Virol 55:402-409

Griffith BP, McCormick SR, Booss J, Hsiung GD (1986) Inbred guinea pig model of intrauterine infection with cytomegalovirus. Am J Pathol 122:112-119

Griffiths PD, Baboonian C (1984) A prospective study of primary cytomegalovirus infection during pregnancy: final report. Br J Obstet Gynaecol 91:307-315

Grundy JE, Shanley JD, Griffiths PD (1987) Is cytomegalovirus interstitial pneumonitis in transplant recipients an immunopathological condition? Lancet 2:996-999

Grundy JE, Lui SF, Super M, Berry NJ, Sweny P, Fernando ON, Moorhead J, Griffiths PD (1988) Symptomatic cytomegalovirus infection in seropositive kidney recipients: reinfection with donor virus rather than reactivation of recipient virus. Lancet 2:132-135

Grundy JE, Lawson KM, MacCormac LP, Fletcher JM, Yong KL (1998) Cytomegalovirus-infected endothelial cells recruit neutrophils by the secretion of C-X-C chemokines and transmit virus by direct neutrophil-endothelial cell contact and during neutrophil transendothelial migration. J Infect Dis 177:1465-1474

Gyorkey F, Melnick JL, Guinn GA, Gyorkey P, DeBakey ME (1984) Herpesviridae in the endothelial and smooth muscle cells of the proximal aorta in arteriosclerotic patients. Exp Mol Pathol 40:328-339

Haerter G, Manfras BJ, de Jong-Hesse Y, Wilts H, Mertens T, Kern P, Schmitt M (2004) Cytomegalovirus retinitis in a patient treated with anti-tumor necrosis factor alpha antibody therapy for rheumatoid arthritis. Clin Infect Dis 39:e88-94

Hahn G, Revello MG, Patrone M, Percivalle E, Campanini G, Sarasini A, Wagner M, Gallina A, Milanesi G, Koszinowski U, Baldanti F, Gerna G (2004) Human cytomegalovirus UL131-128 genes are indispensable for virus growth in endothelial cells and virus transfer to leukocytes. J Virol 78:10023-10033

Hakki M, Geballe AP (2005) Double-stranded RNA binding by human cytomegalovirus pTRS1. J Virol 79:7311-7318

Hakki M, Marshall EE, De Niro KL, Geballe AP (2006) Binding and nuclear relocalization of protein kinase R by human cytomegalovirus TRS1. J Virol 80:11817-11826

Halwachs-Baumann G, Wilders-Truschnig M, Desoye G, Hahn T, Kiesel L, Klingel K, Rieger P, Jahn G, Sinzger C (1998) Human trophoblast cells are permissive to the complete replicative cycle of human cytomegalovirus. J Virol 72:7598-7602

Hamprecht K, Maschmann J, Vochem M, Dietz K, Speer CP, Jahn G (2001) Epidemiology of transmission of cytomegalovirus from mother to preterm infant by breastfeeding. Lancet 357:513-518

Han CP, Tsao YP, Sun CA, Ng HT, Chen SL (1997) Human papillomavirus, cytomegalovirus and herpes simplex virus infections for cervical cancer in Taiwan. Cancer Lett 120:217-221

Handsfield HH, Chandler SH, Caine VA, Meyers JD, Corey L, Medeiros E, McDougall JK (1985) Cytomegalovirus infection in sex partners: evidence for sexual transmission. J Infect Dis 151:344-348

Harris S, Ahlfors K, Ivarsson S, Lemmark B, Svanberg L (1984) Congenital cytomegalovirus infection and sensorineural hearing loss. Ear Hear 5:352-355

Harrison CJ, Caruso N (2000) Correlation of maternal and pup NK-like activity and TNF responses against cytomegalovirus to pregnancy outcome in inbred guinea pigs. J Med Virol 60:230-236

Harrison CJ, Myers MG (1990) Relation of maternal CMV viremia and antibody response to the rate of congenital infection and intrauterine growth retardation. J Med Virol 31:222-228

Hartmann A, Sagedal S, Hjelmesaeth J (2006) The natural course of cytomegalovirus infection and disease in renal transplant recipients. Transplantation 82: S15-S17

Hasan M, Krmpotic A, Ruzsics Z, Bubic I, Lenac T, Halenius A, Loewendorf A, Messerle M, Hengel H, Jonjic S, Koszinowski UH (2005) Selective down-regulation of the NKG2D ligand H60 by mouse cytomegalovirus m155 glycoprotein. J Virol 79:2920-2930

Hassan-Walker AF, Mattes FM, Griffiths PD, Emery VC (2001) Quantity of cytomegalovirus DNA in different leukocyte populations during active infection in vivo and the presence of gB and UL18 transcripts. J Med Virol 64:283-289

Helantera I, Koskinen P, Finne P, Kyllonen L, Salmela K, Gronhagen-Riska C, Lautenschlager I (2006) Persistent cytomegalovirus infection in kidney allografts is associated with inferior graft function and survival. Transpl Int 19:893-900

Hemmings DG, Kilani R, Nykiforuk C, Preiksaitis J, Guilbert LJ (1998) Permissive cytomegalovirus infection of primary villous term and first trimester trophoblasts. J Virol 72:4970-4979

Hendrix MG, Dormans PH, Kitslaar P, Bosman F, Bruggeman CA (1989) The presence of cytomegalovirus nucleic acids in arterial walls of atherosclerotic and nonatherosclerotic patients. Am J Pathol 134:1151-1157

Hendrix MG, Daemen M, Bruggeman CA (1991) Cytomegalovirus nucleic acid distribution within the human vascular tree. Am J Pathol 138:563-567

Herrera GA, Alexander RW, Cooley CF, Luke RG, Kelly DR, Curtis JJ, Gockerman JP (1986) Cytomegalovirus glomerulopathy: a controversial lesion. Kidney Int 29:725-733

Hicks T, Fowler K, Richardson M, Dahle A, Adams L, Pass R (1993) Congenital cytomegalovirus infection and neonatal auditory screening. J Pediatr 123:779-782

Hobom U, Brune W, Messerle M, Hahn G, Koszinowski UH (2000) Fast screening procedures for random transposon libraries of cloned herpesvirus genomes: mutational analysis of human cytomegalovirus envelope glycoprotein genes. J Virol 74:7720-7729

Holland GN (1999) Immune recovery uveitis. Ocul Immunol Inflamm 7:215-221

Holtappels R, Podlech J, Grzimek NK, Thomas D, Pahl-Seibert MF, Reddehase MJ (2001) Experimental preemptive immunotherapy of murine cytomegalovirus disease with CD8 T-cell lines specific for ppM83 and pM84, the two homologs of human cytomegalovirus tegument protein ppUL83 (pp65). J Virol 75:6584-6600

Holtappels R, Grzimek N, Simon C, Thomas D, Dreis D, Reddehase M (2002) Processing and presentation of murine cytomegalovirus pORFm164-derived peptide in fibroblasts in the face of all viral immunosubversive early gene functions. J Virol 76:6044-6053

Holtappels R, Munks MW, Podlech J, Reddehase MJ (2006) CD8 T-cell -based immunotherapy of cytomegalovirus disease in the mouse model of the immunocompromised bone marrow transplantation recipient. In: Reddehase MJ (ed) Cytomegaloviruses: molecular biology and immunology. Caister Academic Press, Norfolk, UK, pp 383-418

Horak DA, Schmidt GM, Zaia JA, Niland JC, Ahn C, Forman SJ (1992) Pretransplant pulmonary function predicts cytomegalovirus-associated interstitial pneumonia following bone marrow transplantation. Chest 102:1484-1490

Horne BD, Muhlestein JB, Carlquist JF, Bair TL, Madsen TE, Hart NI, Anderson JL (2003) Statin therapy interacts with cytomegalovirus seropositivity and high C-reactive protein in reducing mortality among patients with angiographically significant coronary disease. Circulation 107:258-263

Horwitz CA, Henle W, Henle G, Snover D, Rudnick H, Balfour HH, Mazur MH, Watson R, Schwartz B, Muller N (1986) Clinical and laboratory evaluation of cytomegalovirus-induced mononucleosis in previously healthy individuals. Report of 82 cases. Medicine 65:124-134

Hosenpud JD (1999) Coronary artery disease after heart transplantation and its relation to cytomegalovirus. Am Heart J 138:S469-S472

Hosenpud JD, Chou SW, Wagner CR (1991) Cytomegalovirus-induced regulation of major histocompatibility complex class I antigen expression in human aortic smooth muscle cells. Transplantation 52:896-903

Hsich E, Zhou YF, Paigen B, Johnson TM, Burnett MS, Epstein SE (2001) Cytomegalovirus infection increases development of atherosclerosis in apolipoprotein-E knockout mice. Atherosclerosis 156:23-28

Huang ES, Mar EC, Boldogh I, Baskar J (1984) The oncogenicity of human cytomegalovirus. Birth Defects Orig Artic Ser 20:193-211

Hubscher SG, Buckels JA, Elias E, McMaster P, Neuberger J (1991) Vanishing bile-duct syndrome following liver transplantation - is it reversible? Transplantation 51:1004-1010

Hummel M, Zhang Z, Yan S, DePlaen I, Golia P, Varghese T, Thomas G, Abecassis MI (2001) Allogeneic transplantation induces expression of cytomegalovirus immediate-early genes in vivo: a model for reactivation from latency. J Virol 75:4814-4822

Hussain T, Burch M, Fenton MJ, Whitmore PM, Rees P, Elliott M, Aurora P (2007) Positive pretransplantation cytomegalovirus serology is a risk factor for cardiac allograft vasculopathy in children. Circulation 115:1798-1805

Hutto C, Little EA, Ricks R, Lee JD, Pass RF (1986) Isolation of cytomegalovirus from toys and hands in a day care center. J Infect Dis 154:527-530

Iversen AC, Norris PS, Ware CF, Benedict CA (2005) Human NK cells inhibit cytomegalovirus replication through a noncytolytic mechanism involving lymphotoxin-dependent induction of IFN-beta. J Immunol 175:7568-7574

Jacobson MA, Mills J, Rush J, Peiperl L, Seru V, Mohanty PK, Hopewell PC, Hadley WK, Broadus VC, Leoung G et al (1991) Morbidity and mortality of patients with AIDS and first-episode *Pneumocystis carinii* pneumonia unaffected by concomitant pulmonary cytomegalovirus infection. Am Rev Respir Dis 144:6-9

Jin X, Demoitie MA, Donahoe SM, Ogg GS, Bonhoeffer S, Kakimoto WM, Gillespie G, Moss PA, Dyer W, Kurilla MG, Riddell SR, Downie J, Sullivan JS, McMichael AJ, Workman C, Nixon DF (2000) High frequency of cytomegalovirus-specific cytotoxic T-effector cells in HLA-A*0201-positive subjects during multiple viral coinfections. J Infect Dis 181:165-175

Jonjic S, Pavic I, Polic B, Crnkovic I, Lucin P, Koszinowski UH (1994) Antibodies are not essential for the resolution of primary cytomegalovirus infection but limit dissemination of recurrent virus. J Exp Med 179:1713-1717

Jordan MC, Mar VL (1982) Spontaneous activation of latent cytomegalovirus from murine spleen explants. Role of lymphocytes and macrophages in release and replication of virus. J Clin Invest 70:762-768

Kahl M, Siegel-Axel D, Stenglein S, Jahn G, Sinzger C (2000) Efficient lytic infection of human arterial endothelial cells by human cytomegalovirus strains. J Virol 74:7628-7635

Karavellas MP, Lowder CY, Macdonald C, Avila CP Jr, Freeman WR (1998) Immune recovery vitritis associated with inactive cytomegalovirus retinitis: a new syndrome. Arch Ophthalmol 116:169-175

Karavellas MP, Azen SP, MacDonald JC, Shufelt CL, Lowder CY, Plummer DJ, Glasgow B, Torriani FJ, Freeman WR (2001) Immune recovery vitritis and uveitis in AIDS: clinical predictors, sequelae, and treatment outcomes. Retina 21:1-9

Kas-Deelen AM, de Maar EF, Harmsen MC, Driessen C, van Son WJ, The TH (2000) Uninfected and cytomegalic endothelial cells in blood during cytomegalovirus infection: effect of acute rejection. J Infect Dis 181:721-724

Kaur A, Hale CL, Noren B, Kassis N, Simon MA, Johnson RP (2002) Decreased frequency of cytomegalovirus (CMV)-specific CD4+ T lymphocytes in simian immunodeficiency virus-infected rhesus macaques: inverse relationship with CMV viremia. J Virol 76: 3646-3658

Kaur A, Kassis N, Hale CL, Simon M, Elliott M, Gomez-Yafal A, Lifson JD, Desrosiers RC, Wang F, Barry P, Mach M, Johnson RP (2003) Direct relationship between suppression of virus-specific immunity and emergence of cytomegalovirus disease in simian AIDS. J Virol 77:5749-5758

Kempen JH, Min YI, Freeman WR, Holland GN, Friedberg DN, Dieterich DT, Jabs DA (2006) Risk of immune recovery uveitis in patients with AIDS and cytomegalovirus retinitis. Ophthalmology 113:684-694

Kern ER (1999) Animal models for cytomegalovirus infection: murine CMV. In: Zak O, Sande M (eds) Handbook of animal models of infection. Academic Press, London, pp 927-934

Kessler MJ, London WT, Madden DL, Dambrosia JM, Hilliard JK, Soike KF, Rawlins RG (1989) Serological survey for viral diseases in the Cayo Santiago rhesus macaque population. Puerto Rico Health Sci J 8:95-97

Khan N, Bruton R, Taylor GS, Cobbold M, Jones TR, Rickinson AB, Moss PA (2005) Identification of cytomegalovirus-specific cytotoxic T lymphocytes in vitro is greatly enhanced by the use of recombinant virus lacking the US2 to US11 region or modified vaccinia virus Ankara expressing individual viral genes. J Virol 79:2869-2879

Kim SJ, Varghese TK, Zhang Z, Zhao LC, Thomas G, Hummel M, Abecassis M (2005) Renal ischemia/reperfusion injury activates the enhancer domain of the human cytomegalovirus major immediate early promoter. Am J Transplant 5:1606-1613

King A, Hiby SE, Verma S, Burrows T, Gardner L, Loke YW (1997) Uterine NK cells and trophoblast HLA class I molecules. Am J Reprod Immunol (Copenhagen) 37:459-462

Kloppenburg G, de Graaf R, Herngreen S, Grauls G, Bruggeman C, Stassen F (2005) Cytomegalovirus aggravates intimal hyperplasia in rats by stimulating smooth muscle cell proliferation. Microbes Infect 7:164-170

Knight DA, Waldman WJ, Sedmak DD (1999) Cytomegalovirus-mediated modulation of adhesion molecule expression by human arterial and microvascular endothelial cells. Transplantation 68:1814-1818

Koffron A, Varghese T, Hummel M, Yan S, Kaufman D, Fryer J, Leventhal J, Stuart F, Abecassis M (1999) Immunosuppression is not required for reactivation of latent murine cytomegalovirus. Transplant Proc 31:1395-1396

Kohara MM, Blum RN (2006) Cytomegalovirus ileitis and hemophagocytic syndrome associated with use of anti-tumor necrosis factor-alpha antibody. Clin Infect Dis 42:733-734

Komanduri KV, Viswanathan MN, Wieder ED, Schmidt DK, Bredt BM, Jacobson M, McCune JM (1998) Restoration of cytomegalovirus-specific CD4+ T-lymphocyte responses after ganciclovir and highly active antiretroviral therapy in individuals infected with HIV-1. Nature Med 4:953-956

Komanduri KV, Donahoe SM, Moretto WJ, Schmidt DK, Gillespie G, Ogg GS, Roederer M, Nixon DF, McCune JM (2001a) Direct measurement of CD4+ and CD8+ T-cell responses to CMV in HIV-1-infected subjects. Virology 279:459-470

Komanduri KV, Feinberg J, Hutchins RK, Frame RD, Schmidt DK, Viswanathan MN, Lalezari JP, McCune JM (2001b) Loss of cytomegalovirus-specific CD4+ T cell responses in human immunodeficiency virus type 1-infected patients with high CD4+ T cell counts and recurrent retinitis. J Infect Dis 183:1285-1289

Kondo K, Xu J, Mocarski ES (1996) Human cytomegalovirus latent gene expression in granulocyte-macrophage progenitors in culture and in seropositive individuals. Proc Natl Acad Sci U S A 93:11137-11142

Koskinen PK, Nieminen MS, Krogerus LA, Lemstrom KB, Mattila SP, Hayry PJ, Lautenschlager IT (1993) Cytomegalovirus infection and accelerated cardiac allograft vasculopathy in human cardiac allografts. J Heart Lung Transplant 12:724-729

Koskinen PK, Kallio EA, Tikkanen JM, Sihvola RK, Hayry PJ, Lemstrom KB (1999) Cytomegalovirus infection and cardiac allograft vasculopathy. Transplant Infect Dis 1:115-126

Kovacs A, Weber ML, Burns LJ, Jacob HS, Vercellotti GM (1996) Cytoplasmic sequestration of p53 in cytomegalovirus-infected human endothelial cells. Am J Pathol 149:1531-1539

Kovats S, Main EK, Librach C, Stubblebine M, Fisher SJ, DeMars R (1990) A class I antigen, HLA-G, expressed in human trophoblasts. Science 248:220-223

Krmpotic A, Messerle M, Crnkovic-Mertens I, Polic B, Jonjic S, Koszinowski UH (2001) The immunoevasive function encoded by the mouse cytomegalovirus gene m152 protects the virus against T cell control in vivo. J Exp Med 190:1285-1296

Krmpotic A, Busch DH, Bubic I, Gebhardt F, Hengel H, Hasan M, Scalzo AA, Koszinowski UH, Jonjic S (2002) MCMV glycoprotein gp40 confers virus resistance to CD8+ T cells and NK cells in vivo. Nat Immunol 3:529-535

Krmpotic A, Bubic I, Polic B, Lucin P, Jonjic S (2003) Pathogenesis of murine cytomegalovirus infection. Microbes Infect 5:1263-1277

Kropff B, Mach M (1997) Identification of the gene coding for rhesus cytomegalovirus glycoprotein B and immunological analysis of the protein. J Gen Virol 78:1999-2007

Kuhn EM, Stolte N, Matz-Rensing K, Mach M, Stahl-Henning C, Hunsmann G, Kaup FJ (1999) Immunohistochemical studies of productive rhesus cytomegalovirus infection in rhesus monkeys (Macaca mulatta) infected with simian immunodeficiency virus. Vet Pathol 36:51-56

Kurz S, Steffens HP, Mayer A, Harris JR, Reddehase MJ (1997) Latency versus persistence or intermittent recurrences: evidence for a latent state of murine cytomegalovirus in the lungs. J Virol 71:2980-2987

Kurz SK, Rapp M, Steffens HP, Grzimek NK, Schmalz S, Reddehase MJ (1999) Focal transcriptional activity of murine cytomegalovirus during latency in the lungs. J Virol 73:482-494

La Rosa C, Limaye AP, Krishnan A, Longmate J, Diamond DJ (2007) Longitudinal assessment of cytomegalovirus (CMV)-specific immune responses in liver transplant recipients at high risk for late CMV disease. J Infect Dis 195:633-644

Lacey SF, Gallez-Hawkins G, Crooks M, Martinez J, Senitzer D, Forman SJ, Spielberger R, Zaia JA, Diamond DJ (2002) Characterization of cytotoxic function of CMV-pp65-specific CD8+ T-lymphocytes identified by HLA tetramers in recipients and donors of stem-cell transplants. Transplantation 74:722-732

Lang DJ, Kummer JF (1975) Cytomegalovirus in semen: observations in selected populations. J Infect Dis 132:472-473

Lanier LL (1999) Natural killer cells fertile with receptors for HLA-G? Proc Natl Acad Sci USA 96:5343-5345

Lathey JL, Fiscus SA, Rasheed S, Kappes JC, Griffith BP, Elbeik T, Spector SA, Reichelderfer PS (1994) Optimization of quantitative culture assay for human immunodeficiency virus from plasma. Plasma Viremia Group Laboratories of the AIDS Clinical Trials Group (National Institute of Allergy and Infectious Diseases). J Clin Microbiol 32:3064-3067

Lautenschlager I, Soots A, Krogerus L, Kauppinen H, Saarinen O, Bruggeman C, Ahonen J (1997) CMV increases inflammation and accelerates chronic rejection in rat kidney allografts. Transplant Proc 29:802-803

Lautenschlager I, Halme L, Hockerstedt K, Krogerus L, Taskinen E (2006) Cytomegalovirus infection of the liver transplant: virological, histological, immunological, and clinical observations. Transpl Infect Dis 8:21-30

Lawrence RM, Lawrence RA (2004) Breast milk and infection. Clin Perinatol 31:501-528

Le Bouteiller P (2000) HLA-G in the human placenta: expression and potential functions. Biochem Soc Trans 28:208-212

Lemstrom K, Persoons M, Bruggeman C, Ustinov J, Lautenschlager I, Hayry P (1993) Cytomegalovirus infection enhances allograft arteriosclerosis in the rat. Transplant Proc 25:1406-1407

Lemstrom K, Sihvola R, Bruggeman C, Hayry P, Koskinen P (1997) Cytomegalovirus infection-enhanced cardiac allograft vasculopathy is abolished by DHPG prophylaxis in the rat. Circulation 95:2614-2616

Lenfant F, Pizzato N, Liang S, Davrinche C, Le Bouteiller P, Horuzsko A (2003) Induction of HLA-G-restricted human cytomegalovirus pp65 (UL83)-specific cytotoxic T lymphocytes in HLA-G transgenic mice. J Gen Virol 84:307-317

Libby P, Warner SJ, Friedman GB (1988a) Interleukin 1: a mitogen for human vascular smooth muscle cells that induces the release of growth-inhibitory prostanoids. J Clin Invest 81:487-498

Libby P, Warner SJ, Salomon RN, Birinyi LK (1988b) Production of platelet-derived growth factor-like mitogen by smooth-muscle cells from human atheroma. N Engl J Med 318:1493-1498

Limaye AP, Huang ML, Leisenring W, Stensland L, Corey L, Boeckh M (2001) Cytomegalovirus (CMV) DNA load in plasma for the diagnosis of CMV disease before engraftment in hematopoietic stem-cell transplant recipients. J Infect Dis 183:377-382

Lockridge KM, Sequar G, Zhou SS, Yue Y, Mandell CP, Barry PA (1999) Pathogenesis of experimental rhesus cytomegalovirus infection. J Virol 73:9576-9583

Lodoen M, Ogasawara K, Hamerman JA, Arase H, Houchins JP, Mocarski ES, Lanier LL (2003) NKG2D-mediated natural killer cell protection against cytomegalovirus is impaired by viral gp40 modulation of retinoic acid early inducible 1 gene molecules. J Exp Med 197:1245-1253

Loebe M, Schuler S, Zais O et al (1990) Role of cytomegalovirus infection in the development of coronary artery disease in the transplanted heart. J Heart Transplant 9:707-711

Lopez-Botet M, Llano M, Ortega M (2001) Human cytomegalovirus and natural killer-mediated surveillance of HLA class I expression: a paradigm of host-pathogen adaptation. Immunol Rev 181:193-202

Jarvis MA, Borton JA, Keech AM, Wong J, Britt WJ, Magun BE, Nelson JA (2006) Human cytomegalovirus attenuates IL-1B and TNF proinflammatory signaling by inhibition of NFk- B activation. J Virol 80:5588-5596

Manley TJ, Luy L, Jones T, Boeckh M, Mutimer H, Riddell SR (2004) Immune evasion proteins of human cytomegalovirus do not prevent a diverse CD8+ cytotoxic T-cell response in natural infection. Blood 104:1075-1082

Marchant A, Appay V, Van Der Sande M, Dulphy N, Liesnard C, Kidd M, Kaye S, Ojuola O, Gillespie GM, Vargas Cuero AL, Cerundolo V, Callan M, McAdam KP, Rowland-Jones SL, Donner C, McMichael AJ, Whittle H (2003) Mature CD8(+) T lymphocyte response to viral infection during fetal life. J Clin Invest 111:1747-1755

Martelius TJ, Blok MJ, Inkinen KA, Loginov RJ, Hockerstedt KA, Bruggeman CA, Lautenschlager IT (2001) Cytomegalovirus infection, viral DNA, and immediate early-1 gene expression in rejecting rat liver allografts. Transplantation 71:1257-1261

Martelius TJ, Wolff H, Bruggeman CA, Hockerstedt KA, Lautenschlager IT (2002) Induction of cyclo-oxygenase-2 by acute liver allograft rejection and cytomegalovirus infection in the rat. Transpl Int 15:610-614

Martin-Davila P, Fortun J, Gutierrez C, Marti-Belda P, Candelas A, Honrubia A, Barcena R, Martinez A, Puente A, de Vicente E, Moreno S (2005) Analysis of a quantitative PCR assay for CMV infection in liver transplant recipients: an intent to find the optimal cut-off value. J Clin Virol 33:138-144

McCarthy M, Auger D, He J, Wood C (1998) Cytomegalovirus and human herpesvirus-6 trans-activate the HIV-1 long terminal repeat via multiple response regions in human fetal astrocytes. J Neurovirol 4:495-511

McCormick AL, Skaletskaya A, Barry PA, Mocarski ES, Goldmacher VS (2003) Differential function and expression of the viral inhibitor of caspase 8-induced apoptosis (vICA) and the viral mitochondria-localized inhibitor of apoptosis (vMIA) cell death suppressors conserved in primate and rodent cytomegaloviruses. Virology 316:221-233

McDonald K, Rector TS, Braulin EA, Kubo SH, Olivari MT (1989) Association of coronary artery disease in cardiac transplant recipients with cytomegalovirus infection. Am J Cardiol 64:359-362

McLaughlin-Taylor E, Pande H, Forman SJ, Tanamachi B, Li CR, Zaia JA, Greenberg PD, Riddell SR (1994) Identification of the major late human cytomegalovirus matrix protein pp65 as a target antigen for CD8+ virus-specific cytotoxic T lymphocytes. J Med Virol 43:103-110

Mehra MR (2006) Contemporary concepts in prevention and treatment of cardiac allograft vasculopathy. Am J Transplant 6:1248-1256

Melnick JL, Adam E, Debakey ME (1993) Cytomegalovirus and atherosclerosis. Eur Heart J 14:30-38

Melnychuk RM, Smith P, Kreklywich CN, Ruchti F, Vomaske J, Hall L, Loh L, Nelson JA, Orloff SL, Streblow DN (2005) Mouse cytomegalovirus M33 is necessary and sufficient in virus-induced vascular smooth muscle cell migration. J Virol 79:10788-10795

Messerle M, Hahn G, Brune W, Koszinowski UH (2000) Cytomegalovirus bacterial artificial chromosomes: a new herpesvirus vector approach. Adv Virus Res 55:463-478

Meyers JD, Flournoy N, Thomas ED (1982) Nonbacterial pneumonia after allogeneic marrow transplantation: a review of ten years' experience. Rev Infect Dis 4:1119-1132

Meyers JD, Flournoy N, Thomas ED (1986) Risk factors for cytomegalovirus infection after human marrow transplantation. J Infect Dis 153:478-488

Minamishima I, Ueda K, Minematsu T, Minamishima Y, Umemoto M, Take H, Kuraya K (1994) Role of breast milk in acquisition of cytomegalovirus infection. Microbiol Immunol 38:549-552

Miron D, Brosilow S, Felszer K, Reich D, Halle D, Wachtel D, Eidelman AI, Schlesinger Y (2005) Incidence and clinical manifestations of breast milk-acquired Cytomegalovirus infection in low birth weight infants. J Perinatol 25:299-303

Mizuta M, Schuster MG (2005) Cytomegalovirus hepatitis associated with use of anti-tumor necrosis factor-alpha antibody. Clin Infect Dis 40:1071-1072

MMWR (1997) Rubella and congenital rubella syndrome - United States, 1994-1997. MMWR Morb Mortal Wkly Rep 46:350-354

Moreno TN, Fortunato EA, Hsia K, Spector SA, Spector DH (1997) A model system for human cytomegalovirus-mediated modulation of human immunodeficiency virus type 1 long terminal repeat activity in brain cells. J Virol 71:3693-3701

Moretta A, Bottino C, Vitale M, Pende D, Cantoni C, Mingari MC, Biassoni R, Moretta L (2001) Activating receptors and coreceptors involved in human natural killer cell-mediated cytolysis. Annu Rev Immunol 19:197-223

Morris DJ, Martin S, Dyer PA, Hunt L, Mallick NP, Johnson RW (1993) HLA mismatching and cytomegalovirus infection as risk factors for transplant failure in cyclosporin-treated renal allograft recipients. J Med Virol 41:324-327

Mueller C, Hodgson JM, Bestehorn HP, Brutsche M, Perruchoud AP, Marsch S, Roskamm H, Buettner HJ (2003) Previous cytomegalovirus infection and restenosis after aggressive angioplasty with provisional stenting. J Interv Cardiol 16:307-313

Muhlemann K, Miller RK, Metlay L, Menegus MA (1992) Cytomegalovirus infection of the human placenta: an immunocytochemical study. Hum Pathol 23:1234-1237

Muhlestein JB, Horne BD, Carlquist JF, Madsen TE, Bair TL, Pearson RR, Anderson JL (2000) Cytomegalovirus seropositivity and C-reactive protein have independent and combined predictive value for mortality in patients with angiographically demonstrated coronary artery disease. Circulation 102:1917-1923

Munks MW, Pinto AK, Doom CM, Hill AB (2007) Viral interference with antigen presentation does not alter acute or chronic CD8 T cell immunodominance in murine cytomegalovirus infection. J Immunol 178:7235-7241

Murphy E, Yu D, Grimwood J, Schmutz J, Dickson M, Jarvis MA, Nelson JA, Myers RM, Shenk TE (2003) Coding capacity of laboratory and clinical strains of human cytomegalovirus. Proc Natl Acad Sci USA 100:14976-14981

Murray BM, Subramaniam S (2004) Late cytomegalovirus infection after oral ganciclovir prophylaxis in renal transplant recipients. Transpl Infect Dis 6:3-9

Myerson D, Hackman RC, Nelson JA et al (1984) Widespread presence of histologically occult cytomegalovirus. Hum Pathol 15:430-439

Nerheim PL, Meier JL, Vasef MA, Li WG, Hu L, Rice JB, Gavrila D, Richenbacher WE, Weintraub NL (2004) Enhanced cytomegalovirus infection in atherosclerotic human blood vessels. Am J Pathol 164:589-600

Nieto FJ, Adam E, Sorlie P, Farzadegan H, Melnick JL, Comstock GW, Szklo M (1996) Cohort study of cytomegalovirus infection as a risk factor for carotid intimal-medial thickening, a measure of subclinical atherosclerosis. Circulation 94:922-927

Noda S, Aguirre SA, Bitmansour A, Brown JM, Sparer TE, Huang J, Mocarski ES (2006) Cytomegalovirus MCK-2 controls mobilization and recruitment of myeloid progenitor cells to facilitate dissemination. Blood 107:30-38

Noyola DE, Demmler GJ, Nelson CT, Griesser C, Williamson WD, Atkins JT, Rozelle J, Turcich M, Llorente AM, Sellers-Vinson S, Reynolds A, Bale JF Jr., Gerson P, Yow MD (2001) Early predictors of neurodevelopmental outcome in symptomatic congenital cytomegalovirus infection. J Pediatr 138:325-331

Onno M, Pangault C, Le Friec G, Guilloux V, Andre P, Fauchet R (2000) Modulation of HLA-G antigens expression by human cytomegalovirus: specific induction in activated macrophages harboring human cytomegalovirus infection. J Immunol 164:6426-6434

Ozono K, Mushiake S, Takeshima T, Nakayama M (1997) Diagnosis of congenital cytomegalovirus infection by examination of placenta: application of polymerase chain reaction and in situ hybridization. Pediatr Pathol Lab Med 17:249-258

Paavonen T, Mennander A, Lautenschlager I, Mattila S, Hayry P (1993) Endothelialitis and accelerated arteriosclerosis in human heart transplant coronaries. J Heart Lung Transplant 12:117-122

Pannuti CS, Boas LSV, Angelo MJO, Neto VA, Levi GC, de Mendonca JS, de Godoy CVF (1985) Cytomegalovirus mononucleosis in children and adults: differences in clinical presentation. Scand J Infect Dis 17:153-156

Pass RF, Stagno S, Dworsky ME, Smith RJ, Alford CA (1982) Excretion of cytomegalovirus in mothers: observation after delivery of congenitally infected and normal infants. J Infect Dis 146:1-6

Pass RF, Stagno S, Britt WJ, Alford CA (1983) Specific cell mediated immunity and the natural history of congenital infection with cytomegalovirus. J Infect Dis 148:953-961

Pass RF, Little EA, Stagno S, Britt WJ, Alford CA (1987) Young children as a probable source of maternal and congenital cytomegalovirus infection. N Engl J Med 316:1366-1370

Pecorella I, Ciardi A, Garner A, McCartney AC, Lucas S (2000) Postmortem histological survey of the ocular lesions in a British population of AIDS patients. Br J Ophthalmol 84:1275-1281

Penfold ME, Dairaghi DJ, Duke GM, Saederup N, Mocarski ES, Kemble GW, Schall TJ (1999) Cytomegalovirus encodes a potent alpha chemokine. Proc Natl Acad Sci USA 96:9839-9844

Pepose JS, Newman C, Bach MC, Quinn TC, Ambinder RF, Holland GN, Hodstrom PS, Frey HM, Foos RY (1987) Pathologic features of cytomegalovirus retinopathy after treatment with the antiviral agent ganciclovir. Ophthalmology 94:414-424

Percivalle E, Revello MG, Vago L, Morini F, Gerna G (1993) Circulating endothelial giant cells permissive for human cytomegalovirus (HCMV) are detected in disseminated HCMV infections with organ involvement. J Clin Invest 92:663-670

Pereira L, Maidji E, McDonagh S, Tabata T (2005) Insights into viral transmission at the uterine-placental interface. Trends Microbiol 13:164-174

Perlman JM, Argyle C (1992) Lethal cytomegalovirus infection in preterm infants: clinical, radiological, and neuropathological findings. Ann Neurol 31:64-68

Persoons MC, Daemen MJ, Bruning JH, Bruggeman CA (1994) Active cytomegalovirus infection of arterial smooth muscle cells in immunocompromised rats. A clue to herpesvirus-associated atherogenesis? Circ Res 75:214-220

Persoons MC, Stals FS, van dam Mieras MC, Bruggeman CA (1998) Multiple organ involvement during experimental cytomegalovirus infection is associated with disseminated vascular pathology. J Pathol 184:103-199

Petrakopoulou P, Kubrich M, Pehlivanli S, Meiser B, Reichart B, von Scheidt W, Weis M (2004) Cytomegalovirus infection in heart transplant recipients is associated with impaired endothelial function. Circulation 110:11207-11212

Plachter B, Sinzger C, Jahn G (1996) Cell types involved in replication and distribution of human cytomegalovirus. Adv Virus Res 46:195-261

Plotkin SA, Starr SE, Friedman HM et al (1989) Protective effects of Towne cytomegalovirus vaccine against low-passage cytomegalovirus administered as a challenge. J Infect Dis 159:860-865

Podlech J, Holtappels R, Pahl-Seibert MF, Steffens HP, Reddehase MJ (2000) Murine model of interstitial cytomegalovirus pneumonia in syngeneic bone marrow transplantation: persistence of protective pulmonary CD8-T-cell infiltrates after clearance of acute infection. J Virol 74:7496-7507

Poehlmann TG, Schaumann A, Busch S, Fitzgerald JS, Aguerre-Girr M, Le Bouteiller P, Schleussner E, Markert UR (2006) Inhibition of term decidual NK cell cytotoxicity by soluble HLA-G1. Am J Reprod Immunol 56:275-285

Polic B, Hengel H, Krmpotic A, Trgovcich J, Pavic I, Luccaronin P, Jonjic S, Koszinowski UH (1998) Hierarchical and redundant lymphocyte subset control precludes cytomegalovirus replication during latent infection. J Exp Med 188:1047-1054

Pooley RJ Jr, Peterson L, Finn WG, Kroft SH (1999) Cytomegalovirus-infected cells in routinely prepared peripheral blood films of immunosuppressed patients. Am J Clin Pathol 112:108-112

Porath A, Schlaeffer F, Sarov I, Keynan A (1987) Cytomegalovirus mononucleosis - a report of 70 cases in a community hospital. Isr J Med Sci 23:268-273

Potena L, Holweg CT, Chin C, Luikart H, Weisshaar D, Narasimhan B, Fearon WF, Lewis DB, Cooke JP, Mocarski ES, Valantine HA (2006) Acute rejection and cardiac allograft vascular disease is reduced by suppression of subclinical cytomegalovirus infection. Transplantation 82:398-405

Preblud SR, Alford CA (1990) Rubella. In: Remington JS, Klein JO (eds) Infectious diseases of the fetus and newborn infant, 3rd edn. W.B. Saunders, Philadelphia, pp 197-240

Prichard MN, Penfold ME, Duke GM, Spaete RR, Kemble GW (2001) A review of genetic differences between limited and extensively passaged human cytomegalovirus strains. Rev Med Virol 11:191-200

Prod'homme V, Griffin C, Aicheler RJ, Wang EC, McSharry BP, Rickards CR, Stanton RJ, Borysiewicz LK, Lopez-Botet M, Wilkinson GW, Tomasec P (2007) The human cytomegalovirus MHC class I homolog UL18 inhibits LIR-1+ but activates LIR-1-NK cells. J Immunol 178:4473-4481

Racusen LC, Solez K, Colvin RB, Bonsib SM, Castro MC, Cavallo T, Croker BP, Demetris AJ, Drachenberg CB, Fogo AB, Furness P, Gaber LW, Gibson IW, Glotz D, Goldberg JC, Grande J, Halloran PF, Hansen HE, Hartley B, Hayry PJ, Hill CM, Hoffman EO, Hunsicker LG, Lindblad AS, Yamaguchi Y et al (1999) The Banff 97 working classification of renal allograft pathology. Kidney Int 55:713-723

Raeiszadeh M, Webster D, Workman S, Emery VC (2007) The spectrum of CMV associated disease in patients with common variable immunodeficiency. In: 11th International Betaherpesvirus Workshop, Toulouse, France, p 12

Ranga U, Woffendin C, Yang ZY, Xu L, Verma S, Littman DR, Nabel GJ (1997) Cell and viral regulatory elements enhance the expression and function of a human immunodeficiency virus inhibitory gene. J Virol 71:7020-7029

Rao NA, Zhang J, Ishimoto S (1998) Role of retinal vascular endothelial cells in development of CMV retinitis. Trans Am Ophthalmol Soc 96:111-123; discussion 124-116

Rapp M, Messerle M, Buhler B, Tannheimer M, Keil GM, Koszinowski UH (1992) Identification of the murine cytomegalovirus glycoprotein B gene and its expression by recombinant vaccinia virus. J Virol 66:4399-4406

Razonable RR, Rivero A, Rodriguez A, Wilson J, Daniels J, Jenkins G, Larson T, Hellinger WC, Spivey JR, Paya CV (2001) Allograft rejection predicts the occurrence of late-onset cytomegalovirus (CMV) disease among CMV-mismatched solid organ transplant patients receiving prophylaxis with oral ganciclovir. J Infect Dis 184: 1461-1464

Reddehase MJ, Podlech J, Grzimek NK (2002) Mouse models of cytomegalovirus latency: overview. J Clin Virol 25: S23-S36

Redman TK, Britt WJ, Wilcox CM, Graham MF, Smith PD (2002) Human cytomegalovirus enhances chemokine production by lipopolysaccharide-stimulated lamina propria macrophages. J Infect Dis 185:584-590

Reinhardt B, Mertens T, Mayr-Beyrle U, Frank H, Luske A, Schierling K, Waltenberger J (2005) HCMV infection of human vascular smooth muscle cells leads to enhanced expression of functionally intact PDGF beta-receptor. Cardiovasc Res 67:151-160

Reischig T, Jindra P, Svecova M, Kormunda S, Opatrny K Jr, Treska V (2006) The impact of cytomegalovirus disease and asymptomatic infection on acute renal allograft rejection. J Clin Virol 36:146-151

Reusser P, Attenhofer R, Hebart H, Helg C, Chapuis B, Einsele H (1997) Cytomegalovirus specific T-cell immunity in recipients of autologous peripheral blood stem cell or bone marrow transplants. Blood 89:3873-3879

Reusser P, Riddell SR, Meyers JD, Greenberg PD (1991) Cytotoxic T-lymphocyte response to cytomegalovirus after human allogeneic bone marrow transplantation: pattern of recovery and correlation with cytomegalovirus infection and disease. Blood 78:1373-1380

Richardson WP, Colvin RB, Cheeseman SH, Tolkoff-Rubin NE, Herrin JT, Cosimi AB, Collins AB, Hirsch MS, McCluskey RT, Russell PS, Rubin RH (1981) Glomerulopathy associated with cytomegalovirus viremia in renal allografts. N Engl J Med 305:57-63

Ricotta D, Alessandri G, Pollara C, Fiorentini S, Favilli F, Tosetti M, Mantovani A, Grassi M, Garrafa E, Dei Cas L, Muneretto C, Caruso A (2001) Adult human heart microvascular endothelial cells are permissive for non-lytic infection by human cytomegalovirus. CardiovascRes 49:440-448

Rifkind D (1965) Cytomegalovirus infection after renal transplantation. Arch Intern Med 116:554-558

Rinaldo CR Jr, Kingsley LA, Ho M, Armstrong JA, Zhou SY (1992) Enhanced shedding of cytomegalovirus in semen of human immunodeficiency virus-seropositive homosexual men. J Clin Microbiol 30:1148-1155

Ross SA, Novak Z, Ashrith G, Rivera LB, Britt WJ, Hedges S, Schwebke JR, Boppana AS (2005) Association between genital tract cytomegalovirus infection and bacterial vaginosis. J Infect Dis 192:1727-1730

Rubin R (2002) Clinical approach to infection in the compromised host. In: Rubin R, Young LS (eds) Infection in the organ transplant recipient. Kluwer Academic Press, New York, pp 573-679

Rubin RH, Colvin RB (1986) Cytomegalovirus infection in renal transplantation: clinical importance and control. In: Williams GM, Burdick JF, Solez K (eds) Kidney transplant rejection: diagnosis and treatment. Dekker, New York, pp 283-304

Rubin RH, Russell PS, Levin M et al (1979) Summary of a workshop on cytomegalovirus infections during organ transplantation. J Infect Dis 139:728-734

Rue CA, Jarvis MA, Knoche AJ, Meyers HL, DeFilippis VR, Hansen SG, Wagner M, Fruh K, Anders DG, Wong SW, Barry PA, Nelson JA (2004) A cyclooxygenase-2 homologue encoded by rhesus cytomegalovirus is a determinant for endothelial cell tropism. J Virol 78:12529-12536

Rupp B, Ruzsics Z, Sacher T, Koszinowski UH (2005) Conditional cytomegalovirus replication in vitro and in vivo. J Virol 79:486-494

Saederup N, Lin YC, Dairaghi DJ, Schall TJ, Mocarski ES (1999) Cytomegalovirus-encoded beta chemokine promotes monocyte-associated viremia in the host. Proc Natl Acad Sci USA 96:10881-10886

Saez-Lopez C, Ngambe-Tourere E, Rosenzwajg M, Petit JC, Nicolas JC, Gozlan J (2005) Immediate-early antigen expression and modulation of apoptosis after in vitro infection of polymorphonuclear leukocytes by human cytomegalovirus. Microbes Infect 7:1139-1149

Saffert RT, Kalejta RF (2006) Inactivating a cellular intrinsic immune defense mediated by Daxx is the mechanism through which the human cytomegalovirus pp71 protein stimulates viral immediate-early gene expression. J Virol 80:3863-3871

Sagedal S, Hartmann A, Nordal KP, Osnes K, Leivestad T, Foss A, Degre M, Fauchald P, Rollag H (2004) Impact of early cytomegalovirus infection and disease on long-term recipient and kidney graft survival. Kidney Int 66:329-337

Salazar-Mather TP, Lewis CA, Biron CA (2002) Type I interferons regulate inflammatory cell trafficking and macrophage inflammatory protein 1alpha delivery to the liver. J Clin Invest 110:321-330

Scalzo AA, Fitzgerald NA, Wallace CR, Gibbons AE, Smart YC, Burton RC, Shellam GR (1992) The effect of the Cmv-1 resistance gene, which is linked to the natural killer cell gene complex, is mediated by natural killer cells. J Immunol 149:581-589

Scalzo AA, Corbett AJ, Rawlinson WD, Scott GM, Degli-Esposti MA (2007) The interplay between host and viral factors in shaping the outcome of cytomegalovirus infection. Immunol Cell Biol 85:46-54

Schmidt GM, Horak DA, Niland JC, Duncan SR, Forman SJ, Zaia JA (1991) A randomized, controlled trial of prophylactic ganciclovir for cytomegalovirus pulmonary infection in recipients of allogeneic bone marrow transplants; The City of Hope-Stanford-Syntex CMV Study Group. N Eng J Med 324:1005-1011

Schoppel K, Kropff B, Schmidt C, Vornhagen R, Mach M (1997) The humoral immune response against human cytomegalovirus is characterized by a delayed synthesis of glycoprotein-specific antibodies. J Infect Dis 175:533-544

Schoppel K, Schmidt C, Einsele H, Hebart H, Mach M (1998a) Kinetics of the antibody response against human cytomegalovirus-specific proteins in allogeneic bone marrow transplant recipients. J Infect Dis 178:1233-1243

Schoppel K, Schmidt C, Einsele H, Hebart H, Mach M (1998b) Kinetics of the antibody response against human cytomegalovirus-specific proteins in allogeneic bone marrow transplant recipients. J Infect Dis 178:1233-1243

Schrier RD, Song MK, Smith IL, Karavellas MP, Bartsch DU, Torriani FJ, Garcia CR, Freeman WR (2006) Intraocular viral and immune pathogenesis of immune recovery uveitis in patients with healed cytomegalovirus retinitis. Retina 26:165-169

Schust DJ, Tortorella D, Ploegh HL (1999) Viral immunoevasive strategies and trophoblast class I major histocompatibility complex antigens. J Reproduc Immunol 43:243-251

Schust DJ, Tortorella D, Seebach J, Phan C, Ploegh HL (1998) Trophoblast class I major histocompatibility complex (MHC) products are resistant to rapid degradation imposed by the human cytomegalovirus (HCMV) gene products US2 and US11. J Exp Med 188:497-503

Sedmak DD, Knight DA, Vook NC, Waldman JW (1994) Divergent patterns of ELAM-1, ICAM-1, and VCAM-1 expression on cytomegalovirus-infected endothelial cells. Transplantation 58:1379-1385

Sequar G, Britt WJ, Lakeman FD, Lockridge KM, Tarara RP, Canfield DR, Zhou SS, Gardner MB, Barry PA (2002) Experimental coinfection of rhesus macaques with rhesus cytomegalovirus and simian immunodeficiency virus: pathogenesis. J Virol 76:7661-7671

Sester M, Sester U, Gartner B, Heine G, Girndt M, Mueller-Lantzsch N, Meyerhans A, Kohler H (2001) Levels of virus-specific CD4 T cells correlate with cytomegalovirus control and predict virus-induced disease after renal transplantation. Transplantation 71:1287-1294

Sester M, Sester U, Gartner B, Kubuschok B, Girndt M, Meyerhans A, Kohler H (2002a) Sustained high frequencies of specific CD4 T cells restricted to a single persistent virus. J Virol 76:3748-3755

Sester M, Sester U, Gartner BC, Girndt M, Meyerhans A, Kohler H (2002b) Dominance of virus-specific CD8 T cells in human primary cytomegalovirus infection. J Am Soc Nephrol 13:2577-2584

Sharon-Friling R, Goodhouse J, Colberg-Poley AM, Shenk T (2006) Human cytomegalovirus pUL37x1 induces the release of endoplasmic reticulum calcium stores. Proc Natl Acad Sci USA 103:19117-19122

Sharples LD, Tamm M, McNeil K, Higenbottam TW, Stewart S, Wallwork J (1996) Development of bronchiolitis obliterans syndrome in recipients of heart-lung transplantation - early risk factors. Transplantation 61:560-566

Shen CY, Ho MS, Chang SF, Yen MS, Ng HT, Huang ES, Wu CW (1993) High rate of concurrent genital infections with human cytomegalovirus and human papillomaviruses in cervical cancer patients. J Infect Dis 168:449-452

Shen CY, Chang SF, Lin HJ, Ho HN, Yeh TS, Yang SL, Huang ES, Wu CW (1994) Cervical cytomegalovirus infection in prostitutes and in women attending a sexually transmitted disease clinic. J Med Virol 43:362-366

Shen Y, Zhu H, Shenk T (1997) Human cytomagalovirus IE1 and IE2 proteins are mutagenic and mediate "hit-and-run" oncogenic transformation in cooperation with the adenovirus E1A proteins. Proc Natl Acad Sci U S A 94:3341-3345

Shi Y, Tokunaga O (2002) Herpesvirus (HSV-1, EBV and CMV) infections in atherosclerotic compared with non-atherosclerotic aortic tissue. Pathol Int 52:31-39

Simmen KA, Singh J, Luukkonen BG, Lopper M, Bittner A, Miller NE, Jackson MR, Compton T, Fruh K (2001) Global modulation of cellular transcription by human cytomegalovirus is initiated by viral glycoprotein B. Proc Natl Acad Sci U S A 98:7140-7145

Singh N (2005) Late-onset cytomegalovirus disease as a significant complication in solid organ transplant recipients receiving antiviral prophylaxis: a call to heed the mounting evidence. Clin Infect Dis 40:704-708

Sinzger C, Plachter B, Stenglein S, Jahn G (1993) Immunohistochemical detection of viral antigens in smooth muscle, stromal, and epithelial cells from acute human cytomegalovirus gastritis. J Infect Dis 167:1427-1432

Sinzger C, Grefte A, Plachter B, Gouw AS, The TH, Jahn G (1995) Fibroblasts, epithelial cells, endothelial cells and smooth muscle cells are major targets of human cytomegalovirus infection in lung and gastrointestinal tissues. J Gen Virol 76:741-750

Sinzger C, Plachter B, Grefte A, The TH, Jahn G (1996) Tissue macrophages are infected by human cytomegalovirus in vivo. J Infect Dis 173:240-245

Sinzger C, Bissinger AL, Viebahn R, Oettle H, Radke C, Schmidt CA, Jahn G (1999a) Hepatocytes are permissive for human cytomegalovirus infection in human liver cell culture and In vivo. J Infect Dis 180:976-986

Sinzger C, Schmidt K, Knapp J, Kahl M, Beck R, Waldman J, Hebart H, Einsele H, Jahn G (1999b) Modification of human cytomegalovirus tropism through propagation in vitro is associated with changes in the viral genome. J Gen Virol 80:2867-2877

Smith LM, Shellam GR, Redwood AJ (2006) Genes of murine cytomegalovirus exist as a number of distinct genotypes. Virology 352:450-465

Smith PD, Saini SS, Raffeld M, Manischewitz JF, Wahl SM (1992) Cytomegalovirus induction of tumor necrosis factor-alpha by human monocytes and mucosal macrophages. J Clin Invest 90:1642-1648

Smyth RL, Scott JP, Borysiewicz LK, Sharples LD, Stewart S, Wreghitt TG, Gray JJ, Higenbottam TW, Wallwork J (1991) Cytomegalovirus infection in heart-lung transplant recipients: risk factors, clinical associations, and response to treatment. J Infect Dis 164:1045-1050

Snydman DR, Werner BG, Heinze-Lacey B, Berardi VP, Tilney NL, Kirkman RL, Milford EL, Cho SI, Bush HL, Levey AS, Strom TB, Carpenter CE, Levey RH, Harmon WE, Zimmerman CE, Shaprio ME, Steinman T, LoGerfo F, Idelson B, Schroter GPJ, Levin MJ, McIver J, Leszczynski J, Grady GF (1987) Use of cytomegalovirus immune globulin to prevent cytomegalovirus disease in renal transplant recipients. N Engl J Med 317:1049-1054

Soderberg-Naucler C, Fish KN, Nelson JA (1997) Reactivation of latent human cytomegalovirus by allogeneic stimulation of blood cells from healthy donors. Cell 91:119-126

Sohn YM, Oh MK, Balcarek KB, Cloud GA, Pass RF (1991) Cytomegalovirus infection in sexually active adolescents. J Infect Dis 163:460-463

Sohn YM, Park KI, Lee C, Han DG, Lee WY (1992) Congenital cytomegalovirus infection in Korean population with very high prevalence of maternal immunity. J Kor Med Sci 7:47-51

Sola R, Diaz JM, Guirado L, Ravella N, Vila L, Sainz Z, Gich I, Picazo M, Garcia R, Abreu E, Ortiz F, Alcaraz A (2003) Significance of cytomegalovirus infection in renal transplantation. Transplant Proc 35:1753-1755

Soots A, Lautenschlager I, Krogerus L, Saarinen O, Ahonen J (1998) An experimental model of chronic renal allograft rejection in the rat using triple drug immunosuppression. Transplantation 65:42-46

Sorlie PD, Nieto FJ, Adam E, Folsom AR, Shahar E, Massing M (2000) A prospective study of cytomegalovirus, herpes simplex virus 1, and coronary heart disease: the atherosclerosis risk in communities (ARIC) study. Arch Intern Med 160:2027-2032

Span AH, van Dam-Mieras MC, Mullers W, Endert J, Muller AD, Bruggeman CA (1991) The effect of virus infection on the adherence of leukocytes or platelets to endothelial cells. Eur J Clin Invest 21:331-338

Spector SA, Hsia K, Crager M, Pilcher M, Cabral S, Stempien MJ (1999) Cytomegalovirus (CMV) DNA load is an independent predictor of CMV disease and survival in advanced AIDS. J Virol 73:7027-7030

Spector SA, Wong R, Hsia K, Pilcher M, Stempien MJ (1998) Plasma cytomegalovirus (CMV) DNA load predicts CMV disease and survival in AIDS patients. J Clin Invest 101:497-502

Speir E, Modali R, Huang ES, Leon MB, Shawl F, Finkel T, Epstein SE (1994) Potential role of human cytomegalovirus and p53 interaction in coronary restenosis. Science 265:391-394

Srivastava R, Curtis M, Hendrickson S, Burns WH, Hosenpud JD (1999) Strain specific effects of cytomegalovirus on endothelial cells: implications for investigating the relationship between CMV and cardiac allograft vasculopathy. Transplantation 68:1568-1573

Stagno S, Britt WJ (2006) Cytomegalovirus. In: Remington JS, Klein JO (eds) Infectious diseases of the fetus and newborn infant, 6th edn. W.B. Saunders, Philadelphia

Stagno S, Reynolds DW, Tsiantos A, Fucillo DA, Long W, Alford CA (1975) Comparative, serial virologic and serologic studies of symptomatic and subclinical congenital and natally acquired cytomegalovirus infection. J Infect Dis 132:568-577

Stagno S, Reynolds DW, Pass RF, Alford CA (1980) Breast milk and the risk of cytomegalovirus infection. N Engl J Med 302:1073-1076

Stagno S, Pass RF, Dworsky ME, Alford CA (1982a) Maternal cytomegalovirus infection and perinatal transmission. Clin Obstet Gynecol 25:563-576

Stagno S, Pass RF, Dworsky ME, Henderson RE, Moore EG, Walton PD, Alford CA (1982b) Congenital cytomegalovirus infection: the relative importance of primary and recurrent maternal infection. N Eng J Med 306:945-949

Stagno S, Pass RF, Dworsky ME, Alford CA (1983) Congenital and perinatal cytomegaloviral infections. Semin Perinatol 7:31-42

Stals FS (1999) Animal models for cytomegalovirus infection: Rat CMV. In: Zak O, Sande M (eds) Handbook of animal models of infection. Academic Press, London, pp 943-950

Stals FS, Bosman F, van Boven CP, Bruggeman CA (1990) An animal model for therapeutic intervention studies of CMV infection in the immunocompromised host. Arch Virol 114:91-107

Starr SE, Tolpin MD, Friedman HM, Paucker K, Plotkin SA (1979) Impaired cellular immunity to cytomegalovirus in congenitally infected children and their mothers. J Infect Dis 140:500-505

Stassen FR, Vega-Cordova X, Vliegen I, Bruggeman CA (2006) Immune activation following cytomegalovirus infection: more important than direct viral effects in cardiovascular disease? J Clin Virol 35:349-353

Stoica SC, Cafferty F, Pauriah M, Taylor CJ, Sharples LD, Wallwork J, Large SR, Parameshwar J (2006) The cumulative effect of acute rejection on development of cardiac allograft vasculopathy. J Heart Lung Transplant 25:420-425

Streblow DN, Soderberg-Naucler C, Vieira J, Smith P, Wakabayashi E, Ruchti F, Mattison K, Altschuler Y, Nelson JA (1999) The human cytomegalovirus chemokine receptor US28 mediates vascular smooth muscle cell migration. Cell 99:511-520

Streblow DN, van Cleef KW, Kreklywich CN, Meyer C, Smith P, Defilippis V, Grey F, Fruh K, Searles R, Bruggeman C, Vink C, Nelson JA, Orloff SL (2007) Rat cytomegalovirus gene expression in cardiac allograft recipients is tissue specific and does not parallel the profiles detected in vitro. J Virol 81:3816-3826

Swack NS, Hsiung GD (1982) Natural and experimental simian cytomegalovirus infections at a primate center. J Med Primatol 11:169-177

Sylwester AW, Mitchell BL, Edgar JB, Taormina C, Pelte C, Ruchti F, Sleath PR, Grabstein KH, Hosken NA, Kern F, Nelson JA, Picker LJ (2005) Broadly targeted human cytomegalovirus-specific CD4+ and CD8+ T cells dominate the memory compartments of exposed subjects. J Exp Med 202:673-685

Tabiasco J, Rabot M, Aguerre-Girr M, El Costa H, Berrebi A, Parant O, Laskarin G, Juretic K, Bensussan A, Rukavina D, Le Bouteiller P (2006) Human decidual NK cells: unique phenotype and functional properties - a review. Placenta 27 Suppl A:S34-S39

Tarantal AF, Salamat MS, Britt WJ, Luciw PA, Hendrickx AG, Barry PA (1998) Neuropathogenesis induced by rhesus cytomegalovirus in fetal rhesus monkeys (Macaca mulatta). J Infect Dis 177:446-450

Taylor-Wiedeman J, Sissons JG, Borysiewicz LK, Sinclair JH (1991) Monocytes are a major site of persistence of human cytomegalovirus in peripheral blood mononuclear cells. J Gen Virol 72:2059-2064

Taylor-Wiedeman J, Sissons P, Sinclair J (1994) Induction of endogenous human cytomegalovirus gene expression after differentiation of monocytes from healthy carriers. J Virol 68:1597-1604

Taylor PM, Rose ML, Yacoub MH, Pigott R (1992) Induction of vascular adhesion molecules during rejection of human cardiac allografts. Transplantation 54:451-457

Tomazin R, Boname J, Hegde NR, Lewinsohn DM, Altschuler Y, Jones TR, Cresswell P, Nelson JA, Riddell SR, Johnson DC (1999) Cytomegalovirus US2 destroys two components of the MHC class II pathway, preventing recognition by CD4+ T cells. Nat Med 5:1039-1043

Tong CY, Bakran A, Peiris JS, Muir P, Herrington CS (2002) The association of viral infection and chronic allograft nephropathy with graft dysfunction after renal transplantation. Transplantation 74:576-578

Torriani FJ, Freeman WR, Macdonald JC, Karavellas MP, Durand DM, Jeffrey DD, Meylan PR, Schrier RD (2000) CMV retinitis recurs after stopping treatment in virological and immunological failures of potent antiretroviral therapy. AIDS 14:173-180

Tu W, Potena L, Stepick-Biek P, Liu L, Dionis KY, Luikart H, Fearon WF, Holmes TH, Chin C, Cooke JP, Valantine HA, Mocarski ES, Lewis DB (2006) T-cell immunity to subclinical cytomegalovirus infection reduces cardiac allograft disease. Circulation 114:1608-1615

Tumilowicz JJ (1990) Characteristics of human arterial smooth muscle cell cultures infected with cytomegalovirus. In Vitro Cel Dev Biol 26:1144-1150

Tumilowicz JJ, Gawlik ME, Powell BB, Trentin JJ (1985) Replication of cytomegalovirus in human arterial smooth muscle cells. J Virol 56:839-845

Valantine HA (2004) The role of viruses in cardiac allograft vasculopathy. Am J Transplant 4:169-177

Valantine HA, Gao SZ, Menon SG, Renlund DG, Hunt SA, Oyer P, Stinson EB, Brown BW Jr, Merigan TC, Schroeder JS (1999) Impact of prophylactic immediate posttransplant ganciclovir on development of transplant atherosclerosis: a post hoc analysis of a randomized, placebo-controlled study. Circulation 100:61-66

Valchanova RS, Picard-Maureau M, Budt M, Brune W (2006) Murine cytomegalovirus m142 and m143 are both required to block protein kinase R-mediated shutdown of protein synthesis. J Virol 80:10181-10190

Van Dam-Mieras MC, Bruggeman CA, Muller AD, Debie WH, Zwaal RF (1987) Induction of endothelial cell procoagulant activity by cytomegalovirus infection. Thromb Res 47:69-75

Vazquez-Martul E, Mosquera JM, Rivera CF, Hernandez AA, Pertega S (2004) Histological features with clinical impact in chronic allograft nephropathy: review of 66 cases. Transplant Proc 36:770-771

Vliegen I, Duijvestijn A, Grauls G, Herngreen S, Bruggeman C, Stassen F (2004) Cytomegalovirus infection aggravates atherogenesis in apoE knockout mice by both local and systemic immune activation. Microbes Infect 6:17-24

Vochem M, Hamprecht K, Jahn G, Speer CP (1998) Transmission of cytomegalovirus to preterm infants through breast milk. Pediatr Infect Dis J 17:53-58

Vogel P, Weigler BJ, Kerr H, Hendrickx AG, Barry PA (1994) Seroepidemiologic studies of cytomegalovirus infection in a breeding population of rhesus macaques. Lab Anim Sci 44:25-30

Voigt V, Forbes CA, Tonkin JN, Degli-Esposti MA, Smith HR, Yokoyama WM, Scalzo AA (2003) Murine cytomegalovirus m157 mutation and variation leads to immune evasion of natural killer cells. Proc Natl Acad Sci U S A 100:13483-13488

Wagner CS, Riise GC, Bergstrom T, Karre K, Carbone E, Berg L (2007) Increased expression of leukocyte Ig-like receptor-1 and activating role of UL18 in the response to cytomegalovirus infection. J Immunol 178:3536-3543

Waldrop SL, Pitcher CJ, Peterson DM, Maino VC, Picker LJ (1997) Determination of antigen-specific memory/effector CD4+ T cell frequencies by flow cytometry. Evidence for a novel, antigen-specific homeostatic mechanism in HIV-associated immunodeficiency. J Clin Invest 99:1739-1750

Walter EA, Greenberg PD, Gilbert MJ, Finch RJ, Watanabe KS, Thomas ED, Riddell SR (1995) Reconstitution of cellular immunity against cytomegalovirus in recipients of allogeneic bone marrow by transfer of T-cell clones from the donor. N Engl J Med 333:1038-1044

Wang D, Shenk T (2005a) Human cytomegalovirus UL131 open reading frame is required for epithelial cell tropism. J Virol 79:10330-10338

Wang D, Shenk T (2005b) Human cytomegalovirus virion protein complex required for epithelial and endothelial cell tropism. Proc Natl Acad Sci U S A 102:18153-18158

Webster A, Lee CA, Cook DG, Grundy JE, Emery VC, Kernoff PB, Griffiths PD (1989) Cytomegalovirus infection and progression towards AIDS in haemophiliacs with human immunodeficiency virus infection. Lancet 2:63-66

Weller TH (1971a) The cytomegaloviruses: ubiquitous agents with protean clinical manifestations. I. N Eng J Med 285:203-214

Weller TH (1971b) The cytomegaloviruses: ubiquitous agents with protean clinical manifestations. II. N Eng J Med 285:267-274

Westphal M, Lautenschlager I, Backhaus C, Loginov R, Kundt G, Oberender H, Stamm C, Steinhoff G (2006) Cytomegalovirus and proliferative signals in the vascular wall of CABG patients. Thorac Cardiovasc Surg 54:219-226

Whitcup SM, Fortin E, Lindblad AS, Griffiths P, Metcalf JA, Robinson MR, Manischewitz J, Baird B, Perry C, Kidd IM, Vrabec T, Davey RT Jr, Falloon J, Walker RE, Kovacs JA, Lane HC, Nussenblatt RB, Smith J, Masur H, Polis MA (1999) Discontinuation of anticytomegalovirus therapy in patients with HIV infection and cytomegalovirus retinitis. JAMA 282:1633-1637

Wilcox CM, Chalasani N, Lazenby A, Schwartz DA (1998) Cytomegalovirus colitis in acquired immunodeficiency syndrome: a clinical and endoscopic study. Gastrointest Endos 48:39-43

Wiley CA, Nelson JA (1988) Role of human immunodeficiency virus and cytomegalovirus in AIDS encephalitis. Am J Pathol 133:73-81

Willeitner A (2004) Transmission of cytomegalovirus (CMV) through human milk: are new breastfeeding policies required for preterm infants? Adv Exp Med Biol 554:489-494

Wills MR, Ashiru O, Reeves MB, Okecha G, Trowsdale J, Tomasec P, Wilkinson GW, Sinclair J, Sissons JG (2005) Human cytomegalovirus encodes an MHC class I-like molecule (UL142) that functions to inhibit NK cell lysis. J Immunol 175:7457-7465

Wreghitt TG, Hakim M, Gray JJ et al (1988) Cytomegalovirus infections in heart and heart and lung transplant recipients. J Clin Pathol 41:660-667

Wreghitt TG, Abel SJ, McNeil K, Parameshwar J, Stewart S, Cary N, Sharples L, Large S, Wallwork J (1999) Intravenous ganciclovir prophylaxis for cytomegalovirus in heart, heart-lung, and lung transplant recipients. Transplant Int 12:254-260

Wu GD, Shintaku IP, Chien K, Geller SA (1989) A comparison of routine light microscopy, immunohistochemistry and in situ hybridization for the detection of cytomegalovirus in gastrointestinal biopsies. Am J Gastroenterol 84:1517-1520

Wu TC, Hruban RH, Ambinder RF, Pizzorno M, Cameron DE, Baumgartner WA, Reitz BA, Hayward GS, Hutchins GM (1992) Demonstration of cytomegalovirus nucleic acids in the coronary arteries of transplanted hearts. Am J Pathol 140:739-747

Yamamoto AP, Mussi-Pinhata MM, Pinto PC, Figueiredo LT, Jorge SM (2001) Congenital cytomegalovirus infection in preterm and full-term newborn infants from a population with a high seroprevalence rate. Pediatr Infect Dis J 20:188-192

Ye M, Morello CS, Spector DH (2002) Strong CD8 T-cell responses following coimmunization with plasmids expressing the dominant pp89 and subdominant M84 antigens of murine cytomegalovirus correlate with long-term protection against subsequent viral challenge. J Virol 76:2100-2112

Yilmaz S, Koskinen PK, Kallio E, Bruggeman CA, Hayry PJ, Lemstrom KB (1996) Cytomegalovirus infection-enhanced chronic kidney allograft rejection is linked with intercellular adhesion molecule-1 expression. Kidney Int 50:526-537

Yokoyama WM, Scalzo AA (2002) Natural killer cell activation receptors in innate immunity to infection. Microbes Infect 4:1513-1521

Yow MD, Williamson DW, Leeds LJ, Thompson P, Woodward RM, Walmus BF, Lester JW, Six HR, Griffiths PD (1988) Epidemiologic characteristics of cytomegalovirus infection in mothers and their infants. Am J Obstet Gynecol 158:1189-1195

Yu D, Silva MC, Shenk T (2003) Functional map of human cytomegalovirus AD169 defined by global mutational analysis. Proc Natl Acad Sci U S A 100:12396-12401

Zhou YF, Leon MB, Waclawiw MA, Popma JJ, Yu ZX, Finkel T, Epstein SE (1996) Association between prior cytomegalovirus infection and the risk of restenosis after coronary atherectomy. N Engl J Med 335:624-630

Zhou YF, Shou M, Guetta E, Guzman R, Unger EF, Yu ZX, Zhang J, Finkel T, Epstein SE (1999) Cytomegalovirus infection of rats increases the neointimal response to vascular injury without consistent evidence of direct infection of the vascular wall. Circulation 100:1569-1575

Zhu H, Shen Y, Shenk T (1995) Human cytomegalovirus IE1 and IE2 proteins block apoptosis. J Virol 69:7960-7970

Zhu H, Cong JP, Mamtora G, Gingeras T, Shenk T (1998) Cellular gene expression altered by human cytomegalovirus: global monitoring with oligonucleotide arrays. Proc Natl Acad Sci U S A 95:14470-14475

Zhu H, Cong JP, Bresnahan WA, Shenk TE (2002) Inhibition of cyclooxygenase 2 blocks human cytomegalovirus replication. Proc Natl Acad Sci U S A 99:3932-3937

Zhu J, Nieto FJ, Horne BD, Anderson JL, Muhlestein JB, Epstein SE (2001) Prospective study of pathogen burden and risk of myocardial infarction or death. Circulation 103:45-51

Index

A

Adaptation, 67, 68, 74, 75
Adoptive immunotherapy, 320
AICAR, 270–272
Akt, 263–266, 268, 269, 271, 273–276, 283, 287, 288
Alphavirus replicon CMV vaccine, 367, 370
Aly/Ref, 172, 174–177
Amino acid depletion, 270, 271
AMPK, 269–272, 274
Angiogenesis, 392, 397, 399, 402, 405, 407–409, 411
Antigenic peptide, mutation of, 326, 328
Apoptosis, 76, 77
Atherosclerosis, 72, 77
ATP production, 272
ATRX, 125–127
Augmentation of transcription, 122, 124

B

Bacterial artificial chromosome vaccine, 41, 42, 374
Bidirectional gene pair, 317, 321–323

C

CaMKKβ, 270
Canarypox vectored CMV vaccine, 367, 370
Cap-dependent translation, 263–265, 267–269, 272–275
Capsid
   assembly, formation, 190–194
   composition, 189, 190, 192, 196
   DNA packaging, 191, 194–196
   maturation, 188, 191, 194, 195, 198
CCMV, 224, 226–228, 237
CD14+ monocytes, 299–300
CD34+ cells, 299–301, 303–307, 308

Cell
   culture, 64, 65, 67–69, 71, 73, 74, 76, 77
   cycle, 135–137, 141–145
   tropism, 67, 68, 72–78
Cellular stress responses, 263–265, 267, 274
Chemokines, 223, 227, 230, 234, 235
Chromatin, 109, 111, 302, 303, 305–307
Chromatinized viral genomes, 127
Chromatin remodelling, 324–325
Chronic rejection, 397, 446, 447
Cis-acting elements, 3, 4, 139, 176
Cis-complementation, 51–55
Clinical isolates, 1, 4–7, 9, 16, 17, 32, 74, 75, 86, 89, 290, 368, 374, 376, 418
Conditional expression, 55
Constitutive transport element (CTE), 171, 172
Cosmid, 43–45
Costimulation, 341
Counter-selection, 48, 49
CRE. *See* Cytomegalovirus
CREB, 229, 230
CRM1/exportin1, 168, 170, 177
Cross-complementation, 42
Cyclin dependent kinases, 244, 246, 252–255
Cytokines, 77, 91, 92, 146, 306, 307, 334, 398, 399, 402, 404, 406, 407, 411, 425, 435, 444, 445, 448
Cytomegalovirus, 1–4, 16, 21, 22, 29, 30, 41, 63, 68, 78, 221, 231, 232
   AD169 strain, 86, 224, 368, 373
   adenovirus vectored protein, 51, 107, 251, 288, 289, 372–374
   adolescent infection, 364, 365, 367, 422
   breast milk transmission, 70, 363, 421–423, 426
   congenital infection, 127, 298, 362, 365, 377, 384, 388, 392, 427, 428, 432, 437
   DNA vaccine, 367, 371, 372, 374

471

Cytomegalovirus (cont.)
  hyperimmune globulin, 366
  IL-10, 222, 301, 318, 385, 391, 392
  infections in pregnancy, 428
  MVA vaccine, 372, 373
  receptors, 71, 85–89, 91, 92, 95, 96, 126, 168, 207
  replication in utero, 384
  Toledo chimeric vaccine, 367, 368
  Towne vaccine, 367, 368, 370
  transgenic plant vectored vaccine, 372
Cytotrophoblast, 209, 383–392
  cell column, 385, 388, 389
  invasive, 383–386, 389–392, 426, 433, 436
  villous, 383–388, 391

D
Daxx, 107–111
Daxx, ATRX, Sp100, 125
Daxx-/-cells, 124, 126
Daxx-mediated NFkB binding, 125
Dendritic cells, 64, 65, 70, 74, 300, 303, 306
Detectable size, 118
DExD/H box protein, 172
Dissemination, 69, 72, 74, 76, 77
DNA replication, 4, 53, 78, 103, 135, 140, 144, 153–163, 167, 178–180, 243–245, 247, 249, 252, 253, 257, 305

E
4E-BP, 265, 267, 268, 271–275
Effective host defense, 128
EGFR, 87–91, 207, 209, 210, 213, 214, 386–391
eIF4E, 266, 267, 269, 273, 274, 276
Endocytosis, 75, 76
Endothelial cells, 5, 63–68, 72–78, 86, 89, 205, 208, 209, 213, 214, 226, 236, 257, 275, 289, 290, 298, 299, 308, 384–386, 397, 407, 408, 423, 425, 429, 435, 441, 443–445, 448
Endothelial cell tropism, 68, 72, 74–77
Enhancer, 321–324
Entry pathway, 78, 85, 95, 226
Envelope acquisition, 196
Envelope glycoprotein, 85, 86, 91, 95, 207, 208, 210, 373, 375
Epstein Barr Virus EB2, 173, 174
Exon junction complex (EJC), 172
Exon skipping, 129

F
Fatty acid oxidation, 272
Fibroblasts, 63, 65, 67, 68, 71, 73–76, 78
Foreign DNA/foreign protein complexes, 121
FRT, 48, 51, 53–56
Fusion, 73, 75, 76, 86, 87, 90–93, 95, 102, 104, 106, 206, 210, 236, 237

G
gc II complex (gM, gN) vaccine, 86, 90, 373, 438
gc III complex (gH, gL, gO) vaccine, 373
Gene silencing/desilencing, 317, 319, 323–327
Genome
  evolution, 16, 17
  load, 320–321
  ORFs, 1, 5–17, 54–56
  organization, 3, 4
Glucose transport, 272
Glycolysis, 272
Glycoprotein B (gB; gpUL55) vaccine, 86, 91, 208, 369
Glycoproteins, 206–210, 213
GP33, 228
GP78, 228
GPCR, 223, 229, 234, 235
Growth factors, 244, 246, 306, 398, 399, 402, 405–407, 411, 444
Guinea pig cytomegalovirus (GpCMV) vaccination studies, 225–228, 371, 376, 377

H
HDAC, 109, 111, 122, 123, 126–129, 139, 252
Herpes simplex virus ICP27, 173–177
Herpesviruses, 2, 3, 21–24, 29–31, 34, 35, 42, 44–51, 55, 57, 86, 90, 101, 107, 110, 157, 159, 173, 174, 176, 178, 187–189, 192, 196, 197, 206, 208, 222–224, 226–228, 230, 287, 289, 304, 315, 316, 341, 342, 400, 418, 443
HHV-6, 223–228, 230–232, 235, 236
HHV-7, 227, 228, 230, 232
HIF-1, 269
Host Cell Cycle, 243–246
hSPT6, 173–175, 177
Human cytomegalovirus, 1–17, 21–35, 63, 85, 86, 101–111, 133–146
Human cytomegalovirus infections, 439
Hypoxia, 263, 264, 269, 271, 272

I

IE1, 268, 275
  deletion mutant, 124, 125, 128
  disperses ND10, 121, 122
  as drug development candidate, 129
  gene expression, 105, 107–111
  inactivation, 122
IE1491aa and IE2579aa, 283, 287, 288
  prosurvival function, 287, 288
IE2, 268, 275
IE2-p86, 178
IE86 protein, 135–145
IL-10, 318
Immediate-early genes/proteins
  ie1/IE1, 321–323, 325–328
  ie2/IE2, 317, 320–323, 325
  ie3/IE3, 318, 321–326, 328
Immediate early protein 1(IE1) vaccine, 35, 110, 111, 117, 120–129, 134, 139, 154, 249, 250, 275, 320, 321, 323, 325–328, 335, 371, 372, 374, 375
Immediate early protein 2, 250–252
Immediate early transcript environment, 120, 122
Immune
  response, 301, 307
  sensing, 316, 325–327
  suppression, 298, 308
  surveillance, 316, 317
Immunoevasin, 317, 328
Importin-alpha, 178
Importin-beta, 170
Inactive transactivator complex, 126
Inhibition of splicing, 120
Innate immunity, 91, 103
Institute of medicine (IOM), 363, 365
Insulin receptor substrates, 265
Integrin, 87–89, 91, 209, 210, 213, 214, 227, 384–391, 405
  α1β1, 385, 386, 388–391
  αV, 386–391
Interaction between IE1 and PML, 124
Interferon, 334, 345
  pathway, 93, 95, 334, 336
  upregulation, 122, 128, 129
Interstrain differences, 64, 73–76
Intrinsic immunity, 110

K

Karyopherin, 169, 170, 172

L

Laboratory strains, 1, 4–6, 32, 158, 224, 285, 287, 289
Latency, 11, 317–320, 323, 325–327
  checkpoints of, 324, 325, 328
  molecular, 319
  replicative, 316, 319, 320, 323, 325
  transcriptional, 319, 325
Leptomycin B, 171, 173
Leukocytes, 64, 72, 75
Ligand, 226, 229, 230–232, 236
Lipid raft, 89, 94
LKB1, 270
Localization of genomes, 120
Lytic replication, 23, 34, 72, 101–104, 111, 153, 159–161, 163, 430

M

M33, 228, 230, 231, 234, 235
m41, 283
  golgi apparatus, 290
  homologs, 290
  m41-deficient virus, 290
M45, 283
  homologs, 289
  M45-deficient virus, 289
  ribonucleotide reductase, 289
M50, 55, 57
M53, 55, 57
M78, 228, 232, 234–236
Major histocompatibility complex (MHC), 337–340, 342–347
  class I, 317, 326–328
Major-immediate early promoter (MIEP), 302–306
Microtubules, 105, 111
MIE locus, 317, 319–327
miRNA, 22–35
  CCMV, 29, 30
  EBV, 22, 23, 29, 31, 33, 34
  HCMV, 22–24, 27–35
  HSV, 29, 34, 35
  KSHV, 29, 31, 34
  MDV, 30
  MHV-68/76, 24, 26, 33
  rLCV, 29, 31
  SV, 34, 40
mTORC1, 265–275
mTORC2, 266–268, 271, 274
mTOR kinase, 265–267, 269, 270, 272, 273, 275

Murine cytomegalovirus (MCMV), 223–225, 227, 228, 230–232, 234–236
  vaccination studies, 373–377
Mutant(s), 234, 235
  dominant negative, 55, 56
  temperature sensitive, 42

N
Natural killer cells (NK cells), 341–348
ND10-associated proteins, 120–122, 124, 125, 129
Neonatal Fc receptor, FcRn, 384, 386, 388
Neonatal infection, 322
NF-κB, 324
Nuclear defense, 120, 126
Nuclear depots, 120
Nuclear domain 10 (ND10), 107, 120
Nuclear export signal (NES), 170, 171–173, 176, 177, 179
Nuclear localization signal (NLS), 170, 171, 173, 176, 178–180
Nuclear pore complex (NPC), 168–171
Nucleocytoplasmic transport, 168, 169

O
ORF57, 174

P
p70S6 kinase, 267
Pathogenesis, 63, 70, 73, 77, 234
Pathogenetic role, 68, 69
PDK1, 265, 267, 274
Peptide aptamers, 180
Perinatal viral infections, 421
Persistence, 317–320
Physical association between viral genomes and ND10, 120
Physical dimensions of the large viral genome of CMV, 118
PI3K, 265, 266, 268, 269, 271, 274–276, 287, 288
Placenta, 71, 383–386, 388–390, 392, 399, 429, 430
PML, 102, 107–111
PML-NB, 102, 107–111
pp150, 104–106, 111
PP2A, 275
pp65, 104, 106, 109, 111
pp65(ppUL83) vaccine, 371
pp71, 102, 105, 106, 108–111, 118, 119, 126, 142, 248, 249, 302, 336

Pre replication domain, 118, 121, 125
Proliferation, 63, 71, 73
Protease
  function, 190, 192
  maturational, capsid, 188, 191, 195
  structure, 189, 190, 193, 194
Proteasome, 105, 109
Protein-coding ORFs, 6–17
PTEN, 275
pUL26, 105, 110
pUL35, 105, 109, 110
pUL38, 272, 275, 283, 288, 289
  pUL38-mediated survival, 289
  UL38 transcripts, 283
  UL38-deficient virus, 288, 289
pUL47, 104, 106, 107, 110, 111
pUL48, 104, 106, 107, 110, 111
pUL84, 167, 169, 178–180

R
R33, 228, 230, 231, 234, 235
R78, 228, 232, 234, 235
RanGTP, 169–171
Rapamycin, 263, 265–267, 270, 271
Raptor, 266, 271, 273, 275
RCMV, 223–225, 227, 228, 230–232, 234, 235
Reactivation, 298, 300, 302, 305–308, 317–326
Receptor, 85–92, 94, 95, 111, 169–172, 174, 175, 178, 205, 207–210, 212, 214, 221–224, 228, 229, 231, 232, 234, 236, 264, 265, 268, 318, 334
Receptor/ligand, 208–210, 214
Recombinant virus, 326, 328
Recombinase
  Flp, 48, 54–56
  recET, 47
  redαβ, 47, 48
  homologous, 42–48, 52
  site-specific, 53, 54, 56
Recombineering, 47–49
Reconstitution of a single PML isotype, 125
Recurrence, 320, 323–326
Regulation, 142
rh107, 228
rh214, 228
rh215, 228
rh216, 228
rh218, 228
rh220, 228
rh56, 228
RhCMV, 224–226, 228, 230, 236
Rheb-GTP, 266, 267, 270, 271, 275

Ribosomal protein, S6, 267
Rictor, 266, 267, 273, 275
β2.7 RNA, 283
    β2.7-deficient virus, 290
    ATP synthesis, 290
    mitochondrial respiratory complex I, 290

S

Salivary glands, 318–320
Secretome, 398, 405–411
Sensorineural hearing loss (SNHL), 362
Signaling, 226, 227, 229, 230–232, 234–236
Signal transduction, 91–94, 111, 125, 136, 205, 209, 210, 213, 214, 229, 230, 236, 334, 336
Small ubiquitin modifier, 123
Smooth muscle cells, 63–66, 68, 71–73, 77
Sp100, 107–111
Splicing inhibitors, 129
SUMO, 123, 125, 126
SUMO interacting motif (SIM), 126
Syncytiotrophoblast, 383–388, 391

T

TAP, 168, 169, 171, 172, 174, 175, 177
Target cells, 64, 65, 67, 68
T cell
    CD4, 318, 320
    CD8, 316, 317, 320, 325–328
Tegument, 102–111
    acquisition, 196
    proteins, 208, 210–214
    pUL48 deubiquitinylase, 189, 198
TetR, 55, 56
TLR, 209
TNF-α, 324, 325
Toll-like receptor, 85, 92, 209, 334
Trans-complementation, 51–53, 55
Transcription, 134, 137, 139–142, 144, 146
Transmission, 63, 65, 70, 73
Transplantation, 72, 270, 298, 307, 323, 365, 368, 370, 374, 398, 400, 402, 411, 422, 433, 434, 436, 446–448

Tropism, 64, 67–69, 72–78
Tuberous sclerosis complex (TSC), 263–266, 268–272, 274, 275

U

U12, 228, 230–232
U51, 228, 230, 232, 235, 236
UAP56, 169, 172, 173, 175–177
UL128-131, 64, 74, 75
UL24-protein, 74
UL33, 224, 227–231, 234, 236, 237
Ul50, 57
UL69, 167, 169, 173–177, 179
UL78, 224, 227–229, 232, 234, 235, 237
UL84, 8, 12, 135, 153–159, 161–163, 178, 179
US28, 227–230, 232, 234–237

V

Vascular disease, 398–400, 402, 405, 411, 419, 437, 442–448
vICA, 282, 283, 287
    homologs, 287
    procaspase-8, 287
    UL36-deficient virus, 287
    UL36 transcripts, 283
Viral diseases in immunocompromised host, 426, 434, 436
Viral pathogenesis, 418, 419, 426, 430, 431, 439, 441, 445, 447
Virion enzymes, 210, 211
vMIA, 282–287
    BAX and Bcl-xL, 284
    functional domains, 284
    GADD45, 284
    homologs, 286, 287
    UL37 transcripts, 283
    UL37x1-deficient virus, 285, 286

W

Wound healing, 397, 399, 401, 402, 405, 407–411

Current Topics in Microbiology and Immunology

Volumes published since 2002

Vol. 271: **Koehler, Theresa M. (Ed.):** Anthrax. 2002. 14 figs. X, 169 pp. ISBN 3-540-43497-6

Vol. 272: **Doerfler, Walter; Böhm, Petra (Eds.):** Adenoviruses: Model and Vectors in Virus-Host Interactions. Virion and Structure, Viral Replication, Host Cell Interactions. 2003. 63 figs., approx. 280 pp. ISBN 3-540-00154-9

Vol. 273: **Doerfler, Walter; Böhm, Petra (Eds.):** Adenoviruses: Model and Vectors in VirusHost Interactions. Immune System, Oncogenesis, Gene Therapy. 2004. 35 figs., approx. 280 pp. ISBN 3-540-06851-1

Vol. 274: **Workman, Jerry L. (Ed.):** Protein Complexes that Modify Chromatin. 2003. 38 figs., XII, 296 pp. ISBN 3-540-44208-1

Vol. 275: **Fan, Hung (Ed.):** Jaagsiekte Sheep Retrovirus and Lung Cancer. 2003. 63 figs., XII, 252 pp. ISBN 3-540-44096-3

Vol. 276: **Steinkasserer, Alexander (Ed.):** Dendritic Cells and Virus Infection. 2003. 24 figs., X, 296 pp. ISBN 3-540-44290-1

Vol. 277: **Rethwilm, Axel (Ed.):** Foamy Viruses. 2003. 40 figs., X, 214 pp. ISBN 3-540-44388-6

Vol. 278: **Salomon, Daniel R.; Wilson, Carolyn (Eds.):** Xenotransplantation. 2003. 22 figs., IX, 254 pp. ISBN 3-540-00210-3

Vol. 279: **Thomas, George; Sabatini, David; Hall, Michael N. (Eds.):** TOR. 2004. 49 figs., X, 364 pp. ISBN 3-540-00534X

Vol. 280: **Heber-Katz, Ellen (Ed.):** Regeneration: Stem Cells and Beyond. 2004. 42 figs., XII, 194 pp. ISBN 3-540-02238-4

Vol. 281: **Young, John A. T. (Ed.):** Cellular Factors Involved in Early Steps of Retroviral Replication. 2003. 21 figs., IX, 240 pp. ISBN 3-540-00844-6

Vol. 282: **Stenmark, Harald (Ed.):** Phosphoinositides in Subcellular Targeting and Enzyme Activation. 2003. 20 figs., X, 210 pp. ISBN 3-540-00950-7

Vol. 283: **Kawaoka, Yoshihiro (Ed.):** Biology of Negative Strand RNA Viruses: The Power of Reverse Genetics. 2004. 24 figs., IX, 350 pp. ISBN 3-540-40661-1

Vol. 284: **Harris, David (Ed.):** Mad Cow Disease and Related Spongiform Encephalopathies. 2004. 34 figs., IX, 219 pp. ISBN 3-540-20107-6

Vol. 285: **Marsh, Mark (Ed.):** Membrane Trafficking in Viral Replication. 2004. 19 figs., IX, 259 pp. ISBN 3-540-21430-5

Vol. 286: **Madshus, Inger H. (Ed.):** Signalling from Internalized Growth Factor Receptors. 2004. 19 figs., IX, 187 pp. ISBN 3-540-21038-5

Vol. 287: **Enjuanes, Luis (Ed.):** Coronavirus Replication and Reverse Genetics. 2005. 49 figs., XI, 257 pp. ISBN 3-540- 21494-1

Vol. 288: **Mahy, Brain W. J. (Ed.):** Foot-and-Mouth-Disease Virus. 2005. 16 figs., IX, 178 pp. ISBN 3-540-22419X

Vol. 289: **Griffin, Diane E. (Ed.):** Role of Apoptosis in Infection. 2005. 40 figs., IX, 294 pp. ISBN 3-540-23006-8

Vol. 290: **Singh, Harinder; Grosschedl, Rudolf (Eds.):** Molecular Analysis of B Lymphocyte Development and Activation. 2005. 28 figs., XI, 255 pp. ISBN 3-540-23090-4

Vol. 291: **Boquet, Patrice; Lemichez Emmanuel (Eds.):** Bacterial Virulence Factors and Rho GTPases. 2005. 28 figs., IX, 196 pp. ISBN 3-540-23865-4

Vol. 292: **Fu, Zhen F. (Ed.):** The World of Rhabdoviruses. 2005. 27 figs., X, 210 pp. ISBN 3-540-24011-X

Vol. 293: **Kyewski, Bruno; Suri-Payer, Elisabeth (Eds.):** CD4+CD25+ Regulatory T Cells: Origin, Function and Therapeutic Potential. 2005. 22 figs., XII, 332 pp. ISBN 3-540-24444-1

Vol. 294: **Caligaris-Cappio, Federico, Dalla Favera, Ricardo (Eds.):** Chronic Lymphocytic Leukemia. 2005. 25 figs., VIII, 187 pp. ISBN 3-540-25279-7

Vol. 295: **Sullivan, David J.; Krishna Sanjeew (Eds.):** Malaria: Drugs, Disease and Post-genomic Biology. 2005. 40 figs., XI, 446 pp. ISBN 3-540-25363-7

Vol. 296: **Oldstone, Michael B. A. (Ed.):** Molecular Mimicry: Infection Induced Autoimmune Disease. 2005. 28 figs., VIII, 167 pp. ISBN 3-540-25597-4

Vol. 297: **Langhorne, Jean (Ed.):** Immunology and Immunopathogenesis of Malaria. 2005. 8 figs., XII, 236 pp. ISBN 3-540-25718-7

Vol. 298: **Vivier, Eric; Colonna, Marco (Eds.):** Immunobiology of Natural Killer Cell Receptors. 2005. 27 figs., VIII, 286 pp. ISBN 3-540-26083-8

Vol. 299: **Domingo, Esteban (Ed.):** Quasispecies: Concept and Implications. 2006. 44 figs., XII, 401 pp. ISBN 3-540-26395-0

Vol. 300: **Wiertz, Emmanuel J.H.J.; Kikkert, Marjolein (Eds.):** Dislocation and Degradation of Proteins from the Endoplasmic Reticulum. 2006. 19 figs., VIII, 168 pp. ISBN 3-540-28006-5

Vol. 301: **Doerfler, Walter; Böhm, Petra (Eds.):** DNA Methylation: Basic Mechanisms. 2006. 24 figs., VIII, 324 pp. ISBN 3-540-29114-8

Vol. 302: **Robert N. Eisenman (Ed.):** The Myc/Max/Mad Transcription Factor Network. 2006. 28 figs., XII, 278 pp. ISBN 3-540-23968-5

Vol. 303: **Thomas E. Lane (Ed.):** Chemokines and Viral Infection. 2006. 14 figs. XII, 154 pp. ISBN 3-540-29207-1

Vol. 304: **Stanley A. Plotkin (Ed.):** Mass Vaccination: Global Aspects – Progress and Obstacles. 2006. 40 figs. X, 270 pp. ISBN 3-540-29382-5

Vol. 305: **Radbruch, Andreas; Lipsky, Peter E. (Eds.):** Current Concepts in Autoimmunity. 2006. 29 figs. IIX, 276 pp. ISBN 3-540-29713-8

Vol. 306: **William M. Shafer (Ed.):** Antimicrobial Peptides and Human Disease. 2006. 12 figs. XII, 262 pp. ISBN 3-540-29915-7

Vol. 307: **John L. Casey (Ed.):** Hepatitis Delta Virus. 2006. 22 figs. XII, 228 pp. ISBN 3-540-29801-0

Vol. 308: **Honjo, Tasuku; Melchers, Fritz (Eds.):** Gut-Associated Lymphoid Tissues. 2006. 24 figs. XII, 204 pp. ISBN 3-540-30656-0

Vol. 309: **Polly Roy (Ed.):** Reoviruses: Entry, Assembly and Morphogenesis. 2006. 43 figs. XX, 261 pp. ISBN 3-540-30772-9

Vol. 310: **Doerfler, Walter; Böhm, Petra (Eds.):** DNA Methylation: Development, Genetic Disease and Cancer. 2006. 25 figs. X, 284 pp. ISBN 3-540-31180-7

Vol. 311: **Pulendran, Bali; Ahmed, Rafi (Eds.):** From Innate Immunity to Immunological Memory. 2006. 13 figs. X, 177 pp. ISBN 3-540-32635-9

Vol. 312: **Boshoff, Chris; Weiss, Robin A. (Eds.):** Kaposi Sarcoma Herpesvirus: New Perspectives. 2006. 29 figs. XVI, 330 pp. ISBN 3-540-34343-1

Vol. 313: **Pandolfi, Pier P.; Vogt, Peter K. (Eds.):** Acute Promyelocytic Leukemia. 2007. 16 figs. VIII, 273 pp. ISBN 3-540-34592-2

Vol. 314: **Moody, Branch D. (Ed.):** T Cell Activation by CD1 and Lipid Antigens, 2007, 25 figs. VIII, 348 pp. ISBN 978-3-540-69510-3

Vol. 315: **Childs, James, E.; Mackenzie, John S.; Richt, Jürgen A. (Eds.):** Wildlife and Emerging Zoonotic Diseases: The Biology, Circumstances and Consequences of Cross-Species Transmission. 2007. 49 figs. VII, 524 pp. ISBN 978-3-540-70961-9

Vol. 316: **Pitha, Paula M. (Ed.):** Interferon: The 50th Anniversary. 2007. VII, 391 pp. ISBN 978-3-540-71328-9

Vol. 317: **Dessain, Scott K. (Ed.):** Human Antibody Therapeutics for Viral Disease. 2007. XI, 202 pp. ISBN 978-3-540-72144-4

Vol. 318: **Rodriguez, Moses (Ed.):** Advances in Multiple Sclerosis and Experimental Demyelinating Diseases. 2008. XIV, 376. ISBN 978-3-540-73679-9

Vol. 319: **Manser, Tim (Ed.):** Specialization and Complementation of Humoral Immune Responses to Infection. 2008. XII, 174. ISBN 978-3-540-73899-2

Vol. 320: **Paddison, Patrick J.; Vogt, Peter K. (Eds.):** RNA Interference. 2008. VIII, 273. ISBN 978-3-540-75156-4

Vol. 321: **B. Beutler (Ed.):** Immunology, Phenotype First: How Mutations Have Established New Principles and Pathways in Immunology. 2008. ISBN 978-3-540-75202-250

Vol. 322: **Romeo, Tony (Ed.):** Bacterial Biofilms. 2008. XII, 299.
ISBN 978-3-540-75417-6

Vol. 323: **Tracy, Steven; Oberste, M. Steven; Drescher, Kristen M. (Eds.):** Group B Coxsackieviruses. 2008.
ISBN 978-3-540-75545-6

Vol. 324: **Nomura, Tatsuji; Watanabe, Takeshi; Habu, Sonoko (Eds.):** Humanized Mice. 2008. ISBN 978-3-540-75646-0

Vol.325: **Shenk, Thomas E., Stinski, Mark F. (Eds.):** Human Cytomegalovirus. 2008.
ISBN 978-3-540-77348-1

Printing: Krips bv, Meppel, The Netherlands
Binding: Stürtz, Würzburg, Germany